THE NONLINEAR WORKBOOK

THE NONLINEAR WORKBOOK

5th Edition

Chaos,
Fractals,
Cellular Automata,
Genetic Algorithms,
Gene Expression Programming,
Support Vector Machine,
Wavelets,
Hidden Markov Models,
Fuzzy Logic with C++,
Java and SymbolicC++ Programs

Willi-Hans Steeb
International School for Scientific Computing
University of Johannesburg, South Africa

in collaboration with

Yorick Hardy
International School for Scientific Computing, South Africa

Ruedi Stoop
Institute for Neuroinformatics, University of Zürich/ETHZ

World Scientific

NEW JERSEY · LONDON · SINGAPORE · BEIJING · SHANGHAI · HONG KONG · TAIPEI · CHENNAI

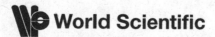

Published by

World Scientific Publishing Co. Pte. Ltd.
5 Toh Tuck Link, Singapore 596224
USA office: 27 Warren Street, Suite 401-402, Hackensack, NJ 07601
UK office: 57 Shelton Street, Covent Garden, London WC2H 9HE

British Library Cataloguing-in-Publication Data
A catalogue record for this book is available from the British Library.

THE NONLINEAR WORKBOOK — Fifth Edition
Chaos, Fractals, Cellular Automata, Genetic Algorithms, Gene Expression Programming, Support Vector Machine, Wavelets, Hidden Markov Models, Fuzzy Logic with C++, Java and SymbolicC++ Programs

ISBN-13 978-981-4335-77-5
ISBN-10 981-4335-77-0
ISBN-13 978-981-4335-78-2 (pbk)
ISBN-10 981-4335-78-9 (pbk)

Printed in Singapore.

Preface

The study of chaos, fractals, cellular automata, neural networks, genetic algorithms and fuzzy logic is one of the most fascinating subjects in science. Most of these fields are interrelated. Chaotic attractors are used in neural networks. Genetic algorithms can be used to train neural networks. Fractals are used in data compression. Neural networks and fuzzy logic are often combined when the input values of the system are not crisp.

In this book we give all the basic concepts in these fields together with the definitions, theorems and algorithms. The algorithms are implemented using C++, Java and SymbolicC++. The level of presentation is such that one can comprehend the subject early on while studying science. There is a balance between practical computation and the underlying mathematical theory.

In chapter 1 we consider one and two-dimensional nonlinear maps. All the relevant quantities to characterize chaotic systems are introduced. Algorithms are given for all the quantities which are used to describe chaos such as invariant density, Liapunov exponent, correlation integral, autocorrelation function, capacity, phase portrait, Poincaré section, Fourier transform, calculations of exact trajectories, fixed points and their stability, etc.. Chaotic repellers and encoding using one-dimensional chaotic maps are also investigated. Newton's method in one and two dimensions is derived. Periodic orbits and topological degree are introduced.

Quite often a dynamical system cannot be modelled by difference equations or differential equations, but an experiment provides a time series. In chapter 2 we consider quantities for the study of chaotic time-series. We also include the Hurst exponent which plays an important role in the study of financial markets. The related Higuchi algorithm is also provided.

In chapter 3 we describe the classification of fixed points in the plane. Furthermore the most important two-dimensional dynamical systems are studied, such as the pendulum, limit cycle systems and a Lotka-Volterra model. Homoclinic orbits are also introduced.

Chapter 4 reviews integrable and chaotic Hamilton systems. Among other concepts we introduce the Lax representation for integrable Hamilton systems, the Poincaré section and the Floquet theory.

In chapter 5 nonlinear dissipative systems are studied. The most famous dissipative system with chaotic behaviour, the Lorenz model, is introduced. We also discuss Hopf bifurcation and hyperchaotic systems.

Nonlinear driven systems play a central role in engineering, in particular in electronics. In most cases the driving force is periodic. Chapter 6 is devoted to these systems. As examples we consider among others the driven pendulum and the driven van der Pol equation. The concept of torsion number is also discussed.

Controlling of chaotic systems is very important in applications in engineering. In chapter 7 we discuss the different concepts of controlling chaos. The Ott-Grebogi-Yorke method for controlling chaotic systems is studied in detail.

Synchronization of chaotic systems is described in chapter 8 and a number of applications are given, such as the coupled Rikitake dynamos.

Fractals have become of increasing interest, not only in art, but also in many different areas of science such as compression algorithms. In chapter 9 we introduce iterated function systems, the Mandelbrot set, the Julia set and the Weierstrass function. The famous Cantor set is considered as an example as well as the Koch curve, fern and the Sierpinski gasket. We also derive the construction of fractals using the Kronecker product of matrices. Grey level maps are also described.

Cellular automata are discrete dynamical systems. We describe in chapter 10 one and two-dimensional cellular automata. The famous game of life with a C++ implementation and the button game with a Java implementation are also considered. The Sznajd model is studied as an application.

Chapter 11 is about integration of differential equations. We describe the Euler method, the Runge-Kutta method, the Lie series technique, symplectic integration, Verlet method, etc.. Furthermore we discuss ghost solutions, invisible chaos and integration in the complex domain.

Optimization problems are studied in chapter 12. We consider the Lagrange multiplier method for optimization problems and also describe an alternative method using differential forms. For problems with inequality constraints the Karush-Kuhn-Tucker condition is provided and the support vector machine is studied. The Kernel-Adatron algorithm, a fast and simple learning procedure for support vector machines, is also implemented. As an application the kernel Fisher discriminant is studied.

Chapter 13 is devoted to neural networks. We introduce the Hopfield algorithm, the Kohonen self-organizing map, the back propagation algorithm and radial basis function networks. One of the applications is the traveling salesman problem. Neural oscillator models are also introduced.

Genetic algorithms are used to solve optimization problems. Chapter 14 is devoted to this technique. We discuss optimization problems with and without constraints. A discussion of bitwise operations is given. We also study simulated annealing.

Gene expression programming is a new genetic algorithm that uses encoded individuals. Gene expression programming individuals are encoded in linear chromosomes which are expressed or translated into expression trees. The linear chromosome is the genetic material that is passed on with modifications to the next generation. Chapter 15 gives an introduction to this technique together with a C++ program. As an alternative to gene expression programming we also describe multi-expression programming together with a C++ program.

Wavelet theory is a form of mathematical transformation, similar to the Fourier transform in that it takes a signal in time domain, and represents it in frequency domain. Wavelet functions are distinguished from other transformations in that they not only dissect signals into their component frequencies, they also vary the scale at which the component frequencies are analyzed. Chapter 16 provides an introduction. Filtering is given as an example application. As examples the Haar wavelet and Daubechies wavelet are studied. Two-dimensional wavelets are also considered.

Discrete Hidden Markov Models are introduced in chapter 17. The forward-backward algorithm, Viterbi algorithm, and Baum-Welch algorithm are described. The application concentrates on speech recognition.

Since its inception 40 years ago the theory of fuzzy sets has advanced in a variety of ways and in many disciplines, not only in science. Chapter 18 is devoted to fuzzy logic. Fuzzy numbers and arithmetic are also considered. Furthermore decision making problems and controlling problems using fuzzy logic are also described. Fuzzy clustering is also included as well as a definition for the fuzzy XOR.

In each chapter we give C++, Java and SymbolicC++ implementations of the algorithms.

Without doubt, this book can be extended. Any comments or suggestions are welcome. The author can be contacted via e-mail:

`steebwilli@gmail.com`
`steeb_wh@yahoo.com`

The web page of the author is:

`http://issc.uj.ac.za`

The International School for Scientific Computing (ISSC) provides certificate courses for these subjects. Please contact the author if you want to do any of these courses.

Contents

Symbol Index

$\emptyset$	empty set
$A \subset B$	the subset A of the set B
$A \cap B$	the intersection of the sets A and B
$A \cup B$	the union of the sets A and B
χ_A	indicator function
$\mathbb{N}$	the set of positive integers: natural numbers
$\mathbb{Z}$	the set of integers
$\mathbb{Q}$	the set of rational numbers
$\mathbb{R}$	the set of real numbers
$\mathbb{R}^+$	nonnegative real numbers
$\mathbb{C}$	the set of complex numbers
$\mathbb{R}^n$	the n-dimensional real linear space
$\mathbb{C}^n$	the n-dimensional complex linear space
i	$:= \sqrt{-1}$
$\Re z$	real part of the complex number z
$\Im z$	imaginary part of the complex number z
$\{0,1\}^n$	n-dimensional binary hypercube
$[a,b]$	closed interval of real-valued numbers between a and b
$[0,1]$	unit interval
S^1	unit circle $\{\,(x,y) : x^2 + y^2 = 1\,\}$
f	mapping (map)
$f \circ g$	composition of mappings $(f \circ g)(x) = f(g(x))$
$f^{(n)}$	n-th iterate of mapping f
$\mathbf{x} \in \mathbb{R}^n$	the element $\mathbf{x}$ of the vector space $\mathbb{R}^n$, column vector
$\mathbf{x}^T$	transpose of $\mathbf{x}$ (row vector)
$\mathbf{x}^T\mathbf{y}$	scalar product in the vector space $\mathbb{R}^n$
t	time: discrete or continuous depending on context
$\mathbf{x}$	dependent variable: discrete systems
$\mathbf{x}^*$	fixed point of maps
$\mathbf{u}$	dependent variable: continuous systems
ρ	invariant density
r	bifurcation parameter, control parameter
$\lVert . \rVert$	norm
tr	trace of a square matrix

det	determinant of a square matrix
I_n	$n \times n$ identity matrix
$[A, B]$	commutator of $n \times n$ matrices A and B
δ_{jk}	Kronecker delta with $\delta_{jk} = 1$ for $j = k$ and $\delta_{jk} = 0$ for $j \neq k$
L	Lagrange function
H	Hamilton function
λ	Lagrange multiplier
λ	eigenvalue
η	learning rate
I	identity operator
$\mathbf{w}$	weight vector (column vector)
$\Delta\mathbf{w}$	small change applied to $\mathbf{w}$
W	weight matrix
$\boldsymbol{\Theta}$	bias vector
$\{\mathbf{x}_k, \mathbf{d}_k\}$	kth training pairs
net	weighted sum or $\mathbf{w}^T\mathbf{x}$
$f(net)$	differentiable activation function, usually a sigmoid function
$f'(net)$	derivative of f with respect to net
$\mu_{\tilde{A}}$	membership function (fuzzy logic)
$\oplus$	XOR bitwise operation
$\otimes$	Kronecker product of matrices
$\wedge$	wedge product (exterior product, Grassmann product)
$L_2(\Omega)$	Hilbert space of square integrable functions
$\ell_2(\mathbb{N})$	Hilbert space of all infinite dimensional sequences

Comments to the Programs

The C++ programs comply to the ANSI C++ standard. Thus they should run under all compilers.

SymbolicC++ version 3 is a symbolic manipulation tool (Y. Hardy, Tan Kiat Shi, W.-H. Steeb [87]) written completely in ANSI C++. It includes classes (abstract data types) to do symbolic and numeric manipulations. The classes include

```
Symbolic, Rational, Verylong, Quaternion,
Derive, Vector, Matrix, Array, Polynomial .
```

The classes from the Standard Template Library are also extensively used and so is the `string` class from C++. The symbolic manipulation is done using the class `Symbolic`.

SymbolicC++ version 3 is available at

`http://issc.uj.ac.za/symbolic/symbolic.html`

We have tested the C++ programs with GCC 4.1.3 and Microsoft Visual Studio.net (VC8).

In C++ graphics do not belong to the standard. Here we use GnuPlot (GNUPLOT Copyright(c) 1986-1993, 1998, Colin Kelly and Thomas Williams) to draw the figures.

The Java programs have been tested with JDK 1.6. The JDK (Java Development Kit) is a product of Sun Microsystems, Inc. The JDK allows us to develop applets that will run in browsers supporting the Java platform 1.6. The Java tools we use are the Java Compiler (`javac`) which compiles programs written in the Java programming language into bytecodes and the Java Interpreter (`java`) that executes Java bytecodes. In other words, it runs programs written in the Java programming language. AppletViewer allows us to run one or more Java applets that are called references in a web page (HTML file) using the APPLET tag. The AppletViewer finds the APPLET tags in the HTML file and runs the applets (in separate windows) as specified by the tags.

Most of the programs are written so that they can be understood by beginners. Thus some of the programs can be improved and written in a more sophisticated manner.

Chapter 1

Nonlinear and Chaotic Maps

1.1 One-Dimensional Maps

In this section we consider nonlinear and chaotic one-dimensional maps (Devaney [47], Arrowsmith and Place [7], Holmgren [98], Collet and Eckmann [37], Gumowski and Mira [77], Ruelle [177], Baker and Gollub [10], van Wyk and Steeb [209])

$$f : S \to S, \qquad S \subset \mathbb{R}.$$

In most cases the set S will be $S = [0,1]$ or $S = [-1,1]$. The one-dimensional map can also be written as a difference equation

$$x_{t+1} = f(x_t), \qquad t = 0,1,2,\ldots \qquad x_0 \in S.$$

Starting from an initial value $x_0 \in S$ we obtain, by iterating the map, the sequence

$$x_0, \quad x_1, \quad x_2, \; \ldots$$

or

$$x_0, \quad f(x_0), \quad f(f(x_0)), \quad f(f(f(x_0))), \; \ldots \; .$$

For any $x_0 \in S$ the sequence of points $x_0, x_1, x_2, \ldots$ is called the forward *orbit* (or forward *trajectory*) generated by x_0. The goal of a dynamical system is to understand the nature of all orbits, and to identify the set of orbits which are periodic, eventually periodic, asymptotic, etc. Thus we want to understand what happens if $t \to \infty$. In some cases the long-time behaviour is quite simple.

Example. Consider the map $f : \mathbb{R}^+ \to \mathbb{R}^+$ with $f(x) = \sqrt{x}$. For all $x_0 \in \mathbb{R}^+$ and $x_0 > 0$ the forward trajectory tends to 1. The fixed points of f are 0 and 1. ♣

Example. Consider the map $f : [0,1] \to [0,1]$, $f(x) = x^2$. If $x_0 = 0$, then $f(x_0) = 0$. Analogously, if $x_0 = 1$, then $f(x_0) = 1$. The points 0 and 1 we call fixed points for this map. For $x \in (0,1)$, the forward trajectory tends to 0. ♣

In most cases the behaviour of a map is much more complex.

Next we introduce some basic definitions for dynamical systems.

Definition. A point $x^* \in S$ is called a *fixed point* of the map f if $f(x^*) = x^*$.

Example. Consider the map $f : [0,1] \to [0,1]$ with $f(x) = 4x(1-x)$. Then $x^* = 3/4$ and $x^* = 0$ are fixed points. ♣

Definition. A point $x^* \in S$ is called a *periodic point* of period n if

$$f^{(n)}(x^*) = x^*$$

where $f^{(n)}$ denotes the n-th iterate of f. The least positive integer n for which $f^{(n)}(x^*) = x^*$ is called the *prime period* of x^*. The set of all iterates of a periodic point form a periodic orbit.

Example. Consider the map $f : \mathbb{R} \to \mathbb{R}$ and $f(x) = x^2 - 1$. Then the points 0 and -1 lie on a periodic orbit of period 2 since $f(0) = -1$ and $f(-1) = 0$. ♣

Definition. A point x^* is *eventually periodic* of period n if x^* is not periodic but there exists $m > 0$ such that

$$f^{(n+i)}(x^*) = f^{(i)}(x^*)$$

for all $i \geq m$. That is, $f^{(i)}(x^*)$ is periodic for $i \geq m$.

Example. Consider the map $f : \mathbb{R} \to \mathbb{R}$ with $f(x) = x^2 - 1$. Then with $x_0 = \sqrt{2}$ we have the orbit $x_1 = 1$, $x_2 = 0$, $x_3 = -1$, $x_4 = 0$, i.e., the orbit is eventually periodic.♣

Definition. Let x^* be a periodic point of prime period n. The point x^* is *hyperbolic* if

$$|(f^{(n)})'(x^*)| \neq 1$$

where $'$ denotes the derivative of $f^{(n)}$ with respect to x.

Example. Consider the map $f_c : \mathbb{R} \to \mathbb{R}$ with $f_c(x) = x^2 + c$ and $c \in \mathbb{R}$. Then the fixed points are $x^*_\pm = 1/2 \pm \sqrt{1/4 - c}$. We have $f'_c \equiv df_c/dx = 2x$ and $df_c(x^*_+)/dx = 1 + \sqrt{1-4c}$. With $c = 1/4$ we have $|f'_c(x^*_+)| = 1$ and the fixed point is non-hyperbolic. ♣

Theorem. Let x^* be a hyperbolic fixed point with $|f'(x^*)| < 1$. Then there is an open interval U about x^* such that if $x \in U$, then

$$\lim_{n \to \infty} f^{(n)}(x) = x^* .$$

Definition. Let M be a differentiable manifold. A mapping $f : M \to M$ is called a *diffeomorphism* if f is a bijection with f and f^{-1} continuously differentiable (of class C^1).

Example. The map $f : \mathbb{R} \to \mathbb{R}$, $f(x) = \sinh(x)$ is a diffeomorphism. ♣

Example. The map $f : \mathbb{R} \to \mathbb{R}$, $f(x) = x^3$ is not a diffeomorphism since its derivative vanishes at 0. ♣

In the following sections we introduce the following concepts important in the study of nonlinear and chaotic maps. The concepts are

1) Fixed points
2) Liapunov exponent
3) Invariant density
4) Autocorrelation functions
5) Moments
6) Fourier transform
7) Bifurcation diagrams
8) Feigenbaum number
9) Symbolic dynamics
10) Chaotic repeller

The one-dimensionial maps we study in the examples are the logistic map, the tent map, the Bernoulli map, the Gauss map, a bungalow-tent map and the circle map. A necessary condition for a one-dimensional map to show chaotic behaviour is that the map is non-invertible.

1.1.1 Exact and Numerical Trajectories

In this section we calculate trajectories for one-dimensional maps. In the first example we consider the map $f : \mathbb{N} \to \mathbb{N}$ defined by

$$f(x) := \begin{cases} x/2 & \text{if } x \text{ is even} \\ 3x+1 & \text{if } x \text{ is odd} \end{cases}$$

where $\mathbb{N}$ denotes the natural numbers. For this map it is conjectured that for all initial values the trajectory finally tends to the periodic orbit `... 4 2 1 4 2 1 ...`. The data type `unsigned long` (4 bytes) in C++ is restricted to the range

```
0...4294967295
```

and the data type `long` (4 bytes) in C++ is restricted to the range

```
-2147483648...+2147483647
```

To check the conjecture for larger initial values we use the abstract data type `Verylong` in SymbolicC++. In this class all arithmetic operators are overloaded. The overloaded operators are

`+, -, *, /, %, +=, -=, *=, /=` .

For example for the initial value 28 we find the sequence

`14, 7, 22, 11, 34, 17, 52, 26, 13, 40, 20, 10, 5, 16, 8, 4, 2, 1, ...`

Thus the orbit is eventually periodic. Two different initial values are considered in the program `trajectory1.cpp`, namely 28 and 998123456789.

```
// trajectory1.cpp

#include <iostream>     // for cout
#include "verylong.h"  // for data type Verylong of SymbolicC++
using namespace std;

int main(void)
{
 unsigned long y = 28;  // initial value
 unsigned long T = 20;  // number of iterations
 unsigned long t;
 for(t=0;t<T;t++)
 { if((y%2)==0) y = y/2; else y = 3*y+1; cout << y << endl; }
 Verylong x("998123456789"); // initial value
 Verylong zero("0"), one("1"), two("2"), three("3");
 T = 350;
 for(t=0;t<T;t++)
 { if((x%two)==zero) x = x/two; else x = three*x + one; cout << x << endl; }
 return 0;
}
```

Java provides a class `BigInteger`. Since operators such as `+, -, *, /, %` cannot be overloaded in Java, Java uses methods to do the arithmetic operations. The methods are

`add(), subtract(), multiply(), divide(), mod()`

where `divide()` provides integer division. The constructor `BigInteger(String val)` translates the decimal `String` representation of a `BigInteger` into a `BigInteger`. The class `BigInteger` also provides the data fields

`BigInteger.ONE   BigInteger.ZERO`

```
// Trajectory1.java

import java.math.*;
```

```
public class Trajectory1
{
 public static void main(String[] args)
 {
 BigInteger X = new BigInteger("998123456789");
 BigInteger TWO = new BigInteger("2");
 BigInteger THREE = new BigInteger("3");
 int T = 350;
 for(int t=0;t<T;t++)
 {
 if((X.mod(TWO)).equals(BigInteger.ZERO)) X = X.divide(TWO);
 else { X = X.multiply(THREE); X = X.add(BigInteger.ONE); }
 System.out.println("X = " + X);
 }
 }
}
```

In the second example we consider the trajectories for the logistic map. The *logistic map* $f : [0,1] \to [0,1]$ is given by $f(x) = 4x(1-x)$. The logistic map can also be written as the difference equation

$$x_{t+1} = 4x_t(1-x_t)$$

where $t = 0, 1, 2, \ldots$ and $x_0 \in [0,1]$. It follows that $x_t \in [0,1]$ for all $t \in \mathbb{N}$. Let $x_0 = 1/3$ be the initial value. Then we find that

$$x_1 = \frac{8}{9}, \quad x_2 = \frac{32}{81}, \quad x_3 = \frac{6272}{6561}, \quad x_4 = \frac{7250432}{43046721}, \quad \ldots$$

The exact solution of the logistic map is given by

$$x_t = \frac{1}{2} - \frac{1}{2}\cos(2^t \arccos(1-2x_0)).$$

In the C++ program `trajectory2.cpp` we evaluate the exact trajectory up to $t = 10$ using the abstract data type `Verylong` of SymbolicC++. For $t = 7$ we find

$$x_7 = \frac{338382616201936779639722410 8032}{343368382029251248465784908 9281}.$$

```
// trajectory2a.cpp

#include <iostream>
#include "verylong.h"  // for data type Verylong
#include "rational.h"  // for data type Rational
using namespace std;

inline void map(Rational<Verylong>& x)
{
```

```
 Rational<Verylong> one("1");       // number 1
 Rational<Verylong> four("4");      // number 4
 Rational<Verylong> x1 = four*x*(one-x);
 x = x1;
}

int main(void)
{
 Rational<Verylong> x0("1/3"); // initial value 1/3
 unsigned long T = 10;         // number of iterations
 Rational<Verylong> x = x0;
 cout << "x[0] = " << x << endl;
 for(unsigned long t=0;t<T;t++)
 { map(x); cout << "x[" << t+1 << "] = " << x << endl; }
 return 0;
}
```

In the C++ program trajectory3.cpp we evaluate the numerical trajectory using the basic data type double. We find that the difference between the exact value and the numerical value for $t = 40$ is

$$x_{40exact} - x_{40approx} = 0.055008 - 0.055015 = -0.000007.$$

```
// trajectory2b.cpp

#include <iostream>
using namespace std;

inline void map(double& x) { double x1 = 4.0*x*(1.0-x); x = x1; }

int main(void)
{
 double x0 = 1.0/3.0;  // initial value
 unsigned long T = 10; // number of iterations
 double x = x0;
 cout << "x[0] = " << x << endl;
 for(unsigned long t=0;t<T;t++)
 { map(x); cout << "x[" << t+1 << "] = " << x << endl; }
 return 0;
}
```

As a third example we consider the *Bernoulli map*. Let $f : [0,1) \to [0,1)$. It is defined by

$$f(x) := 2x \bmod 1 \equiv \text{frac}(2x).$$

The map can be written as the difference equation

$$x_{t+1} = \begin{cases} 2x_t & \text{for } 0 \le x_t < 1/2 \\ (2x_t - 1) & \text{for } 1/2 \le x_t < 1 \end{cases}$$

where $t = 0, 1, 2, \ldots$ and $x_0 \in [0, 1)$. The map admits only one fixed point $x^* = 0$. The fixed point is unstable. Let $x_0 = 1/17$. Then we find the periodic orbit

$$\frac{2}{17}, \quad \frac{4}{17}, \quad \frac{8}{17}, \quad \frac{16}{17}, \quad \frac{15}{17}, \quad \frac{13}{17}, \quad \frac{9}{17}, \quad \frac{1}{17}, \quad \frac{2}{17}, \ldots$$

If x_0 is a rational number in the interval $[0, 1)$, then x_t is either periodic or tends to the fixed point $x^* = 0$. The solution of the Bernoulli map is given by

$$x_t = 2^t x_0 \mod 1$$

where x_0 is the initial value. For almost all initial values the Liapunov exponent is given by $\ln 2$. Every $x_0 \in [0, 1)$ can be written (uniquely) in the *binary representation*

$$x_0 = \sum_{k=1}^{\infty} a_k 2^{-k}, \qquad a_k \in \{0, 1\}.$$

One defines

$$(a_1, a_2, a_3, \ldots) := \sum_{k=1}^{\infty} a_k 2^{-k}$$

and considers the infinite sequence $(a_1, a_2, a_3, \ldots)$. For example, $x_0 = 3/8$ can be represented by the sequence $(0, 1, 1, 0, 0, \ldots)$. Instead of investigating the Bernoulli map we can use the map τ defined by

$$\tau(a_1, a_2, a_3, \ldots) := (a_2, a_3, a_4, \ldots).$$

This map is called the *Bernoulli shift*. In the C++ program `trajectory3.cpp` we find the trajectory of the Bernoulli map using the data type `Rational` and `Verylong` of SymbolicC++. The initial value is 1/17. The orbit is periodic.

```
// trajectory3.cpp

#include <iostream>
#include "verylong.h"  // for data type Verylong
#include "rational.h"  // for data type Rational
using namespace std;

inline void map(Rational<Verylong>& x)
{
 Rational<Verylong> one("1");     // number 1
 Rational<Verylong> two("2");     // number 2
 Rational<Verylong> half("1/2"); // number 1/2
 Rational<Verylong> x1;
 if(x < half) x1 = two*x; else x1 = two*x-one;
 x = x1;
}

int main(void)
```

```
{
 Rational<Verylong> x0("1/17"); // initial value 1/17
 unsigned long T = 10;          // number of iterations
 Rational<Verylong> x = x0;
 cout << "x[0] = " << x << endl;
 for(unsigned long t=0;t<T;t++)
 { map(x); cout << "x[" << t+1 << "] = " << x << endl; }
 return 0;
}
```

As a fourth example we consider the *tent map*. The tent map $f : [0,1] \to [0,1]$ is defined as

$$f(x) := \begin{cases} 2x & \text{if } x \in [0,1/2) \\ 2(1-x) & \text{if } x \in [1/2,1] \end{cases} .$$

The map can also be written as the difference equation

$$x_{t+1} = \begin{cases} 2x_t & \text{if } x_t \in [0,1/2) \\ 2(1-x_t) & \text{if } x_t \in [1/2,1] \end{cases}$$

where $t = 0, 1, 2, \ldots$ and $x_0 \in [0,1]$. Let $x_0 = 1/17$ be the initial value. Then the exact orbit is given by

$$x_0 = \frac{1}{17}, \quad x_1 = \frac{2}{17}, \quad x_2 = \frac{4}{17}, \quad x_3 = \frac{8}{17}, \quad x_4 = \frac{16}{17}, \quad x_5 = \frac{2}{17}, \ldots$$

This is an example of an eventually periodic orbit. If the initial value is a rational number then the orbit is eventually periodic, periodic or tends to a fixed point. For example the initial value $1/16$ tends to the fixed point 1. To find chaotic orbits the initial value must be an irrational number, for example $x_0 = 1/\pi$. The fixed points of the map are given by $x^* = 0$, $x^* = 2/3$. These fixed points are unstable. The map shows fully developed chaotic behaviour. The invariant density is given by $\rho(y) = 1$. For almost all initial values the Liapunov exponent is given by $\lambda = \ln 2$. For the autocorrelation function we find

$$C_{xx}(\tau) = \begin{cases} \frac{1}{12} & \text{for } \tau = 0 \\ 0 & \text{for } \tau \geq 1 \end{cases} .$$

The tent map $f : [0,1] \to [0,1]$ given above and the logistic map $g : [0,1] \to [0,1]$, $g(x) = 4x(1-x)$ are *topologically conjugate* , i.e.

$$f = h \circ g \circ h^{-1}$$

where the homeomorphism $h : [0,1] \to [0,1]$ is given by

$$h(x) = \frac{2}{\pi}\arcsin(\sqrt{x}), \qquad h^{-1}(x) = \frac{1-\cos(\pi x)}{2} .$$

In the C++ program `trajectory4.cpp` we find the trajectory of the tent map with the inital value $1/17$.

```
// trajectory4.cpp

#include <iostream>
#include "verylong.h"
#include "rational.h"
using namespace std;

inline void map(Rational<Verylong>& x)
{
 Rational<Verylong> one("1"), two("2");
 Rational<Verylong> half("1/2"); // number 1/2
 Rational<Verylong> x1;
 if(x < half) x1 = two*x; else x1 = two*(one-x);
 x = x1;
}

int main(void)
{
 Rational<Verylong> x0("1/17"); // initial value 1/17
 unsigned long T = 10;          // number of iterations
 Rational<Verylong> x = x0;
 cout << "x[0] = " << x << endl;
 for(unsigned long t=0;t<T;t++)
 { map(x); cout << "x[" << t+1 << "] = " << x << endl; }
 return 0;
}
```

As a fifth example we consider a *bungalow-tent map.* Our bungalow-tent map $f_r : [0,1] \to [0,1]$ is defined by

$$f_r(x) := \begin{cases} \dfrac{1-r}{r}x & \text{for} \quad x \in [0,r) \\ \dfrac{2r}{1-2r}x + \dfrac{1-3r}{1-2r} & \text{for} \quad x \in [r,1/2) \\ \dfrac{2r}{1-2r}(1-x) + \dfrac{1-3r}{1-2r} & \text{for}\, x \in [1/2, 1-r) \\ \dfrac{1-r}{r}(1-x) & \text{for} \quad x \in [1-r,1] \end{cases}$$

where $r \in (0,1/2)$ is the control parameter. The map is continuous, but not differentiable at the points r, $1-r$ $(r \neq 1/3)$ and $x = 1/2$. The map is piecewise linear. The fixed points are 0 and $1-r$. For $r = 1/3$ we obtain the tent map. The map f_r is a special bungalow-tent map. The intersection point P of the line in the interval $[1/2, 1-r)$ and the line in the interval $[1-r,1]$ lies on the diagonal $y = x$. The invariant density is given by

$$\rho_r(x) = \frac{1}{2-3r}\chi_{[0,1-r]}(x) + \frac{1-2r}{r(2-3r)}\chi_{(1-r,1]}(x)$$

where χ is the *indicator function*, i.e. $\chi_A(x) = 1$ if $x \in A$ and $\chi_A(x) = 0$ if $x \notin A$. Thus the invariant density is constant in the interval $[0, 1-r)$. At $1-r$ the invariant density jumps to another constant value. The Liapunov exponent is given by

$$\lambda(r) = \frac{1-r}{2-3r} \ln\left(\frac{1-r}{r}\right) + \frac{1-2r}{2-3r} \ln\left(\frac{2r}{1-2r}\right).$$

For $r = 1/3$ we obviously obtain $\lambda(1/3) = \ln 2$. This is the Liapunov exponent for the tent map. For $r \to 0$ we obtain $\lambda(r \to 0) = \frac{1}{2}\ln 2$. For $r \to 1/2$ we obtain $\lambda(r \to 1/2) = 0$. $\lambda(r)$ has a maximum for $r = 1/3$ (tent map). Furthermore $\lambda(r)$ is a convex function in the interval $(0, 1/2)$. Thus we have

$$\lambda(r) \leq \ln 2 .$$

The C++ program `trajectory5.cpp` finds the trajectory of the bungalow-tent map for the control parameter $r = 1/7$ and the initial value $x_0 = 1/17$. We find $x_1 = 6/17$, $x_2 = 16/17$, $x_3 = 6/17$. Thus the orbit is eventually periodic.

```
// trajectory5.cpp

#include <iostream>
#include "verylong.h"
#include "rational.h"
using namespace std;

inline void map(Rational<Verylong>& x,Rational<Verylong>& r)
{
 Rational<Verylong> one("1"), two("2"), three("3");
 Rational<Verylong> half("1/2");  // number 1/2
 Rational<Verylong> x1;
 if(x < r) x1 = (one-r)*x/r;
 else if((x >= r) && (x < half))
 x1 = two*r*x/(one-two*r) + (one-three*r)/(one-two*r);
 else if((x >= half) && (x < one-r))
 x1 = two*r*(one-x)/(one-two*r)+(one-three*r)/(one-two*r);
 else if((x <= one) && (x > one-r))
 x1 = (one-r)*(one-x)/r;
 x = x1;
}

int main(void)
{
 Rational<Verylong> x0("1/17"); // initial value 1/17
 Rational<Verylong> r("1/7");   // control parameter 1/7
 unsigned long T = 10;          // number of iterations
 Rational<Verylong> x = x0;
 cout << "x[0] = " << x << endl;
 for(unsigned long t=0;t<T;t++)
 { map(x,r); cout << "x[" << t+1 << "] = " << x << endl; }
```

```
 return 0;
}
```

In the sixth example we consider the *Gauss map*. The Gauss map $f : [0,1] \to [0,1]$ is defined as

$$f(x) := \begin{cases} 0 & \text{if } x = 0 \\ \lfloor 1/x \rfloor & \text{if } x \neq 0 \end{cases}$$

where $\lfloor y \rfloor$ denotes the *fractional part* of y. For example

$$\lfloor 3.2 \rfloor = 0.2, \qquad \lfloor 17/3 \rfloor = \frac{2}{3}.$$

Owing to the definition $x^* = 0$ is a fixed point. Let $x_0 = 23/101$ be the initial value. Then the orbit is given by

$$x_1 = \frac{9}{23}, \quad x_2 = \frac{5}{9}, \quad x_3 = \frac{4}{5}, \quad x_4 = \frac{1}{4}, \quad x_5 = 0$$

where $x_5 = 0$ is a fixed point. The Gauss map possesses an infinite number of discontinuities and is not injective since each $x_0 \in [0,1]$ has countable infinite images. The map admits an infinite number of unstable fixed points and shows chaotic behaviour. For example $x^* = (\sqrt{5}-1)/2$ (golden mean number) is a fixed point, since $x^* = f(x^*)$. The Gauss map preserves the Gauss measure on $[0,1]$ which is given by

$$m(A) := \frac{1}{\ln 2} \int_A \frac{1}{1+x} dx .$$

The periodic points of the Gauss map are the reciprocal of the reduced quadratic irrationals. These numbers are dense in $[0,1)$.

```
// trajectory6.cpp

#include <iostream>
#include "verylong.h"
#include "rational.h"
using namespace std;

inline void map(Rational<Verylong>& x)
{
 Rational<Verylong> zero("0"), one("1");
 Rational<Verylong> x1;
 if(x==zero) return;
 x1 = one/x;
 while(x1 >= one) x1 = x1-one;
 x = x1;
}

int main(void)
{
```

```
 Rational<Verylong> x0("23/101"); // initial value
 unsigned long T = 10;            // number of iterations
 Rational<Verylong> x = x0;
 cout << "x[0] = " << x << endl;
 for(unsigned long t=0;t<T;t++)
 { map(x); cout << "x[" << t+1 << "] = " << x << endl; }
 return 0;
}
```

In the last example we define a fractal signal as the discrete time series

$$x_{2t} = b(1 + x_t), \qquad x_{2t+1} = a(1 + x_t), \qquad t = 0, 1, 2, \ldots$$

where $a < 1$ and $b < 1$. Owing to the first equation the initial value x_0 is given by $x_0 = b/(1 - b)$. This sequence is generated when the elements of the two-scale Cantor set are taken in a definite order. A C++ implementation using templates is

```
// trajectory7.cpp

#include <iostream>
using namespace std;

template <class T> void sequence(T* y,int N,T a,T b)
{
 for(int t=1;t<N;t++)
 {
 if(t%2==0) y[t] = b*(T(1)+y[t/2]);
 else       y[t] = a*(T(1)+y[(t-1)/2]);
 }
}

int main(void)
{
 int N = 20;
 double* y = new double[N];
 double a = 1.0/3.0; double b = 1.0/2.0;
 y[0] = b/(1.0-b);
 sequence(y,N,a,b);
 for(int j=0;j<N;j++) cout << "y[" << j << "] = " << y[j] << endl;
 delete[] y;
 return 0;
}
```

The output from a time series should be stored in a file and then plotted. The next program `datagnu.cpp` shows how to write the output data from an iteration to a file. We consider the logistic map as an example. The output is stored in a file called `timeev.dat`. We use the C++ style for the file manipulation.

```
// datagnu.cpp

#include <fstream>  // for ofstream, close
using namespace std;

int main(void)
{
 ofstream data("timeev.dat"); // filename timeev.dat
 unsigned long T = 100;       // number of iterations
 double x0 = 0.618;           // initial value
 double x1;
 for(unsigned long t=0;t<T;t++)
 { x1 = 4.0*x0*(1.0-x0); data << t << " " << x0 << "\n"; x0 = x1; }
 data.close();
 return 0;
}
```

The data files can now be used to draw a graph of the time evolution using *GNU-plot*. After we entered GNU-plot using the command `gnuplot` the plot command is as follows

```
plot [0:10] 'timeev.dat'
```

This command plots the first eleven points of the time evolution. Furthermore we can create a *postscript file* using the commands:

```
set term postscript default
set output "timeev.ps"
plot 'timeev.dat'
```

The next program shows how to write data to a file and read data from a file using JAVA. We consider the logistic map. The data are stored in a file called `"timeev.dat"`. In the second part of the program we read the data back. In JAVA the filename and the class name which includes the

```
public static void main(String args[])
```

method must coincide (case sensitive). We output data using a `DataOutputStream` that is connected to a `FileOutputStream` via a technique called chaining of stream objects. When the `DataOutputStream` object output is created, its constructor is supplied a `FileOutputStream` object as an argument. The statement creates a `DataOutputStream` object named output associated with the file `timeev.dat`. The argument `"timeev.dat"` is passed to the `FileOutputStream` constructor which opens the file.

`Class DataOutputStream`. A data output stream lets an application write primitive (basic) Java data types to an output stream in a portable way.

`Class FileOutputStream`. A file output stream is an output stream for writing data to File or a FileDescriptor.

The method `void writeDouble(double v)` converts the double argument to a `long` using the `doubleToLongBits()` method in class `Double`, and then writes that `long` value to the underlying output stream as an 8-byte quantity, high byte first. The method `double readDouble()` reads eight input bytes and returns a `double` value.

```
// FileManipulation.java

import java.io.*;
import java.lang.Exception;

public class FileManipulation
{
 public static void main(String args[])
 { DataOutputStream output;
 try
 { output = new DataOutputStream(new FileOutputStream("timeev.dat"));
 int T = 10;
 double x0 = 0.618;  double x1;
 output.writeDouble(x0);
 System.out.println("The output is " + x0);
 for(int t=0;t<T;t++) { x1 = 4.0*x0*(1.0-x0); // logistic map
 System.out.println("The output is " + x1);
 output.writeDouble(x1);
 x0 = x1;
 }
 try { output.flush();  output.close(); }
 catch(IOException e)
 {
 System.err.println("File not closed properly\n" + e.toString());
 System.exit(1);
 }
 }
 catch(IOException e)
 {
 System.err.println("File not opened properly\n" + e.toString());
 System.exit(1);
 }
 System.out.println("\nReading file:");
 try
 {
 FileInputStream fin = new FileInputStream("timeev.dat");
 DataInputStream in = new DataInputStream(fin);
 while(true) System.out.print(in.readDouble() + " ");
 }
 catch(Exception e) { }
```

```
 }
}
```

1.1.2 Fixed Points and Stability

Consider a map $f : [0,1] \to [0,1]$. The *fixed points* are defined as the solutions of the equation

$$f(x^*) = x^* .$$

Assume that the map f is differentiable. Then the *variational equation* of $x_{t+1} = f(x_t)$ is defined as

$$y_{t+1} = \frac{d}{d\epsilon} f(x + \epsilon y)\bigg|_{\epsilon=0, x=x_t, y=y_t} = \frac{df}{dx}(x = x_t) y_t .$$

A fixed point is called *stable* if

$$\left| \frac{df}{dx}(x = x^*) \right| < 1 .$$

Example. Consider the logistic map $f : [0,1] \to [0,1]$, $f(x) = 4x(1-x)$. We have to solve the quadratic equation

$$4x^*(1 - x^*) = x^*$$

to find the fixed points. The fixed points are given by $x_1^* = 0$, $x_2^* = 3/4$. Since

$$\frac{df}{dx} = 4 - 8x$$

we find that the fixed points $x_1^* = 0$ and $x_2^* = 3/4$ are unstable. ♣

In the C++ program `fixpointlog.cpp` we consider the stability of the fixed points for the logistic map $x_{t+1} = 4x_t(1 - x_t)$. We evaluate the variational equation of the logistic equation and determine the stability of the fixed points. We test whether the fixed points of the logistic map $f(x) = 4x(1-x)$ are unstable. We use the header file `derive.h` from `SymbolicC++` to do the differentiation.

```
// fixpointlog.cpp

#include <iostream>
#include <cmath>      // for fabs
#include "verylong.h"
#include "rational.h"
#include "derive.h"
using namespace std;

int main(void)
{
```

```
double x1 = 0.0; double x2 = 3.0/4.0;
Derive<double> C1(1.0);  // constant 1.0
Derive<double> C4(4.0);  // constant 4.0
Derive<double> X1, X2;
X1.set(x1);
Derive<double> R1 = C4*X1*(C1-X1);
double result1 = df(R1);
cout << "result1 = " << result1 << endl;
if(fabs(result1) > 1) cout << "fixpoint x1 unstable " << endl;
X2.set(x2);
Derive<double> R2 = C4*X2*(C1-X2);
double result2 = df(R2);
cout << "result2 = " << result2 << endl;
if(fabs(result2) > 1) cout << "fixpoint x2 unstable ";
return 0;
}
```

1.1.3 Invariant Density

Consider a one-hump fully developed chaotic map $f : [0,1] \to [0,1]$

$$x_{t+1} = f(x_t)$$

where $t = 0, 1, 2, \ldots$. We define the *invariant density* ρ (also called *probability density*) of the iterates, starting from an initial point x_0, by

$$\rho(x) := \lim_{T\to\infty} \frac{1}{T} \sum_{t=0}^{T-1} \delta(x - f^{(t)}(x_0))$$

where $f^{(0)}(x_0) = x_0$ and

$$f^{(1)}(x_0) = f(x_0) = x_1, \ldots \quad , f^{(t)}(x_0) = f^{(t-1)}(f(x_0)) = f(f^{(t-1)}(x_0)) = x_t$$

with $t > 1$. Here δ denotes the *delta function*. Not all starting points $x_0 \in [0,1]$ are allowed in the definition. Those belonging to an unstable cycle must be excluded since we are only interested in the stable chaotic trajectory. For any arbitrary (but integrable in the Lebesgue sense) function g on the unit interval $[0,1]$ the *mean value* of that function along the chaotic trajectory is defined by

$$\langle g(x) \rangle := \lim_{T\to\infty} \frac{1}{T} \sum_{t=0}^{T-1} g(x_t) = \int_0^1 \rho(x) g(x) dx \,.$$

Choosing $g(x) = 1$ we obtain the normalization condition

$$\int_0^1 \rho(x) dx = 1 \,.$$

Since the probability density is independent of the starting point x_0, the expression for ρ can also be written as

$$\rho(x) = \lim_{T\to\infty} \frac{1}{T} \sum_{t=0}^{T-1} \delta(x - x_{t+k}), \qquad k = 0, 1, 2, \ldots.$$

An integral equation for ρ can be derived as follows: Let σ be defined as

$$\sigma(y) := \int_0^1 \delta(y - f^{(k)}(x))\rho(x)dx.$$

Let g be an arbitrary (but integrable in the sense of Lebesgue) function on $[0, 1]$. Then

$$\int_0^1 \sigma(y)g(y)dy = \int_0^1 \int_0^1 \delta(y - f^{(k)}(x))\rho(x)g(y)dydx.$$

Therefore we obtain

$$\int_0^1 \sigma(y)g(y)dy = \lim_{T\to\infty} \frac{1}{T} \sum_{t=0}^{T-1} \int_0^1 \delta(x - f^{(t)}(x_0))g(f^{(k)}(x))dx.$$

Using the properties of the delta function we arrive at

$$\int_0^1 \sigma(y)g(y)dy = \lim_{T\to\infty} \frac{1}{T} \sum_{t=0}^{T-1} g(f^{(t+k)}(x_0)).$$

Hence

$$\int_0^1 \sigma(y)g(y)dy = \lim_{T\to\infty} \frac{1}{T} \sum_{t=0}^{T-1} \int_0^1 \delta(y - f^{(t+k)}(x_0))g(y)dy = \int_0^1 \rho(y)g(y)dy.$$

Since the function g is arbitrarily chosen, we have to set $\sigma(y) = \rho(y)$. Thus the probability density ρ obeys the integral equation

$$\rho(y) = \int_0^1 dx\delta(y - f(x))\rho(x).$$

This equation is called the *Frobenius-Perron integral equation.* This equation has many solutions (Kluiving et al [115]). Among these are the solutions associated with the unstable periodic orbits. If these unstable solutions are left out of consideration and the map is one-hump fully developed chaotic, then there is only one stable chaotic trajectory exploring the unit interval $[0, 1]$ and the Frobenius-Perron equation has a unique solution associated with the chaotic orbit.

Example. Consider the logistic map $f : [0, 1] \to [0, 1]$, $f(x) = 4x(1 - x)$. For the stable chaotic trajectory exploring the unit interval $[0, 1]$ the Frobenius-Perron integral equation has the unique solution

$$\rho(x) = \frac{1}{2\pi\sqrt{x(1-x)}}$$

where

$$\int_0^1 \rho(x)dx = 1$$

and $\rho(x) > 0$ for $x \in [0,1]$. We see that $\rho(x) \to \infty$ for $x \to 0$ and $x \to 1$, respectively. The solution can be found by iteration of

$$\rho_{t+1}(y) = \int_0^1 dx\delta(y - f(x))\rho_t(x)$$

with the initial density $\rho_0(x) = 1$. ♣

In the C++ program `invdensity.cpp` we determine numerically the invariant density for the logistic map $x_{t+1} = 4x_t(1 - x_t)$. We find the histogram for the logistic map. We divide the unit interval $[0,1]$ into 20 bins with bin size 0.05 each. We calculate how many points exist in the intervals $[0.05 \cdot i, 0.05 \cdot (i+1))$, where $i = 0, 1, 2, \ldots, 19$. This gives an approximation for the invariant density defined above. For example, the number of points in the intervals $[0, 0.05)$ and $[0.95, 1.0]$ is much higher than in the other intervals (bins).

```
// invdensity.cpp

#include <iostream>
#include <cmath>      // for floor, sqrt
using namespace std;

void histogram(double* x,int* hist,double T,double xmax,
               double xmin,int n_bins)
{
 double grad = n_bins/(xmax-xmin);
 for(int t=0;t<T;t++) ++hist[((int) floor(grad*(x[t]-xmin)))];
}

int main(void)
{
 int T = 10000;       // number of iterations
 double xmax = 1.0; // length of interval xmax-xmin
 double xmin = 0.0;
 double bin_width = 0.05;
 double* x = new double[T]; // memory allocation
 int n_bins = (int)(xmax-xmin)/bin_width;
 cout << "number of bins = " << n_bins << endl;

 // generating the data for the histogram
 x[0] = (sqrt(5.0)-1.0)/2.0; // initial value
 for(int t=0;t<(T-1);t++) x[t+1] = 4.0*x[t]*(1.0-x[t]);

 int* hist = new int[n_bins]; // memory allocation
 // setting hist[i] to zero
```

```
 for(int i=0;i<n_bins;i++) hist[i] = 0;
 histogram(x,hist,T,xmax,xmin,n_bins);
 for(int i=0;i<n_bins;i++)
 cout << "hist[" << i << "] = " << hist[i] << endl;
 delete[] x;  delete[] hist;
 return 0;
}
```

Consider the *sine map*. The sine map $f : [0,1] \to [0,1]$ is defined by

$$f(x) := \sin(\pi x)\,.$$

The map can also be written as a difference equation

$$x_{t+1} = \sin(\pi x_t), \qquad t = 0, 1, \ldots$$

where $x_0 \in [0,1]$. The fixed points are determined by the solution of the equation

$$x^* = \sin(\pi x^*)\,.$$

The map admits two fixed points. One fixed point is given by $x_1^* = 0$. The other fixed point is determined from $x_2^* = \sin(\pi x_2^*)$ and $x_2^* > 0$. We find $x_2^* = 0.73648\ldots$. The variational equation of the sine map takes the form

$$y_{t+1} = \pi \cos(\pi x_t) y_t, \qquad t = 0, 1, \ldots\,.$$

Both fixed points are unstable. This can be seen by inserting x_1^* and x_2^* into the variational equation.

To find the invariant density for the sine-map we replace in program `invdensity.cpp` the line

```
x[t+1] = 4.0*x[t]*(1.0-x[t]);
```

with

```
x[t+1] = sin(pi*x[t]);
```

and add `const double pi=3.1415927;` in front of this statement. The numerical result suggests that the density for the sine map is quite similar to that of the logistic map.

Example. We find the invariant density for the *bungalow-tent map* $f_r : [0,1] \to [0,1]$

$$f_r(x) := \begin{cases} \dfrac{1-r}{r}x & \text{for} \quad x \in [0,r) \\[2mm] \dfrac{2r}{1-2r}x + \dfrac{1-3r}{1-2r} & \text{for } x \in [r, 1/2) \\[2mm] \dfrac{2r}{1-2r}(1-x) + \dfrac{1-3r}{1-2r} & \text{for } x \in [1/2, 1-r) \\[2mm] \dfrac{1-r}{r}(1-x) & \text{for} \quad x \in [1-r, 1] \end{cases}$$

where $r \in (0, 1/2)$. To find the invariant density exactly we solve the Frobenius-Perron integral equation. The Frobenius-Perron integral equation is given by

$$\rho_r(x) = \int_0^1 \rho_r(y)\delta(x - f_r(y))dy\,.$$

We apply the identities for the *delta function*

$$\delta(cy) \equiv \frac{1}{|c|}\delta(y), \qquad \delta(g(y)) \equiv \sum_n \frac{1}{|g'(y_n)|}\delta(y - y_n)$$

where the sum over n runs over all zeros with multiplicity 1 and $g'(y_n)$ denotes the derivative of g taken at y_n. Taking these identities into account and differentiating in the sense of generalized functions we obtain the invariant density

$$\rho_r(x) = \frac{1}{2-3r}\chi_{[0,1-r]}(x) + \frac{1-2r}{r(2-3r)}\chi_{(1-r,1]}(x)$$

where χ is the *indicator function*, i.e.,

$$\chi_A(x) := \begin{cases} 1 \text{ if } x \in A \\ 0 \text{ if } x \notin A \end{cases}.$$

Thus the invariant density is constant in the interval $[0, 1-r)$. At $1-r$ the invariant density jumps to another constant value. In the calculations we have to consider two domains for x, $[0, 1-r)$ and $[1-r, 1]$. The Liapunov exponent is calculated using

$$\lambda(r) = \int_0^1 \rho_r(x) \ln\left|\frac{df_r}{dx}\right| dx$$

where we differentiated in the sense of generalized functions. Thus we find that the Liapunov exponent as a smooth function of the control parameter r is given by

$$\lambda(r) = \frac{1-r}{2-3r}\ln\left(\frac{1-r}{r}\right) + \frac{1-2r}{2-3r}\ln\left(\frac{2r}{1-2r}\right).$$

For $r = 1/3$ we obviously obtain $\lambda(1/3) = \ln 2$. This is the Liapunov exponent for the tent map. For $r \to 0$ we obtain

$$\lambda(r \to 0) = \frac{1}{2}\ln 2\,.$$

For $r \to 1/2$ we obtain $\lambda(r \to 1/2) = 0$. The Liapunov exponent $\lambda(r)$ has a maximum for $r = 1/3$ (tent map). Furthermore $\lambda(r)$ is a convex function in the interval $(0, 1/2)$. We have $\lambda(r) \leq \ln 2$. The numerical simulation confirms the result for the invariant density, i.e. constant in the interval $[0, 1-r)$ and another constant in the interval $[1-r, 1]$. In the numerical simulation we have to set one of the bins boundary points to $1-r$. ♣

1.1.4 Liapunov Exponent

Here we calculate the *Liapunov exponent* λ for one-dimensional chaotic maps. Consider the one-dimensional map

$$x_{t+1} = f(x_t)$$

where $t = 0, 1, 2, \ldots$ and $x_0 \in [0, 1]$. The *variational equation* (also called the *linearized equation*) of this map takes the form

$$y_{t+1} = \frac{df}{dx}(x_t) y_t$$

with $y_0 \neq 0$. We assumed that f is differentiable. The Liapunov exponent λ is defined as

$$\lambda(x_0, y_0) := \lim_{T \to \infty} \frac{1}{T} \ln \left| \frac{y_T}{y_0} \right| .$$

Example. We calculate the Liapunov exponent for the logistic map. Thus $f(x) = 4x(1 - x)$ and

$$\frac{df}{dx} = 4 - 8x \, .$$

Consequently we obtain the variational equation

$$y_{t+1} = (4 - 8x_t) y_t, \qquad t = 0, 1, \ldots$$

with $y_0 \neq 0$. The exact solution of the logistic map is given by

$$x_t = \frac{1}{2} - \frac{1}{2} \cos(2^t \arccos(1 - 2x_0)).$$

For almost all initial values the Liapunov exponent is given by $\lambda = \ln 2$. ♣

In the C++ program `Liapunov1.cpp` we evaluate the Liapunov exponent by using the variational equation. Overflow occurs if T is made too large. In an alternative method we use nearby trajectories and reset the distance between the two trajectories after each time step. Thus we avoid overflow for large T.

```
// Liapunov1.cpp

#include <iostream>
#include <cmath>        // for fabs, log
using namespace std;

int main(void)
{
 unsigned long T = 200; // number of iterations
 double x = 0.3;        // initial value for logistic map
 double y = 1.0;        // initial value for variational map
 double x1, y1;
 for(unsigned long t=0;t<T;t++)
```

```
 { x1 = x;  y1 = y;
 x = 4.0*x1*(1.0-x1); // logistic map
 y = (4.0-8.0*x1)*y1; // variational map
 }
 // notice that y becomes large very quickly
 double lambda = log(fabs(y))/((double) T); // Liapunov exponent
 cout << "lambda = " << lambda << endl;

 // alternative method
 double eps = 0.001; double xeps, xeps1;
 x = 0.3;  xeps = x-eps;
 // x and xeps are nearby points
 double sum = 0.0;
 T = 1000;
 double distance;
 for(unsigned long t=0;t<T;t++)
 {
 x1 = x;  xeps1 = xeps;
 x = 4.0*x1*(1.0-x1); xeps = 4.0*xeps1*(1.0-xeps1);
 double distance = fabs(x-xeps);
 sum += log(distance/eps);
 xeps = x-eps;
 }
 lambda = sum/((double) T);
 cout << "lambda = " << lambda << endl;
 return 0;
}
```

In the following C++ program we use the Rational, Verylong and Derive class of SymbolicC++ to find an approximation of the Liapunov exponent. The Derive class provides the derivative. Thus the variational equation is obtained via exact differentiation

```
// Liapunov2.cpp
// iteration of logistic equation and variational equation

#include <iostream>
#include <cmath>      // for fabs, log
#include "verylong.h"
#include "rational.h"
#include "derive.h"
using namespace std;

int main(void)
{
 int T = 100;            // number of iterations
 double x = 1.0/3.0;     // initial value
 double x1;
 double y = 1.0;
```

```
 Derive<double> C1(1.0);  // constant 1.0
 Derive<double> C4(4.0);  // constant 4.0
 Derive<double> X;
 cout << "t = 0   x = " << x << "    " << "y = " << y << endl;
 for(int t=1;t<=T;t++)
 { x1 = x; x = 4.0*x1*(1.0-x1);
 X.set(x1);
 Derive<double> Y = C4*X*(C1-X);
 y = df(Y)*y;
 cout << "t = " << t << "    " << "x = " << x << "    "
      << "y = " << y << endl;
 }
 double lambda = log(fabs(y))/((double) T);
 cout << "approximate value for lambda = " << lambda << endl;
 int M = 9;
 Rational<Verylong> u1;
 Rational<Verylong> u("1/3"), v("1");
 Rational<Verylong> K1("1"), K2("4");
 Derive<Rational<Verylong> > D1(K1); // constant 1
 Derive<Rational<Verylong> > D4(K2); // constant 4
 Derive<Rational<Verylong> > U;
 cout << "j = 0   u = " << u << "    " << "v = " << v << endl;
 for(int j=1;j<=M;j++)
 {
 u1 = u; u = K2*u1*(K1-u1);
 U.set(Rational<Verylong>(u1));
 Derive<Rational<Verylong> > V = D4*U*(D1-U);
 v = df(V)*v;
 cout << "j = " << j << "    "
      << "u = " << u << "    " << "v = " << v << endl;
 }
 lambda = log(fabs(double(v)))/((double) M);
 cout << "approximate value for lambda = " << lambda << endl;
 return 0;
}
```

Consider the sine map. The *sine map* $f : [0,1] \to [0,1]$ is defined by

$$f(x) := \sin(\pi x) .$$

The map can be written as the difference equation

$$x_{t+1} = \sin(\pi x_t)$$

where $t = 0, 1, 2, \ldots$ and $x_0 \in [0,1]$. The variational equation of the sine equation is given by

$$y_{t+1} = \frac{df}{dx}(x = x_t) y_t = \pi \cos(\pi x_t) y_t .$$

To find the Liapunov exponent for the sine-map we replace in the C++ program `Liapunov1.cpp` the line

```
x = 4.0*x1*(1.0-x1);  xeps = 4.0*xeps1*(1.0-xeps1);
```

by

```
x = sin(pi*x1);  xeps = sin(pi*xeps1);
```

and add `const double pi = 3.14159;` in front of this statement. For $T = 5000$ we find $\lambda = 0.689$. Thus there is numerical evidence that the sine-map shows chaotic behaviour.

1.1.5 Autocorrelation Function

Consider a one-dimensional difference equation $f : [0,1] \to [0,1]$

$$x_{t+1} = f(x_t)$$

where $t = 0, 1, 2, \ldots$. The *time average* is defined as

$$\langle x_t \rangle := \lim_{T\to\infty} \frac{1}{T} \sum_{t=0}^{T-1} x_t \,.$$

Obviously, $\langle x_t \rangle$ depends on the initial value x_0. The *autocorrelation function* is defined as

$$C_{xx}(\tau) := \lim_{T\to\infty} \frac{1}{T} \sum_{t=0}^{T-1} (x_t - \langle x_t \rangle)(x_{t+\tau} - \langle x_t \rangle)$$

where $\tau = 0, 1, 2, \ldots$. The autocorrelation function depends on the initial value x_0.

Example. For the *logistic map* $f : [0,1] \to [0,1]$, $f(x) = 4x(1-x)$ we find that the time average for almost all initial conditions is given by

$$\langle x_t \rangle = \frac{1}{2} \,.$$

The autocorrelation function is given by

$$C_{xx}(\tau) = \begin{cases} \frac{1}{8} & \text{for} \quad \tau = 0 \\ 0 & \text{otherwise} \end{cases}$$

for almost all initial conditions. ♣

The C++ program `autocorrelation.cpp` calculates the time average and autocorrelation function for the logistic map.

```
// autocorrelation.cpp

#include <iostream>
using namespace std;
```

```
double average(double* x,int T)
{
 double sum = 0.0;
 for(int t=0;t<T;t++) { sum += x[t]; }
 double av = sum/((double) T);
 return av;
}

void autocorr(double* x,double* CXX,int T,int length,double av)
{
 for(int tau=0;tau<length;tau++)
 {
 double C = 0.0;
 double diff = (double) (T-length);
 for(int t=0;t<diff;t++) { C += (x[t]-av)*(x[t+tau]-av); }
 CXX[tau] = C/(diff+1.0);
 } // end for loop tau
}

int main(void)
{
 const int T = 4096;
 double* x = new double[T];
 x[0] = 1.0/3.0;              // initial value
 for(int t=0;t<(T-1);t++) { x[t+1] = 4.0*x[t]*(1.0-x[t]); }
 double av = average(x,T);
 cout << "average value = " << av << endl;
 int length = 11;
 double* CXX = new double[length];
 autocorr(x,CXX,T,length,av);
 delete[] x;
 for(int tau=0;tau<length;tau++)
 cout << "CXX[" << tau << "] = " << CXX[tau] << endl;
 delete[] CXX;
 return 0;
}
```

The output is (exact solution is `0.5`, `CXX[0]=1/8`, `CXX[1]=0`, `CXX[2]=0` etc.)

```
average value = 0.497383
CXX[0] = 0.125707
CXX[1] = 0.00134996
CXX[2] = -0.000105384
CXX[3] = -0.000289099
CXX[4] = 0.00477107
CXX[5] = -0.00186259
CXX[6] = 0.00383531
CXX[7] = -0.00425356
```

```
CXX[8] = -0.00288615
CXX[9] = -0.00110183
CXX[10] = -0.00148765
```

1.1.6 Discrete One-Dimensional Fourier Transform

The *discrete Fourier transform* is an approximation of the continuous Fourier transform. The discrete transform is used when a set of function sample values, $x(t)$, are available at equally spaced time intervals numbered $t = 0, 1, \ldots, T-1$. The discrete Fourier transform maps the given set of function values into a set of uniformly spaced sine waves whose frequencies are numbered $k = 0, 1, \ldots, T-1$, and whose amplitudes are given by

$$\hat{x}(k) = \frac{1}{T}\sum_{t=0}^{T-1} x(t)\exp\left(-i2\pi k\frac{t}{T}\right).$$

This equation can be written as

$$\hat{x}(k) = \frac{1}{T}\sum_{t=0}^{T-1} x(t)\cos\left(2\pi k\frac{t}{T}\right) - \frac{i}{T}\sum_{t=0}^{T-1} x(t)\sin\left(2\pi k\frac{t}{T}\right).$$

The inverse discrete Fourier transformation is given by

$$x(t) = \sum_{k=0}^{T-1} \hat{x}(k)\exp\left(i2\pi t\frac{k}{T}\right).$$

To find the inverse Fourier transformation we use the fact that

$$\sum_{k=0}^{T-1} \exp\left(i2\pi k\frac{(n-m)}{T}\right) = T\delta_{nm}$$

where δ_{nm} denotes the Kronecker symbol.

In the first C++ program (`Fourier.cpp`) we consider the time series

$$x(t) = \cos(2\pi t/T)$$

where $T = 8$ and $t = 0, 1, 2, \ldots, T-1$. We find the discrete Fourier transform $\hat{x}(k)$ $(k = 0, 1, \ldots, T-1)$. We have

$$\hat{x}(k) = \frac{1}{8}\sum_{t=0}^{7} \cos\left(\frac{2\pi t}{8}\right) e^{-i2\pi kt/8}.$$

Using the identity

$$\cos(2\pi t/8) \equiv \frac{e^{i2\pi t/8} + e^{-i2\pi t/8}}{2}$$

we find

$$\hat{x}(k) = \frac{1}{16}\sum_{t=0}^{7}(e^{i2\pi t(1-k)/8} + e^{-i2\pi t(1+k)/8}).$$

Consequently,

$$\hat{x}(k) = \begin{cases} \frac{1}{2} & \text{for} \quad k = 1 \\ \frac{1}{2} & \text{for} \quad k = 7 \\ 0 \text{ otherwise} \end{cases}.$$

```
// fourier.cpp

#include <iostream>
#include <cmath>      // for cos, sin
using namespace std;

int main(void)
{
 const double pi = 3.14159;
 int T = 8;
 double* x = new double[T];
 for(int t=0;t<T;t++) x[t] = cos(2.0*pi*t/((double) T));
 double* rex = new double[T]; double* imx = new double[T];
 for(int k=0;k<T;k++)
 { double cossum = 0.0, sinsum = 0.0;
 for(int t=0;t<T;t++)
 {
 cossum += x[t]*cos(2.0*pi*k*t/((double) T));
 sinsum += x[t]*sin(2.0*pi*k*t/((double) T));
 }
 rex[k] = cossum/((double) T); imx[k] = -sinsum/((double) T);
 }
 // display the output
 for(int k=0;k<T;k++)
 {
 cout << "rex["<< k <<"] = " << rex[k] << "    ";
 cout << "imx["<< k <<"] = " << imx[k] << endl;
 }
 delete[] x;  delete[] rex;  delete[] imx;
 return 0;
}
```

In the next C++ program (`fourierlog.cpp`) we consider the logistic map $x_{t+1} = 4x_t(1-x_t)$, where $t = 0, 1, 2, \ldots$ and $x_0 \in [0, 1]$. We assume that we have a set of T samples from the logistic map, i.e., x_0, x_1, x_2, $\ldots$, x_{T-1}.

```
// fourierlog.cpp

#include <iostream>
```

```
#include <cmath>       // for cos, sin
using namespace std;

int main(void)
{
 const double pi = 3.1415927;
 int T = 256;
 double* x = new double[T];
 x[0] = 0.5;
 for(int t=0;t<(T-1);t++) x[t+1] = 4.0*x[t]*(1.0-x[t]);
 double* rex = new double[T]; double* imx = new double[T];

 for(int k=0;k<T;k++)
 { double cossum = 0.0, sinsum = 0.0;
 for(int t=0;t<T;t++)
 {
 cossum += x[t]*cos(2.0*pi*k*t/((double) T));
 sinsum += x[t]*sin(2.0*pi*k*t/((double) T));
 }
 rex[k] = cossum/((double) T); imx[k] = -sinsum/((double) T);
 }
 // display the output
 for(int k=0;k<T;k++)
 {
 cout << "rex[" << k << "] = " << rex[k] << "    ";
 cout << "imx[" << k << "] = " << imx[k] << endl;
 }
 delete[] x; delete[] rex; delete[] imx;
 return 0;
}
```

1.1.7 Fast Fourier Transform

Let $n \geq 1$. The discrete Fourier transform transforms an n-vector with real components into a complex n-vector. Methods that compute the discrete Fourier transform in $O(N \log N)$ complex floating-point operations are referred to as *fast Fourier transforms*, FFT for short. Based on the odd-even decomposition of a trigonometric polynomial, a problem of size $n = 2^k$ is reduced to two problems of size 2^{k-1}. Subsequently, two problems of size 2^{k-1} are reduced to two problems of size 2^{k-2}. Finally, $n = 2^k$ problems of size 1 are obtained, each of which is solved trivially. Let ω be a *primitive nth root* of 1, i.e.

$$\omega := \exp(2\pi i/n).$$

The matrix F_n denotes the $n \times n$ matrix with entries

$$f_{jk} := \omega^{jk} \equiv e^{2\pi ijk/n}$$

where $0 \le j, k \le n-1$. The discrete Fourier transform of the n-vector

$$P^T = (p_0, p_1, \ldots, p_{n-1})$$

is the product $F_n P$. The components of $F_n P$ are

$$\begin{aligned}
(F_nP)_0 &= \omega^0 p_0 + \omega^0 p_1 + \cdots + \omega^0 p_{n-2} + \omega^0 p_{n-1} \\
(F_nP)_1 &= \omega^0 p_0 + \omega p_1 + \cdots + \omega^{n-2} p_{n-2} + \omega^{n-1} p_{n-1} \\
&\vdots \\
(F_nP)_i &= \omega^0 p_0 + \omega^i p_1 + \cdots + \omega^{i(n-2)} p_{n-2} + \omega^{i(n-1)} p_{n-1} \\
&\vdots \\
(F_nP)_{n-1} &= \omega^0 p_0 + \omega^{n-1} p_1 + \cdots + \omega^{(n-1)(n-2)} p_{n-2} + \omega^{(n-1)(n-1)} p_{n-1} .
\end{aligned}$$

Rewritten in a slightly different form, the ith component is

$$p_{n-1}(\omega^i)^{n-1} + p_{n-2}(\omega^i)^{n-2} + \cdots + p_1\omega^i + p_0 .$$

If we interpret the components of P as coefficients of the polynomial

$$p(x) = p_{n-1}x^{n-1} + p_{n-2}x^{n-2} + \cdots + p_1 x + p_0$$

then the ith component is $p(\omega^i)$ and computing the discrete Fourier transform of P means evaluating the polynomial $p(x)$ at $\omega^0, \omega, \omega^2, \ldots, \omega^{n-1}$, i.e., at each of the nth roots of 1. A *Divide and Conquer algorithm* is as follows. Assume that $n = 2^k$ for some $k \ge 0$. The problem is divided into smaller instances, solve those, and use the solutions to get the solution for the current instance. To evaluate p at n points, we evaluate two smaller polynomials at a subset of the points and then combine the results appropriately. Since $\omega^{n/2} = -1$ we have for $0 \le j \le n/2 - 1$,

$$\omega^{(n/2)+j} = -\omega^j .$$

We order the terms of $p(x)$ with even powers and the terms with odd powers as follows

$$p(x) = \sum_{i=0}^{n-1} p_i x^i \equiv \sum_{i=0}^{n/2-1} p_{2i} x^{2i} + x \sum_{i=0}^{n/2-1} p_{2i+1} x^{2i} .$$

We define

$$p_{even}(x) := \sum_{i=0}^{n/2-1} p_{2i} x^i , \qquad p_{odd}(x) := \sum_{i=0}^{n/2-1} p_{2i+1} x^i .$$

It follows that

$$p(x) = p_{even}(x^2) + x \cdot p_{odd}(x^2), \qquad p(-x) = p_{even}(x^2) - x \cdot p_{odd}(x^2) .$$

To evaluate p at

$$1,\ \omega,\ \ldots,\ \omega^{(n/2)-1},\ -1,\ -\omega,\ \ldots,\ -\omega^{(n/2)-1}$$

it suffices to evaluate p_{even} and p_{odd} at

$$1, \quad \omega^2, \quad \dots \quad , (\omega^{(n/2)-1})^2$$

and then do $n/2$ multiplications (for $x \cdot p_{odd}(x^2)$) and n additions and subtractions. The polynomials p_{even} and p_{odd} can be evaluated recursively by the same scheme. They are polynomials of degree $n/2 - 1$ and will be evaluated at the $n/2th$ roots of unity

$$1, \quad \omega^2, \dots, (\omega^{(n/2)-1})^2 .$$

```
// fft1.cpp

#include <iostream>
#include <cmath>      // for cos, sin
using namespace std;

void p(double wre,double wim,double *re,double *im,
       double &fftre,double &fftim,const int M,int step,int init)
{
 double pre, pim, w2re, w2im;
 if(step==(1 << M))  // << shift operator
 {
 fftre = re[init]*wre-im[init]*wim;
 fftim = im[init]*wre+re[init]*wim;
 return;
 }
 w2re = wre*wre-wim*wim; w2im = 2.0*wre*wim;
 p(w2re,w2im,re,im,pre,pim,M,step<<1,init);      // peven
 fftre = pre; fftim = pim;
 p(w2re,w2im,re,im,pre,pim,M,step<<1,init+step); // podd
 fftre += wre*pre-wim*pim; fftim += wre*pim+wim*pre;
}

void fft(double *re,double *im,double *ftre,double *ftim,const int M)
{
 const double pi = 3.1415927;
 int N = 1 << M;
 double fftre, fftim, wre, wim, w2re, w2im;
 for(int i=0;i<(N>>1);i++)
 {
 wre = cos(i*2.0*pi/N); wim = sin(i*2.0*pi/N);
 w2re = wre*wre-wim*wim; w2im = 2.0*wre*wim;
 p(w2re,w2im,re,im,fftre,fftim,M,2,0);  // peven
 ftre[i] = ftre[i+(N>>1)] = fftre;
 ftim[i] = ftim[i+(N>>1)] = fftim;
 p(w2re,w2im,re,im,fftre,fftim,M,2,1);  // podd
 ftre[i] += wre*fftre-wim*fftim;
 ftre[i+(N>>1)] -= wre*fftre-wim*fftim;
```

```
 ftim[i] += wre*fftim+wim*fftre;
 ftim[i+(N>>1)] -= wre*fftim+wim*fftre;
 }
}

int main(void)
{
 const double pi = 3.1415927;
 const int M = 3;
 int T = 1 << M;
 double* re = new double[T]; double* im = new double[T];
 double* fftre = new double[T]; double* fftim = new double[T];
 for(int i=0;i<T;i++) { re[i] = cos(2.0*i*pi/T); }
 for(int k=0;k<T;k++) { im[k] = 0.0; }
 fft(re,im,fftre,fftim,M);
 for(int k=0;k<T;k++)
 cout << "fftre[" << k << "]=" << fftre[k]/T << endl;
 cout << endl;
 for(int k=0;k<T;k++)
 cout << "fftim[" << k << "]=" << fftim[k]/T << endl;
 delete[] re; delete[] im; delete[] fftre; delete[] fftim;
 return 0;
}
```

A nonrecursive version is given below. We use in place substitution.

```
// FFT2.cpp

#include <iostream>
#include <cmath>      // for sqrt, cos
using namespace std;

// dir = 1 gives the FFT tranform
// dir = -1 gives the inverse FFT transform
// n = 2^m is the length of the time series
// x[] is the real part of the signal
// y[] is the imaginary part of the signal

void FFT(int dir,unsigned long m,double* x,double* y)
{
 unsigned long n, i, i1, j, k, i2, l, l1, l2;
 double c1, c2, tx, ty, t1, t2, u1, u2, z;
 // number of points n = 2^m
 n = 1;
 for(i=0;i<m;i++) n *= 2;
 // bit reversal
 i2 = n >> 1;
 j = 0;
 for(i=0;i<n-1;i++)
```

```
 {
 if(i < j)
 {
 tx = x[i]; ty = y[i];
 x[i] = x[j]; y[i] = y[j]; x[j] = tx; y[j] = ty;
 }
 k = i2;
 while(k <= j) { j -= k; k >>= 1; }
 j += k;
 } // end for loop

 // compute the FFT
 c1 = -1.0; c2 = 0.0; l2 = 1;
 for(l=0;l<m;l++)
 {
 l1 = l2; l2 <<= 1; u1 = 1.0; u2 = 0.0;
 for(j=0;j<l1;j++)
 {
 for(i=j;i<n;i+=l2)
 {
 i1 = i + l1;
 t1 = u1*x[i1]-u2*y[i1]; t2 = u1*y[i1]+u2*x[i1];
 x[i1] = x[i]-t1; y[i1] = y[i]-t2;
 x[i] += t1; y[i] += t2;
 }
 z = u1*c1-u2*c2;
 u2 = u1*c2+u2*c1; u1 = z;
 }
 c2 = sqrt((1.0-c1)/2.0);
 if(dir==1) c2 = -c2;
 c1 = sqrt((1.0+c1)/2.0);
 }
 if(dir==1)
 { for(i=0;i<n;i++) { x[i] /= n; y[i] /= n; } }
} // end function FFT

unsigned long power(unsigned long m)
{
 unsigned long r = 1;
 for(unsigned long i=0;i<m;i++) r *= 2;
 return r;
}

int main(void)
{
 unsigned long m = 3;
 const double pi = 3.1415927;
 unsigned long n = power(m);
```

```
double* x = new double[n];  double* y = new double[n];
unsigned long k;
for(k=0;k<n;k++) { x[k] = cos(2.0*pi*k/n); y[k] = 0.0; }
// call FFT
FFT(1,m,x,y);
for(k=0;k<n;k++) { cout << x[k] << "   " << y[k] << endl; }
// call inverse FFT
cout << "calling inverse FFT" << endl;
FFT(-1,m,x,y);
for(k=0;k<n;k++) { cout << x[k] << "   " << y[k] << endl; }
return 0;
}
```

1.1.8 Logistic Map and Liapunov Exponent for $r \in [3,4]$

We consider the logistic map

$$x_{t+1} = rx_t(1 - x_t)$$

where $t = 0, 1, 2, \ldots$, $x_0 \in [0,1]$ and $r \in [3,4]$. Here r is the bifurcation parameter. Thus the Liapunov exponent depends on r. We evaluate the Liapunov exponent for $r \in [3,4]$. The variational equation is given by

$$y_{t+1} = r(1 - 2x_t)y_t.$$

The *Liapunov exponent* is defined as

$$\lambda(x_0, y_0) := \lim_{T\to\infty} \frac{1}{T} \ln \left| \frac{y_T}{y_0} \right|.$$

The point $r = 3$ is a bifurcation point. The Liapunov exponent is given by $\lambda = 0$ for $r = 3$. In the range $3 < r < 3.5699.....$ we find periodic solutions. The Liapunov exponent is negative. We also find *period doubling*. In the region

$$3.5699... < r < 4$$

we find chaotic behaviour (positive Liapunov exponent) but also periodic windows. For example in the region

$$3.828... < r < 3.842...$$

we have a trajectory with period 3. The Liapunov exponent can be evaluated exactly only for $r = 4$. One finds $\lambda(r = 4) = \ln 2$ for almost all initial values. In the program `lambdaf.cpp` the Liapunov exponent is evaluated for the interval $r \in [3.0, 4.0]$ with step size 0.001.

```
// lambdaf.cpp

#include <fstream>
#include <cmath>    // for fabs, log
```

```
using namespace std;

int main(void)
{
 ofstream data("lambda.dat");
 int T = 10000;          // number of iterations
 double x = 0.618;       // initial value
 double x1;
 double eps = 0.0005;
 double xeps = x-eps;
 double xeps1;
 double r = 3.0;  double sum = 0.0;
 while(r <= 4.0)
 {
 for(int t=0;t<T;t++)
 { x1 = x; x = r*x1*(1.0-x1);
 xeps1 = xeps; xeps = r*xeps1*(1.0-xeps1);
 double distance = fabs(x-xeps);
 sum += log(distance/eps);
 xeps = x-eps;
 }
 double lambda = sum/((double) T);
 data << r << "  " << lambda << "\n";
 sum = 0.0;
 r += 0.001;
 } // end while
 data.close();
 return 0;
}
```

1.1.9 Logistic Map and Bifurcation Diagram

We consider the logistic map

$$x_{t+1} = rx_t(1 - x_t)$$

where $r \in [2, 4]$ and $x_0 \in [0, 1]$. Here r is a bifurcation parameter. We now study the bifurcation diagram. For $r \in [2, 3)$ the fixed point $x^* = 1 - 1/r$ is stable. The fixed point $x^* = 0$ is unstable in the range $(2, 4]$. For $r = 3$ (bifurcation point) the stable fixed point $x^* = 1 - 1/r$ becomes unstable. We find a stable orbit of period 2. With increasing r we find a *period doubling* process with repeated bifurcation from

$$2,\ 4,\ 8, \ldots, 2^n, \ldots .$$

There is a threshold value

$$r_\infty = 3.5699...$$

for the parameter r where the limit 2^n, $n \to \infty$ of the periodicity is reached. For $r = 4$ the logistic map and all its iterates are ergodic and mixing. Within the interval $(r_\infty, 4)$ period triplings $p3^n$ and quadruplings $p4^n$ etc. also occur (so-called *periodic windows*)

In the Java program `Bifurcationlo.java` we display the bifurcation diagram for the interval $r \in [2.0, 4.0]$.

```
// Bifurcationlo.java

import java.awt.*;
import java.awt.Frame;
import java.awt.event.*;
import java.awt.Graphics;

public class Bifurcationlo extends Frame
{
 public Bifurcationlo()
 {
 setSize(600,500);
 addWindowListener(new WindowAdapter()
 { public void windowClosing(WindowEvent event) { System.exit(0); }}); }

 public void paint(Graphics g)
 {
 int xmax = 600; int ymax = 400;
 int j, k, m, n;
 double x, xplot, yplot;
 double r = 2.0;  // bifurcation parameter
 while(r <= 4.0)
 { xplot = xmax*(r-2.0)/2.0; x = 0.5;
 for(j=0;j<400;j++) { x = r*x*(1.0-x); }
 for(k=0;k<400;k++)
 { x = r*x*(1.0-x);
 yplot = ymax*(1.0-x);
 m = (int) Math.round(xplot); n = 50 + (int) Math.round(yplot);
 g.drawLine(m,n,m,n);
 }
 r += 0.0005;
 }  // end while
 }

 public static void main(String[] args)
 { Frame f = new Bifurcationlo(); f.setVisible(true); }
}
```

1.1.10 Random Number Map and Invariant Density

We consider methods for generating a sequence of random fractions, i.e., random real numbers u_t, uniformly distributed between zero and one. Since a computer can represent a real number with only finite accuracy, we shall actually be generating integers x_t between zero and some number m. The fraction

$$u_t = x_t/m, \qquad t = 0, 1, 2, \ldots$$

will then lie between zero and one. Usually m is the word size of the computer, so x_t may be regarded as the integer contents of a computer word with the radix point assumed at the extreme right, and u_t may be regarded as the contents of the same word with the radix point assumed at the extreme left. The most popular random number generators are special cases of the following scheme. We select four numbers

$$\begin{array}{ll} m, \text{the modulus;} & m > 0 \\ a, \text{ the multiplier;} & 0 \leq a < m \\ c, \text{ the increment;} & 0 \leq c < m \\ x_0, \text{the initial value;} & 0 \leq x_0 < m. \end{array}$$

The desired sequence of pseudo-random numbers $x_0, x_1, x_2, \ldots$ is then obtained by the one-dimensional difference equation

$$x_{t+1} = (ax_t + c) \mod m, \qquad t = 0, 1, 2, \ldots \; .$$

This is also called a *linear congruential sequence.* Taking the remainder mod m is somewhat like determining where a ball will land in a spinning roulette wheel.

Example. The sequence obtained when $m = 10$ and $x_0 = a = c = 7$ is

$$7, \quad 6, \quad 9, \quad 0, \quad 7, \quad 6, \quad 9, \quad 0, \ldots .$$

This example shows that the sequence is not always very "random" for all choices of m, a, c, and x_0. ♣

The example also illustrates the fact that congruential sequences always "get into a loop"; i.e., there is ultimately a cycle of numbers which is repeated endlessly. The repeating cycle is called the period. The sequence given above has a period of length 4. A useful sequence will of course have a relatively long period.

In the C++ program we implement a linear congruential sequence.

```
// modulus.cpp

#include <iostream>
using namespace std;

int main(void)
{
 unsigned long a = 7, c = 7;
 unsigned long m = 10; // modulus
 unsigned long x0 = 7; // initial value
 int T = 10;           // number of iterations
```

```
 unsigned long x1;
 for(int t=0;t<T;t++) { x1 = a*x0 + c;
 while(x1 >= m) x1 = x1-m;
 x0 = x1;
 cout << "x[" << t << "] = " << x0 << endl;
 }
 a = 3125; c = 47; m = 2048; // m modulus
 x0 = 3;  // initial value
 T = 12;  // number of iterations
 for(int t=0;t<T;t++)
 { x1 = a*x0 + c;
 while(x1 >= m) x1 = x1-m;
 x0 = x1;
 cout << "x[" << t << "] = " << x0 << endl;
 }
 return 0;
}
```

In the following C++ program we consider the chaotic sequence

$$x_{t+1} = (\pi + x_t)^5 \bmod 1 \equiv \mathrm{frac}(\pi + x_t)^5$$

and ask whether the sequence is uniformly distributed.

```
// random1.cpp

#include <iostream>
#include <cmath>              // for sqrt, fmod
using namespace std;

int main(void)
{
 const double pi = 3.14159;
 int T = 6000;                        // number of iterations
 double* x = new double[T];
 x[0] = (sqrt(5.0)-1.0)/2.0;  // initial value
 for(int t=0;t<(T-1);t++)
 { double r = x[t]+pi; x[t+1] = fmod(r*r*r*r*r,1); }
 const int N = 10;
 double hist[N];
 for(int j=0;j<N;j++) hist[j] = 0.0;
 for(int k=0;k<T;k++)
 hist[(int) floor(N*x[k])] = hist[(int) floor(N*x[k])]+1;
 for(int l=0;l<N;l++)
 cout << "hist[" << l << "] = " << hist[l] << endl;
 return 0;
}
```

1.1.11 Random Number Map and Random Integration

We describe the *Monte Carlo method* for the calculation of integrals. We demonstrate the technique on one-dimensional integrals. Let $f : [0,1] \to [0,1]$ be a continuous function. Consider the integral

$$I = \int_0^1 f(x)dx \,.$$

We choose N number pairs (x_j, y_j) with uniform distribution and define z_j by

$$z_j := \begin{cases} 0 \text{ if } y_j > f(x_j) \\ 1 \text{ if } y_j \le f(x_j). \end{cases}$$

Putting

$$n = \sum_{j=1}^{N} z_j$$

we have $n/N \simeq I$. More precisely, we find

$$I = n/N + O(N^{-1/2}).$$

The accuracy is not very good. The traditional formulas, such as Simpson's formula, are much better. However, in higher dimensions the Monte Carlo technique is favourable, at least if the number of dimensions is ≥ 6. We consider the integral as the mean value of $f(\xi)$ where ξ is uniform. An estimate of the mean value is

$$I \simeq \frac{1}{N} \sum_{j=1}^{N} f(\xi_j) \,.$$

This formula can easily be generalized to higher dimensions.

In the C++ program we use the map $f : [0,1) \to [0,1)$

$$f(x) = (x+\pi)^5 \mod 1$$

as random number generator and evaluate

$$\int_0^1 \sin(x)dx = 0.459697694132.$$

```
// randint.cpp

#include <iostream>
#include <cmath>
using namespace std;

void randval(double* x,double pi)
```

```
{ *x = fmod((*x+pi)*(*x+pi)*(*x+pi)*(*x+pi)*(*x+pi),1); }

int main(void)
{
 const double pi = 3.1415927;
 unsigned long T = 20000;  // number of iterations
 double x = 0.5;           // initial value
 double sum = 0.0;
 for(int t=0;t<T;t++) { randval(&x,pi); sum += sin(x); }
 cout << "The integral is = " << sum/((double) T);
 return 0;
}
```

The output is `0.461403`.

In the Java program we use the same map as in the C++ progam for the random number generator.

```
// Random1.java

class WrappedDouble
{
 WrappedDouble(final double value) { this.value = value; }
 public double value() { return value; }
 public void value(final double newValue) { value = newValue; }
 private double value;
}

class MathUtils
{
 public static void randval(WrappedDouble x)
 { double y = Math.pow(x.value()+Math.PI,5); x.value(y-Math.floor(y)); }
}

class Random1
{
 public static void main(String[] args)
 {
 int n = 20000;
 double sum = 0.0;
 WrappedDouble x = new WrappedDouble(0.5);
 for(int i=0;i<n;++i)
 { MathUtils.randval(x); sum += Math.sin(x.value()); }
 System.out.println("The integral is " + sum/n);
 }
}
```

1.1.12 Circle Map and Rotation Number

The *circle map* is given by

$$x_{t+1} = f(x_t) \equiv x_t + \Omega - \frac{r}{2\pi}\sin(2\pi x_t), \quad t = 0, 1, \ldots$$

which may be regarded as a transformation of the phase of one oscillator through a period of the second one. The map depends on two bifurcation parameters: Ω describes the ratio of undisturbed frequencies while the bifurcation parameter r governs the strength of the nonlinear interaction. The subcritical ($r < 1$) mappings are diffeomorphisms (and thus invertible) whereas the supercritical ones ($r > 1$) are non-invertible and may exhibit chaotic behaviour. The borderline between these two cases consists of the critical circle mappings - homeomorphisms with one (usually cubic) inflection point. This corresponds to $r = 1$ in the family of this map. The dynamics of the map may be characterized by the *rotation number* (also called winding number)

$$\rho := \lim_{T\to\infty} \frac{1}{T}(f^{(T)}(x) - x).$$

When f is invertible, the rotation number is well defined and independent of x. The inverse function f^{-1} does not exist for $r > 1$. For subcritical and critical maps this number does not depend on the initial point x. The dependence $\rho(\Omega)$ is the so-called *devil's staircase*, in which each rational $\rho = p/q$ is represented by an interval of Ω values (which is named the p/q-locking interval). The set of all these intervals has a full measure in the critical case. The locked motion in subcritical and critical cases is represented by a stable periodic orbit of period q. The rotation number is the mean number of rotations per iteration, i.e., the frequency of the underlying dynamical system. If $r = 0$ we obviously find $\rho = \Omega$. Under iteration the variable x_i may converge to a series which is either periodic,

$$x_{i+Q} = x_i + P$$

with rational rotation number $\rho = P/Q$; quasiperiodic, with irrational rotation number $\rho = q$; or chaotic where the sequence behaves irregularly.

```
// circle.cpp

#include <fstream>
#include <cmath>      // for sin
using namespace std;

int main(void)
{
 ofstream data("circle.dat");
 const double pi = 3.1415927;
 int T = 8000;     // number of iterations
 double r = 1.0;  // parameter of map
 double Omega = 0.0;
```

```
while(Omega <= 1.0)
{ double x = 0.3;      // inital value
double x0 = x;
double x1;
for(int t=0;t<T;t++)
{ x1 = x; x = x1+Omega-r*sin(2.0*pi*x1)/(2.0*pi); }
double rho = (x-x0)/((double) T);
data << Omega << "  " << rho << "\n";
Omega += 0.005;
} // end while
data.close();
return 0;
}
```

1.1.13 One-Dimensional Newton Method

Consider the equation $f(x) = 0$ where it is assumed that $f : \mathbb{R} \to \mathbb{R}$ is at least twice differentiable. Let I be some interval containing a root of f. A root is a point $\widetilde{x}$ such that $f(\widetilde{x}) = 0$. We assume that the root is simple (also called multiplicity one). The *Newton method* (Fröberg [64]) can be derived by taking the tangent line to the curve $y = f(x)$ at the point $(x_t, f(x_t))$ corresponding to the current estimate, x_t of the root. The intersection of this line with the x-axis gives the next estimate to the root, x_{t+1}. The gradient of the curve $y = f(x)$ at the point $(x_t, f(x_t))$ is $f'(x_t)$, where $'$ denotes differentiation. The tangent line at this point has the form $y = f'(x)x + b$. Since this passes through $(x_t, f(x_t))$ we see that $b = f(x_t) - x_t f'(x_t)$. Therefore the tangent line is

$$y = f'(x_t)x + f(x_t) - x_t f'(x_t)\,.$$

To determine where this line cuts the x-axis we set $y = 0$. Taking this point of intersection as the next estimate, x_{t+1}, to the root we have

$$0 = f'(x_t)x_{t+1} + f(x_t) - x_t f'(x_t)\,.$$

We obtain the first order difference equation

$$x_{t+1} = x_t - \frac{f(x_t)}{f'(x_t)}, \qquad t = 0, 1, 2, \ldots\,.$$

This is the Newton method. This scheme has the form 'next estimate = current estimate + correction term'. The correction term is $-f(x_t)/f'(x_t)$ and this must be small when x_t is close to the root if convergence is to be achieved. This will depend on the behaviour of $f'(x)$ near the root and, in particular, difficulty will be encountered when $f'(x)$ and $f(x)$ have roots close together. The Newton method is of the form $x_{t+1} = g(x_t)$ with

$$g(x) := x - \frac{f(x)}{f'(x)}\,.$$

The order of the method can be examined. Differentiating this equation leads to

$$g'(x) = \frac{f(x)f''(x)}{(f'(x))^2}.$$

For convergence we require that

$$\left| \frac{f(x)f''(x)}{(f'(x))^2} \right| < 1$$

for all x in some interval I containing the root. Since $f(\widetilde{x}) = 0$, the above condition is satisfied at the root $x = \widetilde{x}$ provided that $f'(\widetilde{x}) \neq 0$. Then provided that the function g is continuous, an interval I must exist in the neighbourhood of the root and over which the condition above is satisfied. Difficulty is sometimes encountered when the interval I is small, because the initial guess must be taken from this interval. This usually arises when $f(x)$ and $f'(x)$ have roots close together, since the correction term is inversely proportional to $f'(x)$.

In the C++ program `Newton.cpp` we consider the function

$$f(x) = x - \sin(\pi x)$$

in the interval $[0.5, 1]$. This means we find the fixed point of the sine-map

$$x_{t+1} = \sin(\pi x_t)$$

in the interval $[0.5, 1]$. We have

$$f'(x) = 1 - \pi \cos(\pi x), \qquad f''(x) = \pi^2 \sin(\pi x).$$

Thus the condition for convergence is satisfied.

```
// Newton.cpp

#include <iostream>
#include <cmath>      // for sin, cos, fabs
using namespace std;

double f(double x)
{ const double pi = 3.1415927; return x-sin(pi*x); }

double fder(double x)  // derivative of f
{ const double pi = 3.1415927; return 1.0-pi*cos(pi*x); }

double newton(double initial,double eps)
{
 double x0, x1; x1 = initial;
 do { x0 = x1; x1 = x0-f(x0)/fder(x0); } while(fabs(x1-x0) > eps);
 return x0;
```

```
}

int main(void)
{
 double initial = 0.5;  double eps = 0.0001;
 double result = newton(initial,eps);
 cout << "result = " << result << endl;
 return 0;
}
```

With the next C++ program we can find all real roots, where n the number of roots we want.

```
// allrootsnewton.cpp

#include <cmath>
#include <iostream>
#include <iterator>
#include <vector>
using namespace std;

double df(double(*f)(double),double x,double h=1e-8)
{ return (f(x+h)-f(x-h))/(2.0*h); }

vector<double> roots(double(*f)(double),int n,double eps)
{
 vector<double> v;
 vector<double>::iterator i;
 double x, delta, m = 1.0;
 while(n > 0)
 { x = m; delta = eps + 1.0;
 while(fabs(delta) >= eps)
 {
 double fx = f(x);
 double dfx = df(f,x);
 for(i=v.begin();i<v.end();i++)
 { dfx = (dfx-fx/(x-*i))/(x-*i); fx  = fx/(x-*i); }
 delta = -fx/dfx;
 x = x + delta;
 }
 v.push_back(x);
 if(m <= x) m = x + 1.0;
 n--;
 }
 return v;
}

double f(double x) { return 4.0*x*x*x-30.0*x*x+54.0*x-18.0; }
double g(double x) { return x*x-1.0; }
```

```
double h(double x) { return sin(x); }

int main(void)
{
 cout.precision(10);
 vector<double> v = roots(f,3,1e-8);
 cout << "[ ";
 copy(v.begin(),v.end(),ostream_iterator<double>(cout," "));
 cout << "]" << endl;

 vector<double> w = roots(g,2,1e-8);
 cout << "[ ";
 copy(w.begin(),w.end(),ostream_iterator<double>(cout," "));
 cout << "]" << endl;

 vector<double> u = roots(h,2,1e-8);
 cout << "[ ";
 copy(u.begin(),u.end(),ostream_iterator<double>(cout," "));
 cout << "]" << endl;
 return 0;
}
```

1.1.14 Feigenbaum's Constant

In a number of mappings which depend on a bifurcation parameter r we find a period doubling cascade. We consider the bifurcation parameter values where period-doubling events occur. The limit of the ratio of distances between consecutive doubling values is *Feigenbaum's constant*. It has the value

$$4.669201609102990671853...$$

One-dimensional maps which show this transition are

$$x_{t+1} = rx_t(1 - x_t), \qquad x_{t+1} = 1 - rx_t^2, \qquad x_{t+1} = x_t^2 + r\,.$$

The C++ program `feigenbaum1.cpp` finds the Feigenbaum constant using the equation $x_{t+1} = x_t^2 + r$. The program shows the constant computed for two doubling cascades. The first one starts with the period 1 cardioid and the second starts with the period 3 cardioid. Newton's method is used to find the root of $x = x^2 + r$ iterated n times.

```
// feigenbaum1.cpp

#include <cstdio>
#include <cstdlib>
#include <cmath>     // for fabs
using namespace std;
```

```
double newton(long n,double c)
{
 double x, x1; double nc = c; double absx = 1.0;
 long i, j; j = 0;
 while((j < 7) && (absx > 1E-13))
 { ++j; x = 0.0; x1 = 0.0;
 for(i=0;i<n;i++) { x1 = 2.0*x1*x+1.0; x = x*x+nc; }
 nc -= x/x1; absx = fabs(x);
 }
 return nc;
}

void go(long n0,double a,double b)
{
 double f = 4.0; double tmp = a;
 double newc = a+(a-b)/f; double oldc = b;
 long n = 2*n0;
 for(int i=0;i<10;++i)
 { newc = newton(n,newc);
 f = (tmp-oldc)/(newc-tmp);
 printf("%.16lf %.16lf %.16lf %.16lf\n",oldc,tmp,newc,f);
 oldc = tmp; tmp = newc;  newc += (newc-oldc)/f;
 n *= 2;
 }
}

int main(void)
{
 double a, b; long n;
 printf("c1                   c2                c3");
 printf("                   f: (c2-c1)/(c3-c2)");
 printf("\n");
 b = 0.0; a = -1.0; n = 2;
 go(n,a,b);
 printf("\n");
 a = -1.7728929033816238; b = -1.7548776662469276;
 n = 6;
 go(n,a,b);
 return 0;
}
```

1.1.15 Symbolic Dynamics

Consider a one-dimensional nonlinear map $f(x,r)$, which maps points from an interval I into the same interval

$$x_{t+1} = f(x_t, r), \qquad x_t \in I$$

where r is a control parameter. The function f may have several monotone branches, divided by turning points, denoted symbolically by C_i (called also *critical points*). For smooth maps the derivative df/dx vanishes at $x = C_i$. The turning points C_i and the end points of the interval I divide I into subintervals I_i. We label each of I_i by a symbol S_i. By iterating the map, we obtain a numerical sequence

$$x_0, \quad x_1 = f(x_0), \quad x_2 = f(x_1), \quad \ldots, \quad x_{t-1} = f(x_{t-2}), \quad x_t = f(x_{t-1}), \ldots.$$

We juxtapose the numerical sequence with a symbolic sequence (Bowen [22], Hao [81], Hao [82]) and call it by the number x_0 which has originated the numerical sequence

$$x_0 = \sigma_0\sigma_1\sigma_2 \ldots \sigma_{n-1}\sigma_n \ldots$$

where σ_i stands for one of the symbols S_j or C_j, depending on whether x_t belongs to the corresponding subinterval or coincides with a turning point. If we want to reverse the numerical sequence, expressing x_0 through x_t, we must indicate which of the monotone branches of f has been used at each iteration. To do so, we attach a subscript σ to f. The symbol σ is chosen by the argument of f

$$x_0, \quad x_1 = f_{\sigma 0}(x_0), \quad x_2 = f_{\sigma_1}(x_1), \quad \ldots, \quad x_t = f_{\sigma_{t-1}}(x_{t-1}), \quad \ldots.$$

Now we are in a position to reverse this sequence. We obtain

$$x_0 = f_{\sigma_0}^{-1} \circ f_{\sigma_1}^{-1} \circ \cdots \circ f_{\sigma_{t-1}}^{-1}(x_t).$$

To simplify the notation, we denote each inverse monotone branch f_σ^{-1} by its subscript, i.e. we define

$$\sigma(y) \equiv f_\sigma^{-1}(y)$$

where σ is one of the symbols S_i.

Example. The logistic map

$$x_{t+1} = 1 - rx_t^2, \qquad x_t \in [-1, 1], \qquad r \in (0, 2)$$

has two *inverse branches*

$$R(y) = \sqrt{(1-y)/r}, \qquad L(y) = -\sqrt{(1-y)/r}\,. \qquad \clubsuit$$

Now we find

$$x_0 = \sigma_0 \circ \sigma_1 \circ \cdots \circ \sigma_{t-1}(x_t).$$

Thus we found the number-symbol-inverse function correspondence.

The logistic map $f(x, r) = 1 - rx^2$ is a unimodal map. Since $df/dx = -2rx$ we find that $C = 0$ is the only critical point. The iterate of the critical point $C = 0$ leads to the rightmost point $f(C)$ on the interval that one can ever reach by iterating the map from any point on the interval. The point $f(C)$ thus starts a special symbolic sequence, called the *kneading sequence* The kneading sequence is named

$f(C)$ and sometimes denoted by K (for kneading). For instance, at the parameter value $r = 1.85$ the logistic map has a kneading sequence

$$K \equiv f(C) = RLLRLRLRRL\cdots$$

The second iterate of C i.e. $f^{(2)}(C)$, gives the leftmost point that one can ever reach by iterating the map twice starting from any point on the interval. All the interesting dynamics takes place on the subinterval

$$[f^{(2)}(C), f(C)]$$

of I. Once a point is in this subinterval, its iterates can never get out. Therefore, this subinterval defines an *invariant dynamical range*. In principle, one can choose an initial point outside this subinterval, but after a trivial transient (in fact, a few iterations), it will fall into the invariant dynamical range. For maps with multiple critical points each C_i leads to a kneading sequence; one collects the dynamical range of all turning points and finds the overall range of interest dynamics. In general one concentrates on the invariant dynamical range only, neglecting trivial transients. Each kneading sequence is represented by a number. This number may be taken as the parameter for the map. This is convenient for maps with multiple critical points. In other words, one can parameterize a map by its independently changing kneading sequences.

In the C++ program using the `Rational` and `Verylong` class of SymbolicC++ we find the kneading sequence for the logistic map with $r = 37/20$.

```
// kneading.cpp

#include <iostream>
#include <string>
#include "rational.h"
#include "verylong.h"
using namespace std;

int main(void)
{
 int n = 12;
 string s = "";
 Rational<Verylong> one("1"), zero("0");
 Rational<Verylong> r("37/20"); // control parameter
 Rational<Verylong> x = zero;   // initial value 0

 if(x < zero) s = s + "L";
 if(x == zero) s = s + "C";
 if(x > zero) s = s + "R";
 for(int i=1;i<(n-1);i++)
 {
 x = one-r*x*x;
```

```
 if(x < zero) s += "L";
 if(x == zero) s += "C";
 if(x > zero) s += "R";
 }
 cout << "symbolic sequence = " << s;
 return 0;
}
```

The output is given by `symbolic sequence = CRLLRLRLRRL`.

1.1.16 Chaotic Repeller

Consider the logistic map

$$x_{t+1} = f(x_t) = rx_t(1 - x_t)$$

where $r > 4.0$, for example $r = 4.1$. Thus, for example, if $x_0 = 0.5$, then $f(x_0) = 1.025 > 1.0$ and $f(f(x_0)) = -0.1050625$. We find that $f^{(n)}(x_0)$ tends to $-\infty$ if $n \to \infty$. Letting $s = r/4 - 1$ the map has a gap of size $\sqrt{s/(1+s)}$. In this gap $f(x) > 1.0$. Initial conditions chosen from this gap maps out of the unit interval $[0,1]$ in one iteration and goes to $-\infty$. Almost all initial conditions in the unit interval eventually escape from it except for a set of Lebesgue measure zero. This set, by construction, is a fractal *Cantor set*. A *chaotic repeller* is a set of points on the attractor that never visit the gap. The chaotic repeller can be used for encoding digital information (Lai [124]).

```
// Repeller.cpp

#include <iostream>
using namespace std;

int main(void)
{
 int count = 0;
 double x0 = 1.0/3.0;  double x1;
 do
 { x1 = 4.1*x0*(1.0-x0); count++; x0 = x1; } while(x0 <= 1.0);
 cout << "x0 = " << x0 << endl;
 cout << "count = " << count << endl;
 return 0;
}
```

1.1.17 Chaos and Encoding

One-dimensional chaotic maps can be used for the communication of information. We consider two techniques of encoding bit strings using chaos: reverse interval mapping and variable bit length encoding (Hardy and Sabatta [86]).

We consider iteration of one-dimensional maps $f : [0,1] \to [0,1]$ of the form,

$$x_{t+1} = rf(x_t), \qquad t = 0, 1, \ldots$$

where $r > 1$ and f is 1 to 1 and monotone on $[0, 0.5]$ and has the properties

$$f(0) = 0, \qquad f(0.5) = 1, \qquad f(1) = 0, \qquad f(x) = f(1-x).$$

Examples of such maps include the

Bell map:	$f_b(x) = (e^{-(x-0.5)^2} - e^{-0.25})/(1 - e^{-0.25})$
Entropy map:	$f_e(x) = -x \log_2 x - (1-x)\log_2(1-x)$
Logistic map:	$f_l(x) = 4x(1-x)$
Tent map:	$f_t(x) = 1 - 2\lvert x - 0.5\rvert$
Sine map:	$f_s(x) = \sin(\pi x)$.

The maps f_l and f_s are good candidates for encoding information. Let f^{-1} denote the inverse of f on $[0, 0.5]$. If we consider the map rf with $r > 1$, there exists two intervals

$$[0, f^{-1}(1/r)], \qquad [1 - f^{-1}(1/r), 1]$$

where $0 \le rf(x) \le 1$. If we now consider points which remain on the unit interval under two iterations of the map, the two intervals are divided into four sub-intervals that remain on the unit interval after two successive maps under the map rf. As we continue in this manner, we construct a Cantor set. Any point on this Cantor set remains on the unit interval under successive iterations of the map, and thus any point not on this set will eventually leave the unit interval. We use orbits that are confined to these intervals to encode messages.

With *reverse interval mapping* one considers a method which encodes messages of the form

$$m = m_1 m_2 \cdots m_n \in \Sigma_2^*$$

where $\Sigma_2 := \{0, 1\}$ and

$$\Sigma_2^* := \Sigma_2 \cup (\Sigma_2 \times \Sigma_2) \cup (\Sigma_2 \times \Sigma_2 \times \Sigma_2) \cup \cdots.$$

Values on the interval $[0, 1]$ are associated with Σ_2 by the map $d : [0, 1] \to \Sigma_2$

$$d(x) = \begin{cases} 0, & 0 \le x \le \frac{1}{2} \\ 1, & \frac{1}{2} < x \le 1 \end{cases}.$$

Initially we begin with any value x_0 outside the interval $[0, 1]$. For example $(1+r)/2$ is a convenient choice. This value marks the beginning of the message and the end of the decoding process. This point has two pre-images under the map rf

$$x_{1,0} = f^{-1}(x_0/r), \qquad x_{1,1} = 1 - f^{-1}(x_0/r).$$

Each of these values lie on either side of $x = 1/2$. Thus we select the value of x_1 from $\{x_{1,0}, x_{1,1}\}$ that satisfies the requirement $d(x_1) = m_1$. We proceed in this manner until the maximum precision of the register storing the value x_i has been reached, or when all of the message has been encoded (i.e. after determining x_n). Thus

$$x_0 := \frac{1+r}{2}$$
$$x_1 := \begin{cases} f^{-1}(x_0/r) & m_1 = 0 \\ 1 - f^{-1}(x_0/r) & m_1 = 1 \end{cases}$$
$$x_2 := \begin{cases} f^{-1}(x_1/r) & m_2 = 0 \\ 1 - f^{-1}(x_1/r) & m_2 = 1 \end{cases}$$
$$\vdots$$

A naive approach is to assume that the register always has adequate precision (i.e. the message must be short enough). This value is then stored and the procedure is repeated. The following C++ function encodes a string of 0s and 1s as a value of type `double`. The function `fi` denotes the inverse f^{-1} of the map f used in the map rf.

```
double encode(string m,double r,double (*fi)(double))
{
 unsigned int i;
 // initial value: 0.5+0.5r > 1, since r > 1
 // consequently m does not lie in [0,1]
 double x = (1.0+r)/2;

 // for each bit 0/1 in the string m="01001..." from left to right
 for(i=0;i<m.length();i++)
 {
 // for a "0" bit choose the pre-image on the left half
 if(m[i]=='0') x = fi(x/r);
 // for a "1" bit choose the pre-image on the right half
 elsex = 1.0-fi(x/r);
 }
 // this is the final pre-image f^(-n)(0.5+0.5r)=fi^(n)(0.5+0.5r)
 // for the sequence of pre-image choices given in m
 return x;
}
```

To decode the message stream we iterate the stored point under the map, at each iteration calculating $m_i = d(x_i)$ to resolve the bit value. This is repeated until the desired number of bits have been retrieved. The message is reconstructed in reverse. The C++ function to decode a value of type `double` to a string of 0s and 1s follows.

```
string decode(double x,double r,double (*f)(double))
{
 // in decoding we work from right to left (opposite to encoding)
 // i.e. every "0" or "1" must be appended on the left: m = "0" + m
 string m;
 // once we leave [0,1] there is no more message to decode
```

```
while((0.0 <= x) && (x <= 1.0))
{
// the left pre-image corresponded to "0"
if(x < 0.5) m = "0" + m;
// the right pre-image corresponded to "1"
else m = "1" + m;
// iterate the map
x = r*f(x);
}
return m;
}
```

Consider the logistic map $f_l(x)$. The arguments below can be extended to any of the other maps listed above. Assume we wish to encode a 16-bit data stream in a single orbit under the logistic map with parameter $r = 1.025$. There are $2^{16} = 65536$ disjoint intervals which remain a subset of $[0, 1]$ under the map $(rf_l)^{(16)}(x)$. These can be found by applying $(rf_l)^{-1}(x)$ to $[0, 1]$, which yields two disjoint intervals. We apply $(rf_l)^{-1}(x)$ to each of the resulting 2 disjoint intervals to obtain 4 disjoint intervals and so on. We apply the inverse $(rf_l)^{-1}(x)$ 16 times. The intervals are numbered from 0 to 65535 according to their relative positions in $[0, 1]$.

In *variable bit length encoding* instead of iterating a single number we iterate an interval

$$[a, b] \subset (f^{-1}(1/r), 0.5)$$

or

$$[a, b] \subset (0.5, 1 - f^{-1}(1/r)).$$

Thus we begin with an interval $[a, b]$ which will iterate out of $[0, 1]$ under rf. A convenient choice in this case is

$$w := 0.1, \quad x_0 := f^{-1}(1/r), \quad a = (1 - w)x_0 + \frac{w}{2}, \quad b = wx_0 + \frac{1 - w}{2}$$

where $0 < w < 0.5$ is a weight for the weighted average of $x_0 = f^{-1}(1/r)$ and 0.5 used to determine a and b. Smaller w is preferred, since it yields a larger initial interval, which will subsequently shrink. In the implementation below we chose $w = 0.1$. One applies the reverse interval mapping for a and b up to a precision limit, given by the interval length $|a - b|$, by determining when the intervals found at each step become smaller in length than $\epsilon > 0$. We need only store a, since b is only used to determine whether we are still outside of the precision limit. If more bits need to be encoded we begin reverse interval mapping with a new initial interval for the remainder of the bits. Repeating this process until all bits are encoded we obtain a sequence of real numbers. Each number should be decoded with reverse interval mapping to obtain that part of the original bit sequence. The number of bits decoded for each real value need not be the same over the sequence, hence the name of the technique. One finds successive intervals $[a_j, b_j]$ as follows

$$a_0 = (1 - w)x_0 + \frac{w}{2}, \qquad b_0 = wx_0 + \frac{1 - w}{2}$$

$$a_1 = \begin{cases} f^{-1}(a_0/r) & m_1 = 0 \\ 1 - f^{-1}(b_0/r) & m_1 = 1 \end{cases}, \quad b_1 = \begin{cases} f^{-1}(b_0/r) & m_1 = 0 \\ 1 - f^{-1}(a_0/r) & m_1 = 1 \end{cases}$$

$$a_2 = \begin{cases} f^{-1}(a_1/r) & m_2 = 0 \\ 1 - f^{-1}(b_1/r) & m_2 = 1 \end{cases}, \quad b_2 = \begin{cases} f^{-1}(b_1/r) & m_2 = 0 \\ 1 - f^{-1}(a_1/r) & m_2 = 1 \end{cases}$$

and continue until $b_n - a_n < \epsilon$. Note that $m_j = 1$ causes a swap in the roles of a_j and b_j so that the absolute value is always given by $b_j - a_j$, and so that the resulting interval lies in $[0.5, 1]$. Once the criteria $b_n - a_n < \epsilon$ is met, we store a_n at the end of the sequence of numbers representing the data and begin once again by redefining a_n and b_n

$$a_n := (1-w)x_0 + \frac{w}{2}, \qquad b_n := wx_0 + \frac{1-w}{2}$$

$$a_{n+1} = \begin{cases} f^{-1}(a_n/r) & m_{n+1} = 0 \\ 1 - f^{-1}(b_n/r) & m_{n+1} = 1 \end{cases}, \quad b_{n+1} = \begin{cases} f^{-1}(b_n/r) & m_{n+1} = 0 \\ 1 - f^{-1}(a_n/r) & m_{n+1} = 1 \end{cases}$$

$$a_{n+2} = \begin{cases} f^{-1}(a_{n+1}/r) & m_{n+2} = 0 \\ 1 - f^{-1}(b_{n+1}/r) & m_{n+2} = 1 \end{cases}, \quad b_{n+2} = \begin{cases} f^{-1}(b_{n+1}/r) & m_{n+2} = 0 \\ 1 - f^{-1}(a_{n+1}/r) & m_{n+2} = 1 \end{cases}$$

until $b_{n+k} - a_{n+k} < \epsilon$ and then continue the process until all the m_j values have been used. Thus the encoding now yields multiple real numbers each representing a portion of the message string, where each real number does not necessarily encode the same number of symbols. A C++ function for the encoding is given below.

```
vector<double> vl_encode(string m,double r,double (*fi)(double),
                         double eps)
{
 double x0, a = 0.0, b = 0.0, w = 0.1;
 vector<double> vd;
 x0 = fi(1.0/r);
 // for each bit 0/1 in the string m="01001..." from left to right
 for(unsigned int i=0;i<m.length();i++)
 {
 // the interval has become too small we start from
 // the initial values again to encode the remaining bits
 if((b-a) < eps)
 {
 // create the interval [a,b] as a subset of (f^(-1)(1/r),0.5)
 a = (1.0-w)*x0+w*0.5;  b = w*x0+(1.0-w)*0.5;
 // push back initializes the next number (appends on the right) in
 // the sequence to a, i.e. we add a number to the end of the sequence
 vd.push_back(a);
}
 // vd.back() is the last number in the sequence which we are computing
 // for a "0" bit choose the pre-image on the left half
 if(m[i]=='0') vd.back() = a;
 // for a "1" bit choose the pre-image on the right half
 else
 {
```

```
vd.back() = 1.0-a;     // since a < 0.5,1-a > 0.5
// after the following two statements we have
// [a,b] -> [1.0-b,1.0-a] i.e. we change halves around 0.5:
// left --> right or right --> left
a = 1.0-b;  b = vd.back();
}
a = fi(a/r); b = fi(b/r); // find the pre-image of [a,b]
}
return vd;
}
```

To decode we simply apply the decoding technique of reverse interval mapping to each real number found in the encoding process as demonstrated in the following C++ function.

```
string vl_decode(vector<double> vd,double r,double (*f)(double))
{
// in decoding we work from right to left (opposite to encoding)
// i.e. every "0" or "1" must be appended on the left: m = "0" + m
string m;
// vd.size() is the number of elements in the sequence
// that are not yet decoded
while(vd.size() != 0)
{
// once we leave [0,1] there is no more message to decode
while((0.0 <= vd.back()) && (vd.back() <= 1.0))
{
// the left pre-image corresponded to "0"
if(vd.back() < 0.5) m = "0" + m;
// the right pre-image corresponded to "1"
else m = "1" + m;
vd.back() = r*f(vd.back());
}
// remove the right most number of the sequence
// still decoding right to left
vd.pop_back();
}
return m;
}
```

1.1.18 Chaotic Data Communication

Data communication utilizing chaotic maps has been studied by various authors (Sushchik et al [201], Steeb and Hardy [199] and references therein). Here we consider the symmetric tent map $f : [-1,+1] \to [-1,+1]$

$$f(x) = \begin{cases} 2x+1 & \text{if } -1 \le x \le 0 \\ -2x+1 & \text{if } 0 \le x \le 1\,. \end{cases}$$

This is a fully chaotic map with Ljapunov exponent $\lambda = \ln(2)$ and invariant density $\rho = 1/2$. The fixed points are $x^* = -1$ and $x^* = 1/3$ which are unstable. The essential part for the data transmission is $f(x) = f(-x)$ and $f : [-1,+1] \to [-1,+1]$. Thus we have the iteration $x_{t+1} = f(x_t)$, where $t = 0, 1, \ldots$ and $x_0 \in [-1,+1]$ is the initial value.

We generate a sequence of length T from the map, i.e. $x_0, x_1, \ldots, x_{T-1}$. This will be the transmitter. Given now the bitstring $\mathbf{b} = (b_0, b_1, \ldots, b_{T-1})$ for the signal of length T, where $b_t \in \{-1,+1\}$. Next we form the transmitted signal $\mathbf{s} = (s_0, s_1, \ldots, s_{T-1})$ via

$$s_t = b_t x_t, \qquad t = 0, 1, \ldots, T-1 .$$

The receiver is now given by

$$y_{t+1} = f(s_t), \qquad t = 0, 1, \ldots, T-2$$

where $y_0 = x_0$ and f is the symmetric tent map given above. The original bitsequence $\mathbf{b}$ can now be found by forming the products

$$s_t y_t, \qquad t = 0, 1, \ldots, T-1 .$$

If $s_t y_t > 0$, then $b_t = 1$ and if $s_t y_t < 0$, then $b_t = -1$. The proof is as follows

$$\begin{aligned} s_t y_t &= s_t f(s_{t-1}) \\ &= s_t f(b_{t-1} x_{t-1}) \\ &= s_t f(x_{t-1}) \quad \text{since } f(x_t) = f(-x_t) \\ &= b_t x_t f(x_{t-1}) \\ &= b_t x_t^2 . \end{aligned}$$

Consequently

$$\text{sign}(s_t y_t) = \text{sign}(b_t x_t^2) = b_t .$$

This scheme does not require a limit on the bit string length, since divergence of the sequence only introduces local errors. The values s_t of the sequence are transmitted and operated on directly by the receiver. This scheme does not rely on the ability to find and control a specific orbit for communication.

Other chaotic maps g with the properties $g : [-1,+1] \to [-1,+1]$ and $g(x) = g(-x)$ can also be used, such as the logistic map

$$g(x) = 1 - 2x^2$$

or the bungalow-tent map $f_r : [-1,1] \to [-1,1]$ $(r \in (0, 1/2))$ desribed above. For $r = 1/3$ we obtain the symmetric tent map.

A C++ implementation using the classes `Verylong` and `Rational` of SymbolicC++ is given by

```
// ChaosComm1.cpp

#include <iostream>
#include "verylong.h"
#include "rational.h"
using namespace std;

int main(void)
{
 int T = 8;
 // transmitter
 Rational<Verylong>* x = new Rational<Verylong>[T];
 x[0] = Rational<Verylong>("1/17"); // initial value
 Rational<Verylong> zero("0");
 Rational<Verylong> one("1");
 Rational<Verylong> two("2");
 cout << x[0];
 for(int t=0;t<T-1;t++) // tent map [-1,1] -> [-1,1]
 {
 if(x[t] <= zero) x[t+1] = two*x[t] + one;
 else x[t+1] = -two*x[t] + one;
 }
 // bitstring
 Rational<Verylong>* b = new Rational<Verylong>[T];
 b[0] = -one; b[1] = one; b[2] = -one; b[3] = -one;
 b[4] = -one; b[5] = one; b[6] = one;  b[7] = -one;
 // signal
 Rational<Verylong>* s = new Rational<Verylong>[T];
 for(int t=0;t<T;t++) { s[t] = b[t]*x[t]; }
 // output of the receiver
 Rational<Verylong>* y = new Rational<Verylong>[T];
 y[0] = x[0];
 for(int t=0;t<T-1;t++)
 {
 if(s[t] <= zero) y[t+1] = two*s[t] + one;
 else y[t+1] = -two*s[t] + one;
 }
 // multiplication
 Rational<Verylong>* bout = new Rational<Verylong>[T];
 for(int t=0;t<T;t++)
 { if(s[t]*y[t] > zero) bout[t] = one; else bout[t] = -one; }
 // display
 for(int t=0;t<T;t++)
 { cout << "bout[" << t << "]=" << bout[t] << endl; }
 delete[] x; delete[] b; delete[] s; delete[] y; delete[] bout;
 return 0;
}
```

Exercises. (1) Consider the nonlinear map $f : [-1,1] \to [-1,1]$

$$f(x) = x(3-4x^2).$$

Show that the fixed points are given by $0, \pm 1/\sqrt{2}$. Study the stability of the fixed points. Show that the exact solution is given by

$$x_t = \sin(3^t \arcsin^{-1}(x_0)), \quad t = 0,1,2,\ldots$$

where x_0 is the initial value.

(2) Consider the nonlinear map $f : [0,1] \to [0,1]$

$$f(x) = 16x(1-x)(1-2x)^2.$$

Show that the fixed points are given by 0, $3/4$, $(5 \pm \sqrt{5})/8$. Study the stability of the fixed points. Show that the exact solution is given by

$$x_t = \sin^2(4^t \arcsin(\sqrt{x_0})).$$

(3) Consider the logistic map

$$x_{t+1} = 4x_t(1-x_t), \qquad t = 0,1,\ldots.$$

Show that the orbit of the initial value

$$x_0 = \sin^2(3\pi/11)$$

has period 5.

(4) Let $r \in \mathbb{R}$. A one-parameter family of functions $f_r : \mathbb{R} \to \mathbb{R}$ is defined by

$$f_r(x) = x^2 - x - r.$$

Find all fixed points and periodic points of prime period two for f_r and establish for which values of r they exist. For all these points establish for which values of r they are attractive or repelling.

1.2 Two-Dimensional Maps

1.2.1 Introduction

Most of the two-dimensional maps (Arrowsmith and Place [7], Devaney [47], Mira [149], Steeb [189], [190]) we consider in this section are diffeomorphisms.

Let U be an open subset of $\mathbb{R}^n$. Then a function $\mathbf{g} : U \to \mathbb{R}$ is said to be of class C^r if it is r-fold continuously differentiable, $1 \leq r \leq \infty$. Let V be an open subset of $\mathbb{R}^m$ and $\mathbf{g} : U \to V$. Given coordinates $(x_1, \ldots, x_n)$ in U and $(y_1, \ldots, y_m)$ in V, $\mathbf{g}$ may be expressed of component functions $g_j : U \to \mathbb{R}$, where

$$y_j = g_j(x_1, \ldots, x_n), \qquad j = 1, \ldots, m.$$

The map $\mathbf{g}$ is called a C^r map if g_j is C^r for each $j = 1, \ldots, m$. The map $\mathbf{g}$ is said to be a *diffeomorphism* if it is a bijection and both $\mathbf{g}$ and $\mathbf{g}^{-1}$ are differentiable mappings. The map $\mathbf{g}$ is called a C^k-diffeomorphism if both $\mathbf{g}$ and $\mathbf{g}^{-1}$ are C^k-maps.

Note that the bijection $\mathbf{g} : U \to V$ is a diffeomorphism if and only if $m = n$ and the *functional matrix* (also called *Jacobian matrix*) of partial derivatives

$$D\mathbf{g}(x_1, \ldots, x_n) := \left(\frac{\partial g_i}{\partial x_j} \right)_{i,j=1}^{n}.$$

For $n = m = 2$ we have the functional matrix

$$\begin{pmatrix} \partial f_1/\partial x_1 & \partial f_1/\partial x_2 \\ \partial f_2/\partial x_1 & \partial f_2/\partial x_2 \end{pmatrix}.$$

If $\mathbf{g}$ satisfies the definition above with $\mathbf{g}$ and $\mathbf{g}^{-1}$ continuous rather than differentiable, then $\mathbf{g}$ is called a *homeomorphism*.

Example. The map $\mathbf{f} : \mathbb{R}^2 \to \mathbb{R}^2$

$$f_1(x_1, x_2) = 1 + x_2 - a x_1^2, \qquad f_2(x_1, x_2) = \frac{1}{2} x_1$$

is a diffeomorphism. The inverse map is given by

$$f_1^{-1}(x_1, x_2) = 2x_2, \qquad f_2^{-1}(x_1, x_2) = x_1 - 1 + 4a x_2^2. \qquad \clubsuit$$

Example. The map $\mathbf{f} : \mathbb{R}^2 \to \mathbb{R}^2$

$$f_1(x_1, x_2) = \sinh(x_2), \qquad f_2(x_1, x_2) = \sinh(x_1)$$

is a diffeomorphism. The inverse map is given by

$$f_1^{-1}(x_1, x_2) = \text{arcsinh}(x_2), \qquad f_2^{-1}(x_1, x_2) = \text{arcsinh}(x_1). \qquad \clubsuit$$

Consider a two-dimensional map $\mathbf{f} : \mathbb{R}^2 \to \mathbb{R}^2$. Then the fixed points are given by the solutions of the equations $f_1(x_1^*, x_2^*) = x_1^*$, $f_2(x_1^*, x_2^*) = x_2^*$. Assume that the functions f_1, f_2 are continuously differentiable. To study the stability of a fixed point $\mathbf{x}^*$ we find the eigenvalues of the functional matrix at the point $\mathbf{x}^*$

$$\begin{pmatrix} \partial f_1(\mathbf{x}^*)/\partial x_1 & \partial f_1(\mathbf{x}^*)/\partial x_2 \\ \partial f_2(\mathbf{x}^*)/\partial x_1 & \partial f_2(\mathbf{x}^*)/\partial x_2 \end{pmatrix}.$$

We have two such eigenvalues, which are either both real or appear as complex conjugate pairs. If $|\lambda| < 1$ for both eigenvalues then the fixed point is stable. If $|\lambda| > 1$ for any of the two eigenvalues, then the fixed point is unstable. If there are two real eigenvalues with one of them $|\lambda| < 1$ and the other one $|\lambda| > 1$ then we have a saddle point.

Let U be an open subset of $\mathbb{R}^n$ and $\mathbf{f} : U \to \mathbb{R}^n$ be a nonlinear diffeomorphism with an isolated fixed point at $\mathbf{x}^* \in U$. The *linearization* of $\mathbf{f}$ at $\mathbf{x}^*$ is given by the $n \times n$ matrix

$$D\mathbf{f}(\mathbf{x}^*) := \left[\frac{\partial f_i}{\partial x_j}\right]_{i,j=1}^{n} |_{\mathbf{x}=\mathbf{x}^*}$$

where $x_1, \ldots, x_n$ are coordinates on U.

Definition. A fixed point $\mathbf{x}^*$ of a diffeomorphism $\mathbf{f}$ is said to be hyperbolic if the map $D\mathbf{f}(\mathbf{x}^*)$ is a hyperbolic, linear diffeomorphism.

Definition. A linear diffeomorphism $A : \mathbb{R}^n \to \mathbb{R}^n$ is said to be *hyperbolic* if it has no eigenvalues with modulus equal to unity.

The following theorems allow us to obtain valuable information from $D\mathbf{f}(\mathbf{x}^*)$.

Theorem. (Hartman-Grobman) Let $\mathbf{x}^*$ be a hyperbolic fixed point of the diffeomorphism $\mathbf{f} : U \to \mathbb{R}^n$. Then there is a neighbourhood $N \subset U$ of $\mathbf{x}^*$ and a neighbourhood $N' \subseteq \mathbb{R}^n$ containing the origin such that $\mathbf{f}|N$ is topologically conjugate to $D\mathbf{f}(\mathbf{x}^*)|N'$.

It follows that there are $4n$ topological types of hyperbolic fixed point for diffeomorphisms $\mathbf{f} : U \to \mathbb{R}^n$.

Theorem. (Invariant Manifold) Let $\mathbf{f} : U \to \mathbb{R}^n$ be a diffeomorphism with a hyperbolic fixed point at $\mathbf{x}^* \in U$. Then on a sufficiently small neighbourhood $N \subseteq U$ of $\mathbf{x}^*$, there exist local stable and unstable manifolds,

$$W^s_{loc}(\mathbf{x}^*) := \{\, \mathbf{x} \in U \mid \mathbf{f}^{(t)}(\mathbf{x}) \to \mathbf{x}^* \text{ as } t \to \infty \,\}$$

$$W^u_{loc}(\mathbf{x}^*) := \{\, \mathbf{x} \in U \mid \mathbf{f}^{(t)}(\mathbf{x}) \to \mathbf{x}^* \text{ as } t \to -\infty \,\}$$

of the same dimensions as E^s and E^u for $D\mathbf{f}(\mathbf{x}^*)$ and tangent to them at $\mathbf{x}^*$.

This theorem allows us to define global stable and unstable manifolds at $\mathbf{x}^*$ by

$$W^s(\mathbf{x}^*) := \bigcup_{m\in\mathbb{Z}^+} \mathbf{f}^{(-m)}(W^s_{loc}(\mathbf{x}^*)), \qquad W^u(\mathbf{x}^*) := \bigcup_{m\in\mathbb{Z}^+} \mathbf{f}^{(m)}(W^u_{loc}(\mathbf{x}^*)).$$

The behaviour of the stable and unstable manifold $W^s(\mathbf{x}^*)$ and $W^u(\mathbf{x}^*)$ reflects in the complexity of the dynamics of the map $\mathbf{f}$. In particular, if $W^s(\mathbf{x}^*)$ and $W^u(\mathbf{x}^*)$ meet transversely at one point, they must do so infinitely many times and a homoclinic tangle results. The theorems given above have an extension to the periodic point of $\mathbf{f}$. Let $\mathbf{x}^*$ belong to a q-cycle of $\mathbf{f}$ then it is said to be a *hyperbolic periodic point* of $\mathbf{f}$ if it is a hyperbolic fixed point of the q-th iterated map $\mathbf{f}^{(q)}$. The orbit of $\mathbf{x}^*$ under $\mathbf{f}$ is referred to as a *hyperbolic periodic orbit* and its topological type is determined by that of the corresponding fixed point of $\mathbf{f}^{(q)}$. Moreover, information about stable and unstable manifolds at each point of the q-cycle can be obtained by applying the theorem to $\mathbf{f}^{(q)}$.

The *Hopf bifurcation theorem* (Marsden and McCraken [142], Iooss and Joseph [103]) for maps in the plane $\mathbf{f}_r : \mathbb{R}^2 \to \mathbb{R}^2$, where r is the bifurcation parameter, is as follows.

Theorem. (Hopf bifurcation theorem) Let $\mathbf{f}(r, \mathbf{x})$ be a one-parameter family of maps in the plane satisfying:
a) An isolated fixed point $\mathbf{x}^*(r)$ exists.
b) The map $\mathbf{f}_r$ is C^k $(k \geq 3)$ in the neighbourhood of $(\mathbf{x}^*(r_0); r_0)$.
c) The Jacobian matrix $D_{\mathbf{x}}\mathbf{f}(\mathbf{x}^*(r); r)$ posseses a pair of complex, simple eigenvalues

$$\lambda(r) = e^{\alpha(r)+i\omega(r)}$$

and $\bar{\lambda}(r)$, such that the critical value $r = r_0$

$$|\lambda(r_0)| = 1, \quad (\lambda(r_0))^3 \neq 1, \quad (\lambda(r_0))^4 \neq 1, \quad \frac{d|\lambda(r)|}{dr}(r = r_0) > 0\,.$$

(Existence) Then there exists a real number $\epsilon_0 > 0$ and a C^{k-1} function such that

$$r(\epsilon) = r_0 + r_1\epsilon + r_3\epsilon^3 + O(\epsilon^4)$$

such that for each $\epsilon \in (0, \epsilon_0]$ the map $\mathbf{f}_r$ has an invariant manifold $H(r)$, i.e. $\mathbf{f}(H(r); r) = H(r)$. The manifold $H(r)$ is C^r diffeomorphic to a circle and consists of points at a distance $O(|r|^{1/2})$ of $\mathbf{x}^*(r)$, for $r = r(\epsilon)$.

(Uniqueness) Each compact invariant manifold close to $\mathbf{x}^*(r)$ for $r = r(\epsilon)$ is contained in $H(r) \cup \{0\}$.

(Stability) If $r_3 < 0$ (respectively $r_3 > 0$) then for $r > 0$ (respectively $r > 0$), the fixed point $\mathbf{x}^*(r(\epsilon))$ is stable (respectively unstable) and for $r > 0$ (respectively $r < 0$) the fixed point $\mathbf{x}^*(r(\epsilon)$ is unstable (respectively stable) and the surroundng manifold $H(r(\epsilon))$ is attracting (respectively repelling). When $r_3 < 0$ (respectively

$r_3 > 0$) the bifurcation at $r = r(\epsilon)$ is said to be *supercritical* (respectively *subcritical*).

Example. Consider the two-dimensional map

$$f_1(x_1, x_2) = rx_1(3x_2 + 1)(1 - x_1), \qquad f_2(x_1, x_2) = rx_2(3x_1 + 1)(1 - x_2)$$

and $r \in \mathbb{R}$. There are fixed points on the diagonal, namely $\mathbf{x}_0^* = (0, 0)$ and for $r \neq 0$

$$\mathbf{x}_1^* = ((r-1)/r, 0), \qquad \mathbf{x}_2^* = (0, (r-1)/r, 0)$$

and on the diagonal ($r \geq 3/4$)

$$\mathbf{x}_3^* = (1/3 - \sqrt{4 - 3/r}/3, 1/3 - \sqrt{4 - 3/r})$$
$$\mathbf{x}_4^* = (1/3 + \sqrt{4 - 3/r}/3, 1/3 + \sqrt{4 - 3/r}).$$

To study Hopf bifurcation we consider the fixed points on the diagonal. Notice that these fixed points exists only for $r \geq 3/4$. For $r > 3/4$ a stable period-2 orbit exists. This period-2 orbit looses stability via a Hopf bifurcation which occurs at $r = r_0$, where $r_0 = 0.957$ and gives rise to a stable limit cycle for $r \in [r_0, r_0 + \delta)$ for some $\delta > 0$. ♣

1.2.2 Phase Portrait

We consider the Hénon map, the Lozi map, the standard map, the Ikeda laser map and a coupled logistic map.

For a two-dimensional map

$$x_{1t+1} = f_1(x_{1t}, x_{2t}), \qquad x_{2t+1} = f_2(x_{1t}, x_{2t})$$

where $t = 0, 1, 2, \ldots$ we can plot the points (x_{1t}, x_{2t}) for $t = 0, 1, 2, \ldots$ in the (x_1, x_2) plane $\mathbb{R}^2$. This is called a *phase portrait.* We also use the notation $x = x_1$ and $y = x_2$.

As the first example we consider the *Hénon map* $\mathbf{f} : \mathbb{R}^2 \to \mathbb{R}^2$ which is given by

$$\mathbf{f}(x, y) := (y + 1 - ax^2, bx)$$

where a and b are bifurcation parameters with $b \neq 0$. The Hénon map is the most studied two-dimensional map with chaotic behaviour. The map can also be written as a system of difference equations

$$x_{t+1} = 1 + y_t - ax_t^2, \qquad y_{t+1} = bx_t$$

where $t = 0, 1, 2, \ldots$. The map is invertible if $b \neq 0$. The inverse map is given by

$$x_t = \frac{1}{b}y_{t+1}, \qquad y_t = x_{t+1} - 1 + \frac{a}{b^2}y_{t+1}^2.$$

In order to visualize the action of the map, note that vertical lines are mapped to horizontal lines, while for $a > 0$ horizontal lines map to parabolas opening to the left.

In the C++ program henon.cpp we evaluate the phase portrait (x_t, y_t) for $a = 1.4$ and $b = 0.3$. The data are written to a file named henon.dat. Then we use GNU plot to display the phase portrait. In the Java program Henon.java the graphics is included and the phase portrait is displayed.

```
// henon.cpp

#include <fstream>
using namespace std;

int main(void)
{
 ofstream data("henon.dat");
 const int T = 2000;            // number of iterations
 double x0 = 0.1, y0 = 0.3;  // initial values
 double x1, y1;
 for(int t=0;t<T;t++)
 { x1 = 1.0+y0-1.4*x0*x0; y1 = 0.3*x0;
 data << x1 << " " << y1 << "\n";
 x0 = x1; y0 = y1;
 }
 data.close();
 return 0;
}
```

The data in the file "henon.dat" can now be used to display the phase portrait using GNU-plot. The command is `plot 'henon.dat' with dots`.

```
// Henon.java

import java.awt.*;
import java.awt.event.*;
import java.awt.Graphics;

public class Henon extends Frame
{
 public Henon()
 {
 setSize(600,500);
 addWindowListener(new WindowAdapter()
 { public void windowClosing(WindowEvent event) { System.exit(0); }}); }

 public void paint(Graphics g)
 {
 double x1, y1;
 double x = 0.0, y = 0.0;  // initial values
```

```
 int T = 4000;                    // number of iterations
 for(int t=0;t<T;t++)
 { x1 = x; y1 = y; x = 1.0+y1-1.4*x1*x1;  y = 0.3*x1;
 int mx = (int) Math.floor(200.0*x+250.0+0.5);
 int ny = (int) Math.floor(200.0*y+150.0+0.5);
 g.drawLine(mx,ny,mx,ny);
 }
 }

 public static void main(String[] args)
 { Frame f = new Henon();  f.setVisible(true); }
}
```

The C++ program `maxmin.cpp` finds the largest and smallest values in the x and y direction of the Hénon attractor for the parameter values $a = 1.4$ and $b = 0.3$.

```
// maxmin.cpp

#include <iostream>
using namespace std;

void maxminxy(double* x,double* y,double& xmax,double& xmin,
              double& ymax,double& ymin,int T)
{
 xmax = x[0]; xmin = x[0];  ymax = y[0]; ymin = y[0];
 for(int t=1;t<T;t++)
 {
 if(x[t] < xmin) xmin = x[t]; if(x[t] > xmax) xmax = x[t];
 if(y[t] < ymin) ymin = y[t]; if(y[t] > ymax) ymax = y[t];
 }
}

int main(void)
{
 unsigned long T = 10000;                 // number of iterations
 double* x = new double[T]; double* y = new double[T];
 x[0] = 1.161094; y[0] = -0.09541356; // initial values
 for(int t=0;t<(T-1);t++)
 { x[t+1] = 1.0+y[t]-1.4*x[t]*x[t]; y[t+1] = 0.3*x[t]; }
 double xmax, xmin, ymax, ymin;
 maxminxy(x,y,xmax,xmin,ymax,ymin,T);
 cout << "xmax= " << xmax << endl; cout << "xmin= " << xmin << endl;
 cout << "ymax= " << ymax << endl; cout << "ymin= " << ymin << endl;
 delete[] x;  delete[] y;
 return 0;
}
```

The *Lozi map* $\mathbf{f} : \mathbb{R}^2 \to \mathbb{R}^2$ is a piecewise linear map which has been introduced to simplify the Hénon map. It is given by

$$\mathbf{f}(x,y) = (1 + y - a|x|, bx)$$

where $b > 0$. It can be shown that if

$$b \in (0,1), \quad a > 0, \quad 2a + b < 4, \quad b < \frac{a^2 - 1}{2a + 1}, \quad a\sqrt{2} > b + 2$$

then there is a hyperbolic fixed point H of saddle type given by

$$x^* = \frac{1}{a + 1 - b}, \qquad y^* = bx^*$$

such that the strange attractor is the closure of their unstable manifold.

In the Java program `Henon.java` for the Hénon model we replace the line

```
x = 1.0+y1-1.4*x1*x1;   y = 0.3*x1;
```

by

```
x = 1.0+y1-a*Math.fabs(x1);   y = b*x1;
```

The parameter values for the Lozi map are $a = 1.7$, $b = 0.4$. For these parameter values the system shows a strange attractor and for these values the conditions given above are satisfied.

The *standard map* is defined as

$$I_{t+1} = I_t + k\sin(\theta_t), \qquad \theta_{t+1} = \theta_t + I_t + k\sin(\theta_t) \equiv \theta_t + I_{t+1}$$

where $0 \leq \theta < 2\pi$. The quantities I, θ are the *action-angle variables.* It can be derived from the Hamilton function

$$H(p_\theta, \theta) = \frac{p_\theta^2}{2} + k\cos\theta \sum_{n\in\mathbb{Z}} \delta(t - n)$$

of a one-dimensional periodically *kicked rotor*, where δ is the delta function. The standard map can be considered as a discrete Hamilton system. We have

$$\det \begin{pmatrix} \frac{\partial f_1}{\partial I} & \frac{\partial f_1}{\partial \theta} \\ \frac{\partial f_2}{\partial I} & \frac{\partial f_2}{\partial \theta} \end{pmatrix} = 1 .$$

This map displays all three types of orbits: periodic cycles, KAM tori and chaotic orbits. The first two types of orbits (the regular ones) dominate the (I, θ) phase space for small k ($k \ll 1$). KAM tori extending over the entire θ interval ($0 \leq \theta < 2\pi$) divide the (I, θ) phase plane into disconnected regions: orbits in one region cannot cross the bounding KAM tori into the other regions, so that the variations in I are bounded. For

$$k \geq k_c \approx 0.9716$$

the bounding KAM tori break, making it possible for chaotic orbits to be unbounded in the I direction. At k_c the last, most robust KAM torus with the winding number

$$w = 2\pi(\sqrt{5} - 1)/2$$

and the other equivalent KAM tori of the standard map with winding numbers

$$\pm w + 2\pi n$$

(n integer) disintegrates. It corresponds to a nonanalytic continuous curve, exhibiting fractal self-similar structure.

The Java program `Standard.java` displays the phase portrait for $k = 0.8$.

```
// Standard.java

import java.awt.*;
import java.awt.event.*;
import java.awt.Graphics;

public class Standard extends Frame
{
 public Standard()
 {
 setSize(600,500);
 addWindowListener(new WindowAdapter()
 { public void windowClosing(WindowEvent event) { System.exit(0); }}); }

 public void paint(Graphics g)
 {
 int T = 10000;                   // number of iterations
 double I = 0.5, theta = 0.8;  // initial values
 double I1, theta1;
 for(int t=0;t<T;t++)
 { I1 = I;  theta1 = theta;  I = I1+0.8*Math.sin(theta1);
 theta = theta1+I1+0.8*Math.sin(theta1);
 if(theta > 2*Math.PI) theta = theta-2.0*Math.PI;
 if(theta < 0.0) theta = theta+2.0*Math.PI;
 int m = (int) Math.floor(90.0*I+200.0+0.5);
 int n = (int) Math.floor(90.0*theta+10.0,0+0.5);
 g.drawLine(n,m,n,m);
 }
 }

 public static void main(String[] args)
```

```
 { Frame f = new Standard(); f.setVisible(true); }
}
```

Optical bistability only represents the simplest amongst the large variety of dynamical behaviours which can occur in passive nonlinear optical cavities. In particular, temporal instabilities leading to various forms of self-oscillations and chaos have been identified. Ikeda predicted the occurrence of period doubling cascades and chaos treating the cavity dynamics by means of a nonlinear mapping of the complex field amplitude. Let z be a complex number. The *Ikeda laser map* $f : \mathbb{C} \to \mathbb{C}$ is given by

$$f(z) = \rho + c_2 z \exp\left(i(c_1 - \frac{c_3}{1+|z|^2})\right).$$

The real bifurcation parameters are ρ, c_1, c_2 and c_3. With $z = x + iy$ and $x, y \in \mathbb{R}$ we can write the map as a system of difference equations

$$x_{t+1} = \rho + c_2(x_t \cos(\tau_t) - y_t \sin(\tau_t)), \quad y_{t+1} = c_2(x_t \sin(\tau_t) + y_t \cos(\tau_t))$$

where

$$\tau_t := c_1 - \frac{c_3}{1 + x_t^2 + y_t^2}$$

and $t = 0, 1, 2, \ldots$.

In the Java program `Ikeda.java` we evaluate the phase portrait (x_t, y_t). The parameter values are $c_1 = 0.4$, $c_2 = 0.9$, $c_3 = 9.0$, $\rho = 0.85$.

```
// Ikeda.java

import java.awt.*;
import java.awt.event.*;
import java.awt.Graphics;

public class Ikeda extends Frame
{
 public Ikeda()
 {
 setSize(400,300);
 addWindowListener(new WindowAdapter()
 { public void windowClosing(WindowEvent event) { System.exit(0); }}); }

 public void paint(Graphics g)
 {
 int T = 20000;                // number of iterations
 double x = 0.5, y = 0.5;   // initial values
 double x1, y1;
 double c1 = 0.4, c2 = 0.9, c3 = 9.0;
 double rho = 0.85;
 for(int t=0;t<T;t++)
 { x1 = x; y1 = y;
```

```
 double taut = c1-c3/(1.0+x1*x1+y1*y1);
 x = rho+c2*x1*Math.cos(taut)-y1*Math.sin(taut);
 y = c2*(x1*Math.sin(taut)+y1*Math.cos(taut));
 int m = (int) Math.floor(90.0*x+200.0+0.5);
 int n = (int) Math.floor(90.0*y+200.0+0.5);
 g.drawLine(m,n,m,n);
 }
 }

 public static void main(String[] args)
 { Frame f = new Ikeda(); f.setVisible(true); }
}
```

The *coupled logistic map* is given by

$$x_{t+1} = rx_t(1-x_t) + e(y_t - x_t), \qquad y_{t+1} = ry_t(1-y_t) + e(x_t - y_t)$$

where $t = 0, 1, 2, \ldots$ and r and e are bifurcation parameters with $1 \le r \le 4$. For $e = 0$ we have two uncoupled logistic equations. The four fixed points are given by

$$(x^*, y^*) = (0,0), \quad (x^*, y^*) = \left(\frac{r-1}{r}, \frac{r-1}{r}\right).$$

$$(x^*, y^*) = (a_+, a_-), \quad (x^*, y^*) = (a_-, a_+)$$

where

$$a_\pm = \frac{1}{2r}\left[(r-1-2e) \pm \sqrt{(r-1-2e)(r-1+2e)}\right].$$

Depending on the initial conditions and the parameter values one can find the following behaviour: (i) orbits tend to a fixed point, (ii) periodic behaviour, (iii) quasiperiodic behaviour, (iv) chaotic behaviour, (v) hyperchaotic behaviour and (vi) x_t and y_t explode, i.e. for a finite time the state variables x_t and (or) y_t tend to infinity. There is numerical evidence of hyperchaos for $r = 3.70$, $e = 0.06$.

In the Java program `Couplog.java` we calculate the phase portrait (x_t, y_t) for these parameter values.

```
// Couplog.java

import java.awt.*;
import java.awt.event.*;
import java.awt.Graphics;

public class Couplog extends Frame
{
 public Couplog()
 {
 setSize(400,300);
 addWindowListener(new WindowAdapter()
```

```
{ public void windowClosing(WindowEvent event) { System.exit(0); }}); }

public void paint(Graphics g)
{
int T = 60000;                   // number of iterations
double r = 3.7, e = 0.06;   // control parameters
double x = 0.1, y = 0.2;    // initial values
double x1, y1;
for(int t=0;t<T;t++)
{ x1 = x; y1 = y;
x = r*x1*(1.0-x1)+e*(y1-x1); y = r*y1*(1.0-y1)+e*(x1-y1);
int m = (int) Math.floor(400.0*x+150.0+0.5);
int n = (int) Math.floor(400.0*y+150.0+0.5);
g.drawLine(m,n,m,n);
}
}

public static void main(String[] args)
{ Frame f = new Couplog(); f.setVisible(true); }
}
```

1.2.3 Fixed Points and Stability

Given a two-dimensional map

$$x_{1t+1} = f_1(x_{1t}, x_{2t}), \qquad x_{2t+1} = f_2(x_{1t}, x_{2t}).$$

The *fixed points* (x_1^*, x_2^*) are defined as the solution of the equations

$$f_1(x_1^*, x_2^*) = x_1^*, \qquad f_2(x_1^*, x_2^*) = x_2^*.$$

The stability of the fixed points is determined by the eigenvalues of the functional matrix taken at the fixed point. This has been described above.

Example. Consider the *Hénon map*

$$x_{t+1} = 1 + y_t - ax_t^2, \qquad y_{t+1} = bx_t$$

where a and b are bifurcation parameters and $t = 0, 1, 2, \ldots$. For $a > 0$ and $1 > b > 0$ the map has two fixed points

$$x^* = \frac{(b-1) \pm \sqrt{(1-b)^2 + 4a}}{2a}, \qquad y^* = bx^*.$$

These fixed points are real for

$$a > a_0 = \frac{(1-b)^2}{4}.$$

The map has been studied in detail for $a = 1.4$ and $b = 0.3$. For these values the fixed points are unstable. There is numerical evidence that the map shows chaos for these parameter values. ♣

In the C++ program we determine the stability of the fixed points for $a = 1.4$ and $b = 0.3$. Since

$$\frac{\partial f_1}{\partial x} = -2ax, \qquad \frac{\partial f_1}{\partial y} = 1, \qquad \frac{\partial f_2}{\partial x} = b, \qquad \frac{\partial f_2}{\partial y} = 0$$

we obtain as solution of the the characteristic equation

$$\lambda_{1,2} = -ax^* \pm \sqrt{b + a^2 x^{*2}}\,.$$

```
// henonstability.cpp

#include <iostream>
#include <cmath>      // for sqrt
using namespace std;

int main(void)
{
 double a = 1.4, b = 0.3; // parameter values
 // first fixed point
 double xf1 = ((b-1.0)+sqrt((1.0-b)*(1.0-b)+4.0*a))/(2.0*a);
 double yf1 = b*xf1;
 double lambda1 = -a*xf1+sqrt(b+a*a*xf1*xf1);
 double lambda2 = -a*xf1-sqrt(b+a*a*xf1*xf1);
 cout << "lambda1 for fixed point 1 = " << lambda1 << endl;
 cout << "lambda2 for fixed point 1 = " << lambda2 << endl;
 // second fixed point
 double xf2 = ((b-1.0)-sqrt((1.0-b)*(1.0-b)+4.0*a))/(2.0*a);
 double yf2 = b*xf2;
 double lambda3 = -a*xf2+sqrt(b+a*a*xf2*xf2);
 double lambda4 = -a*xf2-sqrt(b+a*a*xf2*xf2);
 cout << "lambda3 for fixed point 2 = " << lambda3 << endl;
 cout << "lambda4 for fixed point 2 = " << lambda4 << endl;
 return 0;
}
```

1.2.4 Liapunov Exponents

For a two-dimensional map we have two one-dimensional Liapunov exponents. Consider a system of difference equations

$$x_{t+1} = f_1(x_t, y_t), \qquad y_{t+1} = f_2(x_t, y_t)$$

where we assume that f_1 and f_2 are smooth functions. Then the variational equation is given by

$$u_{t+1} = \frac{\partial f_1}{\partial x}(x = x_t, y = y_t)u_t + \frac{\partial f_1}{\partial y}(x = x_t, y = y_t)v_t$$

$$v_{t+1} = \frac{\partial f_2}{\partial x}(x = x_t, y = y_t)u_t + \frac{\partial f_2}{\partial y}(x = x_t, y = y_t)v_t \,.$$

The maximal one-dimensional Liapunov exponent λ is given by

$$\lambda = \lim_{T\to\infty} \frac{1}{T} \ln(|u_T| + |v_T|) \,.$$

As an example consider the Hénon map described by

$$x_{t+1} = 1 + y_t - ax_t^2 \,, \qquad y_{t+1} = bx_t$$

where a and b are bifurcation parameters and $t = 0, 1, 2, \ldots$. Since

$$f_1(x, y) = 1 + y - ax^2 \,, \qquad f_2(x, y) = bx$$

we obtain the variational equation

$$u_{t+1} = -2ax_t u_t + v_t \,, \qquad v_{t+1} = bu_t \,.$$

For the parameter value $a = 1.4$ and $b = 0.3$ we find from the numerical analysis that the (maximal) one-dimensional Liapunov exponent is given by $\lambda \approx 0.42$. There is numerical evidence that the map shows chaos for these parameter values.

```
// henonliapunov.cpp

#include <iostream>
#include <cmath>   // for fabs, log
using namespace std;

double f1(double x,double y) { double a = 1.4; return 1.0+y-a*x*x; }

double f2(double x,double y) { double b = 0.3; return b*x; }

double vf1(double x,double y,double u,double v)
{ double a = 1.4; return -2.0*a*x*u+v; }

double vf2(double x,double y,double u,double v)
{ double b = 0.3; return b*u; }

int main(void)
{
 int T = 1000;                  // number of iterations
 double x = 0.1, y = 0.2; // initial values
 double u = 0.5, v = 0.5; // initial values
```

```
  double x1, y1, u1, v1;
  for(int t=0;t<T;t++)
  { x1 = x; y1 = y; u1 = u; v1 = v;
  x = f1(x1,y1); y = f2(x1,y1);
  u = vf1(x1,y1,u1,v1);  v = vf2(x1,y1,u1,v1);
  }
  double lambda = log(fabs(u) + fabs(v))/((double) T);
  cout << "lambda = " << lambda << endl;
  return 0;
}
```

1.2.5 Correlation Integral

Dissipative dynamical systems (for example the Hénon map) which exhibit chaotic behaviour often have an attractor in phase space which is strange. Strange attractors are typically characterized by a fractal dimension D which is smaller than the number of degrees of freedom F, $D < F$. Among the fractal dimensions we have the capacity and the Hausdorff dimension. These fractal dimensions have been the most commonly used measure of the strangeness of attractors. Another measure is obtained by considering correlations between points of a long-time series on the attractor. Denote the T points of such a long-time series by

$$\{\mathbf{x}_i\}_{i=1}^T \equiv \{\mathbf{x}(t+i\tau)\}_{i=1}^T$$

where τ is an arbitrary but fixed time increment. The definition of the *correlation integral* (Grassberger and Procaccia [72]) is

$$C(r) := \lim_{T\to\infty} \frac{1}{T^2} \sum_{\substack{i,j=1\\ i\neq j}}^{T} H(r - \|\mathbf{x}_i - \mathbf{x}_j\|)$$

where $H(x)$ is the *Heaviside function*, i.e.

$$H(x) := \begin{cases} 1 \text{ for } x \geq 0 \\ 0 \text{ otherwise} \end{cases}$$

and $\|\ldots\|$ denotes the Euclidean norm. The function $C(r)$ behaves as a power of r for small r

$$C(r) \propto r^\nu .$$

The exponent ν is called the *correlation dimension.* Moreover, the exponent ν is closely related to the capacity D.

In the C++ program `henoncorrelation.cpp` we evaluate the $C(r)$ for the Hénon map

$$x_{t+1} = 1 + y_t - ax_t^2, \qquad y_{t+1} = bx_t$$

with parameter values $a = 1.4$ and $b = 0.3$ and $t = 0, 1, 2, \ldots$. From $C(r)$ we find that $\nu \approx 1.2$.

```
// henoncorrelation.cpp

#include <iostream>
#include <cmath>        // for sqrt
using namespace std;

unsigned long H(double* x,double* y,double r,unsigned long T)
{
 double norm;
 unsigned long sum = 0;
 for(unsigned long i=0;i<T;i++)
 for(unsigned long j=0;j<T;j++)
 {
 if(i != j)
 {
 norm = sqrt((x[i]-x[j])*(x[i]-x[j])+(y[i]-y[j])*(y[i]-y[j]));
 if(r >= norm) sum++;
 }
 }
 return sum;
}

int main(void)
{
 unsigned long T = 20000;                   // number of iterations
 double* x = new double[T];  double* y = new double[T];
 x[0] = 1.161094; y[0] = -0.09541356; // initial values
 for(unsigned long j=0;j<T-1;j++)
 { x[j+1] = 1.0+y[j]-1.4*x[j]*x[j]; y[j+1] = 0.3*x[j]; }
 double r = 0.001;
 while(r <= 0.008)
 { double Cr = H(x,y,r,T);
 cout << "r= " << r << "  " << "Cr= " << Cr/((double)(T*T)) << endl;
 r += 0.001;
 }
 delete[] x; delete[] y;
 return 0;
}
```

1.2.6 Capacity

Let M be a subset of $\mathbb{R}^n$. We assume that the set M is contained in an invariant manifold of some dynamical system (for example the strange attractor of the Hénon model). If N_ϵ is the minimum number of boxes of side ϵ in $\mathbb{R}^n$ needed to cover the set M, the *capacity* (also called *box-counting dimension*) is defined as (Young [225])

$$C := \lim_{\epsilon\to 0} \frac{\ln N_\epsilon}{\ln(1/\epsilon)}.$$

If a set has volume V, the number of boxes of side ϵ needed to cover the set is roughly

$$N_\epsilon \approx V\epsilon^{-C}\,.$$

This equation may be rewritten as

$$\ln N_\epsilon \approx C\ln(1/\epsilon) + \ln V$$

which provides a more practical method of computing the capacity C than the definition, since the latter has a slowly vanishing correction to the capacity $V/\ln(1/\epsilon)$. By plotting

$$\log N_\epsilon \quad \text{versus} \quad \log(1/\epsilon)$$

for decreasing values of ϵ, the formula for C gives the capacity as the asymptotic slope. Let D_H be the Hausdorff dimension. Then we have $D_H \le C$.

In the C++ program `capacity.cpp` we find the capacity for the Hénon map, where $a = 1.4$ and $b = 0.3$. The numerical simulation yields the value $C \approx 1.26$.

```
// capacity.cpp

#include <iostream>
#include <cmath>      // for floor, log
using namespace std;

unsigned int Neps_func(double* x,double* y,unsigned int T,double eps)
{
 unsigned int i, j, k, Nx, Ny, Neps = 0;
 double mx, my, xmin, ymin, xmax, ymax;
 xmin = x[0]; xmax = x[0]; ymin = y[0]; ymax = y[0];
 for(i=1;i<T;i++)
 {
 if(x[i] < xmin) xmin = x[i]; if(x[i] > xmax) xmax = x[i];
 if(y[i] < ymin) ymin = y[i]; if(y[i] > ymax) ymax = y[i];
 }
 Nx = (unsigned)((xmax-xmin)/eps+1.0);
 Ny = (unsigned)((ymax-ymin)/eps+1.0);
 mx = ((double)Nx-1.0)/(xmax-xmin);
 my = ((double)Ny-1.0)/(ymax-ymin);
 unsigned int** box = NULL; box = new unsigned int*[Ny];
 for(j=0;j<Ny;j++) box[j] = new unsigned int[Nx];

 for(i=0;i<Ny;i++)
  for(j=0;j<Nx;j++) box[i][j] = 0;

 for(i=0;i<T;i++)
 {
 k = (unsigned int) floor(mx*(x[i]-xmin)+0.5);
 j = (unsigned int) floor(my*(y[i]-ymin)+0.5);
```

```
 box[j][k] = 1;
 }
 for(i=0;i<Ny;i++)
  for(j=0;j<Nx;j++) Neps += box[i][j];

  for(i=0;i<Ny;i++) delete[] box[i];  delete[] box;
  return Neps;
}

int main(void)
{
 unsigned int T = 200000;
 double* x = new double[T]; double* y = new double[T];
 x[0] = 1.161094;  y[0] = -0.09541356;
 for(int j=0;j<T-1;j++)
 { x[j+1] = 1.0+y[j]-1.4*x[j]*x[j]; y[j+1] = 0.3*x[j]; }
 double eps = 0.01;
 unsigned int Neps = Neps_func(x,y,T,eps);
 cout << "Neps = " << Neps << "   " << "eps = " << eps << endl;
 cout << "log(Neps) = " << log((double) Neps) << "  "
      << "log(1.0/eps) = " << log(1.0/eps) << endl;
 eps = 0.005;
 Neps = Neps_func(x,y,T,eps);
 cout << "Neps = " << Neps << "   " << "eps = " << eps << endl;
 cout << "log(Neps) = " << log((double) Neps) << "  "
      << "log(1.0/eps) = " << log(1.0/eps) << endl;
 eps = 0.002;
 Neps = Neps_func(x,y,T,eps);
 cout << "Neps = " << Neps << "   " << "eps = " << eps << endl;
 cout << "log(Neps) = " << log((double) Neps) << "  "
      << "log(1.0/eps) = " << log(1.0/eps) << endl;
 eps = 0.001;
 Neps = Neps_func(x,y,T,eps);
 cout << "Neps = " << Neps << "   " << "eps = " << eps << endl;
 cout << "log(Neps) = " << log((double) Neps) << "  "
      << "log(1.0/eps) = " << log(1.0/eps) << endl;
 delete[] x; delete[] y;
 return 0;
}
```

1.2.7 Hyperchaos

We consider a system of first order autonomous ordinary difference equations

$$x_{1t+1} = f_1(x_{1t}, x_{2t}), \qquad x_{2t+1} = f_2(x_{1t}, x_{2t}).$$

We assume that f_1 and f_2 are smooth functions. We also assume that the solution (x_{1t}, x_{2t}) is bounded. For certain systems we can find so-called *hyperchaos*. This means the system admits positive two one-dimensional Liapunov exponents $(\lambda_1^I, \lambda_2^I)$ and one positive two-dimensional Liapunov exponent. Next we derive an equation for the two-dimensional Liapunov exponent λ^{II}.

The variational map is given by

$$y_{1t+1} = \frac{\partial f_1}{\partial x_1}(\mathbf{x} = \mathbf{x}_t) y_{1t} + \frac{\partial f_1}{\partial x_2}(\mathbf{x} = \mathbf{x}_t) y_{2t}$$
$$y_{2t+1} = \frac{\partial f_2}{\partial x_1}(\mathbf{x} = \mathbf{x}_t) y_{1t} + \frac{\partial f_2}{\partial x_2}(\mathbf{x} = \mathbf{x}_t) y_{2t} .$$

Let (v_{1t}, v_{2t}) satisfy the variational map, i.e.

$$v_{1t+1} = \frac{\partial f_1}{\partial x_1}(\mathbf{x} = \mathbf{x}_t) v_{1t} + \frac{\partial f_1}{\partial x_2}(\mathbf{x} = \mathbf{x}_t) v_{2t}$$
$$v_{2t+1} = \frac{\partial f_2}{\partial x_1}(\mathbf{x} = \mathbf{x}_t) v_{1t} + \frac{\partial f_2}{\partial x_2}(\mathbf{x} = \mathbf{x}_t) v_{2t} .$$

Let $\{\mathbf{e}_1, \mathbf{e}_2\}$ be the standard basis in $\mathbb{R}^2$, i.e.,

$$\left\{ \mathbf{e}_1 = \begin{pmatrix} 1 \\ 0 \end{pmatrix}, \quad \mathbf{e}_2 = \begin{pmatrix} 0 \\ 1 \end{pmatrix} \right\}.$$

We can write

$$\mathbf{y}_t = y_{1t}\mathbf{e}_1 + y_{2t}\mathbf{e}_2, \qquad \mathbf{v}_t = v_{1t}\mathbf{e}_1 + v_{2t}\mathbf{e}_2.$$

Next we calculate $\mathbf{y}_t \wedge \mathbf{v}_t$, where $\wedge$ denotes the *Graßmann product* (also called *exterior product* or *wedge product*). We find

$$\begin{aligned} \mathbf{y}_t \wedge \mathbf{v}_t &= (y_{1t}\mathbf{e}_1 + y_{2t}\mathbf{e}_2) \wedge (v_{1t}\mathbf{e}_1 + v_{2t}\mathbf{e}_2) \\ &= y_{1t}v_{2t}\mathbf{e}_1 \wedge \mathbf{e}_2 + y_{2t}v_{1t}\mathbf{e}_2 \wedge \mathbf{e}_1 \\ &= (y_{1t}v_{2t} - y_{2t}v_{1t})\mathbf{e}_1 \wedge \mathbf{e}_2 \end{aligned}$$

where we have used the *distributive law* $(a, b, c, d \in \mathbb{R})$

$$(a\mathbf{e}_i + b\mathbf{e}_j) \wedge (c\mathbf{e}_k + d\mathbf{e}_l) = (ac)\mathbf{e}_i \wedge \mathbf{e}_k + (ad)\mathbf{e}_i \wedge \mathbf{e}_l + (bc)\mathbf{e}_j \wedge \mathbf{e}_k + (bd)\mathbf{e}_j \wedge \mathbf{e}_l$$

and that $\mathbf{e}_i \wedge \mathbf{e}_j = -\mathbf{e}_j \wedge \mathbf{e}_i$. It follows that $\mathbf{e}_j \wedge \mathbf{e}_j = 0$. The Graßmann product is also associative. We define

$$w_t := y_{1t}v_{2t} - y_{2t}v_{1t}$$

and evaluate the time evolution of w_t. Since

$$w_{t+1} = y_{1t+1}v_{2t+1} - y_{2t+1}v_{1t+1}$$

we find

$$\begin{aligned} w_{t+1} = &\left(\frac{\partial f_1}{\partial x_1}(\mathbf{x}_t) y_{1t} + \frac{\partial f_1}{\partial x_2}(\mathbf{x}_t) y_{2t} \right) \left(\frac{\partial f_2}{\partial x_1}(\mathbf{x}_t) v_{1t} + \frac{\partial f_2}{\partial x_2}(\mathbf{x}_t) v_{2t} \right) \\ &- \left(\frac{\partial f_2}{\partial x_1}(\mathbf{x}_t) y_{1t} + \frac{\partial f_2}{\partial x_2}(\mathbf{x}_t) y_{2t} \right) \left(\frac{\partial f_1}{\partial x_1}(\mathbf{x}_t) v_{1t} + \frac{\partial f_1}{\partial x_2}(\mathbf{x}_t) v_{2t} \right) . \end{aligned}$$

Consequently

$$w_{t+1} = \left(\frac{\partial f_1}{\partial x_1}(\mathbf{x}_t) \frac{\partial f_2}{\partial x_2}(\mathbf{x}_t) - \frac{\partial f_1}{\partial x_2}(\mathbf{x}_t) \frac{\partial f_2}{\partial x_1}(\mathbf{x}_t) \right) w_t, \quad t = 0, 1, 2, \ldots$$

where the initial value w_0 is given by $w_0 := y_{10}v_{20} - y_{20}v_{10}$. The *two-dimensional Liapunov exponent* is given by

$$\lambda^{II} := \lim_{T \to \infty} \frac{1}{T} \ln |w_T|$$

where λ^{II} depends on the initial values x_{10}, x_{20}, y_{10}, y_{20}, v_{10}, v_{20}. Let λ^I be the maximal one-dimensional Liapunov exponent and let λ^{II} be the maximal two-dimensional Liapunov exponent. If $\lambda^I > 0$ and $\lambda^{II} > 0$ we say that the system shows *hyperchaotic behaviour*. We have

$$\lambda^{II} = \lambda_1^I + \lambda_2^I$$

where λ_1^I and λ_2^I are the two one-dimensional Liapunov exponents.

Example. Consider the coupled logistic equation. It is given by

$$x_{1t+1} = rx_{1t}(1 - x_{1t}) + e(x_{2t} - x_{1t}), \quad x_{2t+1} = rx_{2t}(1 - x_{2t}) + e(x_{1t} - x_{2t})$$

where $t = 0, 1, 2, \ldots$ and r and e are bifurcation parameters with $1 \leq r \leq 4$. For $e = 0$ we have two uncoupled logistic equations. Depending on the initial conditions and the parameter values one can find the following behaviour: (i) orbits tend to a fixed point, (ii) periodic behaviour, (iii) quasiperiodic behaviour, (iv) chaotic behaviour, (v) hyperchaotic behaviour and (vi) x_{1t} and x_{2t} explode, i.e. for a finite time the state variables x_{1t} and (or) x_{2t} tend to infinity. Let

$$f_1(x_1, x_2) = rx_1(1 - x_1) + e(x_2 - x_1), \quad f_2(x_1, x_2) = rx_2(1 - x_2) + e(x_1 - x_2).$$

Thus we find for the variational equation

$$y_{1t+1} = (r(1 - 2x_{1t}) - e)y_{1t} + ey_{2t}, \quad y_{2t+1} = ey_{1t} + (r(1 - 2x_{2t}) - e)y_{2t}.$$

Furthermore we find

$$w_{t+1} = (r^2(1 - 2x_{1t})(1 - 2x_{2t}) - 2re(1 - x_{1t} - x_{2t}))w_t.$$ ♣

In the C++ program **hyperchaos.cpp** we evaluate the one-dimensional and two-dimensional Liapunov exponent for $e = 0.06$ and r in the range $r \in [3.2...3.7]$. For example, for $e = 0.06$ and $r = 3.7$ there is numerical evidence that the system shows hyperchaotic behaviour.

```
// hyperchaos.cpp

#include <fstream>    // for ofstream, close
#include <cmath>      // for fabs, log
using namespace std;
```

```
double f1(double x1,double x2,double r,double e)
{ return r*x1*(1.0-x1)+e*(x2-x1); }

double f2(double x1,double x2,double r,double e)
{ return r*x2*(1.0-x2)+e*(x1-x2); }

double v1(double x1,double x2,double y1,double y2,double r,double e)
{ return (r-2.0*r*x1-e)*y1+e*y2; }

double v2(double x1,double x2,double y1,double y2,double r,double e)
{ return e*y1+(r-2.0*r*x2-e)*y2; }

double varext(double x1,double x2,double w,double r,double e)
{ return (r*r*(1.0-2.0*x1)*(1.0-2.0*x2)-2.0*e*r*(1.0-x1-x2))*w; }

int main(void)
{
 int T = 700;  // number of iterations
 double r, e;  // bifurcation parameters
 double rmin = 3.2, rmax = 3.7;
 r = rmin; e = 0.06;
 ofstream data("lambda.dat");
 while(r <= rmax)
 {
 double x11 = 0.7, x22 = 0.3; // initial values
 double x1, x2;
 for(int t=0;t<10;t++) // remove the transients
 { x1 = x11; x2 = x22; x11 = f1(x1,x2,r,e); x22 = f2(x1,x2,r,e); }

 double y11 = 0.5, y22 = 0.5;
 double w;  double w1 = 0.5;
 double y1, y2;
 for(int t=0;t<T;t++)
 { x1 = x11; x2 = x22; y1 = y11; y2 = y22; w = w1;
 x11 = f1(x1,x2,r,e); x22 = f2(x1,x2,r,e);
 y11 = v1(x1,x2,y1,y2,r,e);  y22 = v2(x1,x2,y1,y2,r,e);
 w1 = varext(x1,x2,w,r,e);
 }
 double lambdaI = log(fabs(y11) + fabs(y22))/((double) T);
 double lambdaII = log(fabs(w1))/((double) T);
 data << r << "  " << lambdaI << "  " << lambdaII << "\n";
 r += 0.005;
 }
 data.close();
 return 0;
}
```

1.2.8 Domain of Attraction

Consider a system of nonlinear first order autonomous difference equations

$$\mathbf{x}_{t+1} = \mathbf{f}(\mathbf{x}_t, \mathbf{r}), \qquad t = 0, 1, 2, \ldots$$

where $\mathbf{r} = (r_1, \ldots, r_p)$ are bifurcation parameters. Assume that the initial values $\mathbf{x}_0$ are given. The behaviour for $t \to \infty$ depends on the bifurcation parameters. In particular one is interested in finding the domain for the bifurcation parameter values $\mathbf{r}$ where the solution escapes to infinity, i.e.,

$$\|\mathbf{x}_t\| \to \infty \qquad \text{for} \quad t \to \infty .$$

The domain of attraction can have a fractal structure. Obviously, the domain also depends on the initial values. Thus one keeps the initial values fixed.

Example. Consider the coupled logistic equation

$$x_{1t+1} = rx_{1t}(1 - x_{1t}) + e(x_{2t} - x_{1t}), \quad x_{2t+1} = rx_{2t}(1 - x_{2t}) + e(x_{1t} - x_{2t})$$

where $t = 0, 1, 2, \ldots$ and r and e are bifurcation parameters with $1 \le r \le 4$. For $e = 0$ we have two uncoupled logistic equations. Depending on the initial conditions and the parameter values one can find the following behaviour for $t \to \infty$: (i) orbits tend to a fixed point, (ii) periodic behaviour, (iii) quasiperiodic behaviour, (iv) chaotic behaviour, (v) hyperchaotic behaviour and (vi) x_{1t} and (or) x_{2t} tend to infinity. ♣

In the C++ program `domain.cpp` we find the escape domain, for the parameter regions

$$e \in [0.04, 0.09] \quad \text{and} \quad r \in [3.5, 4.0] .$$

The initial values are fixed to $x_{10} = 0.7$ and $x_{20} = 0.3$. Obviously, the initial values must be chosen so that $x_{10} \neq x_{20}$ otherwise the coupling term will cancel out and we have two uncoupled logistic maps.

```
// domain.cpp

#include <fstream>    // for ofstream, close
#include <cmath>      // for fabs, log
using namespace std;

double f1(double x1,double x2,double r,double e)
{ return r*x1*(1.0-x1)+e*(x2-x1); }

double f2(double x1,double x2,double r,double e)
{ return r*x2*(1.0-x2)+e*(x1-x2); }

int main(void)
{
```

```
int T = 800;   // number of iterations
double r, e;   // bifurcation parameters
double rmin = 3.6, rmax = 4.0; double emin = 0.04, emax = 0.08;
r = rmin; e = emin;
ofstream data("domain.dat");
while(e <= emax)
{ while(r <= rmax)
{ double x11 = 0.7, x22 = 0.3;
double x1, x2;
for(int t=0;t<=T;t++)
{ x1 = x11; x2 = x22; x11 = f1(x1,x2,r,e); x22 = f2(x1,x2,r,e); }
if((fabs(x11) > 20) || (fabs(x22) > 20))
{ data << r << "  " << e << "\n"; }
r += 0.0005;
}
e += 0.00005;
r = rmin;
}
data.close();
return 0;
}
```

1.2.9 Newton Method in the Complex Domain

In connection with chaotic behaviour and fractals the *Newton method* is considered in the complex domain $\mathbb{C}$. Let f be a differentiable complex valued function. Given an approximate value of z_0 to the solution of $f(z) = 0$ the Newton method finds the next approximation by calculating

$$z_{t+1} = z_t - \frac{f(z_t)}{f'(z_t)}$$

where $t = 0, 1, 2, \ldots$ and $f'(z_t) \neq 0$ is the derivative of f at $z = z_t$. One calls

$$\left\{ \hat{\mathbb{C}} \, : \, g(z) = z - \frac{f(z)}{f'(z)} \right\}$$

the Newton transformation associated with the function f. The general expectation is that a typical orbit $\{ f^{(t)}(z_0) \}$ which starts from an initial guess $z_0 \in \mathbb{C}$, will converge to one of the roots.

Example. Consider the polynomial

$$p(z) = z^4 - 1$$

for $z \in \mathbb{C}$. There are four distinct complex numbers, a_j $(j = 1, 2, 3, 4)$ such that

$$p(a_j) = 0.$$

These are called the *roots*, or the zeros, the polynomial $p(z)$. The roots are given by

$$a_1 = 1, \quad a_2 = -1, \quad a_3 = i, \quad a_4 = -i.$$

In the Newton method we consider the map

$$\left\{ \widehat{\mathbb{C}} : f(z) := z - \frac{p(z)}{p'(z)} \right\}$$

with $p' \equiv dp/dz = 4z^3$. We call f the Newton transformation associated with the function p. One expects that a typical orbit

$$\{ f^{(t)}(z_0) \} = \{ z_0, f(z_0), f(f(z_0)), \ldots \}$$

which starts from an initial value $z_0 \in \mathbb{C}$, will converge to one of the roots of the polynomial p. For the present case we find that the Newton transformation is given by

$$f(z) = \frac{3z^4 + 1}{4z^3}.$$

We expect the orbit of z_0 to converge to one of the numbers a_1, a_2, a_3 or a_4. If we choose z_0 close enough to a_j then it can be proved that

$$\lim_{n \to \infty} f^{(t)}(z_0) = a_j, \qquad \text{for} \quad j = 1, 2, 3, 4.$$

If, on the other hand, z_0 is far away from all of the a_j's, then what happens? Perhaps the orbit of z_0 converges to the root of $p(z)$ closest to z_0? ♣

In the C++ program `complexnewton.cpp` we use the `complex` class of C++.

```
// complexnewton.cpp

#include <iostream>
#include <complex>
using namespace std;

int main(void)
{
 int T = 1000;  // number of iterations
 complex<double> z0(0.4,0.2);
 complex<double> z1 = (3.0*z0*z0*z0*z0+1.0)/(4.0*z0*z0*z0);
 double eps = 0.0001;
 while(abs(z1-z0) > eps)
 { z0 = z1; z1 = (3.0*z0*z0*z0*z0+1.0)/(4.0*z0*z0*z0); }
 cout << "root = " << z1 << endl;
 return 0;
}
```

The iteration provides the root `complex(0,-1)`, i.e. $-i$.

1.2.10 Newton Method in Higher Dimensions

We consider the system

$$f_1(x_1, x_2, \ldots, x_n) = 0, \ \ldots, \ f_n(x_1, x_2, \ldots, x_n) = 0 .$$

We assume that the functions $f_1, \ldots, f_n$ are continuously differentiable. We denote by $\mathbf{x}$ the vector with components $(x_1, x_2, \ldots, x_n)$ and by $\mathbf{f}$ the vector with components $(f_1, f_2, \ldots, f_n)$. Then the system can be written as one vector equation, $\mathbf{f}(\mathbf{x}) = \mathbf{0}$. If we take the gradients of the components, we obtain a function matrix called the *Jacobian matrix* J. It is defined as

$$J(\mathbf{f}(\mathbf{x})) := \left(\frac{\partial f_i}{\partial x_k}\right) \equiv \begin{pmatrix} \frac{\partial f_1}{\partial x_1} & \frac{\partial f_1}{\partial x_2} & \cdots & \frac{\partial f_1}{\partial x_n} \\ \frac{\partial f_2}{\partial x_1} & \frac{\partial f_2}{\partial x_2} & \cdots & \frac{\partial f_2}{\partial x_n} \\ \vdots & & & \\ \frac{\partial f_n}{\partial x_1} & \frac{\partial f_n}{\partial x_2} & \cdots & \frac{\partial f_n}{\partial x_n} \end{pmatrix} .$$

The problem is now to solve the system of equations $\mathbf{f}(\mathbf{x}) = \mathbf{0}$. We suppose that the vector $\mathbf{y}$ is the exact solution and that our present approximation $\mathbf{x}$ can be written as $\mathbf{x} = \mathbf{y} + \mathbf{h}$. We compute $f_i(x_1, \ldots, x_n)$ and call these known values g_i. Hence

$$f_i(y_1 + h_1, y_2 + h_2, \ldots, y_n + h_n) = g_i .$$

The first term is zero by definition, and neglecting higher order terms we obtain with $J_{ik} = (\partial f_i / \partial x_k)_{\mathbf{x}=\mathbf{y}}$:

$$J\mathbf{h} = \mathbf{g} \qquad \text{and} \qquad \mathbf{h} = J^{-1}\mathbf{g} .$$

We suppose that the matrix J is nonsingular. The vector $\mathbf{h}$ found in this way in general does not give us an exact solution. We arrive at the iteration formula

$$\mathbf{x}^{(t+1)} = \mathbf{x}^{(t)} - J^{-1}\mathbf{f}(\mathbf{x}^{(t)}).$$

In one dimension this formula becomes the usual Newton formula. The matrix J must be nonsingular. The matrix J changes from step to step. This suggests a simplification to reduce computational efforts: replace $J(\mathbf{x}^{(t)})$ by $J(\mathbf{x}^{(0)})$.

In the C++ program we consider the case $n = 2$ with

$$f_1(x_1, x_2) = x_1^2 + x_2^2 - 1, \qquad f_2(x_1, x_2) = x_1 - x_2 .$$

The system of equations $f_1(x_1, x_2) = 0$, $f_2(x_1, x_2) = 0$ admits two solutions, namely $(x_1, x_2) = (1/\sqrt{2}, 1/\sqrt{2})$ and $(x_1, x_2) = (-1/\sqrt{2}, -1/\sqrt{2})$.

```
// twonewton.cpp

#include <iostream>
```

```
#include <cmath>        // for fabs
using namespace std;

double g1(double x,double y)
{ return (x-(x*x-y*y+2.0*x*y-1.0)/(2.0*(x+y))); }

double g2(double x,double y)
{ return (y-(-x*x+y*y+2.0*x*y-1.0)/(2.0*(x+y))); }

int main(void)
{
 double x0, y0, x1, y1, eps;
 int t = 0;
 x1 = 3.5, y1 = 20.3;  // initial values
 eps = 0.0005;
 do
 { x0 = x1; y0 = y1; x1 = g1(x0,y0); y1 = g2(x0,y0);
 t++;
 } while((fabs(x0-x1) > eps) && (fabs(y0-y1) > eps));
 cout << "t = " << t << endl;
 cout << "x1 = " << x1 << endl;
 cout << "y1 = " << y1 << endl;
 return 0;
}
```

1.2.11 Ruelle-Takens-Newhouse Scenario

In this route to chaos we have the following sequence when the bifurcation parameter r is changed. A stationary point (fixed point) bifurcates to a periodic orbit, which then bifurcates to a doubly periodic orbit formed by the surface of a torus, which then bifurcates to a system with chaotic behaviour. Newhouse, Ruelle and Takens [153] conjectured that small nonlinearities would destroy triply periodic motion. They proved the following.

Theorem. Let V be a constant vector field on the torus $T^n = \mathbb{R}^n/\mathbb{Z}^n$. If $n \geq 3$ every C^2 neighbourhood of V conatins a vector field V^T with a strange Axiom A attractor. If $n \geq 4$, we may take C^∞ instead of C^2.

A dynamical system is (or satisfies) Axiom A if its nonwandering set
(i) has a hyperbolic structure, and
(ii) is the closure of the set of closed orbits of the system.

We define a point p to be non-wandering if, for all neighbourhoods U of p, $f(U) \setminus U$ is nonempty for arbitrary large $t \in \mathbb{R}$ or $t \in \mathbb{N}$. The term *non-wandering point* is an unhappy one, since not only may the point wander away from its original position but it may never come back again. The set of all non-wandering points of the map

f are called non-wandering set of f.

We consider the map (Lopez-Ruiz and Perez-Garcia [133], [134])

$$x_{t+1} = r(3y_t + 1)x_t(1 - x_t), \qquad y_{t+1} = r(3x_t + 1)y_t(1 - y_t)$$

which shows the Ruelle-Takens-Newhouse transition to chaos. The control parameter is r. We calculate the variational equation symbolically

$$u_{t+1} = r(3y_t + 1)(1 - 2x_t)u_t + 3rx_t(1 - x_t)v_t$$
$$v_{t+1} = 3ry_t(1 - y_t)u_t + r(3x_t + 1)(1 - 2y_t)v_t$$

and then iterate these four equations using the data type `double`. The largest one-dimensional Liapunov exponent is calculated approximately

$$\lambda \approx \frac{1}{T} \ln(|u_T| + |v_T|)$$

where $r = 1.0834$ and T is large. The fixed points of the map are given by the solution of the system

$$r(3y^* + 1)x^*(1 - x^*) = x^*, \qquad r(3x^* + 1)y^*(1 - y^*) = y^*.$$

We find five fixed points

$$x_1^* = \frac{1}{3r}\left(-\sqrt{4r^2 - 3r} + r\right), \qquad y_1^* = \frac{1}{3r}\left(-\sqrt{4r^2 - 3r} + r\right)$$

$$x_2^* = \frac{1}{3r}\left(\sqrt{4r^2 - 3r} + r\right), \qquad y_2^* = \frac{1}{3r}\left(\sqrt{4r^2 - 3r} + r\right)$$

$$x_3^* = (r - 1)/r, \qquad y_3^* = 0$$

$$x_4^* = 0, \quad y_4^* = 0, \quad x_5^* = 0, \quad y_5^* = (r - 1)/r.$$

The fixed points (x_1^*, y_1^*) and (x_2^*, y_2^*) exist only for $r \geq 3/4$.

```
// ruelle.cpp

#include <iostream>
#include <cmath>
#include "symbolicc++.h"
using namespace std;

template <class T> T f(T x,T y,T r)
{ return r*(T(3)*y+T(1))*x*(T(1)-x); }

template <class T> T g(T x,T y,T r)
{ return r*(T(3)*x+T(1))*y*(T(1)-y); }

int main(void)
```

```
{
 int T = 500;    // number of iterations
 double x2, y2, u2, v2;
 Symbolic x("x"), x1("x1"), y("y"), y1("y1"), r("r"),
          u("u"), u1("u1"), v("v"), v1("v1");
 x1 = f(x,y,r); y1 = g(x,y,r);
 cout << "x1 = " << x1 << endl; cout << "y1 = " << y1 << endl;
 u1 = df(x1,x)*u+df(x1,y)*v; // variational equation
 v1 = df(y1,y)*v+df(y1,x)*u; // variational equation
 cout << "u1 = " << u1 << endl; cout << "v1 = " << v1 << endl;
 // initial values
 Equations values = (x==0.3,y==0.4,r==1.0834,u==0.5,v==0.6);
 for(int t=1;t<T;t++)
 {
 x2 = x1[values]; y2 = y1[values]; u2 = u1[values]; v2 = v1[values];
 values = (r==1.0834,x==x2,y==y2,u==u2,v==v2);
 cout << "The Liapunov exponent for t = " << t << " is "
      << log(fabs(double(rhs(values,u)))+fabs(double(rhs(values,v))))/t
      << endl;
 }
 return 0;
}
```

1.2.12 Melnikov Analysis for Maps

Consider an integrable two-dimensional map of the form

$$\mathbf{x}_{t+1} = \mathbf{f}_0(\mathbf{x}_t)\,.$$

Adding a small perturbation to the integrable map we obtain

$$\mathbf{x}_{t+1} = \mathbf{f}_0(\mathbf{x}_t) + \epsilon\mathbf{f}_1(\mathbf{x}_t)$$

where $\epsilon \ll 1$. Assume that for $\epsilon = 0$ (the unperturbed map $\mathbf{f}_0$) there exists a hyperbolic fixed point or saddle point P_0, for which one branch of its stable manifold W_0^s coincides with one branch of its unstable manifold W_0^u. Then for ϵ sufficiently small the perturbed map $\mathbf{f}_0 + \epsilon\mathbf{f}_1$ has a saddle point P_ϵ "close" to P_0 and the stable and unstable manifolds W_0^s and W_0^u of $\mathbf{f}_0+\epsilon\mathbf{f}_1$ are "close" to those of the unperturbed map. Choosing any point $\mathbf{x}_0$ on $W_0^s = W_0^u$ we let $\mathbf{v}_0$ be the unit vector tangent to this manifold at $\mathbf{x}_0$. The first intersection of W_ϵ^s and W_ϵ^u with the normal at $\mathbf{x}_0$ are denoted by $\mathbf{x}_0^s(\epsilon)$ and $\mathbf{x}_0^s(\epsilon)$, repectively. If $\mathbf{v}_t$ is defined as the derivative in the direction of the flow we obtain

$$\mathbf{v}_t = D\mathbf{f}_0^{(t)}(\mathbf{x}_0)\mathbf{v}_0\,.$$

We define

$$\mathbf{y}_t := \left(\frac{d}{d\epsilon}\mathbf{x}_t(\epsilon)\right)_{\epsilon=0}$$

as the derivative towards ϵ. We define the splitting distance as

$$\Delta^u(\epsilon) := (\mathbf{x}_0^u(\epsilon) - \mathbf{x}_0) \wedge \mathbf{v}_0, \qquad \Delta^s(\epsilon) := (\mathbf{x}_0^s(\epsilon) - \mathbf{x}_0) \wedge \mathbf{v}_0$$

where the wedge product is defined by

$$\mathbf{x} \wedge \mathbf{v} := x_1 v_2 - x_2 v_1 .$$

Thus the total splitting distance is

$$\Delta(\epsilon) := \Delta^u(\epsilon) - \Delta^s(\epsilon) .$$

To calculate $\Delta(\epsilon)$ we use the fact that

$$\Delta(\epsilon) = \Delta(0) + \epsilon \left(\frac{d\Delta(\epsilon)}{d\epsilon} \right)_{\epsilon=0} + O(\epsilon^2) = \epsilon \Delta'(0) + O(\epsilon^2) .$$

Now we have

$$\Delta'(0) = \left(\frac{d\Delta(\epsilon)}{d\epsilon} \right)_{\epsilon=0} = \sum_{t=-\infty}^{\infty} \det(D\mathbf{f}_0^{(-t)}(\mathbf{x}_t))\mathbf{f}_1(\mathbf{x}_{t-1}) \wedge \mathbf{v}_t .$$

If the two-dimensional maps are Hamiltonian and therefore area-preserving it follows that $\det(D\mathbf{f}_0^{(t)}) = 1$. Thus

$$\Delta'(0) = \sum_{t=-\infty}^{\infty} \mathbf{f}_1(\mathbf{x}_{t-1}) \wedge \mathbf{v}_t .$$

Consequently the splitting distance can be calculated for a two-dimensional Hamilton map using

$$\Delta(\epsilon) = \epsilon \sum_{t=-\infty}^{\infty} \mathbf{f}_1(\mathbf{x}_{t-1}) \wedge \mathbf{v}_t + O(\epsilon^2) .$$

1.2.13 Periodic Orbits and Topological Degree

We consider the problem of finding the solutions of a system of nonlinear equations of the form

$$\mathbf{f}_n = \mathbf{0}_n$$

where $\mathbf{f}_n = (f_1, f_2, \ldots, f_n) : D_n \subset \mathbb{R}^n \to \mathbb{R}^n$ is a continuously differentiable function from a domain D_n into $\mathbb{R}^n$, $\mathbf{0}_n = (0, 0, \ldots, 0)$ and $\mathbf{x} = (x_1, x_2, \ldots, x_n)$. The system can be written as

$$f_1(x_1, x_2, \ldots, x_n) = 0, \quad f_2(x_1, x_2, \ldots, x_n) = 0, \quad \ldots \quad f_n(x_1, x_2, \ldots, x_n) = 0 .$$

Topological degree theory (Polymilis et al [167]) provides us with information on the existence of solutions of the above system, their number and their nature. Kronecker introduced the concept of topological degree in 1869. Picard in 1892 provided a theorem for computing the exact number of solutions of the system. Numerical methods based on topological degree theory have been applied to numerous dynamical systems.

To define the concept of the topological degree we consider the function $\mathbf{f}_n$ to be continuous on the closure $\overline{D}_n$ of D_n, satisfying also $\mathbf{f}_n(\mathbf{x}) \neq \mathbf{0}_n$ for $\mathbf{x}$ on the boundary $b(D_n)$ of D_n. We also consider the solutions to be simple, i.e. the determinant of the corresponding Jacobian matrix $(J_{\mathbf{f}_n})$ at the solution, to be different from zero. Let sgn be the sign function. Then the *topological degree* of $\mathbf{f}_n$ at $\mathbf{0}_n$ relative to D_n is defined as

$$\deg[\mathbf{f}_n, D_n, \mathbf{0}_n] := \sum_{\mathbf{x}\in\mathbf{f}_n^{-1}(\mathbf{0}_n)} \operatorname{sgn}(\det(J_{\mathbf{f}_n})(\mathbf{x})) = N_+ - N_-$$

where $\det(J_{\mathbf{f}_n})$ is the determinant of the Jacobian matrix of $\mathbf{f}_n$, N_+ the number of roots with $\det(J_{\mathbf{f}_n}) > 0$ and N_- the number of roots with $\det(J_{\mathbf{f}_n}) < 0$. If a nonzero value of $\det[\mathbf{f}_n, D_n, \mathbf{0}_n]$ is obtained then there exists at least one solution of system $\mathbf{f}_n(\mathbf{x}) = \mathbf{0}_n$ within D_n.

To find the topological degree one computates the Kronecker integral. In particular, under the assumption of the definition given above of the topological degree the $\deg[\mathbf{f}_n, D_n, \mathbf{0}_n]$ can be computed as

$$\deg[\mathbf{f}_n, D_n, \mathbf{0}_n] = \frac{\Gamma(n/2)}{2\pi^{n/2}} \int\int_{b(D_n)} \cdots \int \frac{\Sigma_{j=1}^{n} A_j dx_j \cdots dx_{j-1} dx_{j+1} \cdots dx_n}{(f_1^2 + f_2^2 + \cdots + f_n^2)^{n/2}}$$

where

$$A_j := (-1)^{n(j-1)} \det \begin{pmatrix} f_1 & \partial f_1/\partial x_1 & \cdots & \partial f_1/\partial x_{j-1} & \partial f_1/\partial x_{j+1} & \cdots & \partial f_1/\partial x_n \\ f_2 & \partial f_2/\partial x_1 & \cdots & \partial f_2/\partial x_{j-1} & \partial f_2/\partial x_{j+1} & \cdots & \partial f_2/\partial x_n \\ \vdots & \vdots & & \vdots & \vdots & & \vdots \\ f_n & \partial f_n/\partial x_1 & \cdots & \partial f_n/\partial x_{j-1} & \partial f_n/\partial x_{j+1} & \cdots & \partial f_n/\partial x_n \end{pmatrix}$$

and $\Gamma(x)$ is the gamma function. In order to find the number N of solutions of $\mathbf{f}_n = \mathbf{0}_n$ we consider the function

$$\mathbf{f}_{n+1} = (f_1, f_2, \ldots, f_n, f_{n+1})^T : D_{n+1} \subset \mathbb{R}^{n+1} \to \mathbb{R}^{n+1}$$

where

$$f_{n+1} := y \det(J_{\mathbf{f}_n})$$

$\mathbb{R}^{n+1} : x_1, x_2, \ldots, x_n, y$ and D_{n+1} is the product of D_n with a real interval on the y-axis containing $y = 0$. Then the exact number N of the solutions of the equation $\mathbf{f}_n(\mathbf{x}) = \mathbf{0}_n$ is given by

$$N = \deg[\mathbf{f}_{n+1}, D_{n+1}, \mathbf{0}_{n+1}].$$

Example. Consider the case of a set of two equations

$$f_1(x_1, x_2) = 0, \qquad f_2(x_1, x_2) = 0 .$$

We find that the number N of roots in the domain $D_2 = [a, b] \times [c, d]$ is given by

$$N = \frac{1}{2\pi} \int_{b(D_2)} (P_1(x_1, x_2)dx_1 + P_2(x_1, x_2)dx_2) + \delta \int\int_{D_2} \frac{Q(x_1, x_2)dx_1 dx_2}{(f_1^2 + f_2^2 + \epsilon^2 J^2)^{3/2}}$$

where δ is an arbitrary positive value,

$$P_j(x_1, x_2) := \frac{\left(f_1 \frac{\partial f_2}{\partial x_j} - f_2 \frac{\partial f_1}{\partial x_j}\right) \delta J}{(f_1^2 + f_2^2)(f_1^2 + f_2^2 + \delta^2 J^2)^{1/2}}, \quad j = 1, 2$$

and

$$Q(x_1, x_2) = \det \begin{pmatrix} f_1 & \partial f_1/\partial x_1 & \partial f_1/\partial x_2 \\ f_2 & \partial f_2/\partial x_1 & \partial f_2/\partial x_2 \\ J & \partial J/\partial x_1 & \partial J/\partial x_2 \end{pmatrix}$$

with J denoting the determinant of the Jacobian matrix of $\mathbf{f}_2 = (f_1, f_2)$. ♣

1.2.14 JPEG file

JPEG is a lossy compression technique. This means visual information is lost permanently. The key to making JPEG work is choosing what data to throw away. JPEG is the image compression standard developed by the Joint Photographic Experts Group. It works best on natural images (scenes). The JPEG compression can be applied to greyscale images. However, JPEG also compresses color images. For instance, it compresses the red-green-blue parts of a color image as three separate greyscale images - each compressed to a different extent, if desired. JPEG divides up the image into 8 by 8 pixel blocks, and then calculates the two-dimensional discrete cosine transform (DCT) of each block. The two-dimensional *discrete cosine transform* is given by

$$f(k_1, k_2) = 4 \sum_{n_1=0}^{N_1-1} \sum_{n_2=0}^{N_2-1} x(n_1, n_2) \cos\left(\frac{(2n_1 + 1)k_1\pi}{2N_1}\right) \cos\left(\frac{(2n_2 + 1)k_2\pi}{2N_2}\right)$$

where $x(n_1, n_2)$ is an input image with $n_1 = 0, 1, \ldots, N_1 - 1$ and $n_2 = 0, 1, \ldots, N_2 - 1$. The k_1, k_2 are the coordinates in the transform domain, where $k_1 = 0, 1, \ldots, N_1 - 1$ and $k_2 = 0, 1, \ldots, N_2 - 1$. A quantizer rounds off the DCT coefficients according to the quantization matrix. This step produces the lossy nature of JPEG, but allows for large compression ratios. JPEG's compression technique uses a variable length code on these coefficients, and then writes the compressed data stream to an output file `(*.jpg)`. For decompression, JPEG recovers the quantized DCT coefficients from the compressed data stream, takes the inverse transforms and displays the image. Instead of the discrete cosine transform in newer versions discrete wavelets are used.

In the Java program `JPEG1.java` we convert the phase-portrait for the Ikeda-Laser map into a JPEG file `Ikeda.jpg`.

```
// JPEG1.java

import com.sun.image.codec.jpeg.*;
import java.awt.*;
import java.awt.geom.Line2D;
import java.awt.image.BufferedImage;
import java.io.FileOutputStream;

public class JPEG1 extends Frame
{
 BufferedImage bi;
 Graphics2D g2;

 public JPEG1()
 {
 bi = new BufferedImage(400,400,BufferedImage.TYPE_INT_RGB);
 g2 = bi.createGraphics();
 double x = 0.5, y = 0.5; // initial values
 double x1, y1;
 double c1 = 0.4, c2 = 0.9, c3 = 9.0, rho = 0.85;
 int T = 20000;  // number of iterations
 for(int t=0;t<T;t++)
 { x1 = x; y1 = y;
 double taut = c1-c3/(1.0+x1*x1+y1*y1);
 x = rho+c2*x1*Math.cos(taut)-y1*Math.sin(taut);
 y = c2*(x1*Math.sin(taut)+y1*Math.cos(taut));
 double m = 90.0*x+200.0; double n = 90.0*y+200.0;
 g2.draw(new Line2D.Double(m,n,m,n));
 }
 try {
 FileOutputStream jpegOut = new FileOutputStream("Ikeda.jpg");
 JPEGImageEncoder jie = JPEGCodec.createJPEGEncoder(jpegOut);
 jie.encode(bi);
 jpegOut.close();
 System.exit(0);
 }
 catch(Exception e) { }
 } // end constructor JPEG1()

 public static void main(String args[]) { JPEG1 jp = new JPEG1(); }
}
```

Exercises. (1) Consider the linear map $\mathbf{f}:\mathbb{R}^2\to\mathbb{R}^2$

$$\mathbf{f}(x_1,x_2)=(2x_1+x_2,x_1+x_2).$$

If two vectors (x_1,x_2) and (x_1',x_2') represent the same element of $\mathbb{T}^2$, that is, if $(x_1-x_1',x_2-x_2')\in\mathbb{Z}^2$, then $\mathbf{f}(x_1,x_2)$ and $\mathbf{f}(x_1',x_2')$ also represent the same element of $\mathbb{T}^2$. Thus we can define a map

$$\mathbf{F}_{\mathbf{f}}(x_1,x_2)=(2x_1+x_2,x_1+x_2)\mod 1.$$

Show that the map $\mathbf{F}_{\mathbf{f}}$ is invertible. Show that the map $\mathbf{F}_{\mathbf{f}}$ is an automorphism of the abelian group $\mathbb{T}^2=\mathbb{R}^2/\mathbb{Z}^2$.

(2) Consider Baker's map defined by $\mathbf{f}:[0,1]\times[0,1]\to[0,1]\times[0,1]$

$$\mathbf{f}(x_1,x_2):=\begin{cases}(2x_1,x_2/2) & \text{for } 0\le x_1<1/2\\ (2x-1,(x_2+1)/2) & \text{for } 1/2\le x_1\le 1\end{cases}$$

Show that $\mathbf{f}$ is area-preserving. Show that $\mathbf{f}$ is ergodic.

(3) Show that the map $f:\mathbb{R}\to\mathbb{R}$, $f(x)=\exp(x)$ has no fixed points. Find the fixed points of the complex map $f:\mathbb{C}\to\mathbb{C}$

$$f(z)=\exp(z)$$

and study their stability.

Chapter 2

Time Series Analysis

2.1 Introduction

Detecting the existence of deterministic chaos and its characteristics is one of the important studies from the viewpoint of time series analysis on chaos. For quantitative characterization of deterministic chaos, we have several quantities such as the

1) autocorrelation function
2) correlation dimension (Grassberger-Procaccia algorithm)
3) capacity of the attractor
4) Liapunov exponents
5) Hurst exponent
6) complexity

The autocorrelation function, correlation dimension and the capacity (fractal dimension) has been already introduced in chapter 1.

In this chapter we consider the correlation coefficient, Liapunov exponents, the Hurst exponent, and a complexity measure. As for estimating the fractal dimensions, the Grassberger-Procaccia algorithm (Grassberger and Procaccia [72]) has been widely applied to real time series data. The Liapunov exponents and its spectrum are also important statistics to quantify deterministic chaos. Several methods of estimating Liapunov spectra have been proposed (Sano and Sawada [179], Eckmann et al [52], Wolf et al [217], Kantz [111], Sato et al [181], Rosenstein et al [174], Stoop and Meier [200]). Even if the observable is only a single-variable time series in case of observing with enough number of data points, the Liapunov exponent and its spectrum of the original dynamical systems can be estimated with high accuracy. The Hurst exponent (Hurst [102], Peters [165], Steeb and Andrieu [196]) plays a central role in characterizing Brownian motion, pink noise and black noise. The Hurst exponent can also be calculated for chaotic time series.

2.2 Correlation Coefficient

Consider two T-dimensional data vectors

$$\mathbf{x} = (x_0, x_1, \ldots, x_{T-1}), \qquad \mathbf{y} = (y_0, y_1, \ldots, y_{T-1}).$$

The *linear correlation coefficient* r of the two vectors is defined by

$$r := \frac{T\sum_{t=0}^{T-1} x_t y_t - \sum_{t=0}^{T-1} x_t \sum_{t=0}^{T-1} y_t}{\sqrt{(T\sum_{t=0}^{T-1} x_t^2 - (\sum_{t=0}^{T-1} x_t)^2)(T\sum_{t=0}^{T-1} y_t^2 - (\sum_{t=0}^{T-1} y_t)^2)}}$$

and measures the strength of the linear relationship between $\mathbf{x}$ and $\mathbf{y}$. If the $\mathbf{x}$ and $\mathbf{y}$ values are related by

$$\mathbf{y} = c\mathbf{x}$$

where c is any positive constant we find $r = 1$. If the $\mathbf{x}$ and $\mathbf{y}$ values are related by

$$\mathbf{y} = -c\mathbf{x}$$

where c is any positive constant we find $r = -1$. Completely uncorrelated data will give $r = 0$.

The C++ program `correlation.cpp` calculates the correlation coefficient r for a number of data points given by the user. The program first asks the user to enter the number of data. Then the data are entered by the user. Thus the memory for the vectors has to be dynamically allocated.

```
// correlation.cpp

#include <iostream>
#include <cmath>      // for sqrt
using namespace std;

void sums(double* x,double* y,int T,double* a,double* b,
          double* c,double* d,double* e);

int main(void)
{
 unsigned long T;
 cout << "number of data points: ";  cin >> T;
 double* x = new double[T]; double* y = new double[T];

 for(unsigned long t=0;t<T;t++)
 {
 cout << "x[" << t << "] = ";   cin >> x[t];
```

```
 cout << "y[" << t << "] = ";   cin >> y[t];
 }
 double a, b, c, d, e;
 a = 0.0; b = 0.0; c = 0.0; d = 0.0; e = 0.0;
 sums(x,y,T,&a,&b,&c,&d,&e);
 double r = (T*c-a*b)/sqrt((T*d-a*a)*(T*e-b*b));
 delete[] x;  delete[] y;
 cout << " r = " << r << "\n";
 return 0;
} // end main

void sums(double* x,double* y,int T,double* a,double* b,
          double* c,double* d,double* e)
{
 double sum_x = 0.0, sum_y = 0.0;
 double sum_xy = 0.0, sum_x2 = 0.0, sum_y2 = 0.0;
 for(unsigned long t=0;t<T;t++) {
 sum_x += x[t];  sum_y += y[t];
 sum_xy += x[t]*y[t]; sum_x2 += x[t]*x[t]; sum_y2 += y[t]*y[t];
 }
 *a = sum_x; *b = sum_y;
 *c = sum_xy; *d = sum_x2; *e = sum_y2;
}
```

2.3 Liapunov Exponent from Time Series

For a dynamical system, sensitivity to initial conditions is quantified by the Liapunov exponents. Consider two trajectories with nearby initial conditions on an attracting manifold. When the attractor is chaotic, the trajectories diverge, on average, at an exponential rate characterized by the largest one-dimensional Liapunov exponent (Eckmann and Ruelle [51]). This concept is also generalized for the spectrum of one-dimensional Liapunov exponents, λ_j $(j = 1, 2, \ldots, n)$, by considering a small n-dimensional sphere of initial conditions, where n is the number of first order ordinary differential equations used to describe the dynamical system. As time t evolves, the sphere evolves into an ellipsoid whose principal axes expand (or contract) at rates given by the one-dimensional Liapunov exponents. The presence of a positive exponent is sufficient for diagnosing chaos and represents local instability in a particular direction. Note that for the existence of an attractor, the overall dynamics must be dissipative, i.e. globally stable, and the total rate of contraction must outweigh the total rate of expansion. Thus, even when there are several positive one-dimensional Liapunov exponents, the sum across the entire spectrum is negative. In most case a differential equation or difference equation is not given only a data set from an experiment, i.e., a time series.

A large number of authors have discussed the calculation of the spectrum of the one-dimensional Liapunov exponents from time series (Wolf [217], Sano and Sawada

[179], Eckmann et al [52], Sato et al [181], Rosenstein et al [174], Kantz [111]).

There are two types of methods to find Liapunov exponents. One is the Jacobian matrix estimation algorithm (Sano and Sawada [179], Eckmann et al [52]). The Jacobian matrix estimation algorithm can find the whole spectrum of the one-dimensional Liapunov exponents. It involves the least-square-error algorithm and the Gram-Schmidt procedure. Since this algorithm does not have built-in checks against noise, except the fact that the Liapunov spectrum must not depend on the number of near neighbours and the dimension of the reconstructed state space, it would be better to use other methods which have a built-in-check. The method is called the direct method for finding the largest Liapunov exponent. As for estimating largest Liapunov exponents, several algorithms have been already proposed, for example, an algorithm by Wolf et al [217], and its modifications by Sato et al [181], Rosenstein et al [174], Kantz [111], Stoop and Meier [200]. These algorithms can be called a direct method, since they calculate the divergence rates of nearby trajectories and can evaluate whether the orbital instabilities are exponential on t or a power of t.

2.3.1 Jacobian Matrix Estimation Algorithm

We follow in the presentation the work of Eckmann et al [52] and Sano and Sawada [179]. Let

$$x_0, x_1, \ldots, x_{T-1}$$

denote a scalar time series, where the number of data is T. We choose an embedding dimension d_E and construct a d_E dimensional orbit representing the time evolution of the system by the time-delay method. Thus

$$\mathbf{x}_t := (x_t, x_{t+1}, \ldots, x_{t+d_E-1})$$

for $t = 0, 1, \ldots, T - d_E$. Next we have to find the the neighbours of $\mathbf{x}_t$, i.e. the points $\mathbf{x}_\tau$ of the orbit which are contained in an ϵ-neighbourhood centered at $\mathbf{x}_t$

$$\|\mathbf{x}_\tau - \mathbf{x}_t\| \leq \epsilon\,.$$

Thus we have to introduce a norm $\|.\|$. In most cases the Euclidean norm is used. However, using the *max-norm*

$$\|\mathbf{w}\| := \max_{1 \leq k \leq n} |w_k|$$

is more useful in numerical implementations. We now set

$$\mathbf{y}_i := \mathbf{x}_{k_i} - \mathbf{x}_j \quad \text{for} \quad \|\mathbf{x}_{k_i} - \mathbf{x}_j\| \leq \epsilon\,.$$

Thus $\mathbf{y}_i$ is the displacement vector between the vectors $\mathbf{x}_{k_i}$ and $\mathbf{x}_j$. After the evolution of a time interval m, the orbital point $\mathbf{x}_j$ will proceed to $\mathbf{x}_{j+m}$ and neighbouring points $\{\mathbf{x}_{k_i}\}$ to $\{\mathbf{x}_{k_i+m}\}$. The displacement vector $\mathbf{y}_i$ is thereby mapped to

$$\mathbf{z}_i := \mathbf{x}_{k_i+m} - \mathbf{x}_{j+m} \quad \text{for} \quad \|\mathbf{x}_{k_i} - \mathbf{x}_j\| \leq \epsilon\,.$$

If the radius ϵ is small enough for the displacement vectors $\{\mathbf{y}_i\}$ and $\{\mathbf{z}_i\}$ to be regarded as good approximations of tangent vectors in the tangent space, evolution of $\mathbf{y}_i$ to $\mathbf{z}_i$ can be represented by some matrix A_j, as

$$\mathbf{z}_i = A_j \mathbf{y}_i .$$

The matrix A_j is an approximation of the flow map at $\mathbf{x}_j$. Next we have to find the optimal estimation of the linearized flow map A_j from the data sets $\{\mathbf{y}_i\}$ and $\{\mathbf{z}_i\}$. An optimal estimation is the *least-square-error algorithm*, which minimizes the average of the squared error norm between $\mathbf{z}_i$ and $A_j\mathbf{y}_i$ with respect to all components of the matrix A_j as follows

$$\min_{A_j} S = \min_{A_j} \frac{1}{N} \sum_{i=1}^{N} \|\mathbf{z}_i - A_j \mathbf{y}_i\|^2 .$$

Denoting the (k,l) component of matrix A_j by $A_{kl}(j)$ and applying this condition, one obtains $d \times d$ equations to solve,

$$\frac{\partial S}{\partial A_{kl}(j)} = 0 .$$

We obtain the following expression for A_j, namely $A_jV = C$, where the matrices V and C are given by

$$(V)_{kl} = \frac{1}{N} \sum_{i=1}^{n} y_{ik} y_{il}, \qquad (C)_{kl} = \frac{1}{N} \sum_{i=1}^{N} z_{ik} y_{il} .$$

The $d \times d$ matrices V and C are called covariance matrices, and y_{ik} and z_{ik} are the k components of vectors $\mathbf{y}_i$ and $\mathbf{z}_i$, respectively. If $N \geq d$ and there is no degeneracy, the equation $A_jV = C$ has a solution for $a_{kl}(j)$. Since we found the variational equation in the tangent space along the experimentally obtained orbit, the one-dimensional Liapunov exponents can be computed as

$$\lambda_i = \lim_{n\to\infty} \frac{1}{n\tau} \sum_{j=1}^{n} \ln \|A_j \mathbf{e}_i^j\|$$

for $i = 1, \ldots, d$, where A_j is the solution of the linear equation $A_jC = V$ and $\{\,\mathbf{e}_i^j\,\}$ $(i = 1, \ldots, d)$ is a set of basis vectors of the tangent space at $\mathbf{x}_j$. In the numerical procedure, choose an arbitrary set $\{\mathbf{e}_i^j\}$. Operate with the matrix A_j on $\{\,\mathbf{e}_i^j\,\}$ and renormalize $A_j\mathbf{e}_i^j$ to have length 1. Using the Gram-Schmidt procedure, we maintain mutual orthogonality of the basis. We repeat this procedure for n iterations.

2.3.2 Direct Method

The simplest implementation of the *direct method* in the one-dimensional case is as follows. For the illustration we assume that the scalar time series is generated by the *logistic map* $f : [0,1] \to [0,1]$, $f(x) = 4x(1-x)$ with the initial value $x[0] = 0.333333$. We assume that we have 20 data points. Thus the time series given is

```
x[0]  = 0.333333 x[1]  = 0.888888  x[2]  = 0.395063 x[3]  = 0.955953
x[4]  = 0.168427 x[5]  = 0.560239  x[6]  = 0.985485 x[7]  = 0.0572163
x[8]  = 0.21577  x[9]  = 0.676854  x[10] = 0.874891 x[11] = 0.437828
x[12] = 0.984539 x[13] = 0.0608888 x[14] = 0.228725 x[15] = 0.70564
x[16] = 0.830848 x[17] = 0.562158  x[18] = 0.984545 x[19] = 0.060863
```

We start at $t = 0$ with `x[0]=0.333333`. Next we search for the point in the time series which is closest to `x[0]`. This is obviously `x[2] = 0.395063`. Now we calculate the absolute value of the difference of these two points

```
d[0] = |x[0]-x[2]| = 0.061730 .
```

Next we calculate the absolute value difference of the two consecutive points of `x[0]` and `x[2]`, namely `x[1]` and `x[3]`. Thus

```
d1[0] = |x[1]-x[3]| = 0.067065 .
```

Now we repeat this procedure for the point at $t = 1$, `x[1]`. The closest point in the time series to `x[1]` is `x[10]`. Thus

```
d[1] = |x[1]-x[10]| = 0.013997 .
```

For the consecutive points `x[2]` and `x[11]` we obtain

```
d1[1] = |x[2]-x[11]| = 0.042765 .
```

We repeat the procedure up to $t = 18$. The last point is not taken into account since it has no succeeding point. Thus we obtain the following pairs of distances

```
d[0]  = 0.0617301    d1[0]  = 0.0670646
d[1]  = 0.0139979    d1[1]  = 0.0427652
d[2]  = 0.0427652    d1[2]  = 0.0285858
d[3]  = 0.0285858    d1[3]  = 0.107539
d[4]  = 0.0473429    d1[4]  = 0.116615
d[5]  = 0.00191966   d1[5]  = 0.00093984
d[6]  = 0.00093984   d1[6]  = 0.00364669
d[7]  = 0.0036725    d1[7]  = 0.0129551
d[8]  = 0.0129551    d1[8]  = 0.0287863
d[9]  = 0.0287863    d1[9]  = 0.0440425
d[10] = 0.0139979    d1[10] = 0.0427652
d[11] = 0.0427652    d1[11] = 0.0285858
d[12] = 6.6586e-006  d1[12] = 2.5811e-005
d[13] = 0.0036725    d1[13] = 0.0129551
d[14] = 0.0129551    d1[14] = 0.0287863
d[15] = 0.0287863    d1[15] = 0.0440425
d[16] = 0.0440425    d1[16] = 0.12433
d[17] = 0.00191966   d1[17] = 0.00093984
d[18] = 6.6586e-006  d1[18] = 2.5811e-005
```

Now the approximative one-dimensional Liapunov exponent is evaluated as

$$\lambda \approx \frac{1}{19} \sum_{t=0}^{18} \log \left(\frac{d1[t]}{d[t]} \right) .$$

We find $\lambda \approx 0.651626$ which is quite a good agreement with the exact value $\lambda = \ln(2)$ if we take into account that we only had 20 data points. Furthermore we did not introduce an epsilon neighbourhood for each point of the reference trajectory.

```
// Liapunovser1.cpp

#include <iostream>
#include <cmath>      // for fabs
using namespace std;

void find(double* a,int length,int point,double& min,int& position)
{
 int i = 0;
 if(point==0) { min = fabs(a[i]-a[1]); }
 if(point!=0) { min = fabs(a[i]-a[point]); }
 position = i;
 double distance;
 for(i=1;i<(length-1);i++)
 {
 if(i != point)
 {
 distance = fabs(a[i]-a[point]);
 if(distance < min) { min = distance; position = i; } // end if
 } // end if
 } // end for
}

int main(void)
{
 // generate time series
 int T = 20;                       // length of time series
 double* x = new double[T]; // memory allocation
 x[0] = 0.333333;                  // initial value of time series
 for(int t=0;t<(T-1);t++) { x[t+1] = 4.0*x[t]*(1.0-x[t]); }
 double* d = new double[T-1];  // memory allocation
 double* d1 = new double[T-1]; // memory allocation
 int point = 1;
 int position;
 double min;
 for(int t=0;t<(T-1);t++)
 {
 find(x,T,t,min,position);
 d[t] = fabs(x[t]-x[position]);
```

```
 d1[t] = fabs(x[t+1]-x[position+1]);
 }

 for(int t=0;t<(T-1);t++)
 {
 cout << "d[" << t << "] = " << d[t] << "  "
      << "d1[" << t << "] = " << d1[t] << endl;
 }
 double sum = 0.0;
 for(int t=0;t<(T-1);t++) { sum += log(d1[t]/d[t]); }
 double lambda = sum/((double)(T-1));
 cout << "lambda = " << lambda << endl;
 delete[] x;  delete[] d;  delete[] d1;
 return 0;
}
```

The evaluation of the Liapunov exponent can be improved by considering a small ϵ-neighbourhood of each point of the reference trajectory. Then for each point which falls in the ϵ-neighbourhood we calculate the expansion to the next point.

```
// Liapunovser2.cpp

#include <iostream>
#include <cmath>    // for fabs
using namespace std;

int find(double* array,int length,int t,double x,int*& positions,
         int& n_positions,double eps)
{
 for(int s=0;s<length;s++)
 {
 if((fabs(array[s]-x) < eps) && (s != t))
 { positions[n_positions] = s; n_positions++; }
 }
 if(n_positions > 0)
 { cout << "n_positions = " << n_positions << endl; return 1; }
 return 0;
}

int main(void)
{
 // generate time series
 int T = 2000;                 // length of time series
 double* x = new double[T]; // memory allocation
 x[0] = 0.333333;              // initial value of time series
 for(int t=0;t<(T-1);t++) { x[t+1] = 4.0*x[t]*(1.0-x[t]); }
 double eps = 0.005;
 int n_positions = 0;
 int* positions = new int[T-1];
```

```
double sumt = 0.0;
for(int t=0;t<(T-1);t++)
{ cout << "t = " << t << endl;
n_positions = 0;
int result = find(x,T,t,x[t],positions,n_positions,eps);
double sumn = 0.0;
for(int n=0;n<n_positions;n++)
{ if((result==1) && (positions[n] != t))
{ int t1 = positions[n];
double d = fabs(x[t]-x[t1]); double d1 = fabs(x[t+1]-x[t1+1]);
sumn += log(d1/d);
}
}
sumt += sumn/((double) n_positions);
}
double lambda = sumt/((double) T);
cout << "lambda = " << lambda << endl;
delete[] x;  delete[] positions;
return 0;
}
```

The direct method of Wolf et al [217] is a follows. Let

$$x_0, x_1, x_2, \ldots, x_{T-1}$$

be a scalar time series. In the fixed evolution time program the time step

$$\Delta := t_{k+1} - t_k$$

between replacements is held constant and normalized to 1. A d_E-dimensional phase portrait (d_E embedding dimension) is reconstructed with delay coordinates, i.e., a point on the attractor is given by

$$\mathbf{x}_t = (x_t, x_{t+1}, \ldots, x_{t+d_E-1}) .$$

Let $\|.\|$ denote a norm, for example the Euclidian norm, the max norm or sum norm. Using the selected norm we find the nearest neighbour vector to the initial point vector

$$\mathbf{x}_0 = (x_0, \ldots, x_{0+d_E-1}) .$$

We denote the distance between these two points by $d(0)$. At a later time 1, the initial length $d(0)$ will have evolved to length $d'(1)$. The length element is propagated through the attractor for a time short enough so that only small scale attractor structure is likely to be examined. If the evolution time is too large we may see d' shrink as the two trajectories which define it pass through a folding region of the attractor. This would lead to an underestimation of the largest one-dimensional Liapunov exponent λ. We now look for a new data point that satisfies the following two criteria:

(i) its separation, $d(1)$, from the evolved reference point is small,
(ii) and the angular separation between the evolved and replacement elements is small.

If an adequate replacement point cannot be found, we retain the points that were being used. This procedure is repeated until the reference trajectory has traversed the entire data file, at which point we estimate the largest one-dimensional Liapunov exponent as

$$\lambda = \frac{1}{M}\sum_{k=1}^{M} \ln \frac{d'(k)}{d(k-1)}$$

where M is the total number of replacement steps. In the limit of an infinite amount of noise-free data the procedure always provides replacement vectors of infinitesimal magnitude with no orientation error, and λ is obtained by definition.

The algorithm proposed by Kantz [111] evaluates the following quantity

$$S(\tau) = \frac{1}{T}\sum_{t=0}^{T-1} \ln \left(\sum_{k_i=1}^{M} d(\mathbf{x}_t, \mathbf{x}_{k_i}; \tau) \right)$$

where T is the number of data points from the scalar time series, $\mathbf{x}_t$ is a reference point, $\mathbf{x}_{k_i}$ is an ϵ-near neighbour of $\mathbf{x}_t$. M is the number of nearest neighbours and τ is the relative time and $d(\mathbf{x}_t, \mathbf{x}_{k_i}; \tau)$ is the distance between $\mathbf{x}_{t+\tau}$ and $\mathbf{x}_{k_i+\tau}$. If the analyzed time series is produced from nonlinear dynamical systems with a positive largest one-dimensional Liapunov exponent, there is a positive constant slope of the function $S(\tau)$ which corresponds to the largest one-dimensional Liapunov exponent.

In the C++ program `Kantz.cpp` we evaluate the largest Liapunov exponent for data generated from the logistic map. Thus the Euclidean distance is calculated in one dimension. The embedding dimension is set to 1.

```
// Kantz.cpp

#include <iostream>
#include <cmath>     // for sqrt, fabs, log
using namespace std;

int neighbourhood(double* a1,double* a2,int length,double eps)
{
 double d = 0.0;  // distance
 for(int i=0;i<length;i++) { d += (a1[i]-a2[i])*(a1[i]-a2[i]); }
 d = sqrt(d);
 if(d < eps) return 1;
 else return 0;
}

int main(void)
```

```
{
// generate time series
int T = 2048;                    // length of time series
double* x = new double[T]; // memory allocation
x[0] = 0.618;                    // initial value of time series
for(int i=0;i<(T-1);i++) { x[i+1] = 4.0*x[i]*(1.0-x[i]); }

int m = 1;
double** W = NULL; W = new double*[T-m+1];
for(int j=0;j<(T-m+1);j++) { W[j] = new double[m]; }
int count = 0;
for(int j=0;j<(T-m+1);j++)
{
for(int i=0;i<m;i++) { W[j][i] = x[i+count]; }
if(m==1) count++; else count += m-1;
}
double eps = 0.01;  int taurange = 6;
double* S = new double[taurange];
for(int tau=0;tau<taurange;tau++) { S[tau] = 0.0; }
for(int tau=0;tau<taurange;tau++)
{ double sumT = 0.0;
for(int t=0;t<(T-m+1-tau);t++)
{ double sumN = 0.0;  int numbert = 0;
for(int s=0;s<(T-m+1-tau);s++)
{ if(s != t)
{ int result = neighbourhood(W[t],W[s],m,eps);
if(result==1) { numbert++; sumN += fabs(x[t+tau]-x[s+tau]); }
}
} // end s
if(numbert > 0) { sumT += log(sumN); }
} // end t
S[tau] = 1.0/((double)T)*sumT;
} // end tau

for(int tau=0;tau<taurange;tau++)
{ cout << "S[" << tau << "] = " << S[tau] << endl; }
return 0;
}
```

2.4 Hurst Exponent

2.4.1 Introduction

Working extensively on the Nile River Dam Project, Hurst (Hurst [102], Peters [165]) encountered the 847-year record that the Egyptians had kept of the Nile River. Most hydrologists assumed that the inflow into a reservoir was a completely random process. However, in his examination of the Nile's records, Hurst felt that the data did not represent a random structure, though standard statistical methods did not show any correlation between the observations. Thus, Hurst developed a new set of statistical tools to examine data that may not have an underlying Gaussian distribution.

Einstein had done an extensive study on Brownian motion. This study became the main model for random walks in the study of statistics. Einstein discovered that the distance covered by a random particle undergoing random collisions from all sides is directly related to the square root of time. Thus

$$R = kT^{1/2}$$

where R is the distance covered, k is some constant and T is the time index. Using rescaled range analysis (Hurst [102], Peters [165]), Hurst proposed a generalization of Brownian motion that could apply to a broader class of time-series. His generalized equation is

$$R/S = kT^H$$

where R/S = rescaled range (range/standard deviation), T = index for number/time of observations, K = some constant for the time-series, H = *Hurst exponent.*

Thus, Hurst generalized the $T^{1/2}$ law to a T^H law. Analogously, Brownian motion can be generalized to fractal Brownian motion. Fractal Brownian motion exists whenever the Hurst exponent is well-defined.

The R/S value is called the rescaled range and is a dimensionless ratio formed by dividing the range by the standard deviation of the observations. It scales as one increases the time increment by a power law value equal to H. This is the key point in Hurst's analysis: by rescaling, Hurst could compare divers data points, including periods of time that may be separated by many years. In addition, this type of analysis can be used to describe time series that possess no characteristic scale. This equation has a characteristic of fractal geometry: it scales according to a power law. In the lung, for instance, the size of each branch decreases in scale according to an inverse-power law. Likewise, the R/S function increases as a power of H. If the data of the system being measured were independently distributed, or followed a random walk, the equation would fit with Einstein's "T to the one-half" rule, and the value of the Hurst exponent would be 1/2. When Hurst investigated the Nile River he found $H = 0.91$. Thus, the rescaled range was increasing at a faster rate than the square root of time and the system (measured by the changes in water) was covering more "distance" than a purely random process. Thus, the values of the Nile River overflow had to be influencing one another. The river had a memory of past floods.

There are three possibilities for values of H (Peters [165]). If $H = 0.5$ the system follows a random walk. We recover the original scenario of Brownian motion. If not, the observations are not independent; each carries a memory of events which precede it.

- $H = 0.5$

  Independent series. (Brown noise, or Brownian motion) The series is a random walk.

- $0 \leq H < 0.5$

  Antipersistent series. (Pink noise) The system is covering less distance than a random walk. Thus, it has a tendency to reverse itself often. If increasing, it is more likely to be decreasing the next period; if decreasing, it is more likely to be increasing.

- $0.5 < H \leq 1$

  Persistent series. (Black noise) This series covers more distance than a random walk. Thus, if the system increases in one period, it is more likely to keep increasing in the immediately following period. This is called the *Joseph effect*, in that it tends to lead to "seven years" of fortune followed by "seven years" of famine. Such a series also has the potential of sudden catastrophes, the so-called *Noah Effect*.

Thus the Hurst exponent is a useful measure for fractal distributions. There is no characteristic time scale in such a distribution. Hence an exponential, or relative, relation dominates over a polynomial, or absolute, characterization.

The following statements are believed equivalent for a time-series:

1. The Hurst exponent is well-defined for the time-series.
2. The time-series exhibits fractional Brownian motion.
3. The probability distribution is stable (Paretian or Levy).
4. The slope of the log-log R/S graph is constant.

The value $1/H$ is the fractal dimension of the probability space (Mandelbrot [139], Mandelbrot and Wallis [137] and Mandelbrot and Van Ness [138]). The random walk has a fractal dimension (capacity) of $1/0.5 = 2$. Thus it completely fills the phase space. The value $2 - H$ is the fractal dimension of the time-series. The value $2H + 1$ is the rate of decay of the Fourier series. This means the Fourier coefficients decrease in proportion to $1/f^{(2H+1)}$. Estimations of H can be found by taking the slope of the log/log graph of R/S versus T, where

$$\log(R/S) = \log(kT^H) = \log(k) + H\log(T).$$

If there is no long term memory present, scrambling the data should have no effect on this estimate of H. If, however, we destroy the structure by randomizing the data points, the estimate of H should be much lower. Therefore, the Hurst exponent is a meaningful measure of the memory of a system. The R/S statistics for a discrete time series u_t is defined as follows

$$
\begin{aligned}
X(t,\tau) &:= \sum_{i=0}^{t-1}(u_i - \langle u\rangle_\tau) \\
R(\tau) &:= \max_{0\le t\le \tau-1} X(t,\tau) - \min_{0\le t\le \tau-1} X(t,\tau) \\
S(\tau) &:= \left(\frac{1}{\tau}\sum_{t=0}^{\tau-1}(u_t - \langle u\rangle_\tau)^2\right)^{1/2} \\
R/S(\tau) &:= R(\tau)/S(\tau)\,.
\end{aligned}
$$

Regarding the sequence of random numbers u_t as spatial increments in a one-dimensional random walk, then

$$\sum_{t=0}^{\tau-1} u_t$$

is the position of the walker after a time τ. In the quantity $X(t,\tau)$ the mean over the time lag τ

$$\langle u\rangle_\tau := \frac{1}{\tau}\sum_{t=0}^{\tau-1} u_t$$

is subtracted to remove a trend when the expectation value of u_t is not zero. $R(\tau)$ is the self-adjusted range and $R/S(\tau)$ is the self-rescaled self-adjusted range.

2.4.2 Implementation for the Hurst Exponent

In the following we use the notation as in the C++ program. Let

$$u_0,\ u_1,\ \ldots,\ u_{T-1}$$

be a given time series of length T. We divide this time period into a contiguous subperiods of length n, such that $an = T$. We label each subperiod I_j, with $j = 0, 1, \ldots, a-1$. Each element in I_j is labeled $N[j][k]$ such that $k = 0, 1, \ldots, n-1$. Thus N is an $a \times n$ matrix. For each I_j of length n, the average value is defined as

$$E_j := \frac{1}{n}\sum_{k=0}^{n-1} N[j][k]$$

where E_j is the average value of the u_t contained in subperiod I_j of length n. The time series of accumulated departures $X[j][k]$ from the mean value for each subperiod I_j ($j = 0, 1, \ldots, a-1$) is defined as

$$X[j][k] := \sum_{i=0}^{k}(N[j][i] - E_j), \qquad k = 0, 1, \ldots, n-1\,.$$

Thus X is also an $a \times n$ matrix. The range is defined as the maximum minus the minimum value of $X[j][k]$ within each subperiod I_j

$$R_{I_j} := \max_{0 \le k \le n-1} (X[j][k]) - \min_{0 \le k \le n-1} (X[j][k]) .$$

The sample standard deviation calculated for each subperiod I_j ($j = 0, 1, \ldots, a-1$) is

$$S_{I_j} := \left(\frac{1}{n} \sum_{k=0}^{n-1} (N[j][k] - E_j)^2 \right)^{1/2} .$$

Each range, R_{I_j}, is now normalized by dividing by the S_{I_j} corresponding to it. Therefore, the rescaled range for each I_j subperiod is equal to R_{I_j}/S_{I_j}. We had a contiguous subperiods of length n. Therefore, the average R/S value for a fixed length n is defined as

$$(R/S)_n := \frac{1}{a} \sum_{j=0}^{a-1} \frac{R_{I_j}}{S_{I_j}} .$$

The length n is increased to the next higher value, and $(T-1)/n$ is an integer value. We use values of n that include the beginning and ending points of the time series, and steps described above are repeated until $n = (T-1)/2$. We can now apply

$$(R/S)_n = cn^H$$

or

$$\log((R/S)_n) = \log(c) + H \log(n)$$

by performing an ordinary least squares regression on $\log(n)$ as the independent variable and $\log(R/S)_n$ as the dependent variable. The intercept is the estimate for $\log(c)$, the constant. The slope of the equation is the estimate of the Hurst exponent, H. In general, one runs the regression over values of $n \ge 10$. Small values of n produce unstable estimates when sample sizes are small.

In the C++ program `Hurst1.cpp` we generate a time series from the logistic map. The length of the time series is 4096. The smallest length of contiguous subperiods is $n = 16$. Thus $a = 256$ for this case.

```
// Hurst1.cpp

#include <iostream>
#include <cmath>      // for log
using namespace std;

double sum(double* array,int length)
{
 double result = 0.0;
 for(int i=0;i<length;i++) { result += array[i]; }
 return result;
```

```
}

double max(double* array,int length)
{
 double max_value = array[0];
 for(int i=1;i<length;i++)
 { if(max_value < array[i]) max_value = array[i]; }
 return max_value;
}

double min(double* array,int length)
{
 double min_value = array[0];
 for(int i=1;i<length;i++)
 { if(min_value > array[i]) min_value = array[i]; }
 return min_value;
}

int main(void)
{
 // generate time series
 int T = 4096;               // length of time series
 double* u = new double[T]; // memory allocation
 u[0] = 0.618;               // initial value of time series
 for(int i=0;i<(T-1);i++) { u[i+1] = 4.0*u[i]*(1.0-u[i]); }
 int n = 16; // smallest length of contiguous subperiods
 int n_numbers = 0;
 while(n <= T/2) { n_numbers++; n += n; }
 cout << "n_numbers = " << n_numbers << endl;

 n = 16;
 int* n_values = NULL; n_values = new int[n_numbers];
 for(int i=0;i<n_numbers;i++) { n_values[i] = n; n += n; }

 double* RDS = new double[n_numbers];
 n = 16;
 for(int l=0;l<n_numbers;l++)
 {
 int a = T/n;
 double** N = NULL;  N = new double*[a];
 for(int j=0;j<a;j++) { N[j] = new double[n]; }
 int count = 0;
 for(int j=0;j<a;j++)
 {
 for(int i=0;i<n;i++) { N[j][i] = u[i+count]; } count += n;
 }
 double* E = new double[a];
 for(int j=0;j<a;j++) { E[j] = 1.0/((double) n)*sum(N[j],n); }
```

```
double** X = NULL;  X = new double*[a];
for(int j=0;j<a;j++) { X[j] = new double[n]; }
for(int j=0;j<a;j++)
{
for(int k=0;k<n;k++)
{
double temp = 0.0;
for(int i=0;i<=k;i++) { temp += N[j][i]-E[j]; }
X[j][k] = temp;
}
}
double* R = new double[a];
for(int j=0;j<a;j++) { R[j] = max(X[j],n)-min(X[j],n); }

for(int j=0;j<a;j++)
{
for(int k=0;k<n;k++) { N[j][k] = (N[j][k]-E[j])*(N[j][k]-E[j]); }
}
double* S = new double[a];
for(int j=0;j<a;j++) { S[j] = sqrt(1.0/((double) n)*sum(N[j],n)); }
double* D = NULL; D = new double[a];
for(int j=0;j<a;j++) { D[j] = R[j]/S[j]; }
delete[] E; delete[] R; delete[] S;
for(int j=0;j<a;j++) { delete[] N[j]; } delete[] N;
for(int j=0;j<a;j++) { delete[] X[j]; } delete[] X;

RDS[l] = 1.0/((double) a)*sum(D,a);
delete[] D;
n += n;
} // end for loop for l
delete[] u;

for(int i=0;i<n_numbers;i++)
cout << "RDS[" << i << "] = " << RDS[i] << "  " <<
        "n = " << n_values[i] << endl;
cout << endl;
for(int i=0;i<n_numbers;i++)
cout << "logRDS[" << i << "] = " << log(RDS[i]) << "  " <<
        "log(n) = " << log((double) n_values[i]) << endl;
delete[] RDS;
return 0;
}
```

2.4.3 Random Walk

The random walk plays a central role in calculating the Hurst exponent. Hurst random walks are discrete random walks which reverse directions with probability h. Given an undirected, connected graph $G(V, E)$ (V denotes the vertices, E denotes the edges) with $|V| = n$, $|E| = m$ (n number of vertices, m number of edges) a random step in G is a move from some node u to a randomly selected neighbour v. A random walk is a sequence of these random steps starting from some initial node.

The C++ program `RandomWalk.cpp` implements a random walk on a two-dimensional square lattice. No boundary conditions are implemented.

```
// randomwalk.cpp

#include <iostream>
#include <cstdlib>    // for srand(), rand()
using namespace std;

const int n = 24;

void init(char s[n][n])
{
 for(int i=0;i<n;i++)
  for(int j=0;j<n;j++) s[i][j] = '.';
}

void display(char s[n][n])
{
 for(int i=0;i<n;i++)
  { for(int j=0;j<n;j++) cout << s[i][j]; cout << endl; }
}

int main(void)
{
 char s[n][n]; // two-dimensional array of char
 int x = 12, y = 12;
 init(s);
 s[x][y] = '0';  // start cell
 cout << "enter a seed: ";
 unsigned int seed;
 cin >> seed;
 srand(seed);
 int east = 0; int west = 0; int north = 0; int south = 0;

 for(int i=1;i<=n/2;i++)
 {
 switch(rand()%4)
 {
```

```
 case 0: x += 1; west++; break;
 case 1: y += 1; north++; break;
 case 2: x -= 1; east++; break;
 case 3: y -= 1; south++;
 }
 s[x][y] = '$';
 } // end for loop
 s[x][y] = 'E';
 display(s);
 cout << "north = " << north << "\t" << "south = " << south << "\t"
      << "west = " << west << "\t" << "east = " << east;
 return 0;
}
```

The Java program RandomWalk.java also implements a random walk on a two-dimensional square lattice. Cyclic boundary conditions are implemented. The red dot is the initial position. The Java program is an application and an Applet.

```
// RandomWalk.java

import javax.swing.*;
import java.awt.*;
import java.util.Random;

public class RandomWalk extends JApplet
{
 Random r;
 int x0, y0, x, y;
 boolean firstTime;
 int size = 370;

 public void init()
 {
 r = new Random();
 x0 = 150; y0 = 150; x = x0; y = y0;
 firstTime = true;
 }

 public void paint(Graphics g)
 {
 super.paint(g);
 // draw a grid 18 x 18 with each square 20 pixels wide
 for(int i=10;i<size;i+=20)
  for(int j=10;j<size;j+=20) g.drawRect(i,j,20,20);
 g.setColor(Color.RED);
 g.fillOval(x0-3,y0-3,6,6);  // initial position
 // draw a blue dot that represents an ant
 g.setColor(Color.BLUE);
 g.fillOval(x-3,y-3,6,6);
```

```
if(firstTime) walk();
}

private void walk()
{
firstTime = false;
new Thread(new Runnable()
{
int count = 0;
public void run()
{
while(count++ < 400)
{
int d = r.nextInt(4);
switch(d)
{
case 0: y += -20; if(y==-10) y = 350; break;
case 1: x += +20; if(x==370) x = 10; break;
case 2: y += +20; if(y==370) y = 10; break;
case 3: x += -20; if(x==-10) x = 350;
}
repaint();
try { Thread.sleep(200); }
catch(InterruptedException ie)
{ System.err.println("interrupt " + ie.getMessage()); break; }
}
showResults();
}}).start();
}

private void showResults()
{
double distance = (Math.abs(x-x0)+Math.abs(y-y0))/20.0;
JOptionPane.showMessageDialog(null,"effective distance moved = " +
   distance + "blocks!","results",JOptionPane.INFORMATION_MESSAGE);
}

public static void main(String[] args)
{
JApplet applet = new RandomWalk();
JFrame f = new JFrame();
f.setDefaultCloseOperation(JFrame.EXIT_ON_CLOSE);
f.getContentPane().add(applet);
f.setSize(400,400); f.setLocation(200,200);
applet.init();
f.setVisible(true);
} // end main
}
```

Brownian motion (in one dimension) is a random walk on the line where the step length is given by a zero mean Gaussian (normal) probability distribution. Since the steps are independent the cumulative position X is known to satisfy

$$\langle X(t) - X(0)\rangle = 0, \qquad \langle [X(t) - X(0)]^2\rangle^{1/2} \propto |t|^{1/2}$$

so that the standard deviation from the origin grows as $t^{1/2}$. Mandelbrot and Wallis [137] and Mandelbrot and Van Ness [138] introduced *fractional Brownian motion* as a generalization to processes which grow at different rates t^H

$$\langle [X_H(t) - X_H(0)]^2\rangle^{1/2} \propto |t|^H$$

where $0 < H < 1$ is called the Hurst exponent. Successive increments ξ_H of a fractional Brownian motion are called fractional Gaussian noise

$$\xi_H(t) = X_H(t+\delta) - X_H(t)$$

where δ can always be rescaled to one. The autocorrelation funcion which measures the covariance of a data series with itself at some time lag τ is formally defined as

$$C(\tau) := \frac{\langle [\xi_H(t) - \langle \xi_H(t)\rangle][\xi_H(t-\tau) - \langle \xi_H(t-\tau)\rangle]\rangle}{(\langle [\xi_H(t) - \langle \xi_H(t)\rangle]^2\rangle \langle [\xi_H(t-\tau) - \langle \xi_H(t-\tau)\rangle]^2\rangle)^{1/2}}.$$

For a fractional Gaussian noise process the definition yields

$$C(\tau) = \frac{1}{2}(|\tau+1|^{2H} - 2|\tau|^{2H} + |\tau-1|^{2H})$$

which is zero for $H = 1/2$ (except for $\tau = 0$ where the autocorrelation is always one) while for $H \neq 1/2$ and large τ we have

$$\lim_{\tau\to\infty} C(\tau) \propto \tau^{2H-2}.$$

Thus the autocorrelations decay slowly and the resulting fractional Browian motion exhibits long memory effects. Correlations are positive for $H > 1/2$ (persistence) and negative for $H < 1/2$ (antipersistence). As for standard Brownian motion, all fractional Brownian motion are self-affine

$$X_H(at) := a^H X_H(t)$$

meaning that the series appears statistically indentical under scaling at the time axis by some factor a and the displacement X_H by a^H. Thus, fractional Brownian motion lacks any characteristic time scale and when generating or sampling a fractional Brownian motion series, an arbitrary step length of one unit may be used without loss of generality. Self-affine signals can be described by a fractional dimension D (capacity) which is related to the Hurst exponent by $D = 2 - H$ for fractional Brownian motion. The fractal dimension D can be loosely interpreted as the number of dimensions the signal fills up. Let F_H be the Fourier transform of X_H. The power spectrum (defined as the amplitude-squared contributions from the frequencies $\pm f$)

$$S(f) = |F_H(f)|^2 + |F_H(-f)|^2$$

of fractional Brownian motion also demonstrates scaling behaviour. For low frequencies it can be approximated by a *power law*

$$S(f) \sim 1/f^{2H+1}.$$

2.5 Higuchi's Algorithm

Higuchi [95] proposed a method for calculating the fractal dimension from a one-dimensional time series. Consider the time series of length N taken at regular intervals

$$u_1,\ u_2,\ u_3,\ \ldots\ ,u_N\,.$$

Higuchi's algorithm is based on the measure of the mean length of the curve $L(k)$ by using a segment of k-samples as a unit of measure. Let m, k be positive integers with $m = 1, 2, \ldots, k$. Then from the time series we form k new time series as follows

$$u_k^m\ :\ u_m,\ u_{m+k},\ u_{m+2k}, \ldots, u_{m+\lfloor(N-m)/k\rfloor k}$$

where $\lfloor\,\rfloor$ is the Gauss notation (i.e. $\lfloor x\rfloor$ represents an integer not exceeding x). Here m indicates the initial time and k indicates the interval time. Thus for a time interval equal to k we obtain k sets of new time series. For example, in the case of $N = 100$ and $k = 3$ we find

$$\begin{aligned} u_{k=3}^{m=1}\ &:\ u_1,\ u_4,\ u_7, \ldots, u_{100} \\ u_{k=3}^{m=2}\ &:\ u_2,\ u_5,\ u_8, \ldots, u_{98} \\ u_{k=3}^{m=3}\ &:\ u_3,\ u_6,\ u_9, \ldots, u_{99}\,. \end{aligned}$$

One defines the length of the set of the time series $\{\,u_k^m\,\}$ as

$$L_m(k) := \frac{1}{k}\frac{N-1}{(\lfloor(N-m)/k\rfloor)k}\sum_{j=1}^{\lfloor(N-m)/k\rfloor}|u_{m+jk} - u_{m+(j-1)k}|$$

The term $(N-1)/((\lfloor(N-m)/k\rfloor)k)$ is a normalization factor for the time series length of subset time series. Thus $L_m(k)$ represents the normalized sum of the segment length. The length of $L_m(k)$ for the time interval k, $L(k)$, is calculated as the arithmetic mean of the k values $L_m(k)$ for $m = 1, 2, \ldots, k$. If the value is proportional to k^{-D}, i.e.

$$L(k) \propto k^{-D}$$

the time series is fractal-like with the fractal dimension D. If $L(k)$ $(k \geq 1)$ is plotted against k for k ranging from $k = 1, \ldots, k_{max}$ $(k_{max} \ll N)$ on a doubly logarithmic scale, the data should fall on a straight line with slope $-D$.

2.6 Complexity

Many different definitions of *complexity* have been proposed in the literature. Among them are: algorithmic complexity (Kolmogorov-Chaitin), the Lempel-Ziv complexity, the logical depth of Bennet, the effective measure complexity of Grassberger, the complexity of a system based on its diversity, the thermodynamic depth, and a statistical measure of complexity (see Steeb [193] and references therein).

In the following we consider the Lempel-Ziv complexity (Lempel and Ziv [127]). A definition of complexity (Chaitin [27]) of a string (a string of zeros and ones) is given by the number of the bits of the shortest computer program which can generate this string. A general algorithm which determines such a program cannot be given.

Lempel and Ziv [127] have chosen from all possible programs one class that allows only two operations: copying and inserting. For the reconstruction of the given string of length n over a finite alphabet, using these two operations, they have introduced a complexity measure $c(n)$. Here an algorithm can be given. First we give some definitions. A finite nonempty set is called an *alphabet*, and the elements of the set are called *symbols* or *letters*. A *word* (or *string*) over an alphabet A is any finite, possibly empty, sequence of symbols from A, written without punctuation. For example, if

$$A := \{0,1\}$$

then the following are words over A

$$11, \quad 0001, \quad 101010, \quad 0\,.$$

The empty word is the empty sequence of symbols, and is denoted by λ. The *length* of a word S (written $|S|$) is the number of symbols in S, including repetition. Thus the words 11, 0001, 10101, 0, λ have length 2, 4, 5, 1, 0, respectively. The set of all words over A is denoted by A^*. If $A = \{0,1\}$, then

$$A^* = \{\lambda, 0, 1, 00, 01, 10, 11, 000, 001, \dots\}\,.$$

The quantity $S(i,j)$ denotes the substring

$$S(i,j) = s_i s_{i+1} \cdots s_j\,.$$

Definition. A *vocabulary* of a string S, denoted by $v(S)$, is the subset of A^* formed by all the substrings, or words, $S(i,j)$ of S.

Definition. If $S = s_1 s_2 \dots s_m$ and $R = r_1 r_2 \dots r_n$ are words over A, and each s_i and r_j in A, then the *catenation* of S and R is the word

$$s_1 s_2 \cdots s_m r_1 r_2 \cdots r_n$$

denoted by SR.

Obviously we have

(i) $S(RQ) = (SR)Q$
(ii) $S\lambda = \lambda S$
(iii) $|SR| = |S| + |R|$
(iv) $SR \neq RS$ in general

Let us now describe how the Lempel and Ziv complexity is evaluated (Lempel and Ziv [127], Steeb [193]). Given a string of finite length

$$S = s_1 s_2 \cdots s_n \,.$$

The complexity in the sense of Lempel and Ziv of a finite string is evaluated from the point of view of a simple self-delimiting learning machine which, as it scans a given n digit string $S = s_1 s_2 \cdots s_n$ from left to right, adds a new word to its memory every time it discovers a substring of consecutive digits not previously encountered. Thus the calculation of the complexity $c(n)$ proceeds as follows. Assume that a given string $s_1 s_2 \cdots s_n$ has been reconstructed by the program up to the digit s_r and that s_r has been newly inserted, i.e. it was not obtained by simply copying it from $s_1 s_2 \cdots s_{r-1}$. The string up to s_r will be denoted by

$$R := s_1 s_2 \cdots s_r \circ$$

where the $\circ$ indicates that s_r is newly inserted. In order to check whether the rest of R, i.e., $s_{r+1} s_{r+2} \cdots s_n$ can be reconstructed by simple copying or whether one has to insert new digits, we proceed as follows: first, one takes $Q \equiv s_{r+1}$ and asks whether this term is contained in the vocabulary of the string R so that Q can simply be obtained by copying a word of R. This is equivalent to the question of whether Q is contained in the vocabulary $v(RQ\pi)$ of $RQ\pi$ where $RQ\pi$ denotes the string which is composed of R and Q (concatenation) and π means that the last digit has to be deleted. This can be generalized to situations where Q also contains two (i.e., $Q = s_{r+1} s_{r+2}$) or more elements. Let us assume that s_{r+1} can be copied from the vocabulary of R. Then we next ask whether $Q = s_{r+1} s_{r+2}$ is contained in the vocabulary of $RQ\pi$ and so on until Q becomes so large that it can no longer be obtained by copying a word from $v(RQ\pi)$ and one has to insert a new digit. The number c of production steps to create the string S, i.e., the number of newly inserted digits (plus one if the last copy step is not followed by inserting a digit), is used as a measure of the complexity of a given string.

Example. A binary string consisting only of 0's (or 1's) must have the lowest complexity, namely 2, since $0 \circ 000 \cdots$. A string consisting of a sequence of 01's i.e. "010101···01" has complexity 3, since $0 \circ 1 \circ 0101 \cdots 01$. For the binary string 0001101001000101 we find $0 \circ 001 \circ 10 \circ 100 \circ 1000 \circ 101$. ♣

In order to obtain a complexity measure which is independent of the string length we use a normalized complexity measure. For very large strings it makes sense to normalize them. To normalize them we consider the interval $[0, 1]$. The rational numbers in this interval are of Lebesgue measure zero. For almost all numbers in the interval $[0, 1]$ (the irrational numbers) the string of zeros and ones which represents their binary decomposition is not periodic. Therefore almost all strings which correspond to a binary representation of a number $x \in [0, 1]$ should be random and have maximal complexity. The complexity tends to the same value for $n \to \infty$, namely $n/\log_2 n$ for a binary alphabet. For a string with a ternary alphabet we have $n/\log_3 n$. We use this quantity to normalize the complexity $c(n)$. Thus the largest value the normalized complexity can take is equal to 1.

We assumed that the information needed in Lempel and Ziv coding is one unit for each new word. Only then we obtain complexity 3 for an alternating string "0101010$\cdots$", and only then a random string would have complexity $n/\log_2 n$. One is of course free to count complexity in any units one wants. The usual units are bits. In this case the information needed to specify one among n words is $\approx \log_2 n$ instead of 1. Thus the Lempel and Ziv complexity of a fully random binary string is 1 bit/symbol and not $n/\log_2 n$ units. The average Lempel and Ziv complexity per symbol for random strings coincides with the Shannon entropy, and hence the Lempel and Ziv complexity per symbol for the logistic map at fully developed chaos is equal to 1 bit/symbol. This assumes already the limit $n \to \infty$. For finite n there are several sources for logarithmic corrections, one of them being that the Lempel and Ziv complexity is only defined for finite strings in this picture, whence one has to specify also the string length. Thus the Lempel and Ziv complexity for the alternating binary string "0101010$\cdots$" of length n is $\geq \log_2 n$ in this picture, and not finite.

Next we have to find a string from the one-dimensional map. This is done using symbolic dynamics. To construct the symbolic dynamics of a dynamical system, the determination of the partition and the ordering rules for the underlying symbolic sequences is of a crucial importance. In the case of one-dimensional maps, the partition is composed of all the critical points.

```
// Lempelziv.cpp

#include <iostream>
#include <cmath>    // for log
using namespace std;

unsigned int complexity(const unsigned int* S,unsigned int T)
{
 unsigned int c = 1, l = 1;
 do
 {
 unsigned int kmax = 1;
 for(unsigned int i=0;i<l;i++)
 {
 unsigned int k = 0;
 while(S[i+k]==S[l+k]) { ++k; if(l+k >= T-1) return (++c); }
 if(k >= kmax) kmax = k+1;
 }
 ++c;
 l += kmax;
 } while(l < T);
 return c;
```

```
}

int main(void)
{
 unsigned int T = 4000; // length of string
 unsigned int* S = new unsigned int[T];

 // time series from logistic map -> symbolic dynamics
 double x = 0.618; // initial value
 S[0] = 1;         // since x > 0.5

 for(unsigned int t=1;t<T;t++)
 {
 double x1 = x;
 x = 4.0*x1*(1.0-x1);
 if(x >= 0.5) S[t] = 1;
 else S[t] = 0;
 }

 unsigned int sumc = complexity(S,T);
 cout << "sumc = " << sumc << endl;
 double cnorm;
 cnorm = sumc*log((double)T)/(log(2.0)*((double)T));
 cout << "cnorm = " << cnorm;
 delete[] S;
 return 0;
}
```

Chapter 3

Autonomous Systems in the Plane

3.1 Classification of Fixed Points

In this chapter we consider autonomous systems of ordinary differential equations in the plane (Hirsch and Smale [96], Davis [45], Jordan and Smith [106])

$$\frac{du_1}{dt} = f_1(u_1, u_2), \qquad \frac{du_2}{dt} = f_2(u_1, u_2)$$

where $f_1, f_2 \in C^2$. The *fixed points* (u_1^*, u_2^*) (also called *equilibrium points, time-independent solutions*) are given as the solution of the system of equations

$$f_1(u_1^*, u_2^*) = 0, \qquad f_2(u_1^*, u_2^*) = 0.$$

Example. The system (*Van der Pol equation*)

$$\frac{du_1}{dt} = u_2, \qquad \frac{du_2}{dt} = -u_1 + r(1 - u_1^2)u_2, \qquad r \in \mathbb{R}$$

has only the fixed point $(u_1^*, u_2^*) = (0, 0)$. ♣

Example. The dynamical system

$$\frac{du_1}{dt} = u_1(1 - u_1^2 - u_2^2), \qquad \frac{du_2}{dt} = u_2(1 - u_1^2 - u_2^2)$$

has the manifold (circle) $1 - u_1^{*2} - u_2^{*2} = 0$ and the point $(u_1^*, u_2^*) = (0, 0)$ as its fixed points. ♣

Approximation to a nonlinear system by linearizing it at a fixed point is a most important and generally used technique. The *linearized equation* (also called *variational equation*) is given by

$$\frac{dv_1}{dt} = \frac{\partial f_1(\mathbf{u}(t))}{\partial u_1}v_1 + \frac{\partial f_1(\mathbf{u}(t))}{\partial u_2}v_2, \quad \frac{dv_2}{dt} = \frac{\partial f_2(\mathbf{u}(t))}{\partial u_1}v_1 + \frac{\partial f_2(\mathbf{u}(t))}{\partial u_2}v_2\,.$$

If we assume that the only fixed point is at the origin $(0,0)$, then the linearized equation simplifies to

$$\frac{dv_1}{dt} = av_1 + bv_2, \qquad \frac{dv_2}{dt} = cv_1 + dv_2$$

where

$$a := \frac{\partial f_1}{\partial u_1}(0,0), \qquad b := \frac{\partial f_1}{\partial u_2}(0,0), \qquad c := \frac{\partial f_2}{\partial u_1}(0,0), \qquad d := \frac{\partial f_2}{\partial u_2}(0,0) .$$

We expect the solutions of this linearized equation will be geometrically similar to those of the original system near the origin $(0,0)$, an expectation fulfilled in most cases. Non-trivial solutions exist if and only if

$$\det \begin{pmatrix} a-\lambda & b \\ c & d-\lambda \end{pmatrix} = 0 .$$

The *characteristic equation* follows as

$$\lambda^2 - (a+d)\lambda + (ad-bc) = 0 .$$

When this equation has two different roots, λ_1, λ_2, two linearly independent families of solutions are generated. We define $p := a+d$, $q := ad-bc$. Then the characteristic equation takes the form

$$\lambda^2 - p\lambda + q = 0.$$

Let $\Delta := p^2 + 4q$ be the *discriminant.* Then the roots λ_1, λ_2 are given by

$$\lambda_1 = \frac{1}{2}(p + \Delta^{1/2}), \qquad \lambda_2 = \frac{1}{2}(p - \Delta^{1/2}) .$$

The following table lists the possible cases.

(i)	λ_1, λ_2	real, unequal, same sign	$\Delta > 0, q > 0$	node
(ii)	$\lambda_1 = \lambda_2$	(real) $b \neq 0, c \neq 0$	$\Delta = 0, p \neq 0$	inflected node
(iii)	λ_1, λ_2	complex, non-zero real part	$\Delta < 0, p \neq 0$	spiral
(iv)	$\lambda_1 \neq 0, \lambda_2 = 0$		$q = 0$	parallel lines
(v)	λ_1, λ_2	real, different sign	$q < 0$	saddle point
(vi)	λ_1, λ_2	pure imaginary	$q > 0, p = 0$	centre

In classifying the fixed points as in the preceding examples we have taken into account an unproved assumption that the phase paths of the original equation and those of the linearized equation near the fixed point are of the same character. This is true in general for spirals, nodes, and saddle points, but not for a centre.

Example. The linear approximation may predict a centre where the original equation had a spiral. Conversely, the system of differential equations

$$\frac{du_1}{dt} = u_2, \qquad \frac{du_2}{dt} = -u_1^3$$

with the fixed point $(u_1^*, u_2^*) = (0, 0)$ has a centre at the origin, but the linearized system of equations

$$\frac{dv_1}{dt} = v_2, \qquad \frac{dv_2}{dt} = 0$$

has not. ♣

3.2 Homoclinic Orbit

Anharmonic systems with a *homoclinic orbit* play a special role in the study of nonlinear dynamical systems in particular chaotic systems (Lichtenberg and Lieberman [128], Guckenheimer and Holmes [75]). Consider an autonomous system of first order ordinary differential equations $du_j/dt = f_j(\mathbf{u})$, where $j = 1, \ldots, n$ ($n \geq 2$ and f_j are analytic functions.

Definition. A homoclinic orbit is a trajectory of a flow of a first order ordinary differential equations $du_j/dt = f_j(\mathbf{u})$ which joins a saddle fixed point to itself. Thus a homoclinic orbit lies in the intersection of the stable manifold and the unstable manifold of a fixed point.

By definition, these orbits tend to the same fixed point for both $t \to \infty$ and $t \to -\infty$.

Example. Consider the anharmonic system

$$\frac{d^2u}{dt^2} - \frac{u}{2} + u^2 + u^3 = 0.$$

Introducing $u_1 := u$ and $u_2 := du/dt$ we obtain the autonomous system in the plane

$$\frac{du_1}{dt} = u_2, \qquad \frac{du_2}{dt} = \frac{u_1}{2} - u_1^2 - u_1^3.$$

The three fixed points of this system are given by

$$(u_1^*, u_2^*) = (0,0), \quad (u_1^*, u_2^*) = (-(\sqrt{3}+1)/2, 0), \quad (u_1^*, u_2^*) = ((\sqrt{3}-1)/2, 0).$$

The half plane $u_1 > 0$ contains a unique homoclinic orbit $\Gamma(t) = (u_1(t), u_2(t))$ given explicitly by

$$u(t) \equiv u_1(t) = \frac{2e^{t/\sqrt{2}}}{(e^{t/\sqrt{2}} + \frac{2}{3})^2 + 1}.$$

The fixed point $(0, 0)$ is a hyperbolic fixed point and the other two fixed points are elliptic. For the fixed point $(0, 0)$ the eigenvalues of the functional matrix are real and for the two other fixed points the eigenvalues are purely imaginary. ♣

A heteroclinic orbit of a dynamical system $du_j/dt = f_j(\mathbf{u})$ is a path in phase space which joins two different fixed points.

Example. Consider the pendulum in the plane

$$\frac{du_1}{dt} = u_2, \qquad \frac{du_2}{dt} = -\sin(u_1)$$

with the fixed points $(\pi, 0)$ and $(-\pi, 0)$. ♣

In the program Homoclinic.java we apply the Lie series technique to calculate the homoclinic orbit of the differential equation given above.

```
// Homoclinic.java

import java.awt.*;
import java.awt.event.*;
import java.awt.Graphics;

public class Homoclinic extends Frame
{
 public Homoclinic()
 {
 setSize(600,500);
 addWindowListener(new WindowAdapter()
 { public void windowClosing(WindowEvent event) { System.exit(0); }}); }

 public void paint(Graphics g)
 {
 double t = 0.0001, count = 0.0;
 double u11 = -0.001, u22 = 0.001;   // initial values
 double u1, u2; double V2;
 g.drawRect(40,40,500-40,400-40);
 g.drawLine(40,220,500,220);  g.drawLine(400,40,400,400);
 while(count < 42.5)
 {
 u1 = u11; u2 = u22; V2 = u1*(0.5-u1-u1*u1);
 u11 = u1+t*u2+t*t*V2/2.0;
 u22 = u2+t*(u1/2.0-u1*u1-u1*u1*u1)+t*t*u2*(0.5-2.0*u1-3.0*u1*u1)/2.0;
 int m1 = (int) Math.floor(150.0*u1+400.0);
 int n1 = (int) Math.floor(150.0*u2+220.0);
 int m2 = (int) Math.floor(150.0*u11+400.0);
 int n2 = (int) Math.floor(150.0*u22+220.0);
 g.drawLine(m1,n1,m2,n2);
 count += t;
 } // end while
 } // end paint

 public static void main(String[] args)
 { Frame f = new Homoclinic(); f.setVisible(true); }
}
```

3.3 One-Dimensional Pendulum

The equation for the *pendulum* is given by

$$\frac{d^2u}{dt^2} + \omega^2 \sin(u) = 0, \qquad \omega^2 := \frac{g}{L}$$

where L is the length of the pendulum, g the acceleration due to gravity, u is the angular displacement of the pendulum from its position of equilibrium and ω is the frequency. Introducing the quantities $\widetilde{u}(\widetilde{t}(t)) = u(t)$, $\widetilde{t}(t) = \omega t$ we can write the pendulum equation in dimensionless form

$$\frac{d^2\widetilde{u}}{d\widetilde{t}^2} + \sin(\widetilde{u}) = 0\,.$$

We omit the tilde in the following. Introducing the quantity $u_1 = u$ and $u_2 = du/dt = du_1/dt$ we obtain the autonomous first order system

$$\frac{du_1}{dt} = u_2, \qquad \frac{du_2}{dt} = -\sin(u_1)\,.$$

Thus we obtain the fixed points $(n\pi, 0)$, $n \in \mathbb{Z}$. The variational equation is given by

$$\frac{dv_1}{dt} = v_2, \qquad \frac{dv_2}{dt} = -\cos(u_1)v_1\,.$$

Inserting the fixed points $(n\pi, 0)$ into these equations we find

$$\frac{dv_1}{dt} = v_2, \qquad \frac{dv_2}{dt} = \begin{cases} -v_1 \text{ if } n \text{ even} \\ v_1 \text{ if } n \text{ odd} \end{cases}$$

In the first case the eigenvalues are i, $-i$. Thus we have a centre. In the second case the eigenvalues are 1, -1. Thus we have an unstable node.

In the Java program `Pendulum.java` we integrate the dynamical system using a symplectic integrator.

```
// Pendulum.java

import java.awt.*;
import java.awt.event.*;
import java.awt.Graphics;

public class Pendulum extends Frame
{
 public Pendulum()
 {
 setSize(400,300);
 addWindowListener(new WindowAdapter()
 { public void windowClosing(WindowEvent event) { System.exit(0); }}); }
```

```
public void paint(Graphics g)
{
double t = 0.001;
int steps = 20000;
double q1, p1;
double q = 1.0, p = 1.75;  // initial values

for(int i=0;i<steps;i++)
{
q1 = q;  p1 = p;
q = q1+t*p1;  p = p1-t*Math.sin(q);
int mx1 = (int) Math.floor(40.0*q1+250.0+0.5);
int ny1 = (int) Math.floor(40.0*p1+350.0+0.5);
int mx = (int) Math.floor(40.0*q+250.0+0.5);
int ny = (int) Math.floor(40.0*p+350.0+0.5);
g.drawLine(mx1,ny1,mx,ny);
}
}

public static void main(String[] args)
{ Frame f = new Pendulum(); f.setVisible(true); }
}
```

3.4 Limit Cycle Systems

A number of interesting dynamical systems in the plane show stable limit cycle behaviour (Davis [45], Jordan and Smith [106]). The trajectory $(u_1(t), u_2(t))$ approaches asymptotically a periodic orbit. Consider the autonomous system of first order differential equations in the plane

$$\frac{d^2u}{dt^2} + f(u)\frac{du}{dt} + g(u) = 0\,.$$

Let

$$F(u) = \int_0^u f(s)ds\,.$$

Assume that f is an even and g an odd function, both are continuous for all u and g satisfies the Lipschitz condition. Assume further that

$$f(0) < 0, \qquad ug(u) > 0 \quad \text{for} \quad u \neq 0, \qquad F(u) \to \pm\infty \quad \text{if} \quad u \to \infty$$

and that f has a single zero at $u = b$ $(u > 0)$ and is monotonically increasing for $u \geq b$. Then the system

$$\frac{du_1}{dt} = u_2, \qquad \frac{du_2}{dt} = -f(u_1)u_2 - g(u_1)$$

has a stable *limit cycle.*

Example. Consider the *Van der Pol equation*

$$\frac{du_1}{dt} = u_2, \qquad \frac{du_2}{dt} = -u_1 + r(1 - u_1^2)u_2 .$$

It admits only one fixed point at the origin, namely $(0, 0)$. The variational equation is given by

$$\frac{dv_1}{dt} = v_2, \qquad \frac{dv_2}{dt} = -(1 + 2ru_1u_2)v_1 + r(1 - u_1^2)v_2 .$$

Inserting the fixed point $(0, 0)$ yields the linear system

$$\frac{dv_1}{dt} = v_2, \qquad \frac{dv_2}{dt} = -v_1 + rv_2 .$$

The eigenvalues of the matrix

$$A = \begin{pmatrix} 0 & 1 \\ -1 & r \end{pmatrix}$$

of the variational equation are given by

$$\lambda(r) = \frac{r}{2} + \sqrt{-1 + r^2/4}, \qquad \overline{\lambda}(r) = \frac{r}{2} - \sqrt{-1 + r^2/4} .$$

Thus for $r > 0$ the eigenvalues have a positive real part. Hence the fixed point $(0, 0)$ is unstable for $r > 0$. For $r < 0$ the fixed point $(0, 0)$ is stable. We have $\Re\lambda(0)) = 0$ and $\Re(\lambda'(0)) \neq 0$, where $'$ indicates differentiation with respect to r. Thus $(u_1, u_2) = (0, 0)$ undergoes a Hopf bifurcation for $r = 0$ (see section 5.7). ♣

In the C++ program `vdpol.cpp` we find the phase portrait of the Van der Pol equation using the Runge-Kutta technique. The parameter value is $r = 5.0$.

```
// vdpol.cpp

#include <fstream>
using namespace std;

const int N = 2;

void fsystem(double h,double t,double u[N],double hf[N])
{
 double r = 5.0;
 hf[0] = h*u[1]; hf[1] = h*(-u[0]+r*(1.0-u[0]*u[0])*u[1]);
}

void map(double u[N],int steps,double h,double t)
{
 double uk[N];
```

```
 double tk;
 double a[6] = { 0.0, 1.0/4.0, 3.0/8.0, 12.0/13.0, 1.0, 1.0/2.0 };
 double c[6] = { 16.0/135.0, 0.0, 6656.0/12825.0, 28561.0/56430.0,
                 -9.0/50.0, 2.0/55.0 };
 double b[6][5];
 b[0][0] = b[0][1] = b[0][2] = b[0][3] = b[0][4] = 0.0;
 b[1][0] = 0.25; b[1][1] = 0.0; b[1][2] = 0.0; b[1][3] = 0.0; b[1][4] = 0.0;
 b[2][0] = 3.0/32.0; b[2][1] = 9.0/32.0;
 b[2][2] = 0.0; b[2][3] = 0.0; b[2][4] = 0.0;
 b[3][0] = 1932.0/2197.0; b[3][1] = -7200.0/2197.0;
 b[3][2] = 7296.0/2197.0; b[3][3] = b[3][4] = 0.0;
 b[4][0] = 439.0/216.0; b[4][1] = -8.0;
 b[4][2] = 3680.0/513.0; b[4][3] = -845.0/4104.0; b[4][4] = 0.0;
 b[5][0] = -8.0/27.0; b[5][1] = 2.0;
 b[5][2] = -3544.0/2565.0; b[5][3] = 1859.0/4104.0; b[5][4] = -11.0/4.0;
 double f[6][N];
 int i, j, l, k;

 for(i=0;i<steps;i++) { fsystem(h,t,u,f[0]);
 for(k=1;k<=5;k++) { tk = t + a[k]*h;
 for(l=0;l<N;l++) { uk[l] = u[l];
 for(j=0;j<=k-1;j++) uk[l] += b[k][j]*f[j][l];
 }
 fsystem(h,tk,uk,f[k]);
 }
 for(l=0;l<N;l++)
  for(k=0;k<6;k++) u[l] += c[k]*f[k][l];
  }
}

int main(void)
{
 ofstream data;
 data.open("phase_data.dat");
 int steps = 1;
 double h = 0.005;             // step size
 double u[N] = { 0.1, 0.2 };   // initial conditions
 double t = 0.0;
 int i;
 // wait for transients to decay
 for(i=0;i<1000;i++) { t += h; map(u,steps,h,t); }
 t = 0.0;
 for(i=0;i<40000;i++)
 { t += h; map(u,steps,h,t); data << u[0] << " " << u[1] << "\n"; }
 data.close();
 return 0;
}
```

Example. Consider the dynamical system

$$\frac{du_1}{dt} = u_2, \qquad \frac{du_2}{dt} = -u_1 - r\sin(u_2).$$

This nonlinear system shows infinitely many limit cycles. We assume that $r > 0$. Then $(0,0)$ is an unstable fixed point. ♣

In the Java program `LimitCycles.java` we calculate the phase portrait for different initial values. We consider two different initial conditions. One is close to the origin and the other one is between the first and second limit cycle. We apply symplectic integration.

```
// LimitCycles.java

import java.awt.*;
import java.awt.event.*;
import java.awt.Graphics;

public class LimitCycles extends Frame
{
 public LimitCycles()
 {
 setSize(400,300);
 addWindowListener(new WindowAdapter()
 { public void windowClosing(WindowEvent event) { System.exit(0); }}); }

 public void paint(Graphics g)
 {
 double t = 0.005;
 double x1, y1;
 double x = 0.05, y = 0.01; // initial values
 double r = 5.0;            // parameter value
 int T = 10000;             // number of iterations

 for(int i=0;i<T;i++)
 {
 x1 = x;  y1 = y;
 x = x1+t*y1;  y = y1*Math.exp(-t*r*(x*x-1.0))-t*x;
 int mx1 = (int) Math.floor(40.0*x1+250.0+0.5);
 int ny1 = (int) Math.floor(40.0*y1+350.0+0.5);
 int mx = (int) Math.floor(40.0*x+250.0+0.5);
 int ny = (int) Math.floor(40.0*y+350.0+0.5);
 g.drawLine(mx1,ny1,mx,ny);
 } // end for loop
 } // end paint

 public static void main(String[] args)
 { Frame f = new LimitCycles(); f.setVisible(true); }
}
```

3.5 Lotka-Volterra Systems

Lotka-Volterra systems describe the interaction of two (or more) competing species. The simplest model is given by

$$\frac{du_1}{dt} = au_1 - bu_1u_2, \qquad \frac{du_2}{dt} = -cu_2 + bu_1u_2$$

where a, b, c are positive constants. Owing to the term au_1 the species 1 would grow exponentialy. However, owing to the term $-bu_1u_2$ the species 1 will decrease. Similarly, for species 2 the term $-cu_2$ will decrease exponentially but the term bu_1u_2 gives a growing contribution. Thus we expect that the solution is periodic around the fixed point $u_1^* = c/b$, $u_2^* = a/b$. Consider the special case $(a = b = c = 1)$

$$\frac{du_1}{dt} = u_1 - u_1u_2, \qquad \frac{du_2}{dt} = -u_2 + u_1u_2$$

where $u_1(t = 0) > 0$ and $u_2(t = 0) > 0$. Since we assume that $u_1(t) > 0$ and $u_2(t) > 0$ we find that the system has one fixed point at $(u_1^*, u_2^*) = (1, 1)$. The fixed point is a centre. Thus in a neighbourhood of this fixed point the solutions are approximatively circles. Integrating the equation

$$\frac{du_1}{u_1 - u_1u_2} = \frac{du_2}{-u_2 + u_1u_2}$$

yields the constant of motion

$$\ln(u_1) + \ln(u_2) - u_1 - u_2 = C$$

or

$$u_1u_2e^{-u_1}e^{-u_2} = C_1 \, .$$

This constant of motion can be used to check how accurate the integration procedure is.

```
// Lotka.cpp

#include <fstream>
using namespace std;

const int N = 2;

void fsystem(double h,double t,double u[N],double hf[N])
{ hf[0] = h*(u[0]-u[0]*u[1]); hf[1] = h*(-u[1]+u[0]*u[1]); }

void map(double u[N],int steps,double h,double t)
{
 double uk[N]; double tk;
 double a[6] = { 0.0, 1.0/4.0, 3.0/8.0, 12.0/13.0, 1.0, 1.0/2.0 };
 double c[6] = { 16.0/135.0, 0.0, 6656.0/12825.0, 28561.0/56430.0,
```

```
                -9.0/50.0, 2.0/55.0 };
 double b[6][5];
 b[0][0] = b[0][1]= b[0][2] = b[0][3] = b[0][4] = 0.0;
 b[1][0] = 0.25; b[1][1] = 0.0; b[1][2] = 0.0; b[1][3] = 0.0; b[1][4] = 0.0;
 b[2][0] = 3.0/32.0; b[2][1] = 9.0/32.0;
 b[2][2] = 0.0; b[2][3] = 0.0; b[2][4] = 0.0;
 b[3][0] = 1932.0/2197.0; b[3][1] = -7200.0/2197.0;
 b[3][2] = 7296.0/2197.0; b[3][3] = b[3][4] = 0.0;
 b[4][0] = 439.0/216.0; b[4][1] = -8.0;
 b[4][2] = 3680.0/513.0; b[4][3] = -845.0/4104.0; b[4][4] = 0.0;
 b[5][0] = -8.0/27.0; b[5][1] = 2.0;
 b[5][2] = -3544.0/2565.0; b[5][3] = 1859.0/4104.0; b[5][4] = -11.0/4.0;
 double f[6][N];
 int i, j, l, k;
 for(i=0;i<steps;i++) { fsystem(h,t,u,f[0]);
 for(k=1;k<=5;k++) { tk = t + a[k]*h;
 for(l=0;l<N;l++) { uk[l] = u[l];
 for(j=0;j<=k-1;j++) uk[l] += b[k][j]*f[j][l];
 }
 fsystem(h,tk,uk,f[k]);
 }
 for(l=0;l<N;l++)
  for(k=0;k<6;k++) u[l] += c[k]*f[k][l];
 }
}

int main(void)
{
 ofstream data;
 data.open("phase_data.dat");
 int steps = 1;
 double h = 0.001;              // step length
 double u[N] = { 2.0, 1.5 };    // initial conditions
 double t = 0.0;
 for(int i=0;i<20000;i++)
 { t += h; map(u,steps,h,t); data << u[0] << " " << u[1]<< "\n"; }
 data.close();
 return 0;
}
```

In the following program we show that

$$\ln(u_1) + \ln(u_2) - u_1 - u_2 = C$$

is a constant of motion of the Lotka-Volterra system using SymbolicC++.

```
// firstintegral.cpp

#include <iostream>
```

```
#include "symbolicc++.h"
using namespace std;

int main(void)
{
 Symbolic u1("u1"), u2("u2"), t("t");
 Symbolic f, r;
 f = ln(u1[t])+ln(u2[t])-u1[t]-u2[t];
 r = df(f,t);
 cout << "r = " << r << endl;
 r = r.subst(df(u1[t],t),u1[t]-u1[t]*u2[t]);
 r = r.subst(df(u2[t],t),-u2[t]+u1[t]*u2[t]);
 cout << "r = " << r << endl;
 return 0;
}
```

Exercises. (1) Consider the autonomous system

$$\frac{du_1}{dt} = \sin^2\left(\frac{u_1}{2}\right)\cos\left(\frac{u_2}{2}\right), \qquad \frac{du_2}{dt} = \sin^2\left(\frac{u_2}{2}\right)\cos\left(\frac{u_1}{2}\right).$$

Classify all the fixed points of the system. Show that there are connecting heteroclinic orbits.

(2) Consider the autonomous system $du_1/dt = f_1(u_1, u_2)$, $du_2/dt = f_2(u_1, u_2)$ in the plane. Then there are no closed paths in a simply connected region in which

$$\frac{\partial(\rho f_1)}{\partial u_1} + \frac{\partial(\rho f_2)}{\partial u_2}$$

is of one sign, where $\rho(u_1, u_2)$ is any function having continuous first partial derivatives (*Dulac's test*). Show that

$$\frac{du_1}{dt} = u_2, \qquad \frac{du_2}{dt} = -u_1 - u_2 + u_1^2 + u_2^2$$

has no periodic solutions using $\rho(u_1, u_2) = \exp(-2u_1)$.

Chapter 4

Nonlinear Hamilton Systems

4.1 Hamilton Equations of Motion

We consider conservative Hamilton systems (Abraham and Marsden [1], Arnold [5], Arnold and Avez [6], Guillemin and Sternberg [76], MacKay and Meiss [136], Walters [216]) i.e. Hamilton systems that do not depend explicitly on time. Consider a closed classical system with $3N$ degrees of freedom (for example, N particles in a three-dimensional box). The state of such a system is completely specified in terms of a set of $6N$ independent real variables $(\mathbf{p}^N, \mathbf{q}^N)$, where $\mathbf{p}^N$ and $\mathbf{q}^N$ denote the set of vectors

$$\mathbf{p}^N = (\mathbf{p}_1, \mathbf{p}_2, \ldots, \mathbf{p}_N), \qquad \mathbf{q}^N = (\mathbf{q}_1, \mathbf{q}_2, \ldots, \mathbf{q}_N)$$

with $\mathbf{p}_j$ and $\mathbf{q}_j$ are the momentum and position of the ith particle. If the state vector $\mathbf{X}^N = \mathbf{X}^N(\mathbf{p}^N, \mathbf{q}^N)$ is known at one time, then it is completely determined for any other time from Newton's laws. If we can define a Hamilton function, $H(\mathbf{X}, t)$, for the system, then the time evolution of the quantities $\mathbf{p}_j$ and $\mathbf{q}_j$ is given by *Hamilton's equations of motion*

$$\frac{d\mathbf{p}_j}{dt} = -\frac{\partial H}{\partial \mathbf{q}_j}, \qquad \frac{d\mathbf{q}_j}{dt} = \frac{\partial H}{\partial \mathbf{p}_j}, \qquad j = 1, \ldots, N.$$

If the Hamilton function does not depend explicitly on time, then it is a constant of motion

$$H(\mathbf{X}^N) = E$$

where E is the total energy of the system. In this case the dynamical system is called conservative. We associate to the system a $6N$-dimensional phase space, Γ. The state vector $\mathbf{X}^N(\mathbf{p}^N, \mathbf{q}^N)$ then specifies a point in the phase space. As the system evolves in time and its state changes, the system point $\mathbf{X}^N$ traces out a trajectory in Γ-space. Since the subsequent motion of a classical system is uniquely determined from the initial conditions, it follows that no two trajectories in phase space can cross. If they did, one could not uniquely determine the subsequent motion of the trajectory.

Consider the Hamilton system given above and write Hamilton's equations in the form

$$\frac{dq_{11}}{\frac{\partial H}{\partial p_{11}}} = \frac{dq_{12}}{\frac{\partial H}{\partial p_{12}}} = \frac{dq_{13}}{\frac{\partial H}{\partial p_{13}}} = \frac{dq_{21}}{\frac{\partial H}{\partial p_{21}}} = \cdots = \frac{dp_{11}}{-\frac{\partial H}{\partial q_{11}}} = \cdots = \frac{dp_{N3}}{-\frac{\partial H}{\partial q_{N3}}} = \frac{dt}{1}.$$

This equation provides us with $6N - 1$ equations between phase space coordinates which, when solved, give us $6N - 1$ constraints or integrals of the motion

$$f_j(\mathbf{X}^N) = C_j$$

where C_j is a constant. However, these integrals of the motion can be divided into two kinds, isolating and nonisolating. Isolating integrals define a whole surface in the phase space and are important in ergodic theory, while nonisolating integrals do not define a surface and are unimportant. One of the main problems of ergodic theory is to determine how many isolating integrals a given Hamilton system has. An example of an isolating integral is the total energy. For N particles in a box it is probably the only isolating integral.

Only in simple cases these equations can be integrated.

Example. Consider the Hamilton function for the anharmonic oscillator

$$H(p,q) = \frac{1}{2m}p^2 + \frac{m}{2}\omega^2 q^2 + \epsilon\frac{q^4}{4}.$$

Then the equation

$$\frac{dq}{p/m} = -\frac{dp}{m\omega^2 q + \epsilon q^3} = \frac{dt}{1}$$

can be integrated and the solution of the initial value problem can be expressed with Jacobi elliptic functions. ♣

Consider a system for which the only isolating integral of the motion is the total energy and assume that the system has total energy E. Then trajectories in Γ-space (the N-dimensional phase space) which have energy E will be restricted to the energy surface S_E. The energy surface S_E is a $(6N-1)$-dimensional "surface" in phase space. The flow of state points on the energy surface is defined to be ergodic if almost all points $\mathbf{X}^N(\mathbf{p}^N, \mathbf{q}^N)$ on the surface move in such a way that they pass through every small finite neighbourhood R_E on the energy surface. Or, in other words, each point samples small neighbourhoods over the entire surface during the course of its motion (a given point $\mathbf{X}^N(\mathbf{p}^N, \mathbf{q}^N)$ cannot pass through every point on the surface, because a line which cannot intersect itself cannot fill a surface of two or more dimensions). Note that not all points need sample the surface, only "almost all". We can exclude a set of measure zero from this requirement.

Example. For a system of N particles with *central two-body interaction* described by the Hamilton function

$$H(\mathbf{p},\mathbf{q}) = \frac{1}{2}\sum_{k=1}^{N}\sum_{j=1}^{3}\frac{p_{kj}^2}{m_k} + \frac{1}{2}\sum_{\substack{k,l=1\\ k\neq l}}^{N}\sum_{j=1}^{3} V_{kl}(|\mathbf{q_k}-\mathbf{q_l}|)$$

where

$$|\mathbf{q_k}-\mathbf{q_l}| := \sqrt{(q_{k1}-q_{l1})^2+(q_{k2}-q_{l2})^2+(q_{k3}-q_{l3})^2}$$

the first integrals are given by

$$P_j^{kin} = \sum_{k=1}^{N} p_{kj} \qquad \text{momentum}$$

$$J_{ij}^{kin} = \sum_{k=1}^{N}(q_{ki}p_{kj}-q_{kj}p_{ki}) \qquad \text{angular momentum}$$

$$H = \sum_{k=1}^{N}\sum_{j=1}^{3}\frac{p_{kj}^2}{2m_k} + \frac{1}{2}\sum_{\substack{k,l=1\\ k\neq l}}^{N}\sum_{j=1}^{3} V_{kl}(|\mathbf{q_k}-\mathbf{q_l}|) \qquad \text{Hamilton function}$$

$$G_j = tP_j^{kin} - MR_j \qquad \text{centre of gravity}$$

where $j = 1,2,3$ and

$$R_j := \frac{1}{M}\sum_{k=1}^{N} m_k q_{kj}$$

with

$$M := \sum_{k=1}^{N} m_k. \qquad \clubsuit$$

Example. Consider the special case of a Hamilton function ($H : \mathbb{R}^4 \to \mathbb{R}$) with two degrees of freedom

$$H(\mathbf{p},\mathbf{q}) := \frac{1}{2}(p_1^2+p_2^2) + V(q_1,q_2).$$

The Hamilton equations of motion are given by

$$\frac{dq_j}{dt} = \frac{\partial H}{\partial p_j} = p_j, \qquad \frac{dp_j}{dt} = -\frac{\partial H}{\partial q_j} = -\frac{\partial V}{\partial q_j}$$

where $j = 1,2$. $\clubsuit$

In the C++ program using SymbolicC++ we evaluate the Hamilton equations of motion for the *Hénon-Heiles model*

$$H(\mathbf{p},\mathbf{q}) := \frac{1}{2}(p_1^2+p_2^2+q_1^2+q_2^2) + q_1^2q_2 - \frac{1}{3}q_2^3.$$

We find the equations of motion

$$\frac{dq_1}{dt} = p_1, \quad \frac{dq_2}{dt} = p_2, \qquad \frac{dp_1}{dt} = -q_1 - 2q_1q_2, \quad \frac{dp_2}{dt} = -q_2 - q_1^2 + q_2^2.$$

```
// hamiltoneq.cpp

#include <iostream>
#include "symbolicc++.h"
using namespace std;

int main(void)
{
 Symbolic H("H"), q1("q1"), q2("q2"), p1("p1"), p2("p2"),
          pt1, pt2, qt1, qt2;
 // Hamilton function H
 H = (p1*p1+p2*p2+q1*q1+q2*q2)/2+q1*q1*q2-q2*q2*q2/3;
 // Hamilton equations of motion
 pt1 = -df(H,q1); pt2 = -df(H,q2);
 qt1 = df(H,p1);  qt2 = df(H,p2);
 cout << "dp1/dt = " << pt1 << endl;
 cout << "dp2/dt = " << pt2 << endl;
 cout << "dq1/dt = " << qt1 << endl;
 cout << "dq2/dt = " << qt2 << endl;
 return 0;
}
```

Besides conservative Hamilton systems also explicitly time dependent Hamilton systems are studied in chaos theory. An example is the Hamilton function for a pendulum acted upon by a sequence of short kicks

$$H(p,q,t) = \frac{1}{2}p^2 - \frac{1}{2\pi}\sum_{n=-\infty}^{\infty} \cos(2\pi(q-nt)).$$

4.1.1 Hamilton System and Variational Equation

To study the stability of fixed points of a Hamilton systems and calculating the Liapunov exponents of a the Hamilton system we need the variational equation. Consider the phase space

$$M := \mathbb{R}^{2N} = \{(\mathbf{p},\mathbf{q})\}$$

where $\mathbf{p} = (p_1, p_2, \ldots, p_N)$ and $\mathbf{q} = (q_1, q_2, \ldots, q_N)$. From the Hamilton function

$$H(\mathbf{p},\mathbf{q}) = \frac{1}{2}\sum_{j=1}^{N} p_j^2 + V(\mathbf{q})$$

we obtain the Hamilton equation of motion

$$\frac{dq_j}{dt} = \frac{\partial H}{\partial p_j} = p_j, \qquad \frac{dp_j}{dt} = -\frac{\partial H}{\partial q_j} = -\frac{\partial V}{\partial q_j}$$

where $j = 1, 2, \ldots, N$. The *variational equation* is defined as

$$\frac{dy_j}{dt} = y_{j+N}, \qquad \frac{dy_{j+N}}{dt} = -\sum_{i=1}^{N} \frac{\partial^2 V}{\partial q_i \partial q_j} y_i$$

where $j = 1, 2, \ldots, N$. In the C++ program using SymbolicC++ we evaluate the variational equation for the Hénon-Heiles model.

```
// hamiltonvar.cpp

#include <iostream>
#include "symbolicc++.h"
using namespace std;

int main(void)
{
 Symbolic H("H"), q("q",2), p("p",2), u("u",2), v("v",2),
          qt("",2), pt("",2), ut("",2), vt("",2);

 // Hamilton function
 H = (p(0)*p(0)+p(1)*p(1)+q(0)*q(0)+q(1)*q(1))/2
     +q(0)*q(0)*q(1)-q(1)*q(1)*q(1)/3;
 for(int j=0;j<2;j++)
 {
 pt(j) = -df(H,q(j));  qt(j) = df(H,p(j));
 cout << "dp" << j << "/dt = " << pt(j) <<endl;
 cout << "dq" << j << "/dt = " << qt(j) <<endl;
 }
 for(int j=0;j<2;j++)
 {
 ut(j) = v(j);  vt(j) = 0;
 for(int l=0;l<2;l++) { vt(j) += -df(df(H,q(j)),q(l))*u(l); }
 cout << "du" << j << "/dt = " << ut(j) << endl;
 cout << "dv" << j << "/dt = " << vt(j) << endl;
 }
 return 0;
}
```

4.2 Integrable Hamilton Systems

4.2.1 Hamilton Systems and First Integrals

Consider a Hamilton system

$$\frac{dq_j}{dt} = \frac{\partial H}{\partial p_j}, \qquad \frac{dp_j}{dt} = -\frac{\partial H}{\partial q_j}$$

where $j = 1, 2, \ldots, N$. A smooth function $I(\mathbf{p}, \mathbf{q})$ is called a *first integral* of the Hamilton system if

$$\frac{d}{dt} I(\mathbf{p}, \mathbf{q}) = 0.$$

This equation can also be written as

$$\sum_{j=1}^{N} \left(\frac{\partial H}{\partial q_j} \frac{\partial I}{\partial p_j} - \frac{\partial H}{\partial p_j} \frac{\partial I}{\partial q_j} \right) = 0.$$

A number of interesting Hamilton systems have other first integrals besides the Hamilton function.

Example. Consider the three body periodic *Toda lattice* (Toda [204])

$$H(\mathbf{p}, \mathbf{q}) = \frac{1}{2}(p_1^2 + p_2^2 + p_3^2) + \exp(q_1 - q_2) + \exp(q_2 - q_3) + \exp(q_3 - q_1).$$

The first integrals are

$$I_1(\mathbf{p}, \mathbf{q}) = p_1 + p_2 + p_3$$

and

$$I_2(\mathbf{p}, \mathbf{q}) = p_1 p_2 p_3 - p_1 \exp(q_2 - q_3) - p_2 \exp(q_3 - q_1) - p_3 \exp(q_1 - q_2).$$

In the C++ program we show that these functions are first integrals.

```
// hamiltonin.cpp

#include <iostream>
#include "symbolicc++.h"
using namespace std;

int main(void)
{
 Symbolic H("H"), q("q",3), p("p",3), pt("pt",3), qt("qt",3),
          I1, R1, I2, R2;
 int j;

 // Hamilton function
 H = (p(0)*p(0)+p(1)*p(1)+p(2)*p(2))/2
     +exp(q(0)-q(1))+exp(q(1)-q(2))+exp(q(2)-q(0));

 for(j=0;j<3;j++)
 {
 pt(j) = -df(H,q(j)); qt(j) = df(H,p(j));
 cout << "dp" << j << "/dt = " << pt(j) << endl;
 cout << "dq" << j << "/dt = " << qt(j) << endl;
 }
 I1 = p(0)+p(1)+p(2);
```

```
 R1 = 0;
 for(j=0;j<3;j++) R1 += pt(j)*df(I1,p(j))+qt(j)*df(I1,q(j));
 if(R1==0) cout << "I1 is a first integral." << endl;
 else cout << "I1 is not a first integral." << endl;

 I2 = p(0)*p(1)*p(2)-p(0)*exp(q(1)-q(2))-p(1)*exp(q(2)-q(0))
      -p(2)*exp(q(0)-q(1));
 R2 = 0;
 for(j=0;j<3;j++) R2 += pt(j)*df(I2,p(j))+qt(j)*df(I2,q(j));
 if(R2==0) cout << "I2 is a first integral." << endl;
 else cout << "I2 is not a first integral." << endl;
 return 0;
}
```

Example. Consider the Hamilton function

$$H(p_\rho, p_\zeta, \rho, \zeta) = \frac{1}{2}p_\rho^2 + \frac{1}{2}p_\zeta^2 + V(\rho, \zeta)$$

with

$$V(\rho, \zeta) = \frac{1}{2}\rho^2 + \frac{1}{2}\lambda^2\zeta^2 + \frac{\nu^2}{2\rho^2} + \frac{1}{(\rho^2 + \zeta^2)^{1/2}}$$

and

$$p_\rho := \frac{d\rho}{dt}, \qquad \rho_\zeta := \frac{d\zeta}{dt}.$$

This Hamilton function describes the relative motion of two charged particles in a *Paul trap* in the pseudo potential approximation with λ and ν related to the asymmetry of the time average trapping potential and the relative angular momentum, respectively. The Hamilton function describes a single particle moving in two dimensions, ρ and ζ, respectively. The equations of motion derived from the Hamilton function are

$$\frac{d^2\rho}{dt^2} = \frac{\nu^2}{\rho^3} - \rho + \frac{\rho}{(\rho^2 + \zeta^2)^{3/2}},$$

$$\frac{d^2\zeta}{dt^2} = -\lambda^2\zeta + \frac{\zeta}{(\rho^2 + \zeta^2)^{3/2}}.$$

The Hamilton function H is conservative and autonomous. Consequently, E, the total energy of the system, is a constant of the motion. For arbitrary ν two further integrals of motion exist for $\lambda = 2$ and $\lambda = \frac{1}{2}$, respectively. The first integral F, which applies for the case $\lambda = 2$, is given by

$$F\left(\rho, \frac{d\rho}{dt}, \zeta, \frac{d\zeta}{dt}, \nu\right) = \zeta\left(\frac{d\rho}{dt}\right)^2 - \rho\frac{d\zeta}{dt}\frac{d\rho}{dt} + \frac{\zeta}{(\rho^2 + \zeta^2)^{1/2}} - \rho^2\zeta + \frac{\nu^2\zeta}{\rho^2}.$$

For $\lambda = \frac{1}{2}$ a first integral is given by

$$G\left(\rho, \frac{d\rho}{dt}, \zeta, \frac{d\zeta}{dt}, \nu\right) = I_\rho^2 + I_\phi^2 + \nu^2(\rho^2 + \zeta^2),$$

where

$$I_\rho := \frac{\nu^2}{\rho} + \rho\left(\frac{d\zeta}{dt}\right)^2 - \zeta\frac{d\rho}{dt}\frac{d\zeta}{dt} + \frac{\rho}{(\rho^2+\zeta^2)^{1/2}} - \frac{1}{4}\zeta^2\rho$$

and

$$I_\phi := -\frac{\nu}{\rho}\left(\rho\frac{d\rho}{dt} + \zeta\frac{d\zeta}{dt}\right).$$

4.2.2 Lax Pair and Hamilton Systems

A number of Hamilton systems can be written in the form

$$\frac{dL}{dt} = [A, L](t)$$

where L and A are time-dependent $n \times n$ matrices. This is called the *Lax representation* of the Hamilton system. We find that

$$\frac{dL^k}{dt} = [A, L^k](t)$$

and the that $\mathrm{tr}(L^k)$ $(k = 1, 2, \ldots)$ are first integrals.

Example. Consider the Hamilton function (*Toda lattice*)

$$H(\mathbf{p}, \mathbf{q}) = \frac{1}{2}(p_1^2 + p_2^2 + p_3^2) + \exp(q_1 - q_2) + \exp(q_2 - q_3) + \exp(q_3 - q_1).$$

Introducing the quantities

$$a_j := \frac{1}{2}\exp\left(\frac{1}{2}(q_j - q_{j+1})\right), \qquad b_j := \frac{1}{2}p_j$$

and cyclic boundary conditions (i.e., $q_4 \equiv q_1$) we find that the Hamilton equations of motion take the form (with $b_3 = 0$)

$$\frac{da_j}{dt} = a_j(b_j - b_{j+1}), \qquad \frac{db_1}{dt} = -2a_1^2, \qquad \frac{db_2}{dt} = 2(a_1^2 - a_2^2), \qquad \frac{db_3}{dt} = 2a_2^2$$

where $j = 1, 2$. Introducing the matrices (*Lax pair*)

$$L := \begin{pmatrix} b_1 & a_1 & 0 \\ a_1 & b_2 & a_2 \\ 0 & a_2 & b_3 \end{pmatrix}, \qquad A := \begin{pmatrix} 0 & -a_1 & 0 \\ a_1 & 0 & -a_2 \\ 0 & a_2 & 0 \end{pmatrix}$$

the equations of motion can be written as Lax representation. From L we find the first integral as $\mathrm{tr}(L^n)$, where $n = 1, 2, \ldots$, where tr denotes the trace. We obtain

$$\mathrm{tr}L = b_1 + b_2 + b_3, \qquad \mathrm{tr}(L^2) = b_1^2 + b_2^2 + b_3^2 + 2a_1^2 + 2a_2^2 .$$ ♣

In the C++ program `lax.cpp` we show that $\mathrm{tr}(L)$, $\mathrm{tr}(L^2)$ and the determinant of L are first integrals.

```
// lax.cpp

#include <iostream>
#include "symbolicc++.h"
using namespace std;

int main(void)
{
 Symbolic L("L",3,3), A("A",3,3), Lt("Lt",3,3); // Lt=dL/dt
 Symbolic a1("a1"), a2("a2"), b1("b1"), b2("b2"), b3("b3"),
          a1t, a2t, b1t, b2t, b3t;
 L(0,0) = b1; L(0,1) = a1;  L(0,2) = 0;
 L(1,0) = a1; L(1,1) = b2;  L(1,2) = a2;
 L(2,0) = 0;  L(2,1) = a2;  L(2,2) = b3;
 A(0,0) = 0;  A(0,1) = -a1; A(0,2) = 0;
 A(1,0) = a1; A(1,1) = 0;   A(1,2) = -a2;
 A(2,0) = 0;  A(2,1) = a2;  A(2,2) = 0;
 Lt = A*L-L*A;
 cout << "Lt = " << Lt << endl;
 b1t = Lt(0,0); b2t = Lt(1,1); b3t = Lt(2,2);
 a1t = Lt(0,1); a2t = Lt(1,2);
 cout << "b1t = " << b1t << ", b2t = " << b2t
      << ", b3t = " << b3t << endl;
 cout << "a1t = " << a1t << ", a2t = " << a2t << endl;
 cout << endl;

 // I(0),I(1),I(2) are first integrals
 int n = 3;
 Symbolic result;
 Symbolic I("I",n);
 I(0) = L.trace();          cout << "I(0) = " << I(0) << endl;
 I(1) = (L*L).trace();      cout << "I(1) = " << I(1) << endl;
 I(2) = L.determinant(); cout << "I(2) = " << I(2) << endl;
 cout << endl;
 for(int i=0;i<n;i++)
 {
 result = b1t*df(I(i),b1)+b2t*df(I(i),b2) + b3t*df(I(i),b3)
          +a1t*df(I(i),a1)+a2t*df(I(i),a2);
 cout << "result" << i+1 << " = " << result << endl;
 }
 return 0;
}
```

The output is

```
Lt = [-2*a1^(2) -a1*b2+b1*a1 0]
[a1*b1-b2*a1 2*a1^(2)-2*a2^(2) -a2*b3+b2*a2]
[0 a2*b2-b3*a2 2*a2^(2)]
```

```
b1t = -2*a1^(2), b2t = 2*a1^(2)-2*a2^(2), b3t = 2*a2^(2)
a1t = -a1*b2+b1*a1, a2t = -a2*b3+b2*a2
I[0] = b1+b2+b3
I[1] = b1^(2)+2*a1^(2)+b2^(2)+2*a2^(2)+b3^(2)
I[2] = b1*b2*b3-b1*a2^(2)-a1^(2)*b3
result1 = 0
result2 = 0
result3 = 0
```

4.2.3 Floquet Theory

The *Floquet theory* (Iooss and Joseph [103]) has been developed to treat systems of linear differential equations with periodic coefficients. Consider the system of differential equations

$$\frac{d\mathbf{u}}{dt} = A(t)\mathbf{u}(t), \qquad \mathbf{u} = (u_1, u_2, \ldots, u_n)^T$$

where $A(t) = A(t+T)$ for some $T > 0$ and all t. The transition matrix $\Phi(t, t_0)$ is defined by

$$\frac{d\Phi}{dt} = A(t)\Phi$$

where $\Phi(t,t) = I$ for all t and I denotes the $n \times n$ unit matrix. We find that

$$\Phi(t+T, t_0) = \Phi(t, t_0)C$$

where C is a constant matrix. Then there exists a constant matrix RT such that $C = \exp(RT)$, where T is the period. We can prove that

$$\Phi(t_0 + T, t_0) = \exp(RT)\,.$$

Consequently, the state-transition matrix, and the solution to $d\mathbf{u}/dt = A(t)\mathbf{u}$, consists of a periodically modulated exponential matrix function. Therefore the system of linear differential equations $d\mathbf{u}/dt = A(t)\mathbf{u}$ is asymptoticaly stable if the eigenvalues of R all have negative real parts or that

$$\det(I\lambda - \exp(RT)) = 0$$

implies that $|\lambda_j| < 1$ for all $j = 1, 2, \ldots, n$. The eigenvalues of R are the characteristic exponents of $A(t)$, and λ_j are called that characteristic multipliers of $A(t)$. The eigenvalues of R are related to the eigenvalues of $\exp(RT)$ by

$$\rho_j = \frac{\ln \lambda_j}{T}\,.$$

As a special case consider the *Hill equation*

$$\frac{d^2u}{dt^2} + a(t)u = 0$$

where $a(t+T) = a(t)$, $T > 0$ for all t. Then with $u_1 = u$, $u_2 = du/dt$ we have $d\mathbf{u}/dt = A(t)\mathbf{u}$, where

$$A(t) = \begin{pmatrix} 0 & 1 \\ -a(t) & 0 \end{pmatrix}.$$

We find that

$$\lambda_1 \lambda_2 = \exp\left(\int_t^{t+T} \operatorname{tr}(A(s)) ds \right) = 1.$$

The characteristic equation of λ is given by

$$\lambda^2 - \operatorname{tr}(C)\lambda + 1 = 0.$$

Thus if $|\operatorname{tr} C| > 2$ we find that λ_j's are real and the system is stable. If $|\operatorname{tr} C| < 2$ we find that the λ_j's are complex conjugate with $|\lambda_j| = 1$. The system is stable, but not asymptotically stable. If $|\operatorname{tr} C| = 2$ we have $\lambda_1 = \lambda_2 = \pm 1$.

Example. Consider a *Mathieu's equation* with coefficients having period $T = \pi$

$$\frac{d^2 u}{dt^2} + (\delta + 2\epsilon \cos(2t))u = 0$$

where δ and ϵ are constants. Since this is a linear second-order homogeneous differential equation there exists two linear independent solutions $u_1(t)$, $u_2(t)$ which are called a fundamental set of solutions. Any other solution is a linear combination of u_1 and u_2

$$u_1(t+T) = a_{11}u_1(t) + a_{12}u_2(t), \qquad u_2(t+T) = a_{21}u_1(t) + a_{22}u_2(t).$$

The theorem of Floquet states that

$$u_1(t+T) = \lambda_1 u_1(t), \qquad u_2(t+T) = \lambda_2 u_2(t).$$

Consequently

$$u_j(t+nT) = \lambda_j^n u_j(t) \quad \text{for all integers} \quad n.$$

Therefore as $t \to \infty$ we find

$$u_j(t) \to \begin{cases} 0 \text{ if } |\lambda_j| < 1 \\ \infty \text{ if } |\lambda_j| > 1 \end{cases}$$

If $\lambda = +1$, then u has period T, if $\lambda = -1$, then u has period $2T$. ♣

The Java program `Floquet.java` first computes a fundamental set of solutions of the Mathieu equation by numerical integration for two independent initial conditions over one period $T = \pi$. Then it determines the matrix $A = (a_{ij})$ and calculates the eigenvalues λ_1, λ_2. The parameters are set to $\delta = 1$, and $\epsilon = 0.5$. We can find the regions in the (δ, ϵ)-plane which correspond to bounded solutions (*Strutt diagram*). As described above we have $\lambda_1 \lambda_2 = 1$.

```
// Floquet.java

public class Floquet
{
 public static void main(String[] args)
 {
 double delta = 1.0;
 double eps = 0.5;
 double dt = Math.PI/100.0; double dt2 = dt*dt;
 double[] du = new double[2];
 double[][] u = new double[2][3];
 double[][] A = new double[2][2]; double[][] B = new double[2][2];
 u[0][1] = 1.0; u[0][2] = 1.0; // initial value=1, slope=0.0;
 u[1][1] = 0.0; u[1][2] = dt;  // initial value=0, slope=dt
 double[][] uu = new double[2][2];
 uu[0][0] = u[0][1]; uu[0][1] = u[1][1];
 uu[1][0] = (u[0][2]-u[0][1])/dt; uu[1][1] = (u[1][2]-u[1][1])/dt;

 for(int i=0;i<=1;i++)
 {
 double t = dt;
 do
 {
 u[i][0] = u[i][1];  u[i][1] = u[i][2];
 u[i][2] = 2.0*u[i][1]-u[i][0]
           -dt2*(delta+2.0*eps*Math.cos(2.0*t))*u[i][1];
 t += dt;
 }
 while(t <= Math.PI);
 du[i] = (u[i][2]-u[i][0])/(2.0*dt);
 }  // end for loop

 B[0][0] = u[0][1]; B[0][1] = u[1][1];
 B[1][0] = du[0]; B[1][1] = du[1];
 double[][] uui = new double[2][2]; // inverse of uu
 double detuu = uu[0][0]*uu[1][1]-uu[0][1]*uu[1][0];
 uui[0][0] = uu[1][1]/detuu;  uui[0][1] = -uu[0][1]/detuu;
 uui[1][0] = -uu[1][0]/detuu; uui[1][1] = uu[0][0]/detuu;
 A[0][0] = B[0][0]*uui[0][0]+B[0][1]*uui[1][0];
 A[0][1] = B[0][0]*uui[0][1]+B[0][1]*uui[1][1];
 A[1][0] = B[1][0]*uui[0][0]+B[1][1]*uui[1][0];
 A[1][1] = B[1][0]*uui[0][1]+B[1][1]*uui[1][1];
 double D = A[0][1]*A[1][0]-A[0][0]*A[1][1]
          +(A[0][0]+A[1][1])*(A[0][0]+A[1][1])/4.0;
 double lambda1, lambda2;
 if(D >= 0.0)
 {
 lambda1 = (A[0][0]+A[1][1])/2.0+Math.sqrt(D);
```

```
lambda2 = (A[0][0]+A[1][1])/2.0-Math.sqrt(D);
System.out.println("lambda1 = " + lambda1);
System.out.println("lambda2 = " + lambda2);
}
if(D < 0.0)
{
double real = (A[0][0]+A[1][1])/2.0;
double imag = Math.sqrt(-D);
System.out.println("real part lambda1 = " + real);
System.out.println("imaginary part lambda1 = " + imag);
System.out.println("real part lambda1 = " + real);
System.out.println("imaginary part lambda1 = " + -imag);
}
} // end main
}
```

Exercise. Consider the linear system

$$\frac{d\mathbf{u}}{dt} = A(t)\mathbf{u}$$

where $A(t)$ is differentiable $n \times n$ matrix. Consider the fundamental matrix $U(t) = (u_1(t), \ldots, u_n(t))$ consisting of the linear independent solutions $u_j(t)$ with $j = 1, \ldots, n$. Show that

$$\det(U(t)) = \det U(0) \exp\left(\int_0^t \mathrm{tr}(A(\tau))d\tau\right)$$

where tr denotes the trace.

4.3 Chaotic Hamilton Systems

4.3.1 Trajectories and Hénon-Heiles Hamilton Function

Depending on the energy E which is given by the initial values a conservative non-linear Hamilton system could show chaotic behaviour. The set of one-dimensional Liapunov exponents can be used to characterize chaos.

We consider the *Hénon-Heiles Hamilton function*

$$H(\mathbf{p}, \mathbf{q}) = \frac{1}{2}(p_1^2 + p_2^2 + q_1^2 + q_2^2) + q_1^2 q_2 - \frac{1}{3}q_2^3 .$$

The Hamilton equations of motion are given by

$$\frac{dq_1}{dt} = p_1, \qquad \frac{dq_2}{dt} = p_2$$

$$\frac{dp_1}{dt} = -q_1 - 2q_1 q_2, \qquad \frac{dp_2}{dt} = -q_2 - q_1^2 + q_2^2 .$$

Hénon and Heiles studied the bounded motion of orbits for this system. The trajectories move in a four-dimensional phase space but are restricted to a three-dimensional surface because the energy is a constant of motion

$$H(\mathbf{p}, \mathbf{q}) = E\,.$$

One can study a two-dimensional cross-section of the three-dimensional energy surface (see next section). With growing energy the system becomes more and more chaotic. At an energy of $E = 0.16667$ almost no stable motion remains. One could use the one-dimensional Liapunov exponent to characterize chaos.

In the Java program we evaluate the time evolution of q_1 with the initial values $q_1(0) = 0.5$, $q_2(0) = 0.1$, $p_1(0) = 0.1$, and $p_2(0) = 0.1$.

```
// Heilestime.java

import java.awt.*;
import java.awt.event.*;
import java.awt.Graphics;

public class Heilestime extends Frame
{
 public Heilestime()
 {
 setSize(600,500);
 addWindowListener(new WindowAdapter()
 { public void windowClosing(WindowEvent event) { System.exit(0); }}); }

 public void paint(Graphics g)
 {
 g.drawLine(10,10,10,400);  g.drawLine(10,200,630,200);
 g.drawRect(10,10,630,400);
 double t = 0.005; double t2 = t*t;
 double count = 0.0;
 double x11 = 0.5, x22 = 0.1; // initial values
 double x33 = 0.1, x44 = 0.1; // initial values
 double x1, x2, x3, x4;
 do
 {
 x1 = x11; x2 = x22;  x3 = x33; x4 = x44;
 x11 = x1+t*x3+t2*(-x1-2.0*x1*x2)/2.0;
 x22 = x2+t*x4+t2*(-x2-x1*x1+x2*x2)/2.0;
 x33 = x3+t*(-x1-2.0*x1*x2)+t2*(-x3-2.0*x3*x2-2.0*x4*x1)/2.0;
 x44 = x4+t*(-x2-x1*x1+x2*x2)+t2*(-x4-2.0*x1*x3+2.0*x2*x4)/2.0;
 int m = (int)(5.0*count+10.0+0.5);
 int n = (int)(200.0-200.0*x1+0.5);
 int p = (int)(5.0*(count+t)+10.0+0.5);
 int q = (int)(200.0-200.0*x11+0.5);
```

```
g.drawLine(m,n,p,q);
count = count + t;
} while(count < 124.0);
} // end paint

public static void main(String[] args)
{ Frame f = new Heilestime(); f.setVisible(true); }
}
```

4.3.2 Surface of Section Method

The *surface-of-section method* (also called *Poincaré section method*) is particularly suited for Hamilton systems with two degrees of freedom. In this technique one follows the successive crossings of a trajectory through a surface intersecting the energy shell, for example the (p_2, q_2) plane at the point

$$q_1 = 0\,.$$

The position of the system, at a given energy, on such a surface completely specifies its state to within a sign. This is, of course, only true for Hamilton systems with two degrees of freedom which are also quadratic in the momenta. After a large number of crossings a pattern emerges. In the case of quasiperiodic motion the crossing points appear to lie on a smooth curve. This curve is often called an invariant curve. It corresponds to the intersection of the torus, on which the trajectory lies, with the surface of section. Should the trajectory under consideration be one that does not close upon itself (irrational winding number), then the successive points eventually fill up the curve densely. Such orbits are ergodic on the torus. On the other hand, in the case of an orbit being closed, that is, a torus of commensurable frequencies, one only sees a finite number of fixed points. If the trajectory is irregular, no such pattern emerges. One sees a "random splatter" of points that fill up some area.

Consider the Hénon Heiles model. In the Java program we evaluate the surface of section for the energy $E = 1/12$ for two different initial conditions. The initial conditions are

$$q_1(t=0) = 1/\sqrt{8}, \quad q_2(t=0) = 0, \quad p_1(t=0) = 0, \quad p_2(t=0) = 1/\sqrt{24}$$

and

$$q_1(t=0) = 0, \quad q_2(t=0) = 0, \quad p_1(t=0) = 1/\sqrt{6}, \quad p_2(t=0) = 0\,.$$

We find two invariant curves.

```
// HeilesPoincare.java

import java.awt.*;
import java.awt.event.*;
```

```
import java.awt.Graphics;

public class HeilesPoincare extends Frame
{
 public HeilesPoincare()
 {
 setSize(640,480);
 addWindowListener(new WindowAdapter()
 { public void windowClosing(WindowEvent event) { System.exit(0); }}); }

 public void paint(Graphics g)
 {
 g.drawRect(20,20,600,400); g.drawRect(310,20,310,400);
 g.drawLine(20,210,600,210);
 // we identify q1 -> x1, q2 -> x2, p1 -> x3, p2 -> x4
 double t = 0.005;
 double x1[] = new double[3]; double x2[] = new double[3];
 double x3[] = new double[3]; double x4[] = new double[3];
 x1[1] = Math.sqrt(1.0/8.0); x2[1] = 0.0;  // initial condition I
 x3[1] = 0.0; x4[1] = Math.sqrt(1.0/24.0); // initial condition I
 x1[2] = 0.0; x2[2] = 0.0;                 // initial condition II
 x3[2] = Math.sqrt(1.0/6.0); x4[2] = 0.0;  // initial condition II

 for(int j=0;j<2;j++) { double count = 0.0;
 do
 {
 x1[0] = x1[j+1]; x2[0] = x2[j+1];
 x3[0] = x3[j+1]; x4[0] = x4[j+1];
 x1[j+1] = x1[0]+t*x3[0]; x2[j+1] = x2[0]+t*x4[0];
 x3[j+1] = x3[0]-t*(x1[j+1]+2.0*x1[j+1]*x2[j+1]);
 x4[j+1] = x4[0]-t*(x2[j+1]+x1[j+1]*x1[j+1]-x2[j+1]*x2[j+1]);
 if((x1[0]*x1[j+1] < 0.0) && (x1[0] > 0.0))
 {
 double scale = 400.0;
 int m = (int)(scale*x2[j+1]+310.0+0.5);
 int n = (int)(-scale*x4[j+1]+210.0+0.5);
 g.drawLine(m,n,m,n);
 }
 count = count+t;
 } while(count < 4000.0);
 }
 }

 public static void main(String[] args)
 { Frame f = new HeilesPoincare(); f.setVisible(true); }
}
```

Consider the quartic Hamilton function

$$H(\mathbf{p},\mathbf{q}) = \frac{1}{2}(p_1^2+p_2^2) + \frac{1-r}{12}(q_1^4+q_2^4) + \frac{1}{2}q_1^2q_2^2$$

where $r \in [0,1]$. For $r=0$ the system is completely integrable with all motion lying on invariant tori. The second first integral besides the Hamilton function is

$$I(\mathbf{p},\mathbf{q}) = 3p_1p_2 + q_1q_2(q_2+q_2)\,.$$

With increasing r the system becomes more and more chaotic. For the Hamilton equations of motion we find

$$\frac{dq_1}{dt} = p_1, \qquad \frac{dq_2}{dt} = p_2$$

$$\frac{dp_1}{dt} = -\frac{1-r}{3}q_1^3 - q_1q_2^2, \qquad \frac{dp_2}{dt} = -\frac{1-r}{3}q_2^3 - q_1^2q_2.$$

For $r=1$ the system is almost completely chaotic, although there is a very small region with regular behaviour. In this case, i.e. $r=1$, the system can be derived from Yang-Mills theory (Steeb [192]).

In the Java program we evaluate the surface of section for $r=0.15$ and the initial values $p_1(0)=1.5$, $p_2(0)=1.0$, $q_1(0)=1.0$, $q_2(0)=1.0$.

```
// QuarticPotential.java

import java.awt.*;
import java.awt.event.*;
import java.awt.Graphics;

public class QuarticPotential extends Frame
{
 public QuarticPotential()
 {
 setSize(600,500);
 addWindowListener(new WindowAdapter()
 { public void windowClosing(WindowEvent event) { System.exit(0); }}); }

 public void func_system(double h,double t,double u[],double hf[])
 {
 double r = 0.15;
 hf[0] = h*(-(1.0-r)*u[2]*u[2]*u[2]/3.0-u[2]*u[3]*u[3]);
 hf[1] = h*(-(1.0-r)*u[3]*u[3]*u[3]/3.0-u[2]*u[2]*u[3]);
 hf[2] = h*u[0]; hf[3] = h*u[1];
 }

 public void map(double u[],int steps,double h,double t,int N)
 {
```

```
double uk[] = new double[N]; double tk;
double a[] = { 0.0, 1.0/4.0, 3.0/8.0, 12.0/13.0, 1.0, 1.0/2.0 };
double c[] = { 16.0/135.0, 0.0, 6656.0/12825.0, 28561.0/56430.0,
               -9.0/50.0, 2.0/55.0 };
double b[][] = new double[6][5];
b[0][0] = b[0][1]= b[0][2] = b[0][3] = b[0][4] = 0.0;
b[1][0] = 0.25; b[1][1] = 0.0; b[1][2] = 0.0; b[1][3] = 0.0; b[1][4] = 0.0;
b[2][0] = 3.0/32.0; b[2][1] = 9.0/32.0;
b[2][2] = 0.0; b[2][3] = 0.0; b[2][4] = 0.0;
b[3][0] = 1932.0/2197.0; b[3][1] = -7200.0/2197.0;
b[3][2] = 7296.0/2197.0; b[3][3] = b[3][4] = 0.0;
b[4][0] = 439.0/216.0; b[4][1] = -8.0;
b[4][2] = 3680.0/513.0; b[4][3] = -845.0/4104.0; b[4][4] = 0.0;
b[5][0] = -8.0/27.0; b[5][1] = 2.0; b[5][2] = -3544.0/2565.0;
b[5][3] = 1859.0/4104.0; b[5][4] = -11.0/40.0;
double f[][] = new double[6][N];
int i, j, l, k;
for(i=0;i<steps;i++) { func_system(h,t,u,f[0]);
 for(k=1;k<=5;k++) { tk = t+a[k]*h;
  for(l=0;l<N;l++) { uk[l] = u[l];
   for(j=0;j<=k-1;j++) uk[l] += b[k][j]*f[j][l];
}
func_system(h,tk,uk,f[k]);
}
for(l=0;l<N;l++)
 for(k=0;k<6;k++) u[l] += c[k]*f[k][l];
}
}

public void paint(Graphics g)
{
g.drawRect(30,30,570,400);
int steps = 1; int N = 4;
double h = 0.005; double t = 0.0;
double u[] = { 1.5, 1.0, 1.0, 1.0 };  // initial values
for(int i=0;i<20000000;i++)
{ t += h; double temp = u[2]; map(u,steps,h,t,N);
if(((temp*u[2]) < 0.0) && (temp > 0.0))
{
int m = ((int)(200.0*u[3])+300.0); int n = ((int)(250.0*u[1])+600.0);
g.drawLine(m,n,m,n);
}
}
}

public static void main(String[] args)
{ Frame f = new QuarticPotential(); f.setVisible(true); }
}
```

Exercises. (1) Consider equations of motion of the classical three-body problems of the helium atom (one nucleus and two electrons) (Yamamoto and Kaneko [223])

$$
\begin{aligned}
M\frac{d^2\mathbf{r}_0}{dt^2} &= \frac{\mathbf{r}_1-\mathbf{r}_0}{r_{10}^3}2e^2 + \frac{\mathbf{r}_2-\mathbf{r}_0}{r_{20}^3}2e^2 \\
\frac{d^2\mathbf{r}_1}{dt^2} &= -\frac{\mathbf{r}_1-\mathbf{r}_0}{r_{10}^3}2e^2 + \frac{\mathbf{r}_1-\mathbf{r}_2}{r_{12}^3}e^2 \\
\frac{d^2\mathbf{r}_2}{dt^2} &= -\frac{\mathbf{r}_2-\mathbf{r}_0}{r_{20}^3}2e^2 - \frac{\mathbf{r}_1-\mathbf{r}_2}{r_{12}^3}e^2 .
\end{aligned}
$$

The first particle (particle 0) is the nucleus with $2e$ charge and M is the mass ratio of nucleus to electron. The two electrons (particle 1 and 2) have -1 charge. The system of second order differential equations is invariant under the transformation

$$t \to Tt, \qquad \mathbf{r} \to L\mathbf{r}$$

for $T^2/L^3 = 1$, where T and L are units of time and length, respectively. Show that for most initial conditions the trajectories show chaotic transients until one of the electrons escapes to infinity. Show that one can also find stable quasiperiodic motions.

(2) The swinging *Atwood machine* is described by the Hamilton function

$$H_\mu(r,\theta,p_r,p_\theta) = \frac{p_r^2}{2m(1+\mu)} + \frac{p_\theta^2}{2mr^2} + mgr(\mu - \cos\theta)$$

where $\mu > 0$ is a bifurcation parameter. Then the equations of motion are given by

$$
\begin{aligned}
\frac{dr}{dt} &= \frac{\partial H_\mu}{\partial p_r} = \frac{p_r}{m(1+\mu)} \\
\frac{d\theta}{dt} &= \frac{\partial H_\mu}{\partial p_\theta} = \frac{p_\theta}{mr^2} \\
\frac{dp_r}{dt} &= -\frac{\partial H_\mu}{\partial r} = \frac{p_\theta^2}{mr^3} - mg(\mu - \cos\theta) \\
\frac{dp_\theta}{dt} &= -\frac{\partial H_\mu}{\partial \theta} = -mgr\sin\theta .
\end{aligned}
$$

Show that for $\mu = 10$ one can find chaotic behaviour.

(3) Looking for a third integral in the motion of a particle in a galactic potential with cylindrical symmetry Contopoulos et al [40] were lead to the Hamilton function with two degrees of freedom

$$H(\mathbf{p},\mathbf{q}) = \frac{1}{2}(p_1^2+p_2^2) + \frac{1}{2}(q_1^2+q_2^2) - q_1q_2^2 .$$

Study the fixed points of the Hamilton equations of motion.

(4) Consider the dynamics in a hexagonal potential given by the Hamilton function (Zaslavskii et al [226])

$$H(\mathbf{p},\mathbf{q}) = \frac{1}{2}(p_1^2 + p_2^2) + \sum_{j=1}^{3} \cos(q_1 \cos(2\pi j/3) + q_2 \sin(2\pi j/3)) \,.$$

Study the chaotic behaviour of this Hamilton function.

Chapter 5

Nonlinear Dissipative Systems

5.1 Fixed Points and Stability

We consider nonlinear dissipative systems with chaotic behaviour (Arrowsmith and Place [7], Marek and Schreiber [141], Seydel [185], Hirsch and Smale [96], Smale [188], Guckenheimer and Holmes [75]). In particular the Lorenz model (Lorenz [132], Sparrow [187]) is studied. Suppose that the dynamical behaviour of the system is modeled by the solution curves of an autonomous system of first order differential equations (or *dynamical system*)

$$\frac{d\mathbf{u}}{dt} = \mathbf{f}(\mathbf{u}), \qquad \mathbf{f} : U \to \mathbb{R}^n$$

where U is an open subset of $\mathbb{R}^n$. We suppose $\mathbf{f}$ is C^1. In most pratical applications we have div$\mathbf{f} < 0$, where div denotes the divergence. Also in most cases the f_j's are polynomials. If the $f_j : \mathbb{R}^n \to \mathbb{R}$ are analytic functions then the solution of the initial value problem can be expressed as

$$\mathbf{u}(t) = \exp(tV)\mathbf{u}|_{\mathbf{u}\to\mathbf{u}_0}$$

for a sufficiently small t, where V is the vector field

$$V = f_1(\mathbf{u})\frac{\partial}{\partial u_1} + \cdots + f_n(\mathbf{u})\frac{\partial}{\partial u_n}.$$

A point $\mathbf{u}^* \in U$ is called a *fixed point* (also called *equilibrium point* or *stationary point*) of the dynamical system if

$$\mathbf{f}(\mathbf{u}^*) = \mathbf{0}.$$

By uniqueness of solutions, no other solution curve can pass through $\mathbf{u}^*$. Let

$$\Phi_t : U \to \mathbb{R}^n$$

be the flow associated with the dynamical system. The set $U \subset \mathbb{R}^n$ is an open set.

For each $\mathbf{u} \in U$ the map $t \to \Phi(t, \mathbf{u}) = \Phi_t(\mathbf{u})$ is the solution passing through $\mathbf{u}$ when $t = 0$; it is defined for t in some open interval. If $\mathbf{u}^*$ is a fixed point, then

$$\Phi_t(\mathbf{u}^*) = \mathbf{u}^*$$

for all $t \in \mathbb{R}$. Another name for $\mathbf{u}^*$ is a singular point of the vector field $\mathbf{f}$.

Suppose $\mathbf{f}$ is linear. We write $\mathbf{f}(\mathbf{u}) = A\mathbf{u}$, where A is a linear operator (square matrix) on $\mathbb{R}^n$. Then the origin $\mathbf{0} \in \mathbb{R}^n$ is a fixed point. When $\lambda < 0$ is greater than the real parts of the eigenvalues of A, then solutions $\Phi_t(\mathbf{u})$ approach $\mathbf{0}$ exponentially

$$|\Phi_t(\mathbf{u})| \leq Ce^{\lambda t}$$

for some $C > 0$. Now suppose $\mathbf{f}$ is a C^1 vector field with fixed point $\mathbf{0} \in \mathbb{R}^n$. Consider the matrix

$$A := D\mathbf{f}(\mathbf{0}) \equiv (\partial \mathbf{f}/\partial \mathbf{u})(\mathbf{u} = \mathbf{0}).$$

We call it the linear part of $\mathbf{f}$ at $\mathbf{0}$. If all eigenvalues of the matrix A have negative real parts, we call $\mathbf{0}$ a *sink*. More generally, a fixed point $\mathbf{u}^*$ is a sink if all eigenvalues of $D\mathbf{f}(\mathbf{u}^*)$ have negative real parts. We also say the linear flow e^{tA} is a contraction. It can be shown that $\mathbf{0}$ is a sink if and only if every trajectory tends to $\mathbf{0}$ as $t \to \infty$. This is called *asymptotic stability*. It follows that the trajectories approach a sink exponentially.

The following theorem (Hirsch and Smale [96], Arrowsmith and Place [7]) says that a nonlinear sink $\mathbf{u}^*$ behaves locally like a linear sink: nearby solutions approach $\mathbf{u}^*$ exponentially.

Theorem. Let $\mathbf{u}^* \in U$ be a sink of the dynamical system. Suppose every eigenvalue of $D\mathbf{f}(\mathbf{u}^*)$ has real part less than $-c, c > 0$. Then there is a neighbourhood $N \subset U$ of $\mathbf{u}^*$ such that

(a) $\Phi_t(\mathbf{u})$ is defined and in N for all $\mathbf{u} \in N$, $t > 0$.

(b) There is a Euclidean norm on $\mathbb{R}^n$ such that

$$|\Phi_t(\mathbf{u}) - \mathbf{u}^*| \leq e^{-ct}|\mathbf{u} - \mathbf{u}^*|$$

for all $\mathbf{u} \in N$, $t \geq 0$.

(c) For any norm on $\mathbb{R}^n$, there is a constant $B > 0$ such that

$$|\Phi_t(\mathbf{u}) - \mathbf{u}^*| \leq Be^{-ct}|\mathbf{u} - \mathbf{u}^*|$$

for all $\mathbf{u} \in N$, $t \geq 0$.

In particular, $\Phi_t(\mathbf{u}) \to \mathbf{u}^*$ as $t \to \infty$ for all $\mathbf{u} \in N$.

A fixed point is stable if nearby solutions stay nearby for $t \to \infty$. Since in applications of dynamical systems one cannot pinpoint a state exactly, but only approximately, a fixed point must be stable to be physically meaningful. The definition is

Definition. Suppose $\mathbf{u}^* \in U$ is a fixed point of the dynamical system, where $\mathbf{f} : U \to \mathbb{R}^n$ is a C^1 map from an open set U of the vector space $\mathbb{R}^n$ into $\mathbb{R}^n$. Then $\mathbf{u}^*$ is a stable fixed point if for every neighbourhood N of $\mathbf{u}^*$ in U there is a neighborhood N_1 of $\mathbf{u}^*$ in N such that every solution $\mathbf{u}(t)$ with $\mathbf{u}(\mathbf{0})$ in N_1 is defined and in N for all $t > 0$.

Definition. If N_1 can be chosen so that in addition to the properties described in the definition given above, $\lim_{t\to\infty} \mathbf{u}(t) = \mathbf{u}^*$, then $\mathbf{u}^*$ is *asymptotically stable.*

Definition. A fixed point $\mathbf{u}^*$ that is not stable is called *unstable.* This means there is a neighbourhood N of $\mathbf{u}^*$ such that for every neighbourhood N_1 of $\mathbf{u}^*$ in N, there is at least one solution $\mathbf{u}(t)$ starting at $\mathbf{u}(0) \in N_1$, which does not lie entirely in N.

A sink is asymptotically stable and therefore stable. An example of a fixed point that is stable but not asymptotically stable is the origin in $\mathbb{R}^n$ for a linear equation $d\mathbf{u}/dt = A\mathbf{u}$, where A has pure imaginary eigenvalues. The orbits are all ellipses. We have the theorem (Hirsch and Smale [96], Arrowsmith and Place [7])

Theorem. Let $U \subset \mathbb{R}^n$ be open and $\mathbf{f} : U \to \mathbb{R}^n$ continuously differentiable. Suppose $\mathbf{f}(\mathbf{u}^*) = \mathbf{0}$ and $\mathbf{u}^*$ is a stable fixed point of the equation $d\mathbf{u}/dt = \mathbf{f}(\mathbf{u})$. Then no eigenvalue of $D\mathbf{f}(\mathbf{u}^*)$ has a positive real part.

We say that a fixed point $\mathbf{u}^*$ is *hyperbolic* if the derivative $D\mathbf{f}(\mathbf{u}^*)$ has no eigenvalue with real part zero.

Corollary. A hyperbolic fixed point is either unstable or asymptotically stable.

Let

$$\frac{d\mathbf{u}}{dt} = \mathbf{f}(\mathbf{u}), \qquad \mathbf{u} \in U$$

be such that $\mathbf{f}(\mathbf{u}^*) = \mathbf{0}$, $\mathbf{u}^* \in U$, where U is an open subset of $\mathbb{R}^n$. Then the linearization of $d\mathbf{u}/dt = \mathbf{f}(\mathbf{u})$ at $\mathbf{u}^*$ is the linear system of differential equations

$$\frac{d\mathbf{v}}{dt} = D\mathbf{f}(\mathbf{u}^*)\mathbf{v}$$

where the *Jacobian matrix* $D\mathbf{f}(\mathbf{u}^*)$ is the $n \times n$ matrix

$$D\mathbf{f}(\mathbf{u}^*) := \left[\frac{\partial f_i}{\partial u_j}\right]_{i,j=1}^n \Big|_{\mathbf{u}=\mathbf{u}^*}$$

and $\mathbf{v} = (v_1, \ldots; v_n)^T$ are local coordinates at $\mathbf{u}^*$.

Definition. A singular point $\mathbf{u}^*$ of a vector field $\mathbf{f}$ is said to be hyperbolic if no eigenvalue of $D\mathbf{f}(\mathbf{u}^*)$ has a zero real part.

Example. Consider the dynamical system

$$\frac{du_1}{dt} = u_2, \qquad \frac{du_2}{dt} = -u_1 - u_1^3 - ru_2, \quad r \neq 0$$

Then the only fixed point is $(u_1^*, u_2^*) = (0, 0)$. The Jacobian matrix at $(0, 0)$ is

$$J(0,0) = \begin{pmatrix} 0 & 1 \\ -1 & -r \end{pmatrix}$$

with eigenvalues $(-r \pm \sqrt{r^2 - 4})/2$. For all values with $r \neq 0$ the eigenvalues have a nonzero real part. Thus the fixed point $(0, 0)$ is a hyperbolic fixed point. ♣

If $\mathbf{u}^*$ is a singular point of $\mathbf{f}$, then it is a fixed point of the flow of $d\mathbf{u}/dt = \mathbf{f}(\mathbf{u})$. Thus $\mathbf{u}^*$ is a hyperbolic singular point of $\mathbf{f}$ if the flow, $\exp(D\mathbf{f}(\mathbf{u}^*)t)\mathbf{u}$, of the linearisation of $d\mathbf{u}/dt = \mathbf{f}(\mathbf{u})$ is hyperbolic. It is sometimes convenient to distinguish those non-hyperbolic singular points for which $D\mathbf{f}(\mathbf{u}^*)$ has at least one zero eigenvalue. Such points are said to be non-simple.

Theorem. (Hartman-Grobman) Let $\mathbf{u}^*$ be a hyperbolic fixed point of $d\mathbf{u}/dt = \mathbf{f}(\mathbf{u})$ with flow $\Phi_t : U \subseteq \mathbb{R}^n \to \mathbb{R}^n$. Then there is a neighbourhood N of $\mathbf{u}^*$ on which Φ_t is topologically conjugate to the linear flow $\exp(D\mathbf{f}(\mathbf{u}^*)t)\mathbf{u}$.

The *Invariant Manifold Theorem* is also valid for flows. We have

$$W^s_{loc} := \{\, \mathbf{u} \in U \mid \Phi_t(\mathbf{u}) \to \mathbf{u}^* \,\}$$

as $t \to \infty$ and

$$W^u_{loc} := \{\, \mathbf{u} \in U \mid \Phi_t(\mathbf{u}) \to \mathbf{u}^* \,\}$$

as $t \to -\infty$. Global stable and unstable manifolds, W^s and W^u, are defined by taking a union over real $t > 0$ of $\Phi_{-t}(W^s_{loc})$ and $\Phi_t(W^u_{loc})$, respectively.

The following classifies hyperbolic fixed points.

Theorem. Let $\mathbf{u}^*$ be a hyperbolic fixed point of $d\mathbf{u}/dt = \mathbf{f}(\mathbf{u})$ with flow $\Phi_t : U \subseteq \mathbb{R}^n \to \mathbb{R}^n$. Then there is a neighbourhood N of $\mathbf{u}^*$ on which Φ_t is topologically equivalent to the flow of the linear system of differential equations

$$\frac{d\mathbf{u}_s}{dt} = -\mathbf{u}_s, \quad \mathbf{u}_s \in \mathbb{R}^{n_s}$$
$$\frac{d\mathbf{u}_u}{dt} = \mathbf{u}_u, \quad \mathbf{u}_u \in \mathbb{R}^{n_u}$$

where $n_u = n - n_s$. Here n_s is the dimension of the stable eigenspace of $\exp(D\mathbf{f}(\mathbf{u}^*)t)$.

Example. Consider the stability of the fixed points of the Lorenz model. The *Lorenz model* is obtained as an approximation to partial differential equations describing convection in a fluid layer heated below (Bénard problem). The Lorenz model is given by

$$\frac{du_1}{dt} = \sigma(u_2 - u_1)$$
$$\frac{du_2}{dt} = -u_1u_3 + ru_1 - u_2$$
$$\frac{du_3}{dt} = u_1u_2 - bu_3$$

where σ, r and b are positive constants. The fixed points are determined by the system of algebraic equations

$$u_2^* - u_1^* = 0$$
$$-u_1^*u_3^* + ru_1^* - u_2^* = 0$$
$$u_1^*u_2^* - bu_3^* = 0\,.$$

We find that $u_1^* = u_2^* = u_3^* = 0$ is a fixed point for all σ, r and b. This fixed point exits for all parameter values of σ, r and b. If $r < 1$ this fixed point is attracting (a sink). If r becomes larger than 1, this fixed point loses its attracting character (one eigenvalue becomes positive) and two new fixed points appear. For $r > 1$ we find the fixed points

$$u_1^* = u_2^* = \pm\sqrt{b(r-1)}, \qquad u_3^* = r - 1.$$ ♣

The C++ program `lorenzfix.cpp` using SymbolicC++ finds the characteristic equation for the eigenvalues from the variational equations and then determine the stability of the fixed point $(0,0,0)$. The parameter values are $r = 40$, $\sigma = 16$ and $b = 4$. For $r > 1$ the origin $(0,0,0)$ becomes unstable, i.e. one of the three eigenvalues becomes positive (while the other two remain negative).

```
// lorenzfix.cpp

#include <iostream>
#include <cmath>         // for cos, acos, sqrt
#include "symbolicc++.h"
using namespace std;

int main(void)
{
 Symbolic u("u",3), V("V",3), A("A",3,3), s("s"), b("b"), r("r");
 V(0) = s*(u(1)-u(0));                  // Lorenz model
 V(1) = -u(0)*u(2)+r*u(0)-u(1);    // Lorenz model
 V(2) = u(0)*u(1)-b*u(2);              // Lorenz model

 for(int i=0;i<3;i++)
  for(int j=0;j<3;j++) A(i,j) = df(V(i),u(j));
```

```
 Symbolic lambda("lambda");
 Symbolic chareq;
 // characteristic equation
 chareq = det(lambda*A.identity()-A);
 // coefficients
 Symbolic c0 = chareq.coeff(lambda,0);
 Symbolic c1 = chareq.coeff(lambda,1);
 Symbolic c2 = chareq.coeff(lambda,2);
 Symbolic Q = (3*c1-c2*c2)/9;
 Symbolic R = (9*c2*c1-27*c0-2*c2*c2*c2)/54;
 Symbolic D = Q*Q+R*R;
 // parameter values
 Equations values = (s==16.0,b==4.0,r==40.0);
 // fixed point (0,0,0)
 values = (values,u(0)==0.0,u(1)==0.0,u(2)==0.0);
 double q = Q[values];  double rR = R[values];
 double d = D[values];  double nc2 = c2[values];
 if(rR != 0 && q < 0 && d <= 0.0)
 {
 double theta = acos(rR/sqrt(-q*q*q));
 const double PI = 3.1415927;
 double lamb1 = 2.0*sqrt(-q)*cos(theta/3.0)-nc2/3.0;
 double lamb2 = 2.0*sqrt(-q)*cos((theta+2.0*PI)/3.0)-nc2/3.0;
 double lamb3 = 2.0*sqrt(-q)*cos((theta+4.0+PI)/3.0)-nc2/3.0;
 cout << "lamb1 = " << lamb1 << endl;
 cout << "lamb2 = " << lamb2 << endl;
 cout << "lamb3 = " << lamb3 << endl;
 }
 return 0;
}
```

5.2 Trajectories

We evaluate the $u_1(t)$ component of the trajectory for the *Lorenz model* with the initial values

$$u_1(0) = 0.8, \qquad u_2(0) = 0.8, \qquad u_3(0) = 0.8.$$

The parameter values are $r = 40$, $\sigma = 16$ and $b = 4$. For these parameter values there is numerical evidence that the system shows chaotic behaviour.

Lorenz [132] showed that there is an ellipsoid E in $\mathbb{R}^3$ such that every trajectory would eventually cross into the ellipsoid and, once inside, would remain inside. This means the ellipsoid is positively invariant under the system of differential equation. Let $\Phi(\mathbf{u}_0, t) \in \mathbb{R}^3$ be the position in $\mathbb{R}^3$ of the solution of the Lorenz model at time t, starting from $\mathbf{u}_0 \in \mathbb{R}^3$. For a set $S \subset \mathbb{R}^3$ we write $\Phi(S, t)$ for $\{\, \Phi(\mathbf{u}_0, t) \,:\, \mathbf{u}_0 \in S\}$. Denote the three-dimensional volume of S by $vol(S)$. Now the divergence of the right-hand side of the Lorenz model is constant and negative, namely $-(\sigma + b + 1)$.

Thus

$$vol(\Phi(S,t)) = \exp(-t(\sigma+b+1))vol(S)\,.$$

Let E be the positively invariant ellipsoid described above. Since $\Phi(E,t) \subset E$ for t positive, it follows that

$$\Phi(E,t_1) \subset \Phi(E,t_2) \qquad \text{if} \quad t_1 > t_2 > 0\,.$$

Every trajectory tends asymptotically to the limiting set

$$E_\infty = \cap_{t>0}\Phi(E,t)$$

as t tends to infinity. It follows that that this set has volume zero. The shape of E_∞ is extremely complex depending on the parameters.

In the first C++ program `Lie.cpp` we use SymbolicC++ to evaluate symbolically the expansion of the Lie series up to second order with the vector field

$$V = \sigma(u_2 - u_1)\frac{\partial}{\partial u_1} + (-u_1u_3 + ru_1 - u_2)\frac{\partial}{\partial u_2} + (u_1u_2 - bu_3)\frac{\partial}{\partial u_3}\,.$$

Then we iterate numerically the resulting map. In `Lorenztime.java` we use a Runge-Kutta-Fehlberg method to solve the Lorenz system numerically.

```
// lie.cpp

#include <iostream>
#include "symbolicc++.h"
using namespace std;

const int N = 3;
Symbolic u("u",N), ut("ut",N);

// The vector field V
template <class T> T V(const T& ss)
{
 T sum(0);
 for(int i=0;i<N;i++) sum += ut(i)*df(ss,u(i));
 return sum;
}

int main(void)
{
 int i, j;
 Symbolic t("t"), s("s"), b("b"), r("r");
 Symbolic us("",N);
 Equations values;
 // Lorenz model
 ut(0) = s*(u(1)-u(0));
```

```
 ut(1) = -u(1)-u(0)*u(2)+r*u(0);
 ut(2) = u(0)*u(1)-b*u(2);
 // Taylor series expansion up to order 2
 for(i=0;i<N;i++) us(i) = u(i)+t*V(u(i))+0.5*t*t*V(V(u(i)));
 cout << "us =\n" << us << endl;
 // evolution of the approximate solution
 values = (t==0.01,r==40.0,s==16.0,b==4.0,u(0)==0.8,
           u(1)==0.8,u(2)==0.8);
 for(j=0;j<50;j++)
 {
 Equations newvalues = (t==0.01,r==40.0,s==16.0,b==4.0);
 for(i=0;i<N;i++)
 {
 newvalues = (newvalues,u(i)==us(i)[values]);
 cout << newvalues.back() << endl;
 }
 values = newvalues;
 }
 return 0;
}
```

The Runge-Kutta-Fehlberg method is used to solve the Lorenz model numerically. The state variable `u[0]` is displayed as a function of time.

```
// Lorenztime.java

import java.awt.*;
import java.awt.event.*;
import java.awt.Graphics;

public class Lorenztime extends Frame
{
 public Lorenztime()
 {
 setSize(600,500);
 addWindowListener(new WindowAdapter()
 { public void windowClosing(WindowEvent event) { System.exit(0); }}); }

 public void func_system(double h,double t,double u[],double hf[])
 {
 double sigma = 16.0, r = 40.0, b = 4.0;
 hf[0] = h*sigma*(u[1]-u[0]);
 hf[1] = h*(-u[0]*u[2]+r*u[0]-u[1]);
 hf[2] = h*(u[0]*u[1]-b*u[2]);
 }

 public void map(double u[],int steps,double h,double t,int N)
 {
 double uk[] = new double[N];
```

```
double tk;
double a[] = { 0.0, 1.0/4.0, 3.0/8.0, 12.0/13.0, 1.0, 1.0/2.0 };
double c[] = { 16.0/135.0, 0.0, 6656.0/12825.0, 28561.0/56430.0,
               -9.0/50.0, 2.0/55.0 };
double b[][] = new double[6][5];
b[0][0] = b[0][1]= b[0][2] = b[0][3] = b[0][4] = 0.0;
b[1][0] = 1.0/4.0; b[1][1] = 0.0; b[1][2] = 0.0; b[1][3] = 0.0;
b[1][4] = 0.0;
b[2][0] = 3.0/32.0; b[2][1] = 9.0/32.0;
b[2][2] = 0.0; b[2][3] = 0.0; b[2][4] = 0.0;
b[3][0] = 1932.0/2197.0; b[3][1] = -7200.0/2197.0;
b[3][2] = 7296.0/2197.0; b[3][3] = b[3][4] = 0.0;
b[4][0] = 439.0/216.0; b[4][1] = -8.0;
b[4][2] = 3680.0/513.0; b[4][3] = -845.0/4104.0; b[4][4] = 0.0;
b[5][0] = -8.0/27.0; b[5][1] = 2.0;
b[5][2] = -3544.0/2565.0; b[5][3] = 1859.0/4104.0;
b[5][4] = -11.0/40.0;
double f[][] = new double[6][N];
int i, j, l, k;
for(i=0;i<steps;i++) { func_system(h,t,u,f[0]);
for(k=1;k<=5;k++) { tk = t + a[k]*h;
for(l=0;l<N;l++) { uk[l] = u[l];
for(j=0;j<=k-1;j++) uk[l] += b[k][j]*f[j][l];
}
func_system(h,tk,uk,f[k]);
}
for(l=0;l<N;l++)
 for(k=0;k<6;k++) u[l] += c[k]*f[k][l];
}
}

public void paint(Graphics g)
{
g.drawLine(10,10,10,400); g.drawLine(10,200,630,200);
g.drawRect(10,10,630,400);
int steps = 1;  int N = 3;
double h = 0.005;
double t = 0.0;
double u[] = { 0.8, 0.8, 0.8 };  // initial conditions
// wait for transients to decay
for(int i=0;i<1000;i++) { t += h; map(u,steps,h,t,N); }
t = 0.0;
for(int i=0;i<4800;i++) { t += h;
int m = (int)(25.0*t+10.0+0.5); int n = (int)(200.0-6.0*u[0]+0.5);
map(u,steps,h,t,N);
int p = (int)(25.0*(t+h)+10.0+0.5);
int q = (int)(200.0-6.0*u[0]+0.5);
g.drawLine(m,n,p,q);
```

```
 }
 }

 public static void main(String[] args)
 { Frame f = new Lorenztime(); f.setVisible(true); }
}
```

5.3 Phase Portrait

In the program Lorenzphase.java we evaluate the phase portrait $(u_1(t), u_2(t))$ of the Lorenz model for the initial value $u_1(0) = 0.8$, $u_2(0) = 0.8$, $u_3(0) = 0.8$ and the parameter values $\sigma = 16$, $r = 40$ and $b = 4$.

```
// Lorenzphase.java

import java.awt.*;
import java.awt.event.*;
import java.awt.Graphics;

public class Lorenzphase extends Frame
{
 public Lorenzphase()
 {
 setSize(600,500);
 addWindowListener(new WindowAdapter()
 { public void windowClosing(WindowEvent event) { System.exit(0); }}); }

 public void func_system(double h,double t,double u[],double hf[])
 {
 double sigma = 16.0, r = 40.0, b = 4.0;
 hf[0] = h*sigma*(u[1]-u[0]);
 hf[1] = h*(-u[0]*u[2]+r*u[0]-u[1]);
 hf[2] = h*(u[0]*u[1]-b*u[2]);
 }

 public void map(double u[],int steps,double h,double t,int N)
 {
 double uk[] = new double[N];
 double tk;
 double a[] = { 0.0, 1.0/4.0, 3.0/8.0, 12.0/13.0, 1.0, 1.0/2.0 };
 double c[] = { 16.0/135.0, 0.0, 6656.0/12825.0, 28561.0/56430.0,
                -9.0/50.0, 2.0/55.0 };
 double b[][] = new double[6][5];
 b[0][0] = b[0][1]= b[0][2] = b[0][3] = b[0][4] = 0.0;
 b[1][0] = 1.0/4.0; b[1][1] = 0.0; b[1][2] = 0.0; b[1][3] = 0.0;
 b[1][4] = 0.0;
 b[2][0] = 3.0/32.0; b[2][1] = 9.0/32.0;
```

```
b[2][2] = 0.0; b[2][3] = 0.0; b[2][4] = 0.0;
b[3][0] = 1932.0/2197.0; b[3][1] = -7200.0/2197.0;
b[3][2] = 7296.0/2197.0; b[3][3] = b[3][4] = 0.0;
b[4][0] = 439.0/216.0; b[4][1] = -8.0;
b[4][2] = 3680.0/513.0; b[4][3] = -845.0/4104.0;
b[4][4] = 0.0;
b[5][0] = -8.0/27.0; b[5][1] = 2.0;
b[5][2] = -3544.0/2565.0; b[5][3] = 1859.0/4104.0;
b[5][4] = -11.0/40.0;
double f[][] = new double[6][N];
int i, j, l, k;
for(i=0;i<steps;i++) { func_system(h,t,u,f[0]);
for(k=1;k<=5;k++) { tk = t + a[k]*h;
for(l=0;l<N;l++) { uk[l] = u[l];
for(j=0;j<=k-1;j++) uk[l] += b[k][j]*f[j][l];
}
func_system(h,tk,uk,f[k]);
}
for(l=0;l<N;l++)
 for(k=0;k<6;k++) u[l] += c[k]*f[k][l];
}
}

public void paint(Graphics g)
{
g.drawLine(10,10,10,400); g.drawLine(10,200,630,200);
g.drawRect(10,10,630,400);
int steps = 1; int N = 3;
double h = 0.005; double t = 0.0;
double u[] = { 0.8, 0.8, 0.8 };  // initial conditions
// wait for transients to decay
for(int i=0;i<1000;i++) { t += h; map(u,steps,h,t,N); }
t = 0.0;
for(int i=0;i<4800;i++)
{
t += h;
map(u,steps,h,t,N);
int m = (int)(5.0*u[0]+300); int n = (int)(5.0*u[1]+200);
g.drawLine(m,n,m,n);
}
}

public static void main(String[] args)
{ Frame f = new Lorenzphase(); f.setVisible(true); }
}
```

With an appropriate choice of origin the Lorenz model may be rewritten as

$$\begin{aligned}\frac{du_1}{dt} &= \sigma(u_2 - u_1)\\ \frac{du_2}{dt} &= -\sigma u_1 - u_2 - u_1 u_3\\ \frac{du_3}{dt} &= u_1 u_2 - b u_3 - b(r+\sigma)\,.\end{aligned}$$

Here $u_1(t)$ is the rate of rotation of the cylinder, $u_2(t)$ is the temperature difference between opposite sides of the cylinder and $u_3(t)$ measures the deviation from a linear vertical temperature gradient. The positive constants σ, b and r represent respectively the Prandtl number of the fluid (which depends on the viscosity and thermal conductivity), the width to height ratio of the layer, and the fixed temperature difference between the bottom and top of the system. Assume that $\sigma > b+1$. Let

$$u^2(t) := u_1^2(t) + u_2^2(t) + u_3^2(t)\,.$$

Then we find for the time-evolution of $u^2(t)$

$$\frac{1}{2}\frac{d}{dt}u^2 = u_1\frac{du_1}{dt} + u_2\frac{du_2}{dt} + u_3\frac{du_3}{dt} = -\sigma u_1^2 - u_2^2 - b u_3^2 - b u_3(r+\sigma)\,.$$

Thus we have the estimate

$$\begin{aligned}\frac{1}{2}\frac{d}{dt}u^2 &\le -(u_1^2 + u_2^2 + u_3^2) - (b-1)u_3^2 - b u_3(r+\sigma)\\ &\le -u^2 + \frac{b^2(r+\sigma)^2}{4(b-1)}\end{aligned}$$

using that $\sigma > 1$ and the usual estimate for the maximum of a quadratic expression. Then

$$\frac{d}{dt}(u^2(t)e^{2t}) \le \frac{b^2(r+\sigma)^2}{2(b-1)}e^{2t}$$

and integrating yields

$$u^2(t) \le u^2(0)e^{-2t} + \frac{b^2(r+\sigma)^2}{4(b-1)}(1-e^{-2t})\,.$$

It follows that

$$\limsup_{t\to\infty} |u(t)| \le 2\rho_0$$

where $\rho_0 := b(r+\sigma)/(4(b-1)^{1/2})$. Thus $u(t)$ is close to, or inside the ball $B(0, 2\rho_0)$ when t is large. This implies that there is a (maximal) compact set $E \subset B(0, 2\rho_0)$ that is invariant under the solution trajectories.

5.4 Liapunov Exponents

Liapunov exponents provide a meaningful way to characterize the asymptotic behaviour of a nonlinear dynamical system

$$\frac{d\mathbf{u}}{dt} = \mathbf{f}(\mathbf{u}), \qquad \mathbf{u}(0) = \mathbf{u}_0, \qquad \mathbf{u} \in \mathbb{R}^n$$

where $\mathbf{f}$ is continuously differentiable. They provide a generalization of the linear stability analysis of fixed points. For ergodic dynamical systems, the Liapunov exponents are the same for almost all initial conditions $\mathbf{u}_0$ with respect to any invariant measure for the flow. This means their values do not depend on a particular trajectory. For a given solution trajectory $\mathbf{u}(t)$, one considers the *linear variational equation*

$$\frac{dY}{dt} = D\mathbf{f}(\mathbf{u})Y = A(t)Y, \qquad Y(0) = I$$

where $A(t) = (\partial \mathbf{f}/\partial \mathbf{u})$ is the Jacobian matrix at $\mathbf{u}(t)$, Y is an $n \times n$ time-dependent matrix and I is the $n \times n$ identity matrix. Then, for an $n \times n$ fundamental solution matrix $Y(t)$ the symmetric positive definite matrix

$$\Lambda := \lim_{t\to\infty} \Lambda_{\mathbf{u}_0}(t) := \lim_{t\to\infty} (Y^T(t)Y(t))^{1/(2t)}$$

is well-defined, where ^{T} denotes transpose. If

$$\{\, \mathbf{p}_j,\, \mu_j \;:\; j = 1, 2, \ldots, n \,\}$$

denote the eigenvectors and associated eigenvalues of the $n \times n$ matrix Λ such that

$$\Lambda \mathbf{p}_j = \mathbf{p}_j \mu_j \qquad \text{or} \qquad \mathbf{p}_j^T \Lambda \mathbf{p}_j = \mu_j$$

then the one-dimensional Liapunov exponents with respect to the trajectory $\mathbf{u}(t)$ are given by

$$\lambda_j = \log(\mu_j) = \lim_{t\to\infty} \frac{1}{t} \log \|Y(t)\mathbf{p}_j\|$$

where $j = 1, 2, \ldots, n$.

Example. Consider the *Lorenz model*

$$\frac{du_1}{dt} = \sigma(u_2 - u_1)$$
$$\frac{du_2}{dt} = -u_1u_3 + ru_1 - u_2$$
$$\frac{du_3}{dt} = u_1u_2 - bu_3$$

to find the largest one-dimensional Liapunov exponent. The variational equation of the Lorenz model is given by

$$\frac{dv_1}{dt} = \sigma(v_2 - v_1)$$
$$\frac{dv_2}{dt} = (-u_3 + r)v_1 - v_2 - u_1v_3$$
$$\frac{dv_3}{dt} = u_2v_1 + u_1v_2 - bv_3\,.$$

The largest one-dimensional Liapunov exponent is then given by

$$\lambda(u_1(0), u_2(0), u_3(0), v_1(0), v_2(0), v_3(0)) = \lim_{T\to\infty} \frac{1}{T} \ln \|\mathbf{v}(T)\|$$

where $\|\,.\,\|$ denotes any norm in $\mathbb{R}^3$. We choose the norm

$$\|\mathbf{v}\| := |v_1| + |v_2| + |v_3|. \qquad ♣$$

In the C++ program we evaluate the Liapunov exponent for the parameter values $r = 40$, $\sigma = 16$ and $b = 4$. We find $\lambda \approx 1.37$. It is conjectured that the method yields the maximal one-dimensional Liapunov exponent. The accuracy of the one-dimensional Liapunov exponent could be improved when the transients have been decayed.

```
// lorenzliapunov.cpp

#include <iostream>
#include <cmath>          // for log, fabs
#include "symbolicc++.h"
using namespace std;

const int N = 3;
Symbolic u("u",N), ut("ut",N), y("y",N), yt("yt",N);

// The vector field V
template<class T> T V(const T& ss)
{
 T sum(0);
 for(int i=0;i<N;i++) sum += ut(i)*df(ss,u(i));
 return sum;
}

template<class T> T W(const T& ss)
{
 T sum(0);
 for(int i=0;i<N;i++) sum += yt(i)*df(ss,y(i));
 return sum;
}

int main(void)
{
 int i, j;
 Symbolic u("u",N), y("y",N), us("",N), ys("",N),
          t("t"), s("s"), b("b"), r("r");
 Equations v;
 // Lorenz model
 ut(0) = s*(u(1)-u(0));
 ut(1) = -u(1)-u(0)*u(2)+r*u(0);
```

```
ut(2) = u(0)*u(1)-b*u(2);
// variational equations
yt(0) = s*(y(1)-y(0));
yt(1) = (-u(2)+r)*y(0)-y(1)-u(0)*y(2);
yt(2) = (u(1)*y(0)+u(0)*y(1)-b*y(2));
// Taylor series expansion up to order 2
for(i=0;i<N;i++) us(i) = u(i)+t*V(u(i))+t*t*V(V(u(i)))/2;

for(i=0;i<N;i++)
ys(i) = y(i)+t*W(y(i))+t*t*W(W(y(i)))/2;

// Evolution of the approximate solution
v = (t==0.01,r==40.0,s==16.0,b==4.0,
     u(0)==0.8,u(1)==0.8,u(2)==0.8,
     y(0)==0.8,y(1)==0.8,y(2)==0.8);
int iter = 10000;
for(j=0;j<iter;j++)
{
Equations newv = (t==0.01,r==40.0,s==16.0,b==4.0);
for(i=0;i<N;i++) newv = (newv,u(i)==us(i)[v],y(i)==ys(i)[v]);
v = newv;
} // end for loop j

double T = 0.01*iter;
double lambda = log(fabs(double(rhs(v,y(0))))
                +fabs(double(rhs(v,y(1))))
                +fabs(double(rhs(v,y(2)))))/T;
cout << "lambda = " << lambda << endl;
return 0;
}
```

5.5 Generalized Lotka-Volterra Model

The original *Lotka-Volterra model* can be generalized to higher dimensions $n \geq 3$. A three-dimensional generalization of the Lotka-Volterra model is given by

$$\frac{du_1}{dt} = u_1 - u_1 u_2 + c u_1^2 - a u_3 u_1^2$$
$$\frac{du_2}{dt} = -u_2 + u_1 u_2$$
$$\frac{du_3}{dt} = -b u_3 + a u_1 u_3^2$$

where the relations between u_1 and u_2 form the Lotka-Volterra expressions, while the relations between u_1 and u_3 generalize the latter in three dimensions, and where $a, b, c \geq 0$. The parameters a, b and c are bifurcation parameters.

In the Java program `Lotkaphase.java` we evaluate the phase portrait $(u_1(t), u_2(t))$.

```
// Lotkaphase.java

import java.awt.*;
import java.awt.event.*;
import java.awt.Graphics;

public class Lotkaphase extends Frame
{
 public Lotkaphase()
 {
 setSize(600,500);
 addWindowListener(new WindowAdapter()
 { public void windowClosing(WindowEvent event) { System.exit(0); }}); }

 public void func_system(double h,double t,double u[],double hf[])
 {
 double a = 2.9851; double b = 3.0; double c = 7.0;
 hf[0] = h*(u[0]-u[0]*u[1]+c*u[0]*u[0]-a*u[2]*u[0]*u[0]);
 hf[1] = h*(-u[1]+u[0]*u[1]);
 hf[2] = h*(-b*u[2]+a*u[0]*u[0]*u[2]);
 }

 public void map(double u[],int steps,double h,double t,int N)
 {
 double uk[] = new double[N];
 double tk;
 double a[] = { 0.0, 1.0/4.0, 3.0/8.0, 12.0/13.0, 1.0, 1.0/2.0 };
 double c[] = { 16.0/135.0, 0.0, 6656.0/12825.0, 28561.0/56430.0,
                -9.0/50.0, 2.0/55.0 };
 double b[][] = new double[6][5];
 b[0][0] = b[0][1]= b[0][2] = b[0][3] = b[0][4] = 0.0;
 b[1][0] = 1.0/4.0; b[1][1] = 0.0; b[1][2] = 0.0; b[1][3] = 0.0;
 b[1][4] = 0.0;
 b[2][0] = 3.0/32.0; b[2][1] = 9.0/32.0;
 b[2][2] = 0.0; b[2][3] = 0.0; b[2][4] = 0.0;
 b[3][0] = 1932.0/2197.0; b[3][1] = -7200.0/2197.0;
 b[3][2] = 7296.0/2197.0; b[3][3] = b[3][4] = 0.0;
 b[4][0] = 439.0/216.0; b[4][1] = -8.0; b[4][2] = 3680.0/513.0;
 b[4][3] = -845.0/4104.0; b[4][4] = 0.0;
 b[5][0] = -8.0/27.0; b[5][1] = 2.0; b[5][2] = -3544.0/2565.0;
 b[5][3] = 1859.0/4104.0; b[5][4] = -11.0/40.0;
 double f[][] = new double[6][N];
 int i, j, l, k;
 for(i=0;i<steps;i++) { func_system(h,t,u,f[0]);
 for(k=1;k<=5;k++) { tk = t+a[k]*h;
 for(l=0;l<N;l++) { uk[l] = u[l];
 for(j=0;j<=k-1;j++) uk[l] += b[k][j]*f[j][l];
 }
```

```
func_system(h,tk,uk,f[k]);
}
for(l=0;l<N;l++)
 for(k=0;k<6;k++) u[l] += c[k]*f[k][l];
}
}

public void paint(Graphics g)
{
g.drawRect(10,10,630,400);
int steps = 1;  int N = 3;
double h = 0.005;  double t = 0.0;
double u[] = { 1.0, 1.5, 2.5 };  // initial conditions
// wait for transients to decay
for(int i=0;i<1000;i++) { t += h;  map(u,steps,h,t,N); }
t = 0.0;
for(int i=0;i<100000;i++)
{
t += h;  map(u,steps,h,t,N);
int m = (int)(300.0*u[0]-20.0); int n = (int)(280.0*u[1]-60.0);
g.drawLine(m,n,m,n);
}
}

public static void main(String[] args)
{ Frame f = new Lotkaphase(); f.setVisible(true); }
}
```

5.6 Hyperchaotic Systems

For autonomous systems of first-order ordinary differential equations with $n \geq 4$ hyperchaos (Rössler [176], Matsumoto et al [143]) can appear, i.e. the first two largest one-dimensional Liapunov exponents can be positive Consider the autonomous system of differential equations

$$\frac{du_j}{dt} = f_j(\mathbf{u})$$

where $j = 1, 2, 3, 4$. The *variational equation* is given by

$$\frac{dy_i}{dt} = \sum_{j=1}^{4} \frac{\partial f_i}{\partial u_j} y_j\,, \qquad i = 1, 2, 3, 4\,.$$

Assume that $\mathbf{v}(t)$ also satisfies the variational equation. Let $\{\mathbf{e}_1, \mathbf{e}_2, \mathbf{e}_3, \mathbf{e}_4\}$ be the standard basis in the vector space $\mathbb{R}^4$ and

$$\mathbf{y} = \sum_{j=1}^{4} y_j \mathbf{e}_j\,, \qquad \mathbf{v} = \sum_{j=1}^{4} v_j \mathbf{e}_j\,.$$

Then we obtain

$$\mathbf{y} \wedge \mathbf{u} = \left(\sum_{j=1}^{4} y_j \mathbf{e}_j \right) \wedge \left(\sum_{k=1}^{4} u_k \mathbf{e}_k \right) = \sum_{j=1}^{4} \sum_{k=1}^{4} y_j u_k \mathbf{e}_j \wedge \mathbf{e}_k$$

where $\wedge$ denotes the *exterior product*. Using the property of the *exterior product* that

$$\mathbf{e}_i \wedge \mathbf{e}_j = -\mathbf{e}_i \wedge \mathbf{e}_j$$

(and therefore $\mathbf{e}_j \wedge \mathbf{e}_j = 0$) we obtain

$$\begin{aligned} \mathbf{y} \wedge \mathbf{v} = & (y_1 v_2 - y_2 v_1)\, \mathbf{e}_1 \wedge \mathbf{e}_2 + (y_1 v_3 - y_3 v_1)\, \mathbf{e}_1 \wedge \mathbf{e}_3 + (y_1 v_4 - y_4 v_1)\, \mathbf{e}_1 \wedge \mathbf{e}_4 \\ & + (y_2 v_3 - y_3 v_2)\, \mathbf{e}_2 \wedge \mathbf{e}_3 + (y_2 v_4 - y_4 v_2)\, \mathbf{e}_2 \wedge \mathbf{e}_4 + (y_3 v_4 - y_4 v_3)\, \mathbf{e}_3 \wedge \mathbf{e}_4 . \end{aligned}$$

We consider now the time evolution of

$$a_{ij} := y_i v_j - y_j v_i, \qquad i < j.$$

Consequently, using the product rule, we find

$$\frac{da_{ij}}{dt} = \frac{dy_i}{dt} v_j + y_i \frac{dv_j}{dt} - \frac{dy_j}{dt} v_i - y_j \frac{dv_i}{dt} .$$

Therefore we obtain

$$\begin{aligned} \frac{da_{12}}{dt} &= \left(\frac{\partial f_1}{\partial u_1} + \frac{\partial f_2}{\partial u_2} \right) a_{12} - \frac{\partial f_1}{\partial u_3} a_{23} + \frac{\partial f_2}{\partial u_3} a_{13} - \frac{\partial f_1}{\partial u_4} a_{24} + \frac{\partial f_2}{\partial u_4} a_{14} \\ \frac{da_{13}}{dt} &= \left(\frac{\partial f_1}{\partial u_1} + \frac{\partial f_3}{\partial u_3} \right) a_{13} + \frac{\partial f_1}{\partial u_2} a_{23} - \frac{\partial f_1}{\partial u_4} a_{34} + \frac{\partial f_3}{\partial u_4} a_{14} + \frac{\partial f_3}{\partial u_2} a_{12} \\ \frac{da_{14}}{dt} &= \left(\frac{\partial f_1}{\partial u_1} + \frac{\partial f_4}{\partial u_4} \right) a_{14} + \frac{\partial f_1}{\partial u_2} a_{24} + \frac{\partial f_1}{\partial u_3} a_{34} + \frac{\partial f_4}{\partial u_2} a_{12} + \frac{\partial f_4}{\partial u_3} a_{13} \\ \frac{da_{23}}{dt} &= \left(\frac{\partial f_2}{\partial u_2} + \frac{\partial f_3}{\partial u_3} \right) a_{23} + \frac{\partial f_2}{\partial u_1} a_{13} - \frac{\partial f_2}{\partial u_4} a_{34} - \frac{\partial f_3}{\partial u_1} a_{12} + \frac{\partial f_3}{\partial u_4} a_{24} \\ \frac{da_{24}}{dt} &= \left(\frac{\partial f_2}{\partial u_2} + \frac{\partial f_4}{\partial u_4} \right) a_{24} + \frac{\partial f_2}{\partial u_1} a_{14} + \frac{\partial f_2}{\partial u_3} a_{34} - \frac{\partial f_4}{\partial u_1} a_{12} + \frac{\partial f_4}{\partial u_3} a_{23} \\ \frac{da_{34}}{dt} &= \left(\frac{\partial f_3}{\partial u_3} + \frac{\partial f_4}{\partial u_4} \right) a_{34} + \frac{\partial f_3}{\partial u_1} a_{14} + \frac{\partial f_3}{\partial u_2} a_{24} - \frac{\partial f_4}{\partial u_1} a_{13} - \frac{\partial f_4}{\partial u_2} a_{23} . \end{aligned}$$

The two-dimensional Liapunov exponent is given by

$$\lambda^{II} = \lim_{T \to \infty} \frac{1}{T} \ln(\|\mathbf{a}(T)\|)$$

where

$$\|\mathbf{a}(T)\| = |a_{12}(T)| + |a_{13}(T)| + \cdots + |a_{34}(T)| .$$

As a hyperchaotic system we consider the *Rössler model* [176]

$$\frac{du_1}{dt} = -u_2 - u_3$$
$$\frac{du_2}{dt} = u_1 + r_1 u_2 + u_4$$
$$\frac{du_3}{dt} = r_2 + u_1 u_3$$
$$\frac{du_4}{dt} = -r_3 u_3 + r_4 u_4$$

where $r_1 = 0.25$, $r_2 = 2.2$, $r_3 = 0.5$, $r_4 = 0.05$. The two-dimensional Liapunov exponent is calculated in the C++ program.

```
// hyperliapunov.cpp

#include <iostream>
#include <cmath>
#include "symbolicc++.h"
using namespace std;

const int N = 10;
Symbolic u("u",N), ut("ut",N);

// The vector field V
template <class T> T V(const T& ss)
{
 T sum(0);
 for(int i=0;i<N;i++) sum += ut(i)*df(ss,u(i));
 return sum;
}

int main(void)
{
 int i, j;
 Symbolic t("t"), us("",N);
 Equations v;
 double r1 = 1.0/4.0, r2 = 11.0/5.0, r3 = 1.0/20.0, r4 = 3.0/10.0;
 // hyperchaotic model
 ut(0) = -u(1)-u(2);
 ut(1) = u(0)+r1*u(1)+u(3);
 ut(2) = r2+u(0)*u(2);
 ut(3) = r3*u(3)-0.5*u(2);
 ut(4) = r1*u(4)+u(6)+u(7);
 ut(5) = u(0)*u(5)-u(7);
 ut(6) = r3*u(6)-0.5*u(5)-u(8)-u(9);
 ut(7) = (r1+u(0))*u(7)-u(2)*u(4)+u(5)-u(9);
 ut(8) = r4*u(8)+u(6)-0.5*u(7);
 ut(9) = (u(0)+r3)*u(9)+u(2)*u(6);
```

```
// Taylor series expansion up to order 2
for(i=0;i<N;i++) us(i) = u(i)+t*V(u(i))+0.5*t*t*V(V(u(i)));
// Evolution of the approximate solution
// initial values
v = (u(0)==-19.0,u(1)==0.0,u(2)==0.0,u(3)==15.0,u(4)==1.0,
     u(5)==1.0,u(6)==1.0,u(7)==1.0,u(8)==1.0,u(9)==1.0,t==0.01);
int iter = 10000;
for(j=0;j<iter;j++)
{
Equations newv;
newv = (newv,t==0.01);
for(i=0;i<N;i++) newv = (newv,u(i)==us(i)[v]);
v = newv;
} // end for loop j

double T = double(rhs(v,t))*iter;
double lambda =
 log(fabs(double(rhs(v,u(4))))+fabs(double(rhs(v,u(5))))
     +fabs(double(rhs(v,u(6))))+fabs(double(rhs(v,u(7))))
     +fabs(double(rhs(v,u(8))))+fabs(double(rhs(v,u(9)))))/T;
cout << "lambda = " << lambda << endl;
return 0;
}
```

Another hyperchaotic system is given the extended *Chen's chaotic system* [67]

$$\frac{du_1}{dt} = 36(u_2 - u_3)$$
$$\frac{du_2}{dt} = -16u_1 - u_1u_3 + 28u_2 - u_4$$
$$\frac{du_3}{dt} = u_1u_2 - 3u_3$$
$$\frac{du_4}{dt} = u_1 + k$$

where k is the bifurcation parameter. For $k = 0.2$ one has a hyperchaotic attractor.

5.7 Hopf Bifurcation

In the study of nonlinear dynamical systems the *Hopf bifurcation* plays a central role (Marsden and McCracken [142], Chow and Hall [33], Kuznetsov [123]).

Theorem. Let U be an open connected domain in $\mathbb{R}^n$, $c > 0$, and let $\mathbf{f}$ be a real analytic function defined on $U \times [-c, c]$. Consider the autonomous system of first-order ordinary differential equations

$$\frac{d\mathbf{u}}{dt} = \mathbf{f}(\mathbf{u}, r), \qquad \text{where} \quad \mathbf{u} \in U, \quad |r| < c.$$

Suppose there is an analytic, real, vector function $\mathbf{g}$ defined on $[-c, c]$ such that

$$\mathbf{f}(\mathbf{g}(r), r) = \mathbf{0}.$$

Thus one can expand $\mathbf{f}(\mathbf{u}, r)$ about $\mathbf{g}(r)$ in the form

$$\mathbf{f}(\mathbf{u}, r) = L_r \mathbf{u}^* + \mathbf{f}^*(\mathbf{u}^*, r), \qquad \mathbf{u}^* := \mathbf{u} - \mathbf{g}(r)$$

where L_r is an $n \times n$ real matrix which depends only on r, and $\mathbf{f}^*(\mathbf{u}^*, r)$ is the nonlinear part of $\mathbf{f}$. Suppose there exist exactly two complex conjugate eigenvalues $\lambda(r)$, $\bar{\lambda}(r)$ of L_r with the properties

$$\Re(\lambda(0)) = 0 \quad \text{and} \quad \Re(\lambda'(0)) \neq 0 \qquad ('\equiv d/dr).$$

Then there exists a periodic solution $\mathbf{P}(t, \epsilon)$ with period $T(\epsilon)$ with $r = r(\epsilon)$, such that $r(0) = 0$, $\mathbf{P}(t, 0) = \mathbf{g}(0)$ and $\mathbf{P}(t, \epsilon) \neq \mathbf{g}(r(\epsilon))$ for all sufficiently small $\epsilon \neq 0$. Moreover $r(\epsilon)$, $\mathbf{P}(t, \epsilon)$, and $T(\epsilon)$ are analytic at $\epsilon = 0$, and

$$T(0) = \frac{2\pi}{|\Im\alpha(0)|}.$$

These "small" periodic solutions exist for exactly one of three cases: either only for $r > 0$, or only for $r < 0$, or only for $r = 0$.

We apply the Hopf bifurcation to a model for the Belousov-Zhabotinskii reaction.

```
// hopf.cpp

#include <iostream>
#include <cmath>    // for sqrt, acos, cos, fabs, pow
#include "symbolicc++.h"
using namespace std;

int main(void)
{
 int i, j;
 Symbolic u("u",3), V("V",3), b("b"), r("r"), s("s"), A("A",3,3);
 V(0) = s*(u(1)-u(0)*u(1)+u(0)-u(0)*u(1));
 V(1) = u(2)-u(1)-u(0)*u(1);
 V(2) = u(0)-u(2);

 for(i=0;i<3;i++)
  for(j=0;j<3;j++) A(i,j) = df(V(i),u(j));
 Symbolic lambda("lambda");
 Symbolic chareq;
 // characteristic equation
 chareq = det(lambda*A.identity()-A);

 Symbolic c0 = chareq.coeff(lambda,0);
 Symbolic c1 = chareq.coeff(lambda,1);
 Symbolic c2 = chareq.coeff(lambda,2);
 Symbolic Q = (3*c1-c2*c2)/9;
 Symbolic R = (9*c2*c1-27*c0-2*c2*c2*c2)/54;
 Symbolic D = Q*Q*Q+R*R;
```

```
// parameter values s == 2.0
// fixed point (u1,u2,u3) == (0,0,0)
double q =  Q[s==2.0,u(0)==0.0,u(1)==0.0,u(2)==0.0];
double rR =  R[s==2.0,u(0)==0.0,u(1)==0.0,u(2)==0.0];
double d   =  D[s==2.0,u(0)==0.0,u(1)==0.0,u(2)==0.0];
double nc2 = c2[s==2.0,u(0)==0.0,u(1)==0.0,u(2)==0.0];

if(rR != 0 && q < 0 && d <= 0.0)
{
double theta = acos(rR/sqrt(-q*q*q));
double PI = 3.14159;
double lamb1 = 2.0*sqrt(-q)*cos(theta/3.0)-nc2/3.0;
double lamb2 = 2.0*sqrt(-q)*cos((theta+2.0*PI)/3.0)-nc2/3.0;
double lamb3 = 2.0*sqrt(-q)*cos((theta+4.0+PI)/3.0)-nc2/3.0;
cout << "lamb1 = " << lamb1 << endl;
cout << "lamb2 = " << lamb2 << endl;
cout << "lamb3 = " << lamb3 << endl;
}
if(d > 0)
{
double SQRT, T, S;
SQRT = sqrt(d);
if(fabs(rR + SQRT) >= 0.0) S = pow(rR+SQRT,1.0/3.0);
else S = -pow(-rR-SQRT,1.0/3.0);
if(fabs(rR-SQRT) >= 0.0) T = pow(rR-SQRT,1.0/3.0);
else T = -pow(-rR+SQRT,1.0/3.0);
double lamb1 = S+T-nc2/3.0;
double realpart = -(S+T)/2-nc2/3.0;
double imagpart = sqrt(3.0)*(S-T)/2.0;
cout << "lamb1 = " << lamb1 << endl;
cout << "lamb2 = " << realpart << " + i*" << imagpart << endl;
cout << "lamb3 = " << realpart << " - i*" << imagpart << endl;
}
return 0;
}
```

Exercise. An autonomous electronic system is given by

$$
\begin{aligned}
C_0\frac{dv_1}{dt} &= -G_1v_1 + a_1v_1 - a_3v_1^3 + b_1(v_2 - v_1) + b_3(v_2 - v_1)^3 \\
C\frac{dv_2}{dt} &= -i_L - G_2v_2 - b_1(v_2 - v_1) - b_3(v_2 - v_1)^3 \\
L\frac{di_L}{dt} &= v_2
\end{aligned}
$$

where $b_1 > 0$, $b_3 > 0$ and G_1, G_2 are conductances. One sets $\mu = G_1 + b_1 - a_1$, $\delta = G_2 + b_1$, which could be considered as bifurcation parameter. Show that the system can show Hopf bifurcation.

5.8 Time-Dependent First Integrals

A number of dissipative dynamical systems admit *time-dependent first integrals* for a certain choice of the control parameter. As an example we consider the Lorenz model

$$\frac{du_1}{dt} = \sigma(u_2 - u_1)$$
$$\frac{du_2}{dt} = -u_1u_3 + ru_1 - u_2$$
$$\frac{du_3}{dt} = u_1u_2 - bu_3$$

where σ, r and b are positive constants. For example, for $r = 0$, $b = 2$ and σ arbitrary the Lorenz model admits the time-dependent first integral

$$I(\mathbf{u}(t)) = (u_2^2 + u_3^2)\exp(2t).$$

In the program `first2.cpp` we use SymbolicC++ to find the conditions on r, b and σ when we insert the time-dependent first integral into the Lorenz model. We find

$$-2b + 2 = 0, \qquad 2r = 0.$$

Thus σ is arbitrary, $r = 0$ and $b = 1$.

```
// first2.cpp

#include <iostream>
#include "symbolicc++.h"
using namespace std;

int main(void)
{
 Symbolic u("u",3), v("v",4);
 Symbolic term, sum, I, R1, R2;
 Symbolic t("t"), s("s"), b("b"), r("r");
 // Lorenz Model
 v(0) = s*u(1)-s*u(0);
 v(1) = -u(1)-u(0)*u(2)+r*u(0);
 v(2) = u(0)*u(1)-b*u(2);
 v(3) = 1;
 // The ansatz for first integral
 I = (u(1)*u(1)+u(2)*u(2))*exp(2*t);

 sum = 0;
 for(int i=0;i<3;i++) sum += v(i)*df(I,u(i));
 sum += v(3)*df(I,t);
 cout << "sum = " << sum << endl;
 R1 = sum.coeff(u(2),2); R1 = R1/(exp(2*t));
```

```
 cout << "R1 = " << R1 << endl;
 R2 = sum.coeff(u(0),1); R2 = R2.coeff(u(1),1);
 R2 = R2/(exp(2*t));
 cout << "R2 = " << R2 << endl;
 return 0;
}
```

The output is

```
sum = 2*r*u1*u2*exp(2*t)-2*b*u3^(2)*exp(2*t)+2*u3^(2)*exp(2*t)
R1 = -2*b+2
R2 = 2*r
```

Exercises. (1) Find other explicitly time-dependent first integrals for the Lorenz model using the ansatz

$$p(u_1, u_2, u_3)e^{-ct}$$

where p is a polynomial in u_1, u_2, u_3.

(2) The *Rikitake-two-disc dynamo* can be written as

$$\frac{du_1}{dt} = -\mu u_1 + u_3 u_2$$
$$\frac{du_2}{dt} = -\mu u_2 + (u_3 - A)u_1$$
$$\frac{du_3}{dt} = 1 - u_1 u_2$$

with the bifurcation parameters $\mu > 0$ and A. Find the condition on μ and A for explicitly time dependent first integrals.

Chapter 6

Nonlinear Driven Systems

6.1 Introduction

We consider a dynamical nonlinear system described by a system of first order ordinary differential equations

$$\frac{d\mathbf{u}}{dt} = \mathbf{f}(t, \mathbf{u}, r)$$

where $t \in \mathbb{R}$, $\mathbf{u} \in \mathbb{R}^n$, and $r \in \mathbb{R}$ denote the time, an n-dimensional state and a bifurcation parameter, respectively. We assume

$$\mathbf{f} : \mathbb{R} \times \mathbb{R}^n \times \mathbb{R} \to \mathbb{R}^n, \qquad (t, \mathbf{u}, r) \mapsto \mathbf{f}(t, \mathbf{u}, r)$$

is a C^∞ mapping and is periodic in t with period 2π

$$\mathbf{f}(t + 2\pi, \mathbf{u}, r) = \mathbf{f}(t, \mathbf{u}, r).$$

Such dynamical systems have been studied by many authors (Ueda et al [206], Kawakami [113], Ueda and Akamatsu [207]). We assume that the dynamical system has a solution $\mathbf{u}(t) = \Phi(t, \mathbf{v}, r)$ defined on $-\infty < t < \infty$ with every initial condition $\mathbf{v} \in \mathbb{R}^n$ and every $r \in \mathbb{R}$: $\mathbf{u}(0) = \Phi(0, \mathbf{v}, r) = \mathbf{v}$. By the periodic hypothesis given above, we can define a C^∞ diffeomorphism T_r from the state space $\mathbb{R}^n$ into itself

$$T_r : \mathbb{R}^n \to \mathbb{R}^n, \qquad \mathbf{v} \mapsto T_r(\mathbf{v}) = \Phi(2\pi, \mathbf{v}, r)\,.$$

The mapping T_r is often called the *Poincaré mapping* (Ueda et al [206], Kawakami [113], Ueda and Akamatsu [207]). and is used for investigating qualitative properties of the dynamical system given above. If a solution $\mathbf{u}(t) = \Phi(t, P_0, r)$ is periodic with period 2π, then the point P_0 is a *fixed point* of T_r, i.e.

$$T_r(P_0) = P_0\,.$$

If $\mathbf{u}(t)$ is a subharmononic solution of order $1/k$ ($k = 2, 3, \ldots$), i.e., a periodic solution with least period $2k\pi$, then P_0 is a point with period k such that

$$T_r^k(P_0) = P_0$$

and

$$P_0 \neq T_r^{(j)}, \qquad j = 1, 2, \ldots, k-1\,.$$

Hence there are always k points P_0, $P_1 = T_r(P_0)$, $\ldots$, $P_{k-1} = T_r^{(k-1)}(P_0)$ which are all fixed points of T_r^k. An invariant closed curve C such that $T_r(C) = C$ corresponds to quasi-periodic solutions of the dynamical system. Thus in order to study the behaviour of the solutions of the dynamical system, we have only to study the iterates of the Poincaré mapping T_r.

Next we introduce a *hyperbolic fixed point* and give its classification (Ueda et al [206], Kawakami [113], Ueda and Akamatsu [207]). Let $P \in \mathbb{R}^n$ be a fixed point of T_r, then in the neighbourhood of P. Then the mapping T_r may be approximated by its derivative $DT_r(P)$. This is indeed possible if the fixed point is hyperbolic. We call P a hyperbolic fixed point of T_r, if $DT_r(P)$ is hyperbolic, i.e., all the absolute values of the eigenvalues of $DT_r(P)$ are different from unity. We give a topological classification of a hyperbolic fixed point. Let P be a hyperbolic fixed point of T_r and E^u be the intersection of $\mathbb{R}^n$ and the direct sum of the generalized eigenspaces of $DT_r(P)$ corresponding to the eigenvalues μ such that $|\mu| > 1$. Analogously, let E^s be the intersection of $\mathbb{R}^n$ and the direct sum of the generalized eigenspaces of $DT_r(P)$ corresponding to the eigenvalues μ such that $|\mu| < 1$. E^u (or E^s) is called the unstable (or stable) subspace of $DT_r(P)$. Then E^u and E^s have the properties $\mathbb{R}^n = E^u \oplus E^s$ and

$$DT_r(E^u) = E^u, \qquad DT_r(E^s) = E^s$$

$$\dim E^u = \#\{\,\mu_i \mid |\mu_i| > 1\,\}, \qquad \dim E^s = \#\{\,\mu_i \mid |\mu_i| < 1\,\}$$

where μ_i is the eigenvalue of $DT_r(P)$ and $\#\{\quad\}$ indicates the number of elements contained in the set $\{\ \}$. Let

$$L^u := DT_r(P)|_{E^u}\,, \qquad L^s := DT_r(P)|_{E^s}\,.$$

Then the topological type of a hyperbolic fixed point is determined by the $\dim E^u$ (or $\dim E^s$) and the orientation preserving or reversing property of L^u (or L^s). The latter condition is equivalent to the positive or negative sign of $\det L^u$ (or $\det L^s$). One defines two types of hyperbolic fixed points.

Definition. A hyperbolic fixed point P is called

(i) a direct type (abbrev. D-type), if $\det(L^u) > 0$.

(ii) an inversely type (abbrev. I-type), if $\det(L^u) < 0$.

From this definition, at a D-type of fixed point P, L^u is an orientation preserving mapping, whereas at an I-type of fixed point P, L^u is an orientation reversing mapping. If E^u is the empty set, we identify P as a D-type. Secondly, a hyperbolic fixed point is called a positive type (P-type) or a negative type (N-type) according to the evenness or oddness of $\dim E^u$, respectively. The name comes from the index property of a fixed point. Combining these definitions, one studies the following four types of hyperbolic fixed points.

Definition. A hyperbolic fixed point P is called a

(i) PD-type if $\dim E^u$ is even and $\det(L^u) > 0$

(ii) ND-type if $\dim E^u$ is odd and $\det(L^u) > 0$

(iii) PI-type if $\dim E^u$ is even and $\det(L^u) < 0$

(iv) NI-type if $\dim E^u$ is odd and $\det(L^u) < 0$.

Assigning $\dim E^u$ to the above classification, we obtain all the topologically different types of hyperbolic fixed points:

Proposition. Let $T_r : \mathbb{R}^n \to \mathbb{R}^n$ be the Poincaré mapping defined above, then there exist $2n$ topologically different types of hyperbolic fixed points. These types are:

(i) for even-dimensional case: $n = 2m$,

$$_{2k}PD \quad (k = 0, 1, \ldots, m), \qquad _{2k}PI \quad (k = 1, \ldots, m-1)$$

$$_{2k+1}ND \quad (k = 0, 1, 2, \ldots, m) \qquad _{2k+1}NI \quad (k = 0, 1, \ldots, m)$$

(ii) for odd-dimensional case: $n = 2m + 1$,

$$_{2k}PD \quad (k = 0, 1, \ldots, m), \qquad _{2k}PI \quad (k = 1, \ldots, m)$$

$$_{2k+1}ND \quad (k = 0, 1, \ldots, m+1) \qquad _{2k+1}NI \quad (k = 0, 1, \ldots, m)$$

One also uses the notation $^j_kP^m$ denoting a hyperbolic fixed point such that P denotes the type: PD, ND, PI, and NI, m indicates an m-periodic point of T_r, j indicates the number of a set of P^m when there are several sets of P^m, and $k = \dim E^u$. If $m = 1$ or $j = 1$, they will be omitted.

For the two-dimensional case: $n = 2$, we have four types of hyperbolic fixed points: $_0PD$, $_1ND$, $_1NI$, and $_2PD$. These points are called a completely stable, directly unstable, inversely unstable and completely unstable fixed point, respectively. For $n = 3$, we have six types of hyperbolic fixed points: $_0PD$, $_1ND$, $_1NI$, $_2PD$, $_2PI$, and $_3ND$.

A completely stable fixed point, i.e., $\dim E^u = 0$ is often called a *sink*,, whereas a completely unstable fixed point, i.e., $\dim E^u = n$ is called a *source*. Other cases, $1 \le \dim E^u \le n - 1$, are simply called *saddles*.

6.2 Driven Anharmonic Systems

6.2.1 Phase Portrait

We consider the driven anharmonic system

$$\frac{d^2u}{dt^2} + a\frac{du}{dt} + bu + cu^3 = k_1 + k_2\cos(\Omega t).$$

Let $u_1 := u$ and $u_2 := du/dt$. Then the equation can be written as first order system

$$\frac{du_1}{dt} = u_2, \qquad \frac{du_2}{dt} = -au_2 - bu_1 - cu_1^3 + k_1 + k_2\cos(\Omega t)$$

or using $u_3(t) = \Omega t$

$$\frac{du_1}{dt} = u_2, \qquad \frac{du_2}{dt} = -au_2 - bu_1 - cu_1^3 + k_1 + k_2\cos(u_3), \qquad \frac{du_3}{dt} = \Omega.$$

In the Java program Anharmonicph.java we use the values $a = 1$, $b = -10$, $c = 100$, $k_1 = 0$, $k_2 = 1.2$, $\Omega = 3.5$ and find the phase portrait $(u_1(t), u_2(t))$ using the Lie series technique.

```
// Anharmonicph.java

import java.awt.*;
import java.awt.event.*;
import java.awt.Graphics;

public class Anharmonicph extends Frame
{
 public Anharmonicph()
 {
 setSize(600,500);
 addWindowListener(new WindowAdapter()
 { public void windowClosing(WindowEvent event) { System.exit(0); }}); }

 public void paint(Graphics g)
 {
 g.drawRect(40,50,500,440);
 double a, b, c, k1, k2, Omega;
 double u1, u2, u3, u11, u22, u33;
 double tau, count;
 // parameter values
 a = 1.0; b = -10.0; c = 100.0; Omega = 3.5; k1 = 0.0; k2 = 1.2;
 tau = 0.005; count = 0.0;
 // initial values
 u11 = 1.0; u22 = 0.0; u33 = 0.0;

 while(count < 1200.0)
```

```
{
u1 = u11; u2 = u22; u3 = u33;
double vu2 = -a*u2-b*u1-c*u1*u1*u1+k1+k2*Math.cos(u3);
u11 = u1+tau*u2+tau*tau*vu2/2.0;
u22 = u2+tau*vu2+tau*tau*(-b*u2-3.0*c*u1*u1*u2-a*vu2
      -Omega*k2*Math.sin(u3))/2.0;
u33 = Omega*tau+u3;

if(count > 1000.0)  // transients have decayed
{
int m = (int)(250.0*u1+250.0);   int n = (int)(200.0*u2+240.0);
int m1 = (int)(250.0*u11+250.0); int n1 = (int)(200.0*u22+240.0);
g.drawLine(m,n,m,n);
}
count += tau;
}
}

public static void main(String[] args)
{ Frame f = new Anharmonicph(); f.setVisible(true); }
}
```

6.2.2 Poincaré Section

The driven anharmonic system is given by the system

$$\frac{du_1}{dt} = u_2, \qquad \frac{du_2}{dt} = -au_2 - bu_1 - cu_1^3 + k_1 + k_2\cos(\Omega t)$$

where a, b, c, k_1, k_2 and Ω are bifurcation parameters. This system is invariant under the transformation $t \mapsto t + 2\pi n/\Omega$, where $n \in \mathbb{Z}$. Thus we can introduce a *diffeomorphism* on the (u_1, u_2) plane into itself, which can serve to decide numerically whether or not there is chaotic behaviour. Assume that

$$(t, u_1(0), u_2(0)) \mapsto u_1(t, u_1(0), u_2(0)), \quad (t, u_1(0), u_2(0)) \mapsto u_2(t, u_1(0), u_2(0))$$

is a solution of the driven anharmonic system starting from a point

$$\mathbf{p}_0 = (u_1(0), u_2(0))$$

at $t = 0$. Let

$$\mathbf{p}_1 = (u_1(2\pi/\Omega), u_2(2\pi/\Omega)).$$

Thus we have defined a diffeomorphism

$$T_r : \mathbb{R}^2 \mapsto \mathbb{R}^2, \qquad \mathbf{p}_0 \mapsto \mathbf{p}_1$$

where r is a bifurcation parameter.

```
// Anharmonicpo.java

import java.awt.*;
import java.awt.event.*;
import java.awt.Graphics;

public class Anharmonicpo extends Frame
{
 public Anharmonicpo()
 {
 setSize(600,500);
 addWindowListener(new WindowAdapter()
 { public void windowClosing(WindowEvent event) { System.exit(0); }}); }

 public void paint(Graphics g)
 {
 g.drawRect(40,50,500,440);
 double a, b, c, k1, k2, Omega;
 a = 1.0; b = -10.0; c = 100.0; Omega = 3.5; k1 = 0.0; k2 = 1.2;
 double u1, u2, u3, u11, u22, u33, t, count;
 t = 0.005; count = 0.0;
 u11 = 1.0; u22 = 0.0; u33 = 0.0; // initial values
 int iter = 1;
 while(count < 6200.0)
 {
 u1 = u11; u2 = u22; u3 = u33;
 double vu2 = -a*u2-b*u1-c*u1*u1*u1+k1+k2*Math.cos(u3);
 u11 = u1+t*u2+t*t*vu2/2.0;
 u22 = u2+t*vu2+t*t*(-b*u2-3.0*c*u1*u1*u2-a*vu2
       -Omega*k2*Math.sin(u3))/2.0;
 u33 = Omega*t+u3;
 if((iter > 10) && (Math.abs(count-2.0*Math.PI*iter/Omega) < 0.02))
 {
 int m = (int)(200.0*u11+300.0);  int n = (int)(200.0*u22+200.0);
 g.drawLine(m,n,m,n);
 }  // end if
 count += t;
 if(count > 2.0*Math.PI*iter/Omega) iter++;
 }  // end while
 }  // end paint

 public static void main(String[] args)
 { Frame f = new Anharmonicpo(); f.setVisible(true); }
}
```

6.2.3 Liapunov Exponent

The driven anharmonic system is given by

$$\frac{d^2u}{dt^2} + a\frac{du}{dt} + bu + cu^3 = k\cos(\Omega t)$$

where $a > 0$. This equation can be written as an autonomous system

$$\frac{du_1}{dt} = u_2$$
$$\frac{du_2}{dt} = -au_2 - bu_1 - cu_1^3 + k\cos(u_3)$$
$$\frac{du_3}{dt} = \Omega$$

with $u_3(t=0) = 0$. Thus the variational equation

$$\frac{d\mathbf{v}}{dt} = (D_{\mathbf{u}}\mathbf{f})\mathbf{v}$$

is given by

$$\frac{dv_1}{dt} = v_2$$
$$\frac{dv_2}{dt} = -av_2 - bv_1 - 3cu_1^2v_1 - k\sin(u_3)v_3$$
$$\frac{dv_3}{dt} = 0\,.$$

Without loss of generality we can set $v_3(t) = 0$. Then the variational system simplifies to

$$\frac{dv_1}{dt} = v_2, \qquad \frac{dv_2}{dt} = -av_2 - bv_1 - 3cu_1^2v_1\,.$$

The one-dimensional Liapunov exponent follows from

$$\lambda := \lim_{T\to\infty} \frac{1}{T}\ln\|\mathbf{v}(T)\|$$

where

$$\|\mathbf{v}(T)\| = |v_1(T)| + |v_2(T)|.$$

In the Java program **Anharmonicliap.java** we calculate the one-dimensional Liapunov exponent λ for the paramater values $k = 1.2$, $a = 1.0$, $b = -10.0$, $c = 100.0$, $\Omega = 3.5$. We apply the Lie series technique. We find for the one-dimensional Liapunov exponent $\lambda \approx 0.34$.

```
// Anharmonicliap.java

public class Anharmonicliap
{
```

```
public static void main(String[] args)
{
double a, b, c, k, Omega;
double u1, u2, u3, u11, u22, u33;
double v1, v2, v3, v11, v22, v33;
double t, count;
a = 1.0; b = -10.0; c = 100.0; Omega = 3.5; k = 1.2;
t = 0.005; count = 0.0;
double ab1, ab2;
// initial values
u11 = 1.0; u22 = 0.0; u33 = 0.0; v11 = 0.5; v22 = 0.5;

while(count < 1500.0)
{
u1 = u11; u2 = u22; u3 = u33; v1 = v11; v2 = v22;
ab1 = -a*u2-b*u1-c*u1*u1*u1+k*Math.cos(u3);
ab2 = (-b-3.0*c*u1*u1)*v1-a*v2;
u11 = u1+t*u2+t*t*ab1/2.0;
u22 = u2+t*ab1+t*t*(-b*u2-3.0*c*u1*u1*u2
      -a*ab1-Omega*k*Math.sin(u3))/2.0;
u33 = Omega*t+u3;
v11 = v1+t*v2+t*t*ab2/2.0;
v22 = v2+t*ab2+t*t*(-6.0*c*u1*u2*v1-(b+3.0*c*u1*u1)*v2
      -a*ab2/2.0);
count += t;
} // end while
double liap = 1.0/count*Math.log(Math.abs(v11)+Math.abs(v22));
System.out.println("Liapunov exponent = " + liap);
}
}
```

6.2.4 Autocorrelation Function

We consider the driven anharmonic system

$$\frac{d^2u}{dt^2} + a\frac{du}{dt} + bu + cu^3 = k\cos(\Omega t).$$

Let $u_1 := u$ and $u_2 := du/dt$. Then the system can be written as

$$\frac{du_1}{dt} = u_2, \qquad \frac{du_2}{dt} = -au_2 - bu_1 - cu_1^3 + k\cos(\Omega t)\,.$$

The *time average* of $u(t)$ is defined as

$$\langle u \rangle := \lim_{T\to\infty} \frac{1}{T}\int_0^T u(t)dt\,.$$

The *autocorrelation function* of $u(t)$ is defined by

$$C_{uu}(\tau) := \frac{\lim\limits_{T\to\infty}\dfrac{1}{T}\displaystyle\int_0^T (u(t)-\langle u(t)\rangle)(u(t+\tau)-\langle u(t+\tau)\rangle)dt}{\lim\limits_{T\to\infty}\dfrac{1}{T}\displaystyle\int_0^T (u(t)-\langle u(t)\rangle)^2 dt}.$$

The autocorrelation function measures the correlation between subsequent signals. It remains constant or oscillates for regular motion and decays rapidly (mostly with an exponential tail) if the signals become uncorrelated in the chaotic regime. First we generate a time series from integrating the differential equation. The time series is used for calculating the autocorrelation function.

```
// Timeseries.cpp

#include <fstream>
#include <cmath>      // for cos
using namespace std;

void fsystem(double h,double t,double u[],double hf[])
{
 double a = 1.0; double b = -10.0; double c = 100.0;
 double k = 1.2; double Omega = 3.5;
 hf[0] = h*u[1];
 hf[1] = h*(-a*u[1]-b*u[0]-c*u[0]*u[0]*u[0]+k*cos(Omega*t));
}

void map(double u[],int steps,double h,double t)
{
 const int N = 2;
 double uk[N];
 double tk;
 double a[6] = { 0.0, 1.0/4.0, 3.0/8.0, 12.0/13.0, 1.0, 1.0/2.0 };
 double c[6] = { 16.0/135.0, 0.0, 6656.0/12825.0, 28561.0/56430.0,
                 -9.0/50.0, 2.0/55.0 };
 double b[6][5];
 b[0][0] = b[0][1]= b[0][2] = b[0][3] = b[0][4] = 0.0;
 b[1][0] = 1.0/4.0; b[1][1] = 0.0; b[1][2] = 0.0; b[1][3] = 0.0;
 b[1][4] = 0.0;
 b[2][0] = 3.0/32.0; b[2][1] = 9.0/32.0;
 b[2][2] = 0.0; b[2][3] = 0.0; b[2][4] = 0.0;
 b[3][0] = 1932.0/2197.0; b[3][1] = -7200.0/2197.0;
 b[3][2] = 7296.0/2197.0; b[3][3] = b[3][4] = 0.0;
 b[4][0] = 439.0/216.0; b[4][1] = -8.0; b[4][2] = 3680.0/513.0;
 b[4][3] = -845.0/4104.0; b[4][4] = 0.0;
 b[5][0] = -8.0/27.0; b[5][1] = 2.0;
 b[5][2] = -3544.0/2565.0; b[5][3] = 1859.0/4104.0;
 b[5][4] = -11.0/40.0;
```

```
 double f[6][N];
 int i, j, l, k;
 for(i=0;i<steps;i++) { fsystem(h,t,u,f[0]);
 for(k=1;k<=5;k++) { tk = t+a[k]*h;
 for(l=0;l<N;l++) { uk[l] = u[l];
 for(j=0;j<=k-1;j++) uk[l] += b[k][j]*f[j][l];
 }
 fsystem(h,tk,uk,f[k]);
 }
 for(l=0;l<N;l++)
  for(k=0;k<6;k++) u[l] += c[k]*f[k][l];
 }
}

int main(void)
{
 ofstream data;
 data.open("time_ser.dat");
 int steps = 1;
 double h = 0.005;  double t = 0.0;
 double u[2] = { 1.0, 0.0 }; // initial conditions
 int i; // wait for transients to decay
 for(i=0;i<1000;i++) { t += h; map(u,steps,h,t); }
 t = 0.0;
 for(i=0;i<9192;i++)
 { t += h; map(u,steps,h,t); data << u[0] << endl; }
 data.close();
 return 0;
}
```

In the C++ program `correlation3.cpp` we find the autocorrelation functions using the data generated from the program `Timeseries.cpp`.

```
// correlation3.cpp

#include <iostream>
#include <fstream>
using namespace std;

double average(double* series,int T)
{
 double sum = 0.0;
 for(int t=0;t<T;t++) { sum += series[t]; }
 double av = sum/((double)T);
 return av;
}

void autocorrelation(double* series,double* CXX,int T,
                     int length,double av)
```

```
{
 for(int tau=0;tau<length;tau++)
 {
 double C = 0.0;
 double diff = T-length;
 for(int t=0;t<diff;t++) { C += (series[t]-av)*(series[t+tau]-av); }
 CXX[tau] = C/((double) diff+1);
 }
}

int main(void)
{
 int T = 8192;
 double* series = new double[T];
 ifstream data;
 data.open("time_ser.dat");
 for(int t=0;t<T;t++) { data >> series[t]; }
 double av = average(series,T);
 cout << "average value = " << av << endl;
 int length = 200;
 double* CXX = new double[length];
 autocorrelation(series,CXX,T,length,av);
 delete[] series;
 for(int tau=0;tau<length;tau++)
 cout << "CXX[" << tau << "] = " << CXX[tau] << endl;
 delete[] CXX;
 data.close();
 return 0;
}
```

6.2.5 Power Spectral Density

The autocorrelation function of the real scalar signal $u(t)$ is defined by

$$R_{uu}(\tau) = E\left[u(t+\tau)u(t)\right]$$

with

$$E\left[u(t+\tau)u(t)\right] := \int_{-\infty}^{\infty}\int_{-\infty}^{\infty} \rho_{u(t+\tau),u(t)}(x,y)xy\,dx\,dy$$

where $\rho_{u(t+\tau),u(t)}(x,y)$ represents the joint probability density function of $u(t+\tau)$ and $u(t)$. If $u(t)$ is ergodic then the autocorrelation function of $u(t)$ may be equated to

$$R_{uu}(\tau) = \lim_{T\to\infty} \frac{1}{2T}\int_{-T}^{T} u(t+\tau)u(t)\,dt\,.$$

The *power spectral density* $S_{uu}(\omega)$ is now defined as

$$S_{uu}(\omega) := \int_{-\infty}^{\infty} R_{uu}(\tau)e^{-i\omega\tau}\,d\tau\,.$$

It follows that

$$S_{uu}(\omega) := \lim_{T\to\infty} E\left(\frac{1}{2T}\left|\int_{-T}^{T} u(t)e^{-i\omega t}\,dt\right|^2\right)$$

if $u(t)$ is ergodic. The power spectral density can indicate whether the dynamical system is periodic or quasiperiodic. The power spectral density of a periodic system with frequency ω has delta functions at ω and its harmonics 2ω, 3ω, ... A quasiperiodic system with basic frequencies $\omega_1, \ldots, \omega_k$ has delta functions at these positions and also at all linear combinations with integer coefficients.

Exercise. Consider the driven anharmonic oscillator

$$\frac{du_1}{dt} = u_2, \quad \frac{du_2}{dt} = u_1 - u_1^3 + \epsilon(a\cos(\theta) - bu_2), \quad \frac{d\theta}{dt} = 1$$

where $a, b > 0$. Show that this dyanmical system has transverse homoclinic points, for sufficiently small values of ϵ, provided

$$\frac{a}{b} > \frac{4\cosh(\pi/2)}{3(2^{1/2}\pi}\,.$$

6.3 Driven Pendulum

6.3.1 Phase Portrait

The equation for the damped and driven pendulum is given by

$$\frac{d^2u}{dt^2} + a\frac{du}{dt} + b\sin(u) = k\cos(\Omega t).$$

Introducing $u_1 := u$ and $u_2 := du/dt$ we obtain the system

$$\frac{du_1}{dt} = u_2, \quad \frac{du_2}{dt} = -au_2 - b\sin(u_1) + k\cos(\Omega t)\,.$$

For the C++ program we evaluate the phase portrait $(u_1(t), u_2(t))$ for $a = 0.2$, $b = 1$, $k = 1.1$ and $\Omega = 0.8$. We generate `u[0]` and `u[1]` and write it into the file `phase_data.dat`. Then we can use GNUPLOT to display the data.

```
// Pendulumphase.cpp

#include <fstream>
#include <cmath>  // for sin, cos
using namespace std;

void fsystem(double h,double t,double u[],double hf[])
{
 double a = 0.2; double b = 1.0;
 double k = 1.1; double Omega = 0.8;
```

```
 hf[0] = h*u[1]; hf[1] = h*(-a*u[1]-b*sin(u[0])+k*cos(Omega*t));
}

void map(double u[],int steps,double h,double t)
{
 const int N = 2;
 double uk[N]; double tk;
 double a[6] = { 0.0, 1.0/4.0, 3.0/8.0, 12.0/13.0, 1.0, 1.0/2.0 };
 double c[6] = { 16.0/135.0, 0.0, 6656.0/12825.0, 28561.0/56430.0,
                 -9.0/50.0, 2.0/55.0 };
 double b[6][5];
 b[0][0] = b[0][1]= b[0][2] = b[0][3] = b[0][4] = 0.0;
 b[1][0] = 1.0/4.0; b[1][1] = 0.0; b[1][2] = 0.0; b[1][3] = 0.0;
 b[1][4] = 0.0;
 b[2][0] = 3.0/32.0; b[2][1] = 9.0/32.0;
 b[2][2] = 0.0; b[2][3] = 0.0; b[2][4] = 0.0;
 b[3][0] = 1932.0/2197.0; b[3][1] = -7200.0/2197.0;
 b[3][2] = 7296.0/2197.0; b[3][3] = b[3][4] = 0.0;
 b[4][0] = 439.0/216.0; b[4][1] = -8.0;
 b[4][2] = 3680.0/513.0; b[4][3] = -845.0/4104.0; b[4][4] = 0.0;
 b[5][0] = -8.0/27.0; b[5][1] = 2.0;
 b[5][2] = -3544.0/2565.0; b[5][3] = 1859.0/4104.0;
 b[5][4] = -11.0/40.0;
 double f[6][N];
 int i, j, l, k;
 for(i=0;i<steps;i++) { fsystem(h,t,u,f[0]);
 for(k=1;k<=5;k++) { tk = t+a[k]*h;
 for(l=0;l<N;l++) { uk[l] = u[l];
 for(j=0;j<=k-1;j++) uk[l] += b[k][j]*f[j][l];
 }
 fsystem(h,tk,uk,f[k]);
 }
 for(l=0;l<N;l++)
  for(k=0;k<6;k++) u[l] += c[k]*f[k][l];
 }
}

int main(void)
{
 ofstream data;
 data.open("phase_data.dat");
 int steps = 1;
 double h = 0.005; double t = 0.0;
 double u[2] = { 0.8, 0.6 }; // initial conditions
 // wait for transients to decay
 for(int i=0;i<1000;i++) { t += h; map(u,steps,h,t); }
 t = 0.0;
 for(int j=0;j<20000;j++)
```

```
 { t += h; map(u,steps,h,t); data << u[0] << " " << u[1] << "\n"; }
 data.close();
 return 0;
}
```

6.3.2 Poincaré Section

The equation for the damped and driven pendulum is given by

$$\frac{d^2u}{dt^2} + a\frac{du}{dt} + b\sin(u) = k\cos(\Omega t).$$

Introducing $u_1 := u$ and $u_2 := du/dt$ we obtain the system

$$\frac{du_1}{dt} = u_2, \qquad \frac{du_2}{dt} = -au_2 - b\sin u_1 + k\cos(\Omega t).$$

This system is invariant under the translation $t \mapsto t + \frac{2\pi n}{\Omega}$, where $n \in \mathbb{Z}$. Thus we can study the Poincaré map.

In the Java program `PendulumPoincare.java` we evaluate the Poincaré section $(u_1(t), u_2(t))$ for $a = 0.2$, $b = 1$, $k = 1.1$ and $\Omega = 0.8$.

```
// PendulumPoincare.java

import java.awt.*;
import java.awt.event.*;
import java.awt.Graphics;

public class PendulumPoincare extends Frame
{
 public PendulumPoincare()
 {
 setSize(600,500);
 addWindowListener(new WindowAdapter()
 { public void windowClosing(WindowEvent event) { System.exit(0); }}); }

 public void fsystem(double h,double t,double u[],double hf[])
 {
 double a = 0.2, b = 1.0;
 double k = 1.09; double Omega = 0.8;
 hf[0] = h*u[1];
 hf[1] = h*(-a*u[1]-b*Math.sin(u[0])+k*Math.cos(Omega*t));
 }

 public void map(double u[],int steps,double h,double t,int N)
 {
 double uk[] = new double[N];
 double tk;
```

```
double a[] = { 0.0, 1.0/4.0, 3.0/8.0, 12.0/13.0, 1.0, 1.0/2.0 };
double c[] = { 16.0/135.0, 0.0, 6656.0/12825.0, 28561.0/56430.0,
               -9.0/50.0, 2.0/55.0 };
double b[][] = new double[6][5];
b[0][0] = b[0][1]= b[0][2] = b[0][3] = b[0][4] = 0.0;
b[1][0] = 1.0/4.0; b[1][1] = 0.0; b[1][2] = 0.0; b[1][3] = 0.0;
b[1][4] = 0.0;
b[2][0] = 3.0/32.0; b[2][1] = 9.0/32.0;
b[2][2] = 0.0; b[2][3] = 0.0; b[2][4] = 0.0;
b[3][0] = 1932.0/2197.0; b[3][1] = -7200.0/2197.0;
b[3][2] = 7296.0/2197.0; b[3][3] = b[3][4] = 0.0;
b[4][0] = 439.0/216.0; b[4][1] = -8.0;
b[4][2] = 3680.0/513.0; b[4][3] = -845.0/4104.0; b[4][4] = 0.0;
b[5][0] = -8.0/27.0; b[5][1] = 2.0; b[5][2] = -3544.0/2565.0;
b[5][3] = 1859.0/4104.0; b[5][4] = -11.0/40.0;
double f[][] = new double[6][N];
for(int i=0;i<steps;i++) { fsystem(h,t,u,f[0]);
for(int k=1;k<=5;k++) { tk = t+a[k]*h;
for(int l=0;l<N;l++) { uk[l] = u[l];
for(int j=0;j<=k-1;j++) uk[l] += b[k][j]*f[j][l];
}
fsystem(h,tk,uk,f[k]);
}
for(int l=0;l<N;l++)
 for(int k=0;k<6;k++) u[l] += c[k]*f[k][l];
}
}

public void paint(Graphics g)
{
g.drawRect(40,40,500,450);
int steps = 1; int N = 2; int counter = 0;
double h = 0.005;  double Omega = 0.8;  double t = 0.0;
double u[] = { 0.8, 0.6 };  // initial conditions
for(int i=0;i<800000;i++)
{
t += h; map(u,steps,h,t,N);
if((counter>10) && (Math.abs(t-2.0*Math.PI*counter/Omega)<0.02))
{
int m = ((int)(10.0*u[0])+350.0);  int n = ((int)(40.0*u[1])+200.0);
g.drawLine(m,n,m,n);
}
if(t > 2.0*Math.PI*counter/Omega) { counter++; }
}
}

public static void main(String[] args)
{ Frame f = new PendulumPoincare(); f.setVisible(true); }
```

```
}
```

6.4 Parametrically Driven Pendulum

6.4.1 Phase Portrait

We consider the damped parametrically driven pendulum

$$\frac{d^2u}{dt^2} + a\frac{du}{dt} + (1 + k\cos(\Omega t))\sin(u) = 0.$$

This equation can be written as a system of first order differential equations

$$\frac{du_1}{dt} = u_2, \qquad \frac{du_2}{dt} = -au_2 - (1 + k\cos(\Omega t))\sin(u_1).$$

We evaluate the phase portrait $(u_1(t), u_2(t))$. We take into account that

$$-\pi \le u_1 \le \pi.$$

```
// ParametricPhase.java

import java.awt.*;
import java.awt.event.*;
import java.awt.Graphics;

public class ParametricPhase extends Frame
{
 public ParametricPhase()
 {
 setSize(600,500);
 addWindowListener(new WindowAdapter()
 { public void windowClosing(WindowEvent event) { System.exit(0); }}); }

 public void paint(Graphics g)
 {
 g.drawRect(50,30,500,400);
 double a = 0.15, k = 0.94, Omega = 1.56;
 double u1, u2, u3, u11, u22, u33;
 double tau = 0.005; double count = 0.0;
 u11 = 0.1; u22 = 0.2; u33 = 0.0; // initial conditions
 while(count < 1200.0)
 {
 u1 = u11; u2 = u22; u3 = u33;
 double v2 = -a*u2-(1.0+k*Math.cos(u3))*Math.sin(u1);
 u11 = u1+tau*u2+tau*tau*v2/2.0;
 u22 = u2+tau*v2+tau*tau*(-u2*(1.0-k*Math.cos(u3))*Math.cos(u1)
        -a*v2+Omega*k*Math.sin(u3)*Math.sin(u1))/2.0;
```

```
 u33 = Omega*tau+u3;

 if(count > 1000.0)  // transients have decayed
 {
 if(u11 > Math.PI)  { u11 = u11-2.0*Math.PI; }
 if(u11 < -Math.PI) { u11 = u11+2.0*Math.PI; }
 int m = (int)(80.0*u11+300.0); int n = (int)(60.0*u22+200.0);
 g.drawLine(m,n,m,n);
 }
 count += tau;
 } // end while
 } // end paint

 public static void main(String[] args)
 { Frame f = new ParametricPhase(); f.setVisible(true); }
}
```

6.4.2 Poincaré Section

We consider the parametrically driven pendulum

$$\frac{d^2u}{dt^2} + a\frac{du}{dt} + (1 + k\cos(\Omega t))\sin(u) = 0.$$

This equation can be written as a system of first order differential equations

$$\frac{du_1}{dt} = u_2, \qquad \frac{du_2}{dt} = -au_2 - (1 + k\cos(\Omega t))\sin(u_1).$$

This system is invariant under $t \mapsto t + 2\pi n/\Omega$, $n \in \mathbb{Z}$. Thus we can study the Poincaré map of the dynamical system. We take into account that $-\pi \leq u \leq \pi$. In the Java program `ParametricPoincare.java` we use the Lie series technique to integrate the differential equations.

```
// ParametricPoincare.java

import java.awt.*;
import java.awt.event.*;
import java.awt.Graphics;

public class ParametricPoincare extends Frame
{
 public ParametricPoincare()
 {
 setSize(600,500);
 addWindowListener(new WindowAdapter()
 { public void windowClosing(WindowEvent event) { System.exit(0); }}); }

 public void paint(Graphics g)
```

```
{
g.drawRect(50,30,500,400);
double a = 0.15, k = 0.94, Omega = 1.56;
double u1, u2, u3, u11, u22, u33;
double tau = 0.005; double count = 0.0;
u11 = 0.1; u22 = 0.2; u33 = 0.0; // initial conditions
int z = 0;
while(count < 10000.0)
{
u1 = u11; u2 = u22; u3 = u33;
double v2 = -a*u2-(1.0+k*Math.cos(u3))*Math.sin(u1);
u11 = u1+tau*u2+tau*tau*v2/2.0;
u22 = u2+tau*v2+tau*tau*(-u2*(1.0-k*Math.cos(u3))*Math.cos(u1)
      -a*v2+Omega*k*Math.sin(u3)*Math.sin(u1))/2.0;
u33 = Omega*tau+u3;
if(count > 1000.0)  // transients have decayed
{
if(u11 > Math.PI)  { u11 = u11-2.0*Math.PI; }
if(u11 < -Math.PI) { u11 = u11+2.0*Math.PI; }
if(Math.abs(count-2.0*Math.PI*z/Omega) < 0.02)
{
int m = (int)(80.0*u11+300.0); int n = (int)(60.0*u22+200.0);
g.drawLine(m,n,m,n);
}
}
count += tau;
if(count > 2.0*Math.PI*z/Omega) { z++; }
}  // end while
}  // end paint

public static void main(String[] args)
{ Frame f = new ParametricPoincare(); f.setVisible(true); }
}
```

6.5 Driven Van der Pol Equation

6.5.1 Phase Portrait

Consider the *driven van der Pol equation* (Parlitz and Lauterborn [159])

$$\frac{d^2u}{dt^2} + a(u^2-1)\frac{du}{dt} + u = k\cos(\Omega t)$$

or equivalently written as first order system

$$\frac{du_1}{dt} = u_2, \quad \frac{du_2}{dt} = -a(u_1^2-1)u_2 - u_1 + k\cos u_3, \quad \frac{du_3}{dt} = \Omega$$

where $u_3(t=0)=0$. We evaluate the phase portrait $(u_1(t), u_2(t))$ for the parameter values $a = 5.0$, $k = 5.0$, $\Omega = 2.466$. For these parameter values it is conjectured that the system shows chaotic behaviour. We apply the Runge-Kutta technique to integrate the differential equations.

```
// VdpPhase.java

import java.awt.*;
import java.awt.event.*;
import java.awt.Graphics;

public class VdpPhase extends Frame
{
 public VdpPhase()
 {
 setSize(600,500);
 addWindowListener(new WindowAdapter()
 { public void windowClosing(WindowEvent event) { System.exit(0); }}); }

 public void fsystem(double h,double t,double u[],double hf[])
 {
 double a = 5.0; double k = 5.0; double Omega = 2.466;
 hf[0] = h*u[1];
 hf[1] = h*(a*(1.0-u[0]*u[0])*u[1]-u[0]+k*Math.cos(Omega*t));
 }

 public void map(double u[],int steps,double h,double t,int N)
 {
 double uk[] = new double[N];
 double tk;
 double a[] = { 0.0, 1.0/4.0, 3.0/8.0, 12.0/13.0, 1.0, 1.0/2.0 };
 double c[] = { 16.0/135.0, 0.0, 6656.0/12825.0, 28561.0/56430.0,
                -9.0/50.0, 2.0/55.0 };
 double b[][] = new double[6][5];
 b[0][0] = b[0][1]= b[0][2] = b[0][3] = b[0][4] = 0.0;
 b[1][0] = 1.0/4.0; b[1][1] = 0.0; b[1][2] = 0.0; b[1][3] = 0.0;
 b[1][4] = 0.0;
 b[2][0] = 3.0/32.0; b[2][1] = 9.0/32.0;
 b[2][2] = 0.0; b[2][3] = 0.0; b[2][4] = 0.0;
 b[3][0] = 1932.0/2197.0; b[3][1] = -7200.0/2197.0;
 b[3][2] = 7296.0/2197.0; b[3][3] = b[3][4] = 0.0;
 b[4][0] = 439.0/216.0; b[4][1] = -8.0;
 b[4][2] = 3680.0/513.0; b[4][3] = -845.0/4104.0; b[4][4] = 0.0;
 b[5][0] = -8.0/27.0; b[5][1] = 2.0;
 b[5][2] = -3544.0/2565.0; b[5][3] = 1859.0/4104.0;
 b[5][4] = -11.0/40.0;
 double f[][] = new double[6][N];
 int i, j, l, k;
```

```
for(i=0;i<steps;i++) { fsystem(h,t,u,f[0]);
for(k=1;k<=5;k++) { tk = t+a[k]*h;
for(l=0;l<N;l++) { uk[l] = u[l];
for(j=0;j<=k-1;j++) uk[l] += b[k][j]*f[j][l];
}
fsystem(h,tk,uk,f[k]);
}
for(l=0;l<N;l++)
 for(k=0;k<6;k++) u[l] += c[k]*f[k][l];
}
}

public void paint(Graphics g)
{
g.drawRect(20,40,550,400);
int steps = 1;  int N = 2;
double h = 0.005;  double t = 0.0;
double u[] = { 0.8, 0.6 };  // initial conditions
// wait for transients to decay
for(int i=0;i<1000;i++) { t += h; map(u,steps,h,t,N); }
t = 0.0;
for(int i=0;i<100000;i++)
{ t += h;
map(u,steps,h,t,N);
int m = (int)(100.0*u[0]+350.0); int n = (int)(15.0*u[1]+220.0);
g.drawLine(m,n,m,n);
}
}

public static void main(String[] args)
{ Frame f = new VdpPhase(); f.setVisible(true); }
}
```

6.5.2 Liapunov Exponent

We consider the driven van der Pol equation described in section 6.5.1. To find the largest one-dimensional Liapunov exponent we calculate the variational system. The variational system of the driven van der Pol equation is given by

$$\frac{dv_1}{dt} = v_2, \quad \frac{dv_2}{dt} = -(2au_1u_2 + 1)v_1 - a(u_1^2 - 1)v_2 - k\sin(u_3)v_3, \quad \frac{dv_3}{dt} = 0\,.$$

Without loss of generality we can set $v_3 = 0$. Then we arrive at

$$\frac{dv_1}{dt} = v_2, \qquad \frac{dv_2}{dt} = -(2au_1u_2 + 1)v_1 - a(u_1^2 - 1)v_2\,.$$

The one-dimensional Liapunov exponent λ follows from

$$\lambda := \lim_{T\to\infty} \frac{1}{T} \ln(|v_1(T)| + |v_2(T)|).$$

In the C++ program `vdpliapunov.cpp` we evaluate the one-dimensional Liapunov exponent for the parameter values $a = 5.0$, $k = 5.0$, $\Omega = 2.466$.

```
// vdpliapunov.cpp

#include <iostream>
#include <cmath>
#include "symbolicc++.h"
using namespace std;

const int N = 3;
Symbolic u("u",N), ut("ut",N), y("y",N-1), yt("yt",N-1);

// The vector field V
template <class T> T V(const T& ss)
{
 T sum(0);
 for(int i=0;i<N;i++) sum += ut(i)*df(ss,u(i));
 return sum;
}

template <class T> T W(const T& ss)
{
 T sum(0);
 for(int i=0;i<(N-1);i++) sum += yt(i)*df(ss,y(i));
 return sum;
}

int main(void)
{
 int i, j;
 Symbolic t("t"), a("a"), k("k"), om("om"), y("y",2),
          us("us",N), ys("ys",N-1);
 Equations v;
 // driven van der Pol
 ut(0) = u(1);
 ut(1) = a*(1.0-u(0)*u(0))*u(1)-u(0)+k*cos(u(2));
 ut(2) = om;
 // variational equations
 yt(0) = y(1);
 yt(1) = -(2.0*u(0)*u(1)+1.0)*y(0)-a*(u(0)*u(0)-1.0)*y(1);
 // Taylor series expansion up to order 2
 for(i=0;i<N;i++) us(i) = u(i)+t*V(u(i))+t*t*V(V(u(i)))/2;
 for(i=0;i<(N-1);i++) ys(i) = y(i)+t*W(y(i))+t*t*W(W(y(i)))/2;
 // evolution of the approximate solution
 v = (t==0.01,a==5.0,k==5.0,om==2.466,
      u(0)==1.5504,u(1)==-0.71214,u(2)==0.0,y(0)==0.5,y(1)==0.5);
 int iter = 10000;
```

```
 for(j=0;j<iter;j++)
 {
 Equations newv = (t==0.01,a==5.0,k==5.0,om==2.466);
 for(i=0;i<N;i++) newv = (newv,u(i)==us(i)[v]);
  for(i=0;i<N-1;i++) newv = (newv,y(i)==ys(i)[v]);
  v = newv;
 } // end for loop j

 double T = 0.01*iter;
 double lambda = log(fabs(double(rhs(v,y(0))))
                     +fabs(double(rhs(v,y(1)))))/T;
 cout << "lambda = " << lambda << endl;
 return 0;
}
```

6.6 Parametrically and Externally Driven Pendulum

The parametrically and externally driven pendulum is given by

$$\frac{d^2u}{dt^2} + a\frac{du}{dt} + (1 + k_2\cos(\Omega_2 t))\sin(u) = k_1\cos(\Omega_1 t).$$

We set

$$u_1(t) := u(t), \qquad u_2(t) := \frac{du_1}{dt}, \qquad u_3(t) := \Omega_1 t, \qquad u_4(t) := \Omega_2 t\,.$$

Thus the second order differential equation can be written as the autonomous system of first order ordinary differential equations

$$\frac{du_1}{dt} = u_2, \qquad \frac{du_2}{dt} = -au_2 - (1 + k_2\cos(u_4))\sin(u_1) + k_1\cos(u_3)$$

$$\frac{du_3}{dt} = \Omega_1, \qquad \frac{du_4}{dt} = \Omega_2$$

where $u_3(t=0) = 0$ and $u_4(t=0) = 0$. In the Java program we evaluate the phase portrait $(u_1(t), u_2(t))$.

```
// ParametricExtPhase.java

import java.awt.*;
import java.awt.event.*;
import java.awt.Graphics;

public class ParametricExtPhase extends Frame
{
 public ParametricExtPhase()
```

```
{
setSize(600,500);
addWindowListener(new WindowAdapter()
{ public void windowClosing(WindowEvent event) { System.exit(0); }}); }

public void fsystem(double h,double t,double u[],double hf[])
{
double a = 0.1; double k1 = 0.3; double k2 = 0.3;
double Omega1 = 0.618034; double Omega2 = 1.618034;
hf[0] = h*u[1];
hf[1] = h*(-a*u[1]-(1.0+k2*Math.cos(Omega2*t))*Math.sin(u[0])
       +k1*Math.cos(Omega1*t));
}

public void map(double u[],int steps,double h,double t,int N)
{
double uk[] = new double[N];
double tk;
double a[] = { 0.0, 1.0/4.0, 3.0/8.0, 12.0/13.0, 1.0, 1.0/2.0 };
double c[] = { 16.0/135.0, 0.0, 6656.0/12825.0, 28561.0/56430.0,
               -9.0/50.0, 2.0/55.0 };
double b[][] = new double[6][5];
b[0][0] = b[0][1]= b[0][2] = b[0][3] = b[0][4] = 0.0;
b[1][0] = 1.0/4.0; b[1][1] = 0.0; b[1][2] = 0.0; b[1][3] = 0.0;
b[1][4] = 0.0;
b[2][0] = 3.0/32.0; b[2][1] = 9.0/32.0;
b[2][2] = 0.0; b[2][3] = 0.0; b[2][4] = 0.0;
b[3][0] = 1932.0/2197.0; b[3][1] = -7200.0/2197.0;
b[3][2] = 7296.0/2197.0; b[3][3] = b[3][4] = 0.0;
b[4][0] = 439.0/216.0; b[4][1] = -8.0;
b[4][2] = 3680.0/513.0; b[4][3] = -845.0/4104.0; b[4][4] = 0.0;
b[5][0] = -8.0/27.0; b[5][1] = 2.0;
b[5][2] = -3544.0/2565.0; b[5][3] = 1859.0/4104.0;
b[5][4] = -11.0/40.0;
double f[][] = new double[6][N];
int i, j, l, k;
for(i=0;i<steps;i++) { fsystem(h,t,u,f[0]);
for(k=1;k<=5;k++) { tk = t+a[k]*h;
for(l=0;l<N;l++) { uk[l] = u[l];
for(j=0;j<=k-1;j++) uk[l] += b[k][j]*f[j][l];
}
fsystem(h,tk,uk,f[k]);
}
for(l=0;l<N;l++)
 for(k=0;k<6;k++) u[l] += c[k]*f[k][l];
}
}
```

```
 public void paint(Graphics g)
 {
 g.drawLine(20,200,620,200); g.drawRect(20,40,600,360);
 int steps = 1;
 int N = 2;
 double h = 0.005;
 double t = 0.0;
 double u[] = { 1.0, 0.0 };  // initial conditions
 // wait for transients to decay
 for(int i=0;i<2000;i++) { t += h; map(u,steps,h,t,N); }
 t = 0.0;
 for(int i=0;i<50000;i++)
 { t += h;
 map(u,steps,h,t,N);
 int mx = (int)(150.0*u[0]+250.0); int my = (int)(150.0*u[1]+200.0);
 t += h;
 map(u,steps,h,t,N);
 int nx = (int)(150.0*u[0]+250.0); int ny = (int)(150.0*u[1]+200.0);
 g.drawLine(mx,my,nx,ny);
 }
 }

 public static void main(String[] args)
 { Frame f = new ParametricExtPhase(); f.setVisible(true); }
}
```

6.7 Torsion Numbers

For a driven oscillator such as the driven Duffing equation

$$\frac{d^2u}{dt^2} + a\frac{du}{dt} + u^3 = k\cos(\omega t)$$

one finds that for a given periodic orbit nearby starting trajectories are more or less twisted around it. The twisting behaviour determines which bifurcations are possible (Parlitz and Lauterborn [158]). A torsion by 2π, for example, excludes periodic doubling, because the eigenvalues of the linearization of the Poincaré map are positive. The torsion number of a periodic orbit is a quantity that counts the (average) number of twists of nearby trajectories around the closed orbit. We have the following definition for torsion numbers. Consider the second order differential equation

$$\frac{d^2u}{dt^2} + g(u, du/dt) = h(t)$$

where $h(t+T) = h(t)$ and T is the period. Using

$$u_1 = u, \quad u_2 = du/dt, \quad u_3 = t/T$$

we can cast it into a first order system of differential equations

$$\frac{du_1}{dt} = u_2, \qquad \frac{du_2}{dt} = -g(u_1, u_2) + h(Tu_3), \qquad \frac{du_3}{dt} = 1/T = \omega/(2\pi).$$

This system generates a flow $\{\Phi^t\}$ on the phase space $\mathbb{R}^2 \times S^1$. The *Poincaré map* $P = \Phi^T|_\Sigma$ is defined on the cross section

$$\Sigma := \{(u_1, u_2, u_3) \in \mathbb{R}^2 \times S^1 \,:\, u_3 = c = \text{const}\} \cong \mathbb{R}^2$$

Let γ be a period-m orbit through

$$\mathbf{u}^* = (u_{1f}, u_{2f}, u_{3f} = c) \in \mathbb{R}^2 \times S^1$$

described by the solution $\mathbf{u}(t)$ of the system. Let $\mathbf{u}_f = (u_{1f}, u_{2f}) \in \mathbb{R}^2$ be a fixed point of $P^{(m)}$ associated with the period-m orbit of γ we are considering. The variational equation of the system is given by

$$\frac{d\mathbf{v}}{dt} = A(\mathbf{u})\mathbf{v}$$

where the 3×3 matrix A is given by

$$\begin{pmatrix} 0 & 1 & 0 \\ -\partial g/\partial u_1 & -\partial g/\partial u_2 & T\partial h/\partial u_3 \\ 0 & 0 & 0 \end{pmatrix}.$$

Since the third component has a trivial solution $v_3 = \text{const}$ we can reduce the variational system to a two-dimensional system of differential equations. If $\mathbf{u}_0$ is restricted to

$$\Sigma := \{\mathbf{u} \in \mathbb{R}^2 \times S^1 \,:\, u_3 = 0\}$$

(i.e. $\mathbf{v}_0 = (v_{10}, v_{20}, v_{30})$) this reduced variational equation is given by

$$\frac{dv_1}{dt} = v_2, \qquad \frac{dv_2}{dt} = -\frac{\partial g}{\partial u}v_1 - \frac{\partial g}{\partial \dot{u}}v_2.$$

Any solution $\mathbf{v}(t) = (v_1(t), v_2(t))$ of this reduced variational equation describes a spiral-like curve in the phase space $(v, \dot{v})$. The number of torsions can be counted by

1) considering $\mathbf{v}(t) = (v_1(t), v_2(t))$ in polar coordinates $(r(t), \phi(t))$

$$\cos\phi(t) = \frac{v_2(t)}{r(t)}, \quad \sin\phi(t) = \frac{v_1(t)}{r(t)}, \quad r(t) = \sqrt{v_1^2(t) + v_2^2(t)}$$

2) computing the angular velocity

$$\frac{d\phi}{dt} = \frac{v_1 dv_2/dt - v_2 dv_1/dt}{v_1^2 + v_2^2}$$

3) calculating the average angular velocity

$$\Omega(\gamma) := \lim_{t\to\infty} \frac{1}{t} \left| \int_0^t \frac{d\phi(t')}{dt'} dt' \right|$$

4) counting the number of torsions during one perdiod $T_m = mT = 2\pi m/\omega$ of the oscillation

$$n := \frac{T_m}{2\pi}\Omega(\gamma) = \frac{\Omega(\gamma)}{\omega_m} = m\frac{\Omega(\gamma)}{\omega}$$

where

$$\Omega(\gamma) := \lim_{t\to\infty} \frac{1}{t} \left| \int_0^t \frac{v_1 du_2/dt - v_2 du_1/dt}{v_1^2 + v_2^2} dt \right| .$$

$\Omega(\gamma)$ is independent of the initial condition $\mathbf{v}(0)$ for the solution $\mathbf{v}(t)$ of the variational equation and may be viewed as the orbit dependent eigenfrequency of the oscillator. If $\lambda_{1,2} = re^{i\varphi}$, $r > 0$ are complex eigenvalues of the derivative $DP^{(m)}$ of the iterated Poincare map $P^{(m)}$ (or the Floquet multiplier) then the phase factor φ is given by $\varphi = T_m\Omega(\gamma) \bmod 2\pi$. The torsion number is measued in 2π.

Example. Consider the driven linear harmonic oscillator

$$\frac{d^2u}{dt^2} + a\frac{du}{dt} + \omega_0^2 u = k\cos(\omega t)$$

where $a > 0$. For $t \to \infty$ the solutions converge to the periodic solution

$$u(t) = \frac{k}{\sqrt{(\omega_0^2 - \omega^2)^2 + a\omega^2}} \cos(\omega t - q)$$

where

$$q := \operatorname{arccot}((\omega_0^2 - \omega^2)/(a\omega)) .$$

Thus the Poincaré map of this system has a single stable fixed point $\mathbf{u}_f$. The time evolution of a perturbation (initial value in the Poincaré cross section located at

$$u_3 = 0 = t : \mathbf{u}_p = \mathbf{u}_f + (v_1, \dot{v}_1)$$

describing the behaviour of a trajectory in the neighbourhood of the closed orbit is given by the variational equation

$$\frac{d^2v}{dt^2} + a\frac{dv}{dt} + \omega_0^2 v = 0 .$$

For $a < 2\omega_0$ the solution of this linear differential equation with constant coefficients is given by

$$v(t) = e^{-at/2}\left(v_0\cos(\widetilde{\omega}t) + \frac{\dot{v}_0 + av_0/2}{\widetilde{\omega}}\sin(\widetilde{\omega}t)\right)$$

where

$$\widetilde{w} := \sqrt{\omega_0^2 - a^2/4} .$$

The frequency $\widetilde{\omega}$ of the oscillations of the perturbation is independent of the excitation frequency ω and the excitation amplitude k. The rotation angle of the neighbourhood trajectory around the closed orbit depends only on the frequency relation $\widetilde{\omega}/\omega$. For $a < 2\omega_0$ the torsion number is given by $n = \widetilde{\omega}/\omega$. For $a \geq 2\omega_0$ the torsion number is always 0. ♣

For a nonlinear driven oscillator the torsion number must be found numerically. Parlitz and Lauterborn [158] studied the driven Toda oscillator

$$\frac{d^2u}{dt^2} + a\frac{du}{dt} + e^u - 1 = a\cos(\omega t)$$

with the corresponding variational equation

$$\frac{d^2v}{dt^2} + a\frac{dv}{dt} + e^u v = 0\,.$$

The *torsion number* can also be introduced (Uezu [208]) for an autonomous first order system

$$\frac{d\mathbf{u}}{dt} = \mathbf{f}(\mathbf{u})$$

where $\mathbf{u} = (u_1, u_2, u_3)$. A periodic solution $\mathbf{u}_p(t)$ of this autonomous system is characterized topologically by its knot type, torsion numbers n_i and relative torsion number r_i. A solution $\mathbf{v}(t)$ of the linearized differential equation (variational equation) around $\mathbf{u}_p(t)$ is given by

$$\mathbf{v}(t) = \exp\left(\int_0^t \frac{\partial \mathbf{f}(\mathbf{u}_p(s))}{\partial \mathbf{u}} ds\right)\mathbf{v}_0 \equiv S(t)\mathbf{v}_0$$

where $\exp(\ldots)$ is an ordered exponential. Let λ_i be a real eigenvalue of $S(T)$ (T is the period of $\mathbf{u}_p(t)$) and $\mathbf{e}_i$ be the eigenvectors belonging to λ_i with $|\lambda_1| \geq |\lambda_2|$ and $\lambda_0 \equiv 1$. Then

$$\mathbf{v}_i(t) = S(t)\mathbf{e}_i$$

with $i = 1, 2$. The *link number* between two loops C_1 and C_2 in $\mathbb{R}^3$ is given by the *Gauss integral*

$$L(C_1, C_2) := -\frac{1}{4\pi}\int_{C_1}\int_{C_2} \frac{(\mathbf{r}_2 - \mathbf{r}_1, d\mathbf{r}_1 \times d\mathbf{r}_2)}{\|\mathbf{r}_1 - \mathbf{r}_2\|^3}$$

where $\|\ \|$ denotes the Euclidean norm, $(\ ,\ )$ the inner product (scalar product in $\mathbb{R}^3$) and $\times$ denotes the vector product in $\mathbb{R}^3$. Using this notation, n_i and r_i are defined as follows

$$2n_i := L([\mathbf{u}_p(t); 0 \leq t < T], \mathbf{u}_p(t) + \chi\mathbf{v}_i(t)/\|\mathbf{v}_i(t)\|; 0 \leq t < 2T])$$

and

$$r_i := \frac{1}{2\pi}\int_0^T \frac{\alpha_{i2}(t)d\alpha_{i3}(t)/dt - (d\alpha_{i2}(t)/dt)\alpha_{i3}(t)}{\alpha_{i2}^2(t) + \alpha_{i3}^2(t)}dt$$

where we define

$$\alpha_{ij}(t) := (\mathbf{v}_i(t), \mathbf{g}_j(t))$$

with

$$\mathbf{g}_1 := \frac{\dot{\mathbf{u}}_p(t)}{\|\dot{\mathbf{u}}_p(t)\|}, \qquad \mathbf{g}_2 := \frac{\dot{\mathbf{u}}_p(t) \times \ddot{\mathbf{u}}_p(t)}{\|\dot{\mathbf{u}}_p(t) \times \ddot{\mathbf{u}}_p(t)\|}, \qquad \mathbf{g}_3 := \mathbf{g}_1 \times \mathbf{g}_2 .$$

Thus n_i describes the torsion of the tangent space around the periodic orbit $\mathbf{u}_p(t)$ and r_i is the rotation number around the direction of the velocity vector $d\mathbf{u}_p(t)/dt$ in the moving coordinate system.

Exercise. (1) Find the torsion number (numerically) for the *Rikitake two-disc dynamo*

$$\frac{du_1}{dt} = -\mu u_1 + u_3 u_2$$

$$\frac{du_2}{dt} = -\mu u_2 + (u_3 - a)u_1$$

$$\frac{du_3}{dt} = 1 - u_1 u_2 .$$

Chapter 7

Controlling of Chaos

7.1 Introduction

In this chapter we discuss different strategies for controlling classical chaos (Ott et al [156], Kapitaniak [112], Schöll and Schuster [183], Zhang et al [227]). Controlling chaos plays a central role in engineering in particular in electronics and mechanical systems. Roughly speaking, there are two kinds of ways to control chaos: feedback control and non-feedback control.

The frame of a chaotic attractor is given by infinitely many unstable periodic orbits. The task is to use the unstable periodic orbits to control chaos. We describe the Ott-Grebogi-Yorke method which uses this fact. By applying small, judiciously chosen temporal perturbations to an accessible control parameter of the dissipative system, they demonstrated that an original chaotic trajectory can be converted to a desired fixed point or periodic orbit. The Ott-Grebogi-Yorke method [157] belongs to feedback control. We apply this method to one and two-dimensional maps. The maps under consideration are the logistic map and the Hénon map. Controlling can also be achieved by coupling periodic modulations to the control parameter. We study the Lorenz model with this technique in section 7.3. This is a non-feedback control. Resonant perturbation and control is considered in section 7.4 and applied to a driven anharmonic oscillator.

7.2 Ott-Yorke-Grebogi Method

7.2.1 One-Dimensional Maps

The basic idea of the Ott-Yorke-Grebogi method [157] for stabilizing unstable periodic orbits embedded in a chaotic attractor can be understood by considering a simple model system. We consider the *logistic map*

$$x_{t+1} = f(x_t, r) = rx_t(1 - x_t) \qquad r \in [3, 4]$$

where $x_0 \in [0,1]$ and thus $x_t \in [0,1]$. The logistic map develops chaos via the period-doubling bifurcation route. The period-doubling cascade accumulates at $r = r_\infty \approx 3.57...$, after which chaos can arise. Consider the case $r = 3.8$. The system is apparently chaotic for this value of r and the chaotic attractor is contained in the interval $[0,1]$. The chaotic attractor contains an infinite number of unstable periodic orbits embedded within it and they are dense in it. For example, a fixed point ($r = 3.8$)

$$x^* = 1 - \frac{1}{r} = 0.7368...$$

and a period-2 orbit,

$$x(1) \approx 0.3737..., \qquad x(2) \approx 0.8894...$$

where

$$x(1) = f[x(2)], \qquad x(2) = f[x(1)] .$$

We find $x(1)$ and $x(2)$ by solving the quartic equation (second iterate)

$$x^* = r^2 x^*(1 - x^*)(1 - rx^*(1 - x^*)) .$$

Besides the solutions $x^* = 0$, $x^* = 1 - 1/r$ we obtain $x(1)$ and $x(2)$ as fixed points of the second iterate of the map.

Suppose we want to avoid chaos at $r = 3.8$. In particular, we want trajectories resulting from a randomly chosen initial condition x_0 to be as close as possible to the period-2 orbit assuming that this period-2 orbit gives the best system performance. We can choose the desired asymptotic state of the map to be any of the infinite number of unstable periodic orbits, if that periodic orbits gives the best system performance. To achieve this goal, we suppose that the parameter r can be finely tuned in a very small range around the value $r_0 = 3.8$, namely, we allow r to vary in the range $[r_0 - \delta, r_0 + \delta]$, where $\delta \ll 1$. Due to the ergodicity of the chaotic attractor, the trajectory that begins from an arbitrary value of x_0 will fall, with probability one, into the neighbourhood of the desired period-2 orbit at some later time. The trajectory would diverge quickly from the period-2 orbit if we do not intervene. We program the parameter perturbations in such a way that the trajectory stays in the neighbourhood of the period-2 orbit for as long as the control is present. The small parameter perturbations will be time-dependent in general. The logistic map in the neighbourhood of a periodic orbit can be approximated by a linear equation expanded around the periodic orbit. Let the target period-m orbit to be controlled be $x(i)$, $i = 1, \ldots, m$, where $x(i+1) = f[x(i)]$ and $x(m+1) \equiv x(1)$. Assume that at time t, the trajectory falls into the neighbourhood of the ith component of the period-m orbit. The linearized dynamics in the neighbourhood of the $(i+1)$th component is then

$$x_{t+1} - x(i+1) = \left.\frac{\partial f(x,r)}{\partial x}\right|_{x=x(i),r=r_0} (x_t - x(i)) + \left.\frac{\partial f(x,r)}{\partial r}\right|_{x=x(i),r=r_0} \Delta r_t$$

where the partial derivatives are evaluated at $x = x(i)$ and $r = r_0$. Thus we obtain

$$x_{t+1} - x(i+1) = r_0(1-2x(i))(x_t - x(i)) + x(i)(1-x(i))\Delta r_t \,.$$

We require x_{t+1} to stay in the neighbourhood of $x(i+1)$. Hence, we set

$$|x_{t+1} - x(i+1)| = 0 \,.$$

From this condition it follows that

$$\Delta r_t = r_0 \frac{(2x(i)-1)(x_t - x(i))}{x(i)(1-x(i))} \,.$$

This equation holds only when the trajectory x_t enters a small neighbourhood of the period-m orbit, i.e., when $|x_t - x(i)| \to 0$. Hence, the required parameter perturbation Δr_t is small. Let the length of a small interval defining the neighbourhood around each component of the period-m orbit be 2ϵ. In general, the required maximum parameter perturbation δ is proportional to ϵ. Since ϵ can be chosen to be arbitrarily small, δ also can be made arbitrarily small. However, the average transient time before a trajectory enters the neighbourhood of the target periodic orbit depends on ϵ (or δ). A larger δ will mean a shorter average time for the trajectory to be controlled. Of course, if δ is too large, nonlinear effects become important and the linear control strategy might not work. When the trajectory is outside the neighbourhood of the target periodic orbit, we do not apply any parameter perturbation and the system evolves at its nominal parameter value r_0. We usually set $\Delta r_t = 0$ when $\Delta r_t > \delta$. The parameter perturbations Δr_t depend on x_t and are therefore time-dependent.

In the following C++ program `control1.cpp` we control the fixed point $x^* = 1 - 1/r$, where $r = 3.8$ for the logistic equation.

```
// control1.cpp

#include <iostream>
#include <cmath>      // for fabs
using namespace std;

int main(void)
{
 double x = 0.28;          // initial value
 double r0 = 3.8;          // control parameter
 double T = 650;           // number of iterations
 double eps = 0.001;       // neighbourhood
 double xf = 1.0-1.0/r0; // fixed point
 double x1;
 int count = 0;
 double r = r0;
 for(int t=0;t<T;t++)
```

```
 {
 x1 = x; x = r*x1*(1.0-x1);
 count++;
 if(fabs(x-xf) < eps)
 {
 cout << "in eps neighbourhood" << endl;
 cout << "count = " << count << endl;
 double delta = r0*(2.0*xf-1.0)*(x-xf)/(xf*(1.0-xf));
 cout << "delta = " << delta << endl;
 r = r0+delta;  // change to control parameter for controlling
 cout << "x = " << x << endl;
 }  // end if
 }  // end for loop
 return 0;
}
```

In the following C++ program `control2.cpp` we control a period two-orbit for the logistic map with $r = 3.8$.

```
// control2.cpp

#include <iostream>
#include <cmath>      // for sqrt, fabs
using namespace std;

int main(void)
{
 double x = 0.28;       // initial value
 double r0 = 3.8;       // control parameter
 double T = 353;        // number of iterations
 double eps = 0.001;    // neighbourhood
 int count = 0;
 double x1;
 // periodic points
 double ab = (1.0+1.0/r0);
 double xp1 = ab/2.0+sqrt(-ab/r0+ab*ab/4.0);
 double xp2 = ab/2.0-sqrt(-ab/r0+ab*ab/4.0);
 cout << "xp1 = " << xp1 << endl; // periodic point 0.88942
 cout << "xp2 = " << xp2 << endl; // periodic point 0.373738

 double r = r0;
 for(int t=0;t<T;t++)
 {
 x1 = x; x = r*x1*(1.0-x1);
 count++;
 if((fabs(x-xp1) < eps) || (fabs(x-xp2) < eps))
 {
 cout << "in eps neighbourhood of xp1 or xp2" << endl;
 cout << "count = " << count << endl;
```

```
if(fabs(x-xp1) < eps)
{
double delta1;
delta1 = r0*(2.0*xp1-1.0)*(x-xp1)/(xp1*(1.0-xp1));
cout << "delta1 = "<< delta1 << endl;
r = r0+delta1;
cout << "x = " << x << endl;
}
else
{
double delta2;
delta2 = r0*(2.0*xp2-1.0)*(x-xp2)/(xp2*(1.0-xp2));
cout << "delta2 = " << delta2 << endl;
r = r0+delta2;
cout << "x = " << x << endl;
}
}
} // end for loop
return 0;
}
```

7.2.2 Systems of Difference Equations

Here we consider a system of difference equations (Ott et al [157])

$$\mathbf{x}_{t+1} = \mathbf{f}(\mathbf{x}_t, r)$$

where $\mathbf{x}_t \in \mathbb{R}^n$, $r \in \mathbb{R}$ and $\mathbf{f}$ is sufficiently smooth in both variables. Here, r is considered a real parameter which is available for external adjustment but is restricted to lie in some small interval $|r - r_0| < \delta$ around a nominal value r_0. We assume that the nominal system (i.e., for $r = r_0$) contains a chaotic attractor. We vary the control parameter r with time t in such a way that for almost all initial conditions in the basin of the chaotic attractor, the dynamics of the system converge onto a desired time periodic orbit contained in the attractor. The control strategy is the following. We find a stabilizing local feedback control law which is defined on a neighbourhood of the desired periodic orbit. This is done by considering the first order approximation of the system at the chosen unstable periodic orbit. Here we assume that this approximation is stabilizable. Since stabilizability is a generic property of linear systems, this assumption is quite reasonable. The ergodic nature of the chaotic dynamics ensures that the state trajectory eventually enters into the neighbourhood. Once inside, we apply the stabilizing feedback control law in order to steer the trajectory towards the desired orbit.

Thus the control of unstable periodic orbits $\mathbf{x}_{P,i} \in \mathbb{R}^n$ with $i = 1, \ldots, P$, where P denotes the period length, can be achieved by small, exactly determined variations of the system parameter during each iteration t given by the control $r_t \in \mathbb{R}$. Hence we have to consider the parametrised system

$$\mathbf{x}_{t+1} = \mathbf{f}(\mathbf{x}_t, r_t).$$

By changing r_t slightly the periodic points are also shifted slightly, i.e., $\mathbf{x}_{P,i}(r_t)$ for $i = 1, \ldots, P$.

We describe the method as applied to the stabilization of fixed points (i.e., period one orbits) of the map $\mathbf{f}$. The consideration of periodic orbits of period larger than one is straightforward. Let $\mathbf{x}^*(r)$ denote an unstable fixed point on the attractor. For values of r close to r_0 and in the neighbourhood of the fixed point $\mathbf{x}^*(r_0)$ the map can be approximated by the linear map

$$\mathbf{x}_{t+1} - \mathbf{x}^*(r_0) = A[\mathbf{x}_t - \mathbf{x}^*(r_0)] + B(r - r_0)$$

where A is an $n \times n$ Jacobian matrix and B is an n-dimensional column vector

$$A := D_{\mathbf{x}}\mathbf{f}(\mathbf{x}, r), \qquad B := D_r\mathbf{f}(\mathbf{x}, r).$$

The partial derivatives are evaluated at $\mathbf{x} = \mathbf{x}^*(r_0)$ and $r = r_0$. We now introduce the time-dependence of the parameter r by assuming that it is a linear function of the variable $\mathbf{x}_t$ of the form

$$r - r_0 = -K^T(\mathbf{x}_t - \mathbf{x}^*(r_0)).$$

The $1 \times n$ matrix K^T is to be determined so that the fixed point $\mathbf{x}^*(r_0)$ becomes stable. We obtain

$$\mathbf{x}_{t+1} - \mathbf{x}^*(r_0) = (A - BK^T)(\mathbf{x}_t - \mathbf{x}^*(r_0))$$

which shows that the fixed point will be stable provided the $n \times n$ matrix

$$A - BK^T$$

is asymptotically stable: that is, all its eigenvalues have modulus smaller than unity. The solution to the problem of the determination of K^T, such that the eigenvalues of the matrix $A - BK^T$ have specified values, is well known from control systems theory and is called the *pole placement technique*.

Example. Consider the *Hénon map* in the form,

$$x_{1t+1} = a + bx_{2t} - x_{1t}^2, \qquad x_{2t+1} = x_{1t}.$$

For $b = 0.3$ and $a = \bar{a} = 1.4$ there is an unstable saddle point contained in the chaotic attractor. This fixed point is given by

$$x_1^* = -c + \sqrt{c^2 + a}, \qquad x_2^* = x_1^*$$

where $c = (1-b)/2$ and $a \geq -c^2$. We obtain

$$A = D_{\mathbf{x}}\mathbf{f}(\mathbf{x}) = \begin{pmatrix} -2x_1 & b \\ 1 & 0 \end{pmatrix}, \qquad B = D_r\mathbf{f}(\mathbf{x}) = \begin{pmatrix} 1 \\ 0 \end{pmatrix}.$$

The eigenvalues and eigenvectors of A are given by

$$\lambda_s = -x_1 + \sqrt{b + x_1^2}, \quad \begin{pmatrix} \lambda_s \\ 1 \end{pmatrix}, \qquad \lambda_u = -x_1 - \sqrt{b + x_1^2}, \quad \begin{pmatrix} \lambda_u \\ 1 \end{pmatrix}.$$

The 2×2 matrix $A - BK^T$ is given by

$$A - BK^T = \begin{pmatrix} -2x_1^* - k_1 & b - k_2 \\ 1 & 0 \end{pmatrix}. \quad \clubsuit$$

```
// Controlhenon.cpp

#include <iostream>
#include <cmath>        // for sqrt
using namespace std;

int main(void)
{
 double T = 3380;                           // number of iterations
 double x = -0.760343, y = 1.40007;   // initial values
 double a = 1.4, b = 0.3;                   // parameter values
 double x1, y1; double a0 = a;
 // fixed point (fixed point is unstable)
 double c = (1.0-b)/2.0;
 double xf = -c+sqrt(c*c+a);
 double yf = xf;
 cout << "xf = " << xf << endl;  // 0.883896
 cout << "yf = " << yf << endl;  // 0.883896
 double k1 = -0.1, k2 = 1.3;
 int counter = 0;
 double eps = 0.005;  // eps neighbourhood
 for(int t=0;t<T;t++)
 { x1 = x;  y1 = y;
 x = a0-x1*x1+b*y1;  y = x1;
 counter++;
 if((fabs(x-xf) < eps) && (fabs(y-yf) < eps))
 {
 cout << "in eps neighbourhood" << endl;
 cout << "counter = " << counter << endl;
 a0 = a-k1*(x-xf)-k2*(y-yf);
 cout << "x = " << x << endl; cout << "y = " << y << endl;
 }
 }
 cout << "x = " << x << endl; cout << "y = " << y << endl;
 return 0;
}
```

In the following C++ program we stabilize a fixed point of the Henon model using the OGY-method.

```
// henonogy.cpp

#include <iostream>
#include <cmath>     // for sqrt
using namespace std;

double f(double p,double x,double y,double b)
{ return (p-x*x+b*y); }

double fp(double a,double b) // fixed point
{ return ((b-1.0)/2.0 + sqrt((1.0-b)*(1.0-b)/4.0+a)); }

double Es(double x,double b)  // stable eigenvalue
{ return (-x+sqrt(x*x+b)); }

double Eu(double x,double b)  // unstable eigenvalue
{ return (-x-sqrt(x*x+b)); }

int main(void)
{
 double a0 = 1.4, b = 0.3;          // parameters
 double xinit = 0.6, yinit = 0.7; // initial values
 const double delta = 0.01;         // maximum allowed perturbation
 double xi = xinit; double yi = yinit;
 int i = 0;  // indentation
 double x_ = xinit; int k = 0;
 double at;   // control parameter
 double xf = fp(a0,b);  double yf = xf; // fixed point
 double lambdas = Es(xf,b);  // eigenvalue stable
 double lambdau = Eu(xf,b);  // eigenvalue unstable
 double fu = sqrt(lambdau*lambdau+1.0)/(lambdau-lambdas);
 double fu1 = fu;
 double fu2 = -lambdas*fu;         // eigenvectors
 cout << "fixed point x,y: " << xf ;
 cout << "   ,  " << yf <<  endl;
 cout << "eigenvalues: " << lambdas;
 cout << "   ,  " << lambdau << endl;
 do
 {
 double deltap = -lambdau*(fu1*(xi-xf)+fu2*(yi-yf))/fu1;
 if(fabs(deltap) < delta)
 {
 at = a0+deltap;
 x_ = xi;  xi = f(at,xi,yi,b);   yi = x_;
 cout << "x" << i;   cout << " : " << xi;
```

```
 cout << "  y" << i; cout << " : " << yi;
 cout << "  p :" << at-a0 << endl;
 i++; k++;
 }
 else
 { i++; k = 0; x_ = xi; xi = f(a0,xi,yi,b); yi = x_; }
 }
 while(k<=15);
 return 0;
}
```

7.3 Time-Delayed Feedback Control

The time-delayed feedback control (or Pyragas method) ([170]) makes use of a control signal obtained from the difference between the current state and the state of the system delayed by one period of an unstable orbit, so that the signal vanishes when the target orbit is stabilized. The method does not require exact knowledge of the form of the periodic orbit or the system equations. For maps consider, as a simple example, the logistic map $f_r(x) = rx(1-x)$, where $r \in [3,4]$. Then we have

$$x_{t+1} = rx_t(1-x_t) + K(x^* - f_r(x_t))$$

where x^* is the desired fixed point. We recall that the fixed points of the logistic map f_r are given by $x^* = 0$ and $x^* = 1 - 1/r$. In the C++ program we consider the case $r = 4$ and the desired fixed point is $x^* = 3/4$.

```
// DFC.cpp

#include <iostream>
using namespace std;

int main(void)
{
 double x1, x0 = 0.1; double r = 4.0;
 double eps = 0.02;
 double K = 1.0 + 1.0/(2.0-r) + eps;
 double xf = 1.0 - 1.0/r;
 int T = 400;
 for(int t=0;t<T;t++)
 {
 x1 = r*x0*(1.0-x0) + K*(xf-r*x0*(1.0-x0));
 cout << "x1 = " << x1 << endl;
 x0 = x1;
 }
 return 0;
}
```

Pyragas [169] has proposed a method for controlling chaos based on continuous-time perturbation using feedback. Later Chen and Dong [30] provided a rigorous basis for this control scheme. Consider an $(n+m)$-dimensional nonlinear dynamical system of the form

$$\frac{d\mathbf{u}}{dt} = \mathbf{f}(\mathbf{u}, \mathbf{v}, t), \qquad \frac{d\mathbf{v}}{dt} = \mathbf{g}(\mathbf{u}, \mathbf{v}, t) + \mathbf{w}(\mathbf{u}, \mathbf{v}, t)$$

where $\mathbf{u} \in \mathbb{R}^n$ $(n > 0)$ and $\mathbf{v} \in \mathbb{R}^m$ $(m > 0)$ with a control input $\mathbf{w} \in L_1(S \times \mathbb{R}^+)$ to the second subsystem with $S \subseteq \mathbb{R}^n \times \mathbb{R}^m$. Here $L_1(S \times \mathbb{R}^+)$ denotes the set of all Lebesgue integrable functions on $S \times \mathbb{R}^+$. Suppose that the two nonlinear vector-valued functions

$$\mathbf{f} : \mathbb{R}^n \times \mathbb{R}^m \times \mathbb{R}^+ \to \mathbb{R}^n, \qquad \mathbf{g} : \mathbb{R}^n \times \mathbb{R}^m \times \mathbb{R}^+ \to \mathbb{R}^m$$

are integrable and that the system has a unique solution trajectory $(\mathbf{u}(t), \mathbf{v}(t))$ in S and $t \geq t_0 \geq 0$, for any given initial point $(\mathbf{u}_0, \mathbf{v}_0, t_0) \in I \subset S \times \mathbb{R}^+$. Selfcontrolling feedback then refers to a control input $\mathbf{w}$ (satisfying the above requirements) which achieves stabilization of an existing unstable periodic trajectory of the system. Chen and Dong [30] presented some general results for dynamical systems of the above form.

Example. Consider the driven anharmonic oscillator

$$\frac{du_1}{dt} = u_2, \qquad \frac{du_2}{dt} = -p_1 u_1 - u_1^3 - p u_2 + q\cos(\omega t)$$

which can show periodic orbits and chaotic behaviour depending on the parameters. Assume that $(u_1^*(t), u_2^*(t))$ is a periodic solution of the system. By adding a linear correction term to the driven anharmonic oscillator we have in matrix notation

$$\begin{pmatrix} du_1/dt \\ du_2/dt \end{pmatrix} = \begin{pmatrix} u_2 \\ -p_1 u_1 - u_1^3 - p u_2 + q\cos(\omega t) \end{pmatrix} - \begin{pmatrix} K_{11} & K_{12} \\ K_{21} & K_{22} \end{pmatrix} \begin{pmatrix} u_1 - u_1^* \\ u_2 - u_2^* \end{pmatrix}$$

where the 2×2 matrix on the right-hand side is called the feedback gain matrix. It follows that

$$\begin{aligned} \frac{du_1}{dt} &= -K_{11}u_1 + (1 - K_{12})u_2 + (K_{11}u_1^* + K_{12}u_2^*) \\ \frac{du_2}{dt} &= -(K_{21} + p)u_1 - u_1^3 - (K_{22} + p)u_2 + (K_{21}u_1^* + K_{22}u_2^*) + q\cos(\omega t)\,. \end{aligned}$$

If the solution $(u_1(t), u_2(t))$ coincides with $(u_1^*(t), u_2^*(t))$ then the controlled driven anharmonic oscillator yields exactly the driven anharmonic oscillator, since then the control objective is satisfied and no correction is required. However, if the solution $(u_1(t), u_2(t))$ differs from $(u_1^*(t), u_2^*(t))$ then there is a linear correcting term added to each equation in the driven anharmonic oscillator which attempts to stabilize $(u_1^*(t), u_2^*(t))$. Next we show that we can find a feedback gain matrix K for the controller which stabilizes the periodic solution $(u_1^*(t), u_2^*(t))$. The Jacobian matrix corresponding to the controlled driven anharmonic oscillator at the point $(u_1^*(t), u_2^*(t))$ is given by

$$J_c(u_1^*, u_1^*) = \begin{pmatrix} -K_{11} & 1 - K_{12} \\ -(K_{21} + p_1) - 3u_1^{*2} & -(K_{22} + p) \end{pmatrix}.$$

The characteristic equation $\det(sI_2 - J_c) = 0$ of the controlled driven anharmonic oscillator linearized about the periodic solution $(u_1^*(t), u_2^*(t))$ is given by

$$s^2 + (p + K_{11} + K_{22})s + (K_{11}(p + K_{22}) + (1 - K_{12})(K_{21} + p_1 + 3u_1^{*2})) = 0$$

which is required to have all its roots located in the open left-hand s-plane in order for the controlled system to be stable. A necessary and sufficient condition for this to be the case for a second order system is that all coefficients of the polynomial $\det(sI_2 - J_c)$ have the same sign

$$p + K_{11} + K_{22} > 0$$
$$K_{11}(p + K_{22}) + (1 - K_{12})(K_{21} + p_1 + 3u_1^{*2}) > 0.$$

Consider the special case $K_{11} = K_{22} = 0$. Since $p > 0$ the first inequality is satisfied and the second inequality takes the form

$$(1 - K_{12})(K_{21} + p_1 + 3u_1^{*2}) > 0.$$

If we also assume that $K_{12} = 0$ then we have

$$K_{21} > -p_1 - 3u_1^{*2}, \quad t \geq 0.$$

Next we have to linearize the driven anharmonic oscillator about $(u_1^*(t), u_2^*(t))$ and then add the correction term. We obtain

$$\begin{pmatrix} d\tilde{u}_1/dt \\ d\tilde{u}_2/dt \end{pmatrix} = \begin{pmatrix} 0 & 1 \\ -p_1 - 3u_1^{*2} & -p \end{pmatrix} \begin{pmatrix} \tilde{u}_1 \\ \tilde{u}_2 \end{pmatrix} + \begin{pmatrix} 0 \\ w \end{pmatrix}$$

which is completely controllable. Therefore the controlled driven anharmonic oscillator is locally controllable by the feedback

$$w(t) = -K_{21}(u_1(t) - u_1^*(t)).$$

In the linearized controlled driven anharmonic oscillator we have

$$\tilde{u}_1 := u_1 - u_1^*, \quad \tilde{u}_2 := u_2 - u_2^*.$$

7.4 Small Periodic Perturbation

Control of chaos of dissipative systems with a control parameter can be realized by coupling weak periodic oscillations to the control parameter. We consider the

Lorenz equations

$$\frac{du_1}{dt} = \sigma(u_2 - u_1)$$
$$\frac{du_2}{dt} = ru_1 - u_1u_2 - u_2$$
$$\frac{du_3}{dt} = u_1u_2 - bu_3 \,.$$

One chooses the parameters of Lorenz equations where the system shows chaotic behaviour, for example $\sigma = 10$, $b = 0.4$ and $r = 80$. To couple the periodic oscillations to the control parameter, we replace the parameter r in the Lorenz system with

$$r \to r + k\cos(\Omega t)$$

where Ω is the modulation frequency and k is the modulation amplitude. Thus we consider the time-dependent system

$$\frac{du_1}{dt} = \sigma(u_2 - u_1)$$
$$\frac{du_2}{dt} = (r + k\cos(\Omega t))u_1 - u_1u_2 - u_2$$
$$\frac{du_3}{dt} = u_1u_2 - bu_3 \,.$$

After controlling with the frequency $\Omega = 6.28318$, we obtained period one, period two and period four orbits for the values $k = 2.11$, $k = 3$ and $k = 36.8$, respectively.

```
// ControlLorenz.java

import java.awt.*;
import java.awt.event.*;
import java.awt.Graphics;

public class ControlLorenz extends Frame
{
 public ControlLorenz()
 {
 setSize(600,500);
 addWindowListener(new WindowAdapter()
 { public void windowClosing(WindowEvent event) { System.exit(0); }}); }

 public void fsystem(double h,double t,double u[],double hf[])
 {
 double sigma = 10.0, r = 80.0, b = 0.4;
 double Omega = 6.28318; double k = 2.0;
 hf[0] = h*sigma*(u[1]-u[0]);
 hf[1] = h*(-u[0]*u[2]+(r+k*Math.cos(Omega*t))*u[0]-u[1]);
 hf[2] = h*(u[0]*u[1]-b*u[2]);
```

```
}

public void map(double u[],int steps,double h,double t,int N)
{
double uk[] = new double[N];
double tk;
double a[] = { 0.0, 1.0/4.0, 3.0/8.0, 12.0/13.0, 1.0, 1.0/2.0 };
double c[] = { 16.0/135.0, 0.0, 6656.0/12825.0, 28561.0/56430.0,
               -9.0/50.0, 2.0/55.0 };
double b[][] = new double[6][5];
b[0][0] = b[0][1]= b[0][2] = b[0][3] =b[0][4] = 0.0;
b[1][0] = 1.0/4.0; b[1][1] = 0.0; b[1][2] = 0.0; b[1][3] = 0.0;
b[1][4] = 0.0;
b[2][0] = 3.0/32.0; b[2][1] = 9.0/32.0;
b[2][2] = 0.0; b[2][3] = 0.0; b[2][4] = 0.0;
b[3][0] = 1932.0/2197.0; b[3][1] = -7200.0/2197.0;
b[3][2] = 7296.0/2197.0; b[3][3] = b[3][4] = 0.0;
b[4][0] = 439.0/216.0; b[4][1] = -8.0;
b[4][2] = 3680.0/513.0; b[4][3] = -845.0/4104.0; b[4][4] = 0.0;
b[5][0] = -8.0/27.0; b[5][1] = 2.0; b[5][2] = -3544.0/2565.0;
b[5][3] = 1859.0/4104.0; b[5][4] = -11.0/40.0;
double f[][] = new double[6][N];
for(int i=0;i<steps;i++)
{ fsystem(h,t,u,f[0]);
for(int k=1;k<=5;k++)
{ tk = t + a[k]*h;
for(int l=0;l<N;l++)
{ uk[l] = u[l];
for(int j=0;j<=k-1;j++) uk[l] += b[k][j]*f[j][l];
}
fsystem(h,tk,uk,f[k]);
}
for(int l=0;l<N;l++)
 for(int k=0;k<6;k++) u[l] += c[k]*f[k][l];
}
}

public void paint(Graphics g)
{
g.drawRect(40,40,600,400);
int steps = 1;   int N = 3;
double h = 0.005;   double t = 0.0;
double u[] = { 0.8, 0.8, 0.8 };  // initial conditions
// wait for transients to decay
for(int i=0;i<4000;i++)  { t += h; map(u,steps,h,t,N); }
t = 0.0;
for(int i=0;i<20000;i++)
{ t += h; map(u,steps,h,t,N);
```

```
int m = (int)(10.0*u[2]-400); int n = (int)(10.0*u[0]+200);
g.drawLine(m,n,m,n);
}
}

public static void main(String[] args)
{ Frame f = new ControlLorenz(); f.setVisible(true); }
}
```

7.5 Resonant Perturbation and Control

Another method to control a nonlinear dynamical system is to follow a prescribed goal dynamics (Hübler [101]). If the experimental dynamics is described by

$$\frac{d\mathbf{u}}{dt} = \mathbf{f}(\mathbf{u}, \mathbf{p}(t) + \mathbf{F}(t))$$

where both the set of parameters $\mathbf{p}$ and the driving force $\mathbf{F}$ depend only on time t, then the limiting behaviour of $\mathbf{u}(t)$ can be made equal to a given goal dynamics $\mathbf{y}(t)$, i.e.,

$$|\mathbf{u}(t) - \mathbf{y}(t)| \to 0 \quad \text{as} \quad t \to \infty$$

where

$$\frac{d\mathbf{y}}{dt} = \mathbf{g}(\mathbf{y}(t), t)$$

by an appropiate choice of $\mathbf{F}$. Complete entrainment occurs if both sets of flows are made equal, i.e.,

$$\mathbf{f}(\mathbf{y}, \mathbf{p}(t) + \mathbf{F}(t)) = \mathbf{g}(\mathbf{y}(t), t)$$

and if the special solution $\mathbf{u}(t) = \mathbf{y}(t)$ is stable. Whether convergence occurs, i.e. $|\mathbf{u}(t) - \mathbf{y}(t)| \to 0$ as $t \to \infty$, depends on the particular choice of $\mathbf{F}$ and the initial conditions, i.e., $\mathbf{u}(0)$. The regions of the $\mathbf{u}$ state space such that convergence occur are called entrainment regions. The applied controls are not typically small and convergence to the goal is not assured.

Chapter 8

Synchronization of Chaos

8.1 Introduction

It is possible to synchronize two chaotic systems by coupling them (Pikovsky et al [166], González-Miranda [71], Mosekilde et al [151]). Different types of synchronization have been considered. Many types of synchronization have been studied: chaotic synchronization (Pecora and Carroll [160]), adaptive synchronization (Wang and Ge [214]), phase synchronization (van Wyk and Steeb [209]), and generalized synchronization (Brown and Kocarev [24]). One, called master-slave, or unidirectional coupling, is made up of an element, called drive system, one or several components of its state vector are transmitted to a second part, called the response system. The other type of synchronization is the mutual coupling. Synchronization is studied in section 8.2. Phase coupled chaotic systems are studied in section 8.3. We also consider synchronization of coupled Rikitake two-disc dynamos.

8.2 Synchronization of Chaos

8.2.1 Synchronization Using Control

To synchronize two chaotic systems that we call A and B, we assume that some parameter of one system (say B) is externally adjustable (Lai and Grebogi [125]). Suppose that some state variables of both systems A and B can be measured. Some dynamical variables of two systems are measured, based on which temporal-parameter perturbations are calculated and applied to the system B. We assume that before the synchronization, some information about the geometrical structure of the chaotic attractor (e.g., the Jacobian matrices along a long chaotic trajectory that practically covers the whole attractor) has been obtained. Based on this measurement and our knowledge about the system (we can, for example, observe and learn the system first), when it is determined that the state variables of A and B are close, we calculate a small parameter perturbation based on the Ott-Grebogi-Yorke algorithm and apply it to system B.

Two systems can then be synchronized, although their trajectories are still chaotic. Under the influence of external noise, there is a finite probability that two already synchronized trajectories may lose synchronization. However, with probability one (due to the ergodicity of chaotic trajectories), after a finite amount of transient time, the trajectories of A and B will get close and can then be synchronized again. In this sense, the synchronization method is robust against small external noise.

We consider two almost identical chaotic systems (Lai and Grebogi [125]) that can be described by two-dimensional maps on the Poincaré surface of section

$$\mathbf{x}_{t+1} = \mathbf{f}(\mathbf{x}_t, r_0)\ [A], \qquad \mathbf{y}_{t+1} = \mathbf{f}(\mathbf{y}_t, r)\ [B]$$

where $\mathbf{x}_t, \mathbf{y}_t \in \mathbb{R}^2$, $\mathbf{f}$ is a smooth function in its variables, r_0 for system A is a fixed parameter value, and r for system B is an external controllable parameter. For the purpose of synchronization, we require that the dynamics should not be substantially different for system A and B. In other words, any parameter perturbations should be small. Thus, we require

$$|r - r_0| < \delta$$

where δ is a small number defining the range of parameter variation. Suppose that two systems start with different initial conditions. In general, the resulting chaotic trajectories are completely uncorrelated. However, due to ergodicity, with probability one two trajectories can get arbitrarily close to each other at some later time n_c. Without control, two trajectories will separate from each other exponentially again. The objective is to program the parameter r in such a way that $|\mathbf{y}_t - \mathbf{x}_t| \to 0$ for $n \geq n_c$, which means that A and B are synchronized for $t \geq t_c$.

The linearized dynamics in the neighbourhood of the "target" trajectory $\{\mathbf{x}_t\}$ is

$$\mathbf{y}_{t+1} - \mathbf{x}_{t+1}(r_0) = J(\mathbf{y}_t - \mathbf{x}_t(r_0)) + V(\Delta r)_t$$

where $r_t := r_0 + (\Delta r)_t$, $(\Delta r)_t \leq \delta$, J is the 2×2 Jacobian matrix, and V is a two-dimensional column vector

$$J := D_{\mathbf{y}}\mathbf{f}(\mathbf{y}, r)|_{\mathbf{y}=\mathbf{x}, r=r_0}, \qquad V := D_r\mathbf{f}(\mathbf{y}, r)|_{\mathbf{y}=\mathbf{x}, r=r_0}.$$

A property of a chaotic trajectory is the existence of both stable and unstable directions at almost each trajectory point.

Let $\mathbf{e}_{s(t)}$ and $\mathbf{e}_{u(t)}$ be the stable and unstable unit vectors at $\mathbf{x}_t$ and $\mathbf{f}_{s(t)}$ and $\mathbf{f}_{u(t)}$ be two unit vectors that satisfy

$$\mathbf{f}_{u(t)} \cdot \mathbf{e}_{u(t)} = \mathbf{f}_{s(t)} \cdot \mathbf{e}_{s(t)} = 1, \qquad \mathbf{f}_{u(t)} \cdot \mathbf{e}_{s(t)} = \mathbf{f}_{s(t)} \cdot \mathbf{e}_{u(t)} = 0 \cdot$$

where $\cdot$ denotes the scalar product. To stabilize $\{\mathbf{y}_t\}$ around $\{\mathbf{x}_t\}$, we require the next iteration of $\mathbf{x}_t$ after falling into a small neighbourhood around $\mathbf{x}_t$ to lie on the stable direction at $\mathbf{x}_{(t+1)}(r_0)$, i.e.,

$$[\mathbf{y}_{t+1} - \mathbf{x}_{(t+1)}(r_0)] \cdot \mathbf{f}_{u(t+1)} = 0.$$

We obtain the following expression for the parameter perturbations

$$(\Delta r)_t = \frac{(J[\mathbf{y}_t - \mathbf{x}_t(r_0)]) \cdot \mathbf{f}_{u(t+1)}}{-V \cdot \mathbf{f}_{u(t+1)}}.$$

If $(\Delta r)_t > \delta$, we set $(\Delta r)_t = 0$. Note that $\mathbf{f}_{u(t)}$ can be calculated in terms of J.

Example. We illustrate the synchronization algorithm by using the standard Hénon map

$$x_{1t+1} = a - x_{1t}^2 + 0.3x_{2t}, \qquad x_{2t+1} = x_{1t}$$

where a is our control parameter. Consider two such Hénon systems

$$x_{1t+1} = a - x_{1t}^2 + 0.3x_{2t}, \qquad x_{2t+1} = x_{1t}$$

$$y_{1t+1} = a - y_{1t}^2 + 0.3y_{2t}, \qquad y_{2t+1} = y_{1t}$$

where $t = 0, 1, 2, \ldots$. We obtain

$$J = \begin{pmatrix} -2x_1 & b \\ 1 & 0 \end{pmatrix}, \qquad V = \begin{pmatrix} 1 \\ 0 \end{pmatrix}.$$

The eigenvalues and corresponding eigenvectors of the 2×2 matrix J are given by

$$\lambda_s = -x_1 + \sqrt{b + x_1^2}, \quad \begin{pmatrix} \lambda_s \\ 1 \end{pmatrix}, \qquad \lambda_u = -x_1 - \sqrt{b + x_1^2}, \quad \begin{pmatrix} \lambda_u \\ 1 \end{pmatrix}.$$

One has a fixed parameter value ($a = a_0 = 1.4$) which serves the target and in the other system we adjust a in a small range (1.39,1.41). At time $t = 0$, we start two systems with different initial conditions. The two systems then move on completely uncorrelated chaotic trajectories. At a certain time step, the trajectory points of the uncoupled two systems come close to each other within an epsilon neighbourhood. When this occurs, we turn on the parameter perturbations. ♣

```
// SynchronHenon.java

import java.awt.*;

public class SynchronHenon
{
 public static void main(String[] args)
 {
 double x1, x11, x2, x22, y1, y11, y2, y22;
 x1 = 0.5; x2 = -0.8; y1 = 0.001; y2 = 0.001; // initial values
 int T = 2140; // number of iterations
 double b = 0.3, a = 1.4, a0 = 1.4;
 for(int t=0;t<T;t++)
 {
 x11 = x1; x22 = x2; y11 = y1; y22 = y2;
 x1 = a0-x11*x11+b*x22; x2 = x11;
```

```
y1 = a-y11*y11+b*y22; y2 = y11;
double distance = Math.abs(x1-y1)+Math.abs(x2-y2);
double epsilon = 0.01; double range = 0.01;
if(distance < epsilon)
{
System.out.println("");
System.out.println("inside epsilon neighbourhood: t = " + t);
double t1 = a0-x1*x1+b*x2;
double t2 = x1;
double lams = -t1+Math.sqrt(t1*t1+b);
double lamu = -t1-Math.sqrt(t1*t1+b);
double f1 = Math.sqrt(lamu*lamu+1.0)/(lamu-lams);
double f2 = -f1*lams;
double deltaa = -((-2.0*x1*(y1-x1)+b*(y2-x2))*f1+(y1-x1)*f2)/f1;
System.out.println("deltaa = " + deltaa);
if(Math.abs(deltaa) < range) a = a0+deltaa;
System.out.println("a = " + a);
System.out.print("x1 = " + x1); System.out.println(" x2 = " + x2);
System.out.print("y1 = " + y1); System.out.println(" y2 = " + y2);
}
}
}
}
```

8.2.2 Synchronizing Subsystems

Pecora and Carroll [160] showed that certain subsystems of chaotic systems can be made to synchronize by linking them with common signals. The criterion for this is the sign of the one-dimensional sub-Liapunov exponents. The method is as follows. Consider a chaotic system described by the autonomous system of first-order differential equations with n state vectors as

$$\frac{d\mathbf{u}}{dt} = \mathbf{f}(\mathbf{u}) .$$

We divide the dynamical system into two subsystems

$$\mathbf{u}^T = (\mathbf{v}, \mathbf{w})^T$$

$$\frac{d\mathbf{v}}{dt} = \mathbf{g}(\mathbf{v}, \mathbf{w}), \qquad \frac{d\mathbf{w}}{dt} = \mathbf{h}(\mathbf{v}, \mathbf{w})$$

where

$$\mathbf{v} = (u_1, u_2, \ldots, u_m), \quad \mathbf{g} = (f_1, f_2, \ldots, f_m), \quad \mathbf{w} = (u_{m+1}, u_{m+2}, \ldots, u_n)$$

and

$$\mathbf{h} = (f_{m+1}, f_{m+2}, \ldots, f_n) .$$

Now we create a new subsystem $\mathbf{w}'$ identical to the $\mathbf{w}$ system by replacing the set of variables $\mathbf{w}$ for the corresponding $\mathbf{w}'$ in the function $\mathbf{h}$. This yields the dynamical systems

$$\frac{d\mathbf{v}}{dt} = \mathbf{g}(\mathbf{v}, \mathbf{w}), \qquad \frac{d\mathbf{w}}{dt} = \mathbf{h}(\mathbf{v}, \mathbf{w}), \qquad \frac{d\mathbf{w}'}{dt} = \mathbf{h}(\mathbf{v}, \mathbf{w}') .$$

Let

$$\Delta \mathbf{w} := \mathbf{w}' - \mathbf{w} .$$

The subsystem components $\mathbf{w}$ and $\mathbf{w}'$ will synchronize only if $\Delta\mathbf{w} \to 0$ as $t \to \infty$. In the infinitesimal limit this leads to the variational equation for the subsystem

$$\frac{d\mathbf{y}}{dt} = D_{\mathbf{w}}\mathbf{h}(\mathbf{v}(t), \mathbf{w}(t))\mathbf{y}$$

where $D_{\mathbf{w}}\mathbf{h}$ is the Jacobian matrix of the $\mathbf{w}$ subsystem vector field with respect to $\mathbf{w}$ only. The behaviour of the variational system depends on the one-dimensional Liapunov exponents of the $\mathbf{w}$ subsystem.

We have the following theorem (Pecora and Carroll [160])

Theorem. The subsystems $\mathbf{w}$ and $\mathbf{w}'$ will synchronize only if the one-dimensional Liapunov exponents of the $\mathbf{w}$ subsystem are all negative.

The above theorem is a necessary, but not sufficient, condition for synchronization. It says nothing about the set of initial conditions in which $\mathbf{w}'$ will synchronize with $\mathbf{w}$.

A chaotic system, for example, has at least one positive one-dimensional Liapunov exponent. For the response system of the subsystem to synchronize with the drive system, all of its one-dimensional Liapunov exponents must be less than zero.

Example. Consider the *Rössler model*

$$\frac{du_1}{dt} = u_2 - u_3, \qquad \frac{du_2}{dt} = u_1 + au_2, \qquad \frac{du_3}{dt} = b + u_3(u_1 - c)$$

where a, b and c are positive constants. The slave system is

$$\frac{du_1'}{dt} = -u_2 - u_3', \qquad \frac{du_3'}{dt} = b + u_3'(u_1' - c) .$$

Thus in the Rössler system it is possible to use the u_2 component to drive a (u_1', u_3') response system and attain synchronization with the (u_1, u_3) components of the driving system. ♣

In the Java program we integrate all five equations as an autonomous system of differential equations using a Runge-Kutta-Fehlberg method. We display the phase-portrait for (u_1, u_3) and (u_1', u_3').

```
// Rossler.java

import java.awt.*;
import java.awt.event.*;
import java.awt.Graphics;

public class Rossler extends Frame
{
 public Rossler()
 {
 setSize(600,500);
 addWindowListener(new WindowAdapter()
 { public void windowClosing(WindowEvent event) { System.exit(0); }}); }

 public void fsystem(double h,double t,double u[],double hf[])
 {
 double a = 0.2, b = 0.2, c = 9.0;
 hf[0] = h*(-u[1]-u[2]);  hf[1] = h*(u[0]+a*u[1]);
 hf[2] = h*(b+u[2]*(u[0]-c));  hf[3] = h*(-u[1]-u[4]);
 hf[4] = h*(b+u[4]*(u[3]-c));
 }

 public void map(double u[],int steps,double h,double t,int N)
 {
 double uk[] = new double[N];
 double tk;
 double a[] = { 0.0, 1.0/4.0, 3.0/8.0, 12.0/13.0, 1.0, 1.0/2.0 };
 double c[] = { 16.0/135.0, 0.0, 6656.0/12825.0, 28561.0/56430.0,
               -9.0/50.0, 2.0/55.0 };
 double b[][] = new double[6][5];
 b[0][0] = b[0][1]= b[0][2] = b[0][3] = b[0][4] = 0.0;
 b[1][0] = 0.25; b[1][1] = 0.0; b[1][2] = 0.0; b[1][3] = 0.0;
 b[1][4] = 0.0;
 b[2][0] = 3.0/32.0; b[2][1] = 9.0/32.0;
 b[2][2] = 0.0; b[2][3] = 0.0; b[2][4] = 0.0;
 b[3][0] = 1932.0/2197.0; b[3][1] = -7200.0/2197.0;
 b[3][2] = 7296.0/2197.0; b[3][3] = b[3][4] = 0.0;
 b[4][0] = 439.0/216.0; b[4][1] = -8.0;
 b[4][2] = 3680.0/513.0; b[4][3] = -845.0/4104.0; b[4][4] = 0.0;
 b[5][0] = -8.0/27.0; b[5][1] = 2.0; b[5][2] = -3544.0/2565.0;
 b[5][3] = 1859.0/4104.0; b[5][4] = -11.0/40.0;
 double f[][] = new double[6][N];
 for(int i=0;i<steps;i++) { fsystem(h,t,u,f[0]);
 for(int k=1;k<=5;k++) { tk = t+a[k]*h;
 for(int l=0;l<N;l++) { uk[l] = u[l];
 for(int j=0;j<=k-1;j++) uk[l] += b[k][j]*f[j][l];
 }
 fsystem(h,tk,uk,f[k]);
```

```
}
for(int l=0;l<N;l++)
 for(int k=0;k<6;k++) u[l] += c[k]*f[k][l];
}
}

public void paint(Graphics g)
{
g.drawRect(20,40,560,480);
int steps = 1;  int N = 5;
double h = 0.002;  double t = 0.0;
double u[] = { 0.8, 0.4, 0.8, 1.1, 2.3 };  // initial conditions
// wait for transients to decay
for(int i=0;i<4000;i++)  { t += h; map(u,steps,h,t,N); }
t = 0.0;
for(int i=0;i<40000;i++)
{ t += h;
map(u,steps,h,t,N);
int m1 = (int)(5.0*u[0]+100.0); int n1 = (int)(5.0*u[2]+200.0);
g.drawLine(m1,n1,m1,n1);
int m2 = (int)(5.0*u[3]+300.0); int n2 = (int)(5.0*u[4]+200.0);
g.drawLine(m2,n2,m2,n2);
}
}

public static void main(String[] args)
{ Frame f = new Rossler(); f.setVisible(true); }
}
```

8.3 Synchronization of Coupled Dynamos

We consrider the synchronization of two identical coupled Rikitake two-disc dynamos (Xiang-Jun et al [220]). The two coupled Rikitake two-disc dynamos are given by

$$\frac{du_1}{dt} = -\mu u_1 + u_2(u_3+\alpha)$$
$$\frac{du_2}{dt} = -\mu u_2 + u_1(u_3-\alpha)$$
$$\frac{du_3}{dt} = 1 - u_1 u_2$$

and

$$\frac{dv_1}{dt} = -\mu v_1 + v_2(v_3+\alpha) + w_1$$
$$\frac{dv_2}{dt} = -\mu v_2 + v_1(v_3-\alpha) + w_2$$
$$\frac{dv_3}{dt} = 1 - v_1 v_2 + w_3$$

where μ, α are positive constants and w_1, w_2, w_3 are three control inputs. The control functions w_1, w_2, w_3 are determined so that the two identical Rikitake dynamos (with the same parameters μ and α) are synchronized, where the initial conditions for the two dynamical systems can be different. We define the state error as

$$e_1 := v_1 - u_1, \quad e_2 := v_2 - u_2, \quad e_3 := v_3 - u_3 .$$

Subtracting the two dynamical systems yields

$$\frac{de_1}{dt} = -\mu e_1 + \alpha e_2 + u_2 e_3 + v_3 e_2 + w_1$$
$$\frac{de_2}{dt} = -\alpha e_1 - \mu e_2 + u_3 e_1 + v_1 e_3 + w_2$$
$$\frac{de_3}{dt} = -u_2 e_1 - v_1 e_2 + w_3 .$$

Consider the *Liapunov function*

$$V(e_1, e_2, e_3) = \frac{1}{2}(e_1^2 + e_2^2 + e_3^2) .$$

We choose the active control functions w_1, w_2, w_3 as

$$w_1 = -v_3 e_2, \quad w_2 = -u_3 e_1, \quad w_3 = -e_3 .$$

Thus the controller is independent of the positive parameters μ and α. Calculating the time derivative of V and inserting the equations of motions for e_1, e_2, e_3 yields

$$\frac{dV}{dt} = e_1 \frac{de_1}{dt} + e_2 \frac{de_2}{dt} + e_3 \frac{de_3}{dt} = -(\mu e_1^2 + \mu e_2^2 + e_3^2) .$$

Since μ is positive it follows that dV/dt is negative definite. Since V is a positive definite function, we find from the *Liapunov direct method* that the fixed point (equilibrium point)

$$(e_1^*, e_2^*, e_3^*) = (0, 0, 0)$$

for the dynamical system for e_1, e_2, e_3 is asymptotically stable. This implies that

$$(e_1(t), e_2(t), e_3(t)) \to (0, 0, 0)$$

as $t \to \infty$. Thus synchronization is achieved under the given controller.

8.4 Phase Coupled Systems

In this section the stability of a class of coupled identical autonomous systems of first order nonlinear ordinary differential equations is investigated. These couplings play a central role in synchronization of chaotic systems and can be applied in electronic circuits (Schuster et al [184], van Wyk and Steeb [209], Hardy et al [83]). As applications we consider two coupled Van der Pol equations and two coupled logistic maps. When the uncoupled system admits a first integral we study whether a first integral exists for the coupled system. Finally the relation of the Liapunov exponents of the uncoupled and coupled system are discussed.

Consider the autonomous system of first order ordinary differential equations

$$\frac{d\mathbf{u}}{dt} = \mathbf{f}(\mathbf{u}), \qquad \mathbf{u} = (u_1, u_2, \ldots, u_n)^T.$$

We assume that the functions $f_j : \mathbb{R}^n \to \mathbb{R}$ are analytic. Assume that $\mathbf{u}^*$ is a fixed point of this system. The variational equation of the dynamical system is given by

$$\frac{d\mathbf{y}}{dt} = \frac{\partial \mathbf{f}}{\partial \mathbf{u}}(\mathbf{u}(t))\mathbf{y}$$

where $\partial\mathbf{f}/\partial\mathbf{u}$ is the Jacobian matrix. Inserting the fixed point $\mathbf{u}^*$ into the Jacobian matrix results in an $n \times n$ matrix

$$A := \frac{\partial \mathbf{f}}{\partial \mathbf{u}}(\mathbf{u}^*)$$

with constant coefficients. The eigenvalues $\lambda_1, \ldots, \lambda_n$ of this matrix determine the stability of the fixed point $\mathbf{u}^*$. Furthermore the eigenvalues provide information as to whether Hopf bifurcations can occur. In this case we assume that $\mathbf{f}$ depends on a (bifurcation) parameter. Moreover, the variational system is used to find the one-dimensional Liapunov exponents.

In controlling the chaos of the autonomous sytem the coupling of two identical systems according to

$$\frac{d\mathbf{u}}{dt} = \mathbf{f}(\mathbf{u}) + c(\mathbf{v} - \mathbf{u}), \qquad \frac{d\mathbf{v}}{dt} = \mathbf{f}(\mathbf{v}) + c(\mathbf{u} - \mathbf{v})$$

plays a central role. Here $n \geq 3$ and $c \in \mathbb{R}$. First we realize that $(\mathbf{u}^*, \mathbf{v}^*)$ with $\mathbf{v}^* = \mathbf{u}^*$ is a fixed point of the coupled system if $\mathbf{u}^*$ is a fixed point of the system $d\mathbf{u}/dt = \mathbf{f}(\mathbf{u})$. Inserting the fixed point $(\mathbf{u}^*, \mathbf{u}^*)$ into the Jacobian matrix yields a $2n \times 2n$ matrix. We now show that the eigenvalues of this $2n \times 2n$ matrix can be found from the eigenvalues of the $n \times n$ matrix given by A. Then from the $2n$ eigenvalues of the uncoupled system we can determine the stability of the fixed point $(\mathbf{u}^*, \mathbf{u}^*)$ for the phase coupled system.

To find these eigenvalues we prove the following theorem.

Theorem. Let A be an $n \times n$ matrix over the real numbers. Let $\lambda_1, \ldots, \lambda_n$ be the eigenvalues of A. Define the $2n \times 2n$ matrix M as

$$M := \begin{pmatrix} A - cI & cI \\ cI & A - cI \end{pmatrix}$$

where I is the $n \times n$ identity matrix. Then the eigenvalues of M are given by

$$\lambda_1, \lambda_2, \ldots, \lambda_n, \quad \lambda_1 - 2c, \lambda_2 - 2c, \ldots, \lambda_n - 2c.$$

Proof. There exists an $n \times n$ orthogonal matrix Q such that

$$Q^T A Q = D + U$$

where $D := \mathrm{diag}(\lambda_1, \lambda_2, \ldots, \lambda_n)$ and U is a strictly upper-triangular $n \times n$ matrix. Let

$$P := \begin{pmatrix} Q & 0 \\ Q & Q \end{pmatrix}$$

where 0 is the $n \times n$ zero matrix. It follows that the inverse is given by

$$P^{-1} = \begin{pmatrix} Q^T & 0 \\ -Q^T & Q^T \end{pmatrix}.$$

giving

$$P^{-1} M P = \begin{pmatrix} D + U & cI \\ 0 & (D - 2cI) + U \end{pmatrix}.$$

Now $P^{-1}MP$ is an upper-triangular matrix of which the entries on the diagonal are

$$\lambda_1, \ldots, \lambda_n, \quad \lambda_1 - 2c, \ldots, \lambda_n - 2c$$

the eigenvalues of M. This proves the theorem.

Example. Consider the *Van der Pol equation*

$$\frac{du_1}{dt} = u_2, \qquad \frac{du_2}{dt} = r(1 - u_1^2)u_2 - u_1.$$

Then $\mathbf{u}^* = (0, 0)$ is a fixed point. The eigenvalues of the functional matrix for this fixed point are given by

$$\lambda_{1,2} = r/2 \pm \sqrt{r^2/4 - 1}.$$

The uncoupled system shows a Hopf bifurcation. We find a Hopf bifurcation when the characteristic exponents cross the imaginary axis. A stable limit cycle is born. If we consider the coupling we find that the eigenvalues of the coupled system are given by

$$\mu_1 = \lambda_1, \quad \mu_2 = \lambda_2, \quad \mu_3 = \lambda_1 - 2c, \quad \mu_4 = \lambda_2 - 2c.$$

♣

The theorem stated above also applies to coupled maps. For example it can be applied to

$$\mathbf{x}_{t+1} = \mathbf{f}(\mathbf{x}_t) + c(\mathbf{y}_t - \mathbf{x}_t), \qquad \mathbf{y}_{t+1} = \mathbf{f}(\mathbf{y}_t) + c(\mathbf{x}_t - \mathbf{y}_t).$$

Example. Consider the logistic map $f : [0, 1] \to [0, 1]$ with $f(x) = rx(1 - x)$. The map f admits the fixed points $x_1^* = 0$ and $x_2^* = (r - 1)/r$. We consider the fixed point x_2^*. Then we find that the Jacobian matrix (which is a 1×1 matrix) at this fixed point is given by

$$A = \lambda = 2 - r\,.$$

Consequently the eigenvalues for the coupled system are $2 - r$ and $2 - r - 2c$. ♣

Consider now the Liapunov exponents of the coupled system. To find a relation between the Liapunov exponents of the coupled system and uncoupled system we consider the time-evolution of

$$\mathbf{\Theta}(t) := \mathbf{u}(t) - \mathbf{v}(t)\,.$$

We call $\mathbf{\Theta}$ the *phase difference*. For the time evolution of $\mathbf{\Theta}$ it follows that

$$\frac{d\mathbf{\Theta}}{dt} = \frac{d\mathbf{u}}{dt} - \frac{d\mathbf{v}}{dt} = \mathbf{f}(\mathbf{u}) - \mathbf{f}(\mathbf{v}) - 2c\mathbf{\Theta}\,.$$

Using a Taylor expansion for $\mathbf{f}(\mathbf{u})$ and $\mathbf{f}(\mathbf{v})$ and the fact that

$$\frac{\partial \mathbf{f}}{\partial \mathbf{u}}(\mathbf{u}(t)) = \frac{\partial \mathbf{f}}{\partial \mathbf{v}}(\mathbf{u}(t))$$

we obtain

$$\frac{d\mathbf{\Theta}}{dt} = (A(t) - 2cI)\mathbf{\Theta} + O(\mathbf{\Theta}^2), \qquad A(t) := \frac{\partial \mathbf{f}}{\partial \mathbf{u}}(\mathbf{u}(t))$$

where $O(\mathbf{\Theta}^2)$ indicates higher order terms in $\mathbf{\Theta}$. Integrating this equation while neglecting the higher order terms yields

$$\mathbf{\Theta}(t) = \left(e^{-2ct} T \exp\left(\int_0^t A(s)ds\right)\right)\mathbf{\Theta}(0)$$

where T is the time ordering operator. The eigenvalues μ_j $(j = 1, 2, \ldots, n)$ of

$$\lim_{t\to\infty}\left(T \exp\left(\int_0^t A(s)ds\right)\right)$$

are related to the Liapunov exponents λ_j $(j = 1, 2, \ldots, n)$ via $\lambda_j = \ln|\mu_j|$. We find

$$\langle|\mathbf{\Theta}(t)|\rangle \propto e^{(\lambda_1 - 2c)t}$$

where the average is taken over all initial conditions $\mathbf{u}(0)$ and all directions of $\mathbf{\Theta}(0)$ and λ_1 is the largest one-dimensional Liapunov exponent. The equation tells us that for

$$2c > \lambda_1$$

both systems stay in phase. Consequently, they have the λ_1 of the uncoupled system. The two systems get out of phase at the value $c^* = \lambda_1/2$. Thus c^* provides the largest one-dimensional Liapunov exponent.

In the Java program we consider two phase-coupled Lorenz models. The parameter values for the Lorenz model are $\sigma = 16$, $r = 40$ and $b = 4$. For these values we have numerical evidence that the system shows chaotic behaviour. The largest one-dimensional Liapunov exponent is given by $\lambda = 1.37$. Thus for $c = 0.75$ we find that the systems stay in phase, since $2c > \lambda$. Thus the phase portrait (u_1, v_1) is a straight line. For $c = 0.45$ we have $2c < \lambda$ and the systems are out of phase. Thus the phase portrait is no longer a straight line. We display the phase portrait for different values of c.

```
// Phasecoupling.java

import java.awt.*;
import java.awt.event.*;
import java.awt.Graphics;

public class Phasecoupling extends Frame
{
 public Phasecoupling()
 {
 setSize(600,500);
 addWindowListener(new WindowAdapter()
 { public void windowClosing(WindowEvent event) { System.exit(0); }}); }

 public void fsystem(double h,double t,double u[],double hf[])
 {
 double sigma = 16.0, r = 40.0, b = 4.0;
 double c = 0.65;   // coupling constant change to c = 0.70
 hf[0] = h*(sigma*(u[1]-u[0])+c*(u[3]-u[0]));
 hf[1] = h*(-u[0]*u[2]+r*u[0]-u[1]+c*(u[4]-u[1]));
 hf[2] = h*(u[0]*u[1]-b*u[2]+c*(u[5]-u[2]));
 hf[3] = h*(sigma*(u[4]-u[3])+c*(u[0]-u[3]));
 hf[4] = h*(-u[3]*u[5]+r*u[3]-u[4]+c*(u[1]-u[4]));
 hf[5] = h*(u[3]*u[4]-b*u[5]+c*(u[2]-u[5]));
 }

 public void map(double u[],int steps,double h,double t,int N)
 {
 double uk[] = new double[N];
 double tk;
 double a[] = { 0.0, 1.0/4.0, 3.0/8.0, 12.0/13.0, 1.0, 1.0/2.0 };
 double c[] = { 16.0/135.0, 0.0, 6656.0/12825.0, 28561.0/56430.0,
                -9.0/50.0, 2.0/55.0 };
 double b[][] = new double[6][5];
 b[0][0] = b[0][1]= b[0][2] = b[0][3] = b[0][4] = 0.0;
 b[1][0] = 0.25; b[1][1] = 0.0; b[1][2] = 0.0; b[1][3] = 0.0;
 b[1][4] = 0.0;
 b[2][0] = 3.0/32.0; b[2][1] = 9.0/32.0;
 b[2][2] = 0.0; b[2][3] = 0.0; b[2][4] = 0.0;
```

```
 b[3][0] = 1932.0/2197.0; b[3][1] = -7200.0/2197.0;
 b[3][2] = 7296.0/2197.0; b[3][3] = b[3][4] = 0.0;
 b[4][0] = 439.0/216.0; b[4][1] = -8.0;
 b[4][2] = 3680.0/513.0; b[4][3] = -845.0/4104.0; b[4][4] = 0.0;
 b[5][0] = -8.0/27.0; b[5][1] = 2.0; b[5][2] = -3544.0/2565.0;
 b[5][3] = 1859.0/4104.0; b[5][4] = -11.0/40.0;
 double f[][] = new double[6][N];
 for(int i=0;i<steps;i++) { fsystem(h,t,u,f[0]);
 for(int k=1;k<=5;k++) { tk = t+a[k]*h;
 for(int l=0;l<N;l++) { uk[l] = u[l];
 for(int j=0;j<=k-1;j++) uk[l] += b[k][j]*f[j][l];
 }
 fsystem(h,tk,uk,f[k]);
 }
 for(int l=0;l<N;l++)
  for(int k=0;k<6;k++) u[l] += c[k]*f[k][l];
 }
 }

 public void paint(Graphics g)
 {
 g.drawRect(20,30,500,400);
 int steps = 1;  int N = 6;
 double h = 0.005;  double t = 0.0;
 // initial conditions
 double u[] = { 0.8, 0.8, 0.8, 1.0, 0.4, 0.3 };
 // wait for transients to decay
 for(int i=0;i<2000;i++) { t += h; map(u,steps,h,t,N); }
 t = 0.0;
 for(int i=0;i<6800;i++) { t += h;
 map(u,steps,h,t,N);
 int m = (int)(5.0*u[0]+300.0); int n = (int)(5.0*u[3]+200.0);
 g.drawLine(m,n,m,n);
 }
 }

 public static void main(String[] args)
 { Frame f = new Phasecoupling(); f.setVisible(true); }
}
```

Exercises. (1) The nonlinear differential equation

$$\frac{du}{dt} = u - u^3$$

has attractors at $u = \pm 1$ and a repeller at $u = 0$. Study the phase-coupled system

$$\begin{aligned}\frac{du}{dt} &= u - u^3 + c(v - u)\\ \frac{dv}{dt} &= v - v^3 + c(u - v)\,.\end{aligned}$$

(2) The dynamical system

$$\frac{du_1}{dt} = u_2, \qquad \frac{du_2}{dt} = -u_1 - r\sin(u_2)$$

admits an infinite number of *limit cycles*. For $r > 0$ the fixed point $(0, 0)$ is stable and for $r < 0$ the fixed point $(0, 0)$ is unstable. Study the phase-coupled system

$$\begin{aligned}\frac{du_1}{dt} &= u_2 + c(v_1 - u_1)\\ \frac{du_2}{dt} &= -u_1 - r\sin(u_2) + c(v_2 - u_2)\\ \frac{dv_1}{dt} &= v_2 + c(u_1 - v_1)\\ \frac{dv_2}{dt} &= -v_1 - r\sin(u_2)\,.\end{aligned}$$

(3) Study the coupled maps

$$\begin{aligned}x_{1,t+1} &= rx_{1,t}(1 - x_{1,t}) + c(x_{1,t} - x_{2,t})\\ x_{2,t+1} &= rx_{2,t}(1 - x_{2,t}) + c(x_{2,t} - x_{3,t})\\ x_{3,t+1} &= rx_{3,t}(1 - x_{3,t}) + x(x_{3,t} - x_{1,t})\,.\end{aligned}$$

Chapter 9

Fractals

9.1 Introduction

Mandelbrot [139] introduced the term 'fractal' (from the latin *fractus*, meaning 'broken') to characterize spatial or temporal phenomena that are continuous but not differentiable. Unlike more familiar Euclidean constructs, every attempt to split a fractal into smaller pieces results in the resolution of more structure. Fractal objects and processes are therefore said to display 'self-invariant' (self-similar or self-affine) properties. Self-similar objects are isotropic upon rescaling, whereas rescaling of self-affine objects is directing-dependent (anisotropic). Thus the trace of particulate Brownian motion in two-dimensional space is self-similar, whereas a plot of the x-coordinate of the particle as a function of time is self-affine.

Fractal properties include scale independence, self-similarity, complexity, and infinite length or detail (Falconer [56], Peitgen and Richter [162], Barnsley [12], Barnsley [13], Edgar [53], Peitgen et al [163]). Fractal structures do not have a single length scale, while fractal processes (time series) cannot be characterized by a single time scale. Nonetheless, the necessary and sufficient conditions for an object (or process) to possess fractal properties have not been formally defined. Indeed, fractal geometry has been described as a collection of examples, linked by a common point of view, not an organized theory. Fractal theory offers methods for describing the inherent irregularity of natural objects. In fractal analysis, the Euclidean concept of 'length' is viewed as a process. This process is characterized by a constant parameter D known as the fractal (or fractional) dimension.

Fractals are all around us, in the shape of a mountain range or in the windings of a coast line. The most convincing arguments in favour of the study of fractals is their beauty (Peitgen and Richter [162]). Furthermore the domain of attraction in chaotic systems has in many cases a fractal structure.

There are several fractal dimensions described in the literature (Mandelbrot [139], Falconer [56], Peitgen and Richter [162], Barnsley [12], Barnsley [13], Edgar [53], Peitgen et al [163]). The two most used are the *Hausdorff dimension* and *capacity*. The definition of the Hausdorff dimension is as follows. Let X be a subset of $\mathbb{R}^n$. A cover of X is a (possibly infinite) collection of balls, the union of which contains X. The diameter of a cover $\mathcal{A}$ is the maximum diameter of the balls in $\mathcal{A}$. For d, ϵ, we define

$$\alpha(d,\epsilon) := \inf_{\substack{\mathcal{A}=\text{ cover of } X \\ \text{diam}\,\mathcal{A}\le\epsilon}} \sum_{A\in\mathcal{A}} (\text{diam A})^d$$

and

$$\alpha(d) := \lim_{\epsilon\to 0} \alpha(d,\epsilon)\,.$$

It can be shown that there is a unique d_0 such that

$$d < d_0 \to \alpha(d) = \infty$$

$$d > d_0 \to \alpha(d) = 0\,.$$

This d_0 is defined to be the Hausdorff dimension of X, written $HD(X)$.

The capacity C is a special case of the Hausdorff dimension. In this case the covers consist of balls of uniform size ϵ. We put a grid on $\mathbb{R}^n$ and count the number of boxes that intersect the set X. Suppose we have such an ϵ-grid. Let $N(\epsilon)$ = the number of boxes that intersect the set X. Then $N(\epsilon)\cdot\epsilon^d$ would seem to be a simplified version of $\alpha(d,\epsilon)$ and the desired d can sometimes be computed as

$$C := \lim_{\epsilon\to 0} \frac{\ln N(\epsilon)}{\ln\left(\frac{1}{\epsilon}\right)}\,.$$

This number C is called capacity. We have the inequality $C(X) \le HD(X)$.

The fractal dimension can be viewed as a relative measure of complexity, or as an index of the scale-dependency of a pattern. The fractal dimension is a summary statistic measuring overall 'complexity'. Like many summary statistics (e.g. mean), it is obtained by 'averaging' variation in data structure. In doing so, information is necessarily lost. Excellent summaries of basic concepts of fractal geometry can be found in the books (Mandelbrot [139], Falconer [56], Peitgen and Richter [162], Barnsley [12], Barnsley [13], Edgar [53], Peitgen et al [163]).

Definition. A fractal is by definition a set for which the Hausdorff dimension strictly exceeds the topological dimension.

9.2 Iterated Function System

9.2.1 Introduction

To define an iterated function system (Barnsley [12], Barnsley [13]) we introdue the definition of a complete metric space. Let X be a non-empty set. A real-valued function d defined on $X \times X$, i.e. ordered pairs of elements in X, is called a *metric* or a distance function in X, iff it satisfies, for every $a, b, c \in X$, the following axioms

(i) $d(a, b) \geq 0$ and $d(a, a) = 0$
(ii) $d(a, b) = d(b, a)$ symmetry
(iii) $d(a, c) \leq d(a, b) + d(b, c)$
(iv) If $a \neq b$, then $d(a, b) > 0$.

The nonnegative real number $d(a, b)$ is called the distance from a to b.

Example. Let $X = \mathbb{R}$, then $d(a, b) := |a - b|$ is a metric. ♣

Example. Let $X = \mathbb{R}^n$, then $d(\mathbf{x}, \mathbf{y}) := \sqrt{(x_1 - y_1)^2 + \cdots + (x_n - y_n)^2}$ provides a metric. ♣

Let d be a metric on a non-empty set X. The topology $\mathcal{T}$ on X generated by the class of open spheres in X is called the metric topology. The set X together with the topology $\mathcal{T}$ induced by the metric d is called a *metric space* and is denoted by (X, d).

A hyperbolic *iterated function system* consists of a complete metric space (X, d) together with a finite set of contraction mappings

$$w_n : X \to X, \quad n = 1, 2, \ldots, N$$

with respective contractivity factors s_n, for $n = 1, 2, \ldots, N$.

The notation for the iterated function system is $\{ X : w_n,\ n = 1, 2, \ldots, N \}$ and its contractivity factor is $s := \max\{ s_n : n = 1, 2, \ldots, N \}$. Let (X, d) be a complete metric space. Let $(\mathcal{H}(X), h(d))$ denote the corresponding space of nonempty compact subsets, with the Hausdorff metric $h(d)$. The *Hausdorff metric* is defined by as follows. Let $A, B \in \mathcal{H}(X)$. We define

$$d(A, B) := \max\{ d(x, B) : x \in A \}$$

where

$$d(x, B) := \min \{ d(x, y) : y \in B \}.$$

Here $d(A, B)$ is called the distance from the set $A \in \mathcal{H}(X)$ to the set $B \in \mathcal{H}(X)$. If $A, B, C \in \mathcal{H}(X)$, then

$$d(A \cup B, C) = d(A, C) \vee d(B, C)$$

where $d(A,C) \vee d(B,C)$ denotes the maximum of the two real numbers $d(A,C)$ and $d(B,C)$.

Definition. Let (X,d) be a complete metric space. Then the Hausdorff distance between points A and B in $\mathcal{H}(X)$ is defined as

$$h(A,B) = d(A,B) \vee d(B,A).$$

The following theorem (Barnsley [12], Barnsley [13]) summarizes the main result about a hyperbolic iterated function system.

Theorem. Let $\{X : w_n,\ n = 1,2,\ldots,N\}$ be a hyperbolic iterated function system with contractivity factor s. Then the transformation $W : \mathcal{H}(X) \to \mathcal{H}(X)$ defined by

$$W(B) := \bigcup_{n=1}^{N} w_n(B)$$

for all $B \in \mathcal{H}(X)$, is a contraction mapping on the complete metric space $(\mathcal{H}(X), h(d))$ with contractivity factor s. That is

$$h(W(B), W(C)) \leq s \cdot h(B,C)$$

for all $B, C \in \mathcal{H}(X)$. Its unique fixed point, $A \in \mathcal{H}(X)$, obeys

$$A = W(A) = \bigcup_{n=1}^{N} w_n(A)$$

and is given by

$$A = \lim_{n\to\infty} W^{(n)}(B)$$

for any $B \in \mathcal{H}(X)$.

Definition. The fixed point $A \in \mathcal{H}(X)$ is called the attractor of the iterated function system.

9.2.2 Cantor Set

The standard *Cantor set* is described by the following iterated function system

$$\{\mathbb{R} : w_1,\ w_2\}$$

where

$$w_1(x) := \frac{1}{3}x, \qquad w_2(x) := \frac{1}{3}x + \frac{2}{3}.$$

This is an iterated function system. Starting from the unit interval $[0,1]$ we have

$$w_1([0,1]) = [0,1/3], \qquad w_2([0,1]) = [2/3,1]$$

$$w_1([0,1/3]) = [0,1/9], \quad w_2([0,1/3]) = [2/3,7/9]$$
$$w_1([2/3,1]) = [2/9,1/3], \qquad w_2([2/3,1]) = [8/9,1]$$

etc.. Thus the Cantor set is an example of a fractal. The construction of the standard Cantor set (also called Cantor middle third set or ternary Cantor set) is as follows. We denote the open interval

$$((3r-2)/3^n, (3r-1)/3^n)$$

by $I_{n,r}$, where $r = 1, 2, \ldots, 3^{n-1}$. We set

$$G_n := \bigcup_{r=1}^{3^{n-1}} I_{n,r}, \qquad G := \bigcup_{n=1}^{\infty} G_n.$$

The set $C = [0,1]\backslash G$ is called the standard Cantor set.

In other words: let $I = [0,1]$ be the unit interval. We construct the standard Cantor set in the following way. We set

$$E_0 = [0,1]$$

$$E_1 = \left[0, \frac{1}{3}\right] \cup \left[\frac{2}{3}, 1\right]$$

$$E_2 = \left[0, \frac{1}{9}\right] \cup \left[\frac{2}{9}, \frac{1}{3}\right] \cup \left[\frac{2}{3}, \frac{7}{9}\right] \cup \left[\frac{8}{9}, 1\right]$$

etc.. This means we remove the open middle third, i.e. the interval $(\frac{1}{3}, \frac{2}{3})$. In the second step we remove the pair of intervals $(\frac{1}{9}, \frac{2}{9})$ and $(\frac{7}{9}, \frac{8}{9})$. Then we continue removing middle thirds in this fashion. Then E_k is the union of 2^k disjoint compact intervals and

$$C := \bigcap_{k=0}^{\infty} E_k.$$

The standard Cantor set has cardinal c, is perfect, nowhere dense and has Lebesgue measure zero. Every $x \in C$ can be written as

$$x = a_1 3^{-1} + a_2 3^{-2} + a_3 3^{-3} + \cdots \qquad \text{where } a_j \in \{0, 2\}.$$

The corresponding *Cantor function* ψ is also called the *Devil's staircase.* It satisfies the functional equations

$$\psi(x/3) = \frac{1}{2}\psi(x)$$
$$\psi(1 - x/3) = 1 - \frac{1}{2}\psi(x), \quad (0 \le x \le 1)$$
$$\psi((1-x)/3 + 2x/3) = \frac{1}{2}$$

and is a continuous nondecreasing function with $\psi(0) = 0$, $\psi(1) = 1$.

In the C++ program we generate the first few steps in the construction of the Cantor set.

```
// cantor.cpp

#include <iostream>
#include "rational.h"
#include "verylong.h"
#include "vector.h"
using namespace std;

const Rational<Verylong> a = Rational<Verylong>("0"); // lower limit
const Rational<Verylong> b = Rational<Verylong>("1"); // upper limit

class Cantor
{
 private:
  Vector<Rational<Verylong> > CS;
  int currentSize;
 public:
  Cantor(int);              // constructor
  Cantor(const Cantor&);  // copy constructor
  int step();
  void run();
  friend ostream& operator << (ostream&,const Cantor&);
};

Cantor::Cantor(int nStep) : CS((int)pow(2.0,nStep+1)),currentSize(2)
 { CS[0] = a; CS[1] = b; }

Cantor::Cantor(const Cantor& s) : CS(s.CS),currentSize(s.currentSize) { }

int Cantor::step()
{
 int i, newSize;
 static Rational<Verylong> three(3), tt(2,3);
 static int maxSize = CS.size();
 if(currentSize < maxSize)
 {
 for(i=0;i<currentSize;i++) CS[i] /= three;
 newSize = currentSize+currentSize;
 for(i=currentSize;i<newSize;i++) CS[i] = CS[i-currentSize]+tt;
 currentSize = newSize;
 return 1;
 }
 return 0;
```

```
}

void Cantor::run() { while(step() != 0); }

ostream& operator << (ostream& s,const Cantor& c)
{
 for(int i=0;i<c.currentSize;i+=2)
 { s << "[" << c.CS[i] << " "; s << c.CS[i+1] << "] "; }
 return s;
}

int main(void)
{
 const int nStep = 6;  // number of steps in construction
 Cantor C(nStep);
 cout << C << endl;
 for(int i=0;i<nStep;i++) { C.step(); cout << C << endl; }
 return 0;
}
```

9.2.3 Heighway's Dragon

The *Heighway dragon* is a set in the plane $\mathbb{R}^2$ (Edgar [53]). It is the attractor of an iterated function system. The iterated function system is given by

$$w_1(\mathbf{x}) = \begin{pmatrix} 1/2 & -1/2 \\ 1/2 & 1/2 \end{pmatrix} \mathbf{x}, \qquad w_2(\mathbf{x}) = \begin{pmatrix} -1/2 & -1/2 \\ 1/2 & -1/2 \end{pmatrix} \mathbf{x} + \begin{pmatrix} 1 \\ 0 \end{pmatrix}$$

where $\mathbf{x} \in \mathbb{R}^2$, i.e. $\mathbf{x} = (x_1, x_2)^T$. We have a hyperbolic IFS with each map being a similitude of ratio $r < 1$. Therefore the similarity dimension, d, of the unique invariant set of the IFS is the solution to

$$\sum_{k=1}^{2} r^d = 1 \Longrightarrow d = \frac{\ln(1/2)}{\ln(1/\sqrt{2})} = \frac{\ln(2)}{\ln(\sqrt{2})} = 2$$

where $r = 1/\sqrt{2}$.

The Heighway's dragon can also be constructed as follows. The first approximation P_0 is a line segment of length 1. The next approximation is P_1. It results from P_0 by replacing the line segment by a polygon with two segments, each of length $1/\sqrt{2}$, joined at a right angle. The two ends are the same as before. There are two choices of how this can be done. We choose the left side. For P_2, each line segment in P_1 is replaced by a polygon with two segments, each having length $1/\sqrt{2}$ times the length of the segment that is replaced. The choices alternate between left and right, starting with left, counting from the endpoint of the bottom. The dragon is the limit P of this sequence, P_n, of polygons. Closely related to the Heighway dragon is the *Lévy dragon.*

The Java program asks the user to enter the n-th step in the construction of the dragon.

```
// Dragon.java

import java.awt.*;
import java.awt.event.*;

public class Dragon extends Frame implements ActionListener
{
 public Dragon()
 {
 addWindowListener(new WindowAdapter()
 { public void windowClosing(WindowEvent event) { System.exit(0); }});

 drawButton.addActionListener(this);
 setTitle("Dragon");
 Panel parameterPanel = new Panel();
 parameterPanel.setLayout(new GridLayout(2,1));
 Panel nStepsPanel = new Panel();
 nStepsPanel.add(new Label("no of steps = "));
 nStepsPanel.add(nStepsField);
 Panel buttonPanel = new Panel();
 buttonPanel.add(drawButton);
 parameterPanel.add(nStepsPanel); parameterPanel.add(buttonPanel);
 add("North",parameterPanel);     add("Center",dragonAttractor);
 setSize(400,400); setVisible(true);
 }

 public static void main(String[] args) { new Dragon(); }

 public void actionPerformed(ActionEvent action)
 {
 if(action.getSource()==drawButton)
 dragonAttractor.setSteps(Integer.parseInt(nStepsField.getText()));
 System.out.println(Integer.parseInt(nStepsField.getText()));
 }
 TextField nStepsField = new TextField("6",6);
 Button drawButton = new Button("Draw");
 DragonAttractor dragonAttractor = new DragonAttractor();
}

class DragonAttractor extends Canvas
{
 private int n, scaling;
 public DragonAttractor() { n = 6; }
```

```
public void dragonr(int x1,int y1,int x2,int y2,int x3,int y3,int n)
{
if(n==1)
{
Graphics g = getGraphics();
g.drawLine(x1+scaling,y1+scaling,x2+scaling,y2+scaling);
g.drawLine(x2+scaling,y2+scaling,x3+scaling,y3+scaling);
}
else
{
int x4 = (x1+x3)/2; int y4 = (y1+y3)/2;
int x5 = x3+x2-x4; int y5 = y3+y2-y4;
dragonr(x2,y2,x4,y4,x1,y1,n-1); dragonr(x2,y2,x5,y5,x3,y3,n-1);
}
}

public void paint(Graphics g)
{
Dimension size = getSize();
scaling = Math.min(size.width,size.height)/4;
int xorig = scaling;  int yorig = scaling;
int x1 = xorig+scaling;  int y1 = yorig;
int x2 = xorig;  int y2 = yorig-scaling;
int x3 = xorig-scaling;  int y3 = yorig;
dragonr(x1,y1,x2,y2,x3,y3,n);
}

public void setSteps(int nSteps) { n = nSteps; repaint(); }
}
```

9.2.4 Sierpinski Gasket

The *Sierpinski gasket* (Edgar [53]) is the attractor of an iterated function system with $X = \mathbb{R}^2$ and

$$w_1\begin{pmatrix} x_1 \\ x_2 \end{pmatrix} := \begin{pmatrix} 0.5 & 0 \\ 0 & 0.5 \end{pmatrix}\begin{pmatrix} x_1 \\ x_2 \end{pmatrix}$$

$$w_2\begin{pmatrix} x_1 \\ x_2 \end{pmatrix} := \begin{pmatrix} 0.5 & 0 \\ 0 & 0.5 \end{pmatrix}\begin{pmatrix} x_1 \\ x_2 \end{pmatrix} + \begin{pmatrix} 0.5 \\ 0 \end{pmatrix}$$

$$w_3\begin{pmatrix} x_1 \\ x_2 \end{pmatrix} := \begin{pmatrix} 0.5 & 0 \\ 0 & 0.5 \end{pmatrix}\begin{pmatrix} x_1 \\ x_2 \end{pmatrix} + \begin{pmatrix} 1/4 \\ \sqrt{3}/4 \end{pmatrix}.$$

Thus $N = 3$. One of the ways to construct the Sierpinski gasket is by starting with a triangle. Then splitting the triangle into four triangles and removing the open middle one. The same procedure then applies to the remaining three triangles. The term trema refers to the removed pieces.

In the Java program we display the p-th step in the construction of the Sierpinski gasket. We set $p = 4$.

```
// Sierpinski.java

import java.awt.*;
import java.awt.event.*;
import java.awt.Graphics;

public class Sierpinski extends Frame
{
 public Sierpinski()
 {
 setSize(600,500);
 addWindowListener(new WindowAdapter()
 { public void windowClosing(WindowEvent event) { System.exit(0); }}); }

 public void paint(Graphics g)
 {
 g.drawRect(40,40,600,400);
 int p, n, n1, l, k, m, u1l, u2l, v1l, v2l, xl;
 double u1, u2, v1, v2, a, h, s, x, y;
 p = 4; // step in the construction
 double T[] = new double[p];
 a = Math.sqrt(3.0);
 for(m=0;m<=p;m++)
 {
 for(n=0;n<=((int) Math.exp(m*Math.log(3.0)));n++)
 { n1 = n;
 for(l=0;l<=(m-1);l++) { T[l] = n1%3; n1 = n1/3; }
 x = 0.0; y = 0.0;
 for(k=0;k<=(m-1);k++)
 { double temp = Math.exp(k*Math.log(2.0));
 x += Math.cos((4.0*T[k]+1.0)*Math.PI/6.0)/temp;
 y += Math.sin((4.0*T[k]+1.0)*Math.PI/6.0)/temp;
 }
 u1 = x+a/(Math.exp((m+1.0)*Math.log(2.0)));
 u2 = x-a/(Math.exp((m+1.0)*Math.log(2.0)));
 v1 = y-1.0/(Math.exp((m+1.0)*Math.log(2.0)));
 v2 = y+1.0/(Math.exp(m*Math.log(2.0)));
 xl = (int)(100.0*x+300.0+0.5);
 u1l = (int)(100.0*u1+300.0+0.5); u2l = (int)(100.0*u2+300.0+0.5);
 v1l = (int)(100.0*v1+300.0+0.5); v2l = (int)(100.0*v2+300.0+0.5);
 g.drawLine(u1l,v1l,xl,v2l);  g.drawLine(xl,v2l,u2l,v1l);
 g.drawLine(u2l,v1l,u1l,v1l);
 }
 }
 }
```

```
public static void main(String[] args)
{ Frame f = new Sierpinski(); f.setVisible(true); }
}
```

9.2.5 Koch Curve

The *Koch curve* is a so-called monster curve (Koch [117]). Koch constructed his curve in 1904 as an example of a non-differentiable curve, that is, a continuous curve that does not have a tangent at any of its points. It is of infinite length. However it encloses a simply connected region of finite area in the x_1x_2-plane. The method of constructing the Koch curve is as follows. Take an equilateral triangle and trisect each of it sides; then, on the middle segment of each side, construct equilateral triangles whose interiors lie external to the region enclosed by the base triangle and delete the middle segments of the base triangle. This basic construction is then repeated on all of the sides of the resulting curve, and so on. The curve is defined so that the areas of all triangles lie inside. The perimeter is the length of this curve. If $r = 0$ denotes the stage of evolution when only the base triangle of unit side is there, $r = 1$ denotes the stage when three triangles of side $\frac{1}{3}$ have been added, etc.. Thus the perimeter P_r of the Koch curve at the r-th stage is

$$P_r = (4/3)^r P_0 \,.$$

The perimeter grows unboundedly as r increases, but the area A_r included by the curve at the r-th stage is

$$A_r = A_0 \left(1 + \frac{3}{4} \sum_{i \in \{1,2,\ldots,r\}} 4^i 3^{-2i} \right).$$

As $r \to \infty$ we obtain $A_r \to \frac{8}{5} A_0$. The number of the straight line segments making up the curve at the r-th stage is four times that of the line segments making up the curve at the $(r-1)$-th stage, but the segments become three times smaller in size. The similarity dimension, the capacity and the Hausdorff dimension of the Koch curve are given by

$$D = \frac{\ln(4)}{\ln(3)} \,.$$

The topological dimension is equal to one.

The iterated function system for the Koch curve is given by $X = \mathbb{R}^2$ and

$$w_1(\mathbf{x}) := \begin{pmatrix} 1/3 & 0 \\ 0 & 1/3 \end{pmatrix} \mathbf{x}$$

$$w_2(\mathbf{x}) := \begin{pmatrix} 1/6 & -\sqrt{3}/6 \\ \sqrt{3}/6 & 1/6 \end{pmatrix} \mathbf{x} + \begin{pmatrix} 1/3 \\ 0 \end{pmatrix}$$

$$w_3(\mathbf{x}) := \begin{pmatrix} 1/6 & \sqrt{3}/6 \\ -\sqrt{3}/6 & 1/6 \end{pmatrix} \mathbf{x} + \begin{pmatrix} 1/2 \\ \sqrt{3}/6 \end{pmatrix}$$

$$w_4(\mathbf{x}) := \begin{pmatrix} 1/3 & 0 \\ 0 & 1/3 \end{pmatrix} \mathbf{x} + \begin{pmatrix} 2/3 \\ 0 \end{pmatrix}.$$

The fixed invariant set of this IFS is the Koch curve. We have a hyperbolic IFS with each map being a similitude of ration $r < 1$. Therefore the similarity dimension, d, of the unique invariant set of the IFS is the solution to

$$\sum_{k=1}^{4} r^d = 1 \Longrightarrow d = \frac{\ln(1/4)}{\ln(r)} = \frac{\ln(4)}{\ln(3)} = 1.2619...$$

where $r = 1/3$. The length of the intermediate curve at the n-th iteration of the construction is $(4/3)^n$, where $n = 0$ denotes the original straight line segment. Therefore the length of the Koch curve is infinite. Moreover, the length of the curve between any two points on the curve is also infinite since there is a copy of the Koch curve between any two points.

Instead of starting from an equilateral triangle, one can also start from the unit interval. This is done in the Java program `Koch.java`.

```
// Koch.java

import java.awt.*;
import java.awt.event.*;
import java.awt.Graphics;

public class Koch extends Frame
{
 public Koch()
 {
 setSize(600,500);
 addWindowListener(new WindowAdapter()
 { public void windowClosing(WindowEvent event) { System.exit(0); }}); }

 public int power(int a,int n)
 {
 int t = 1;
 for(int i=1;i<=n;i++) { t = t*a; }
 return t;
 }

 public void paint(Graphics g)
 {
 g.drawRect(20,40,500,400);
 int p, m, u, v, u1, v1, a, b;
 double h, s, x, y, x1, y1;
 p = 3;  // step in the construction
 int kmax = 100;
 int T[] = new int[kmax];
```

```
 m = 0; h = 1.0/((double) power(3,p));
 x = 0.0; y = 0.0;
 for(int n=1;n<=(power(4,p)-1);n++)
 {
 m = n;
 {
 for(int l=0;l<=(p-1);l++) { T[l] = m%4;  m = m/4; }
 s = 0.0;
 for(int k=0;k<=(p-1);k++) { s += ((double)((T[k]+1)%3-1)); }
 x1 = x; y1 = y;
 x += Math.cos(Math.PI*s/3.0)*h; y += Math.sin(Math.PI*s/3.0)*h;
 u = (int)(300*x+100+0.5); v = (int)(300*y+150+0.5);
 u1 = (int)(300*x1+100+0.5); v1 = (int)(300*y1+150+0.5);
 g.drawLine(u,v,u1,v1);
 }
 }
 }

 public static void main(String[] args)
 { Frame f = new Koch(); f.setVisible(true); }
}
```

9.2.6 Fern

Iterated functions systems can also be used to create fern like patterns. The Fern iterated function system was found by M. Barnsley (Barnsley [12], Barnsley [13]). It generates a Black Spleenwort fern. The fern is created using the following four affine transformations

$$w_j(\mathbf{x}) = \begin{pmatrix} r_j\cos(\alpha_j) & -s_j\sin(\beta_j) \\ r_j\sin(\alpha_j) & s_j\cos(\beta_j) \end{pmatrix} \begin{pmatrix} x_1 \\ x_2 \end{pmatrix} + \begin{pmatrix} a_{1j} \\ a_{2j} \end{pmatrix}$$

for $j = 1, 2, 3, 4$. For the four transformations we have

$$w_1(\mathbf{x}) := \begin{pmatrix} 0 & 0 \\ 0 & 0.16 \end{pmatrix} \begin{pmatrix} x_1 \\ x_2 \end{pmatrix} + \begin{pmatrix} 0 \\ 0 \end{pmatrix}$$

$$w_2(\mathbf{x}) := \begin{pmatrix} 0.85 & 0.04 \\ -0.04 & 0.85 \end{pmatrix} \begin{pmatrix} x_1 \\ x_2 \end{pmatrix} + \begin{pmatrix} 0 \\ 1.6 \end{pmatrix}$$

$$w_3(\mathbf{x}) := \begin{pmatrix} 0.2 & -0.26 \\ 0.23 & 0.22 \end{pmatrix} \begin{pmatrix} x_1 \\ x_2 \end{pmatrix} + \begin{pmatrix} 0 \\ 1.6 \end{pmatrix}$$

$$w_4(\mathbf{x}) := \begin{pmatrix} -0.15 & 0.28 \\ 0.26 & 0.24 \end{pmatrix} \begin{pmatrix} x_1 \\ x_2 \end{pmatrix} + \begin{pmatrix} 0 \\ 0.44 \end{pmatrix}.$$

The matrix norms of the matrices are smaller than 1. Thus the maps w_1, w_2, w_3 and w_4 are all contractions. Now w_1 maps the entire fern onto the fern with the two lowest branches and the lowest part of the stem removed. The fixed point of w_1 is the tip of the fern. w_2 transforms the entire fern to the branch on the lower right; it shrinks, narrows, and flips. Similarly, w_3 transforms the entire fern to the branch on the lower left. Finally, w_4 maps the entire fern into a line segment on the x_2 axis; its fixed point is the origin, and it is used to generate the stem.

In our Java program (Applet) we modify the coefficients of the four maps for a better graphical display.

```
// Fern.java

import java.awt.*;
import java.applet.*;
import java.awt.Graphics;

public class Fern extends Applet
{
 double mxx, myy, bxx, byy;

 public void plot(Graphics g)
 {
 double x, y, xn, yn, r;
 int pex, pey;
 int max = 15000;     // number of iterations
 x = 0.5; y = 0.0;    // starting point
 convert(0.0,1.0,1.0,-0.5);
 setBackground(Color.white); g.setColor(Color.black);
 for(int i=0;i<=max;i++)
 {
 r = Math.random();  // generate a random number
 if(r <= 0.02) { xn = 0.5; yn = 0.27*y; } // map 1
 else if((r>0.02) && (r<=0.17)) // map2
 { xn = -0.139*x+0.263*y+0.57; yn =  0.246*x+0.224*y-0.036; }
 else if((r>0.17) && (r<=0.3))  // map3
 { xn = 0.17*x-0.215*y+0.4086; yn = 0.222*x+0.176*y+0.0893; }
 else // map4
 { xn = 0.781*x+0.034*y+0.1075; yn = -0.032*x+0.739*y+0.27; }
 x = xn; y = yn;
 pex = (int)(mxx*x+bxx); pey = (int)(myy*y+byy);
 g.drawLine(pex,pey,pex,pey);  // output to screen
 }
 }

 void convert(double xiz,double ysu,double xde,double yinf)
 {
 double maxx, maxy, xxfin, xxcom, yyin, yysu;
 maxx = 600; maxy = 450;
 xxcom = 0.15*maxx; xxfin = 0.75*maxx;
 yyin = 0.8*maxy; yysu = 0.2*maxy;
 mxx = (xxfin-xxcom)/(xde-xiz);
```

```
 bxx = 0.5*(xxcom+xxfin-mxx*(xiz+xde));
 myy = (yyin-yysu)/(yinf-ysu);
 byy = 0.5*(yysu+yyin-myy*(yinf+ysu));
 }

 public void paint(Graphics g)
 { plot(g); showStatus("Done"); }
}
```

9.2.7 Grey Level Maps

An n-map iterated function system with grey level maps is defined as

1. The iterated function system $\mathbf{w} = \{w_1, w_2, \ldots, w_n\}$ where each $w_i : X \to X$ is a contraction and

2. the grey level component $\Phi = \{\phi_1, \phi_2, \ldots, \phi_n\}$ where each $\phi_i : \mathbb{R} \to \mathbb{R}$ is Lipshitz.

Here (X, d) is a complete metric space, the pixel space. In many cases we have $X = [0,1] \times [0,1]$ (unit square) with the Euclidean metric. Associated with an iterated function system with grey level maps is an operator T, the fractal transform, whose action on a suitable space of function $\mathcal{F}(X)$ is given by

$$T(f)(x) = \sum_{i=1}^{n} \phi_i(f(w_i^{-1}(x))), \qquad f \in \mathcal{F}(X).$$

It is assumed that both w_i and ϕ_i are affine. Hence

$$w_i(x) = s_i x + a_i, \qquad \phi_i(t) = \alpha_i t + \beta_i .$$

Another special case of an iterated function system with grey level maps operator is the following

$$T(f)(x) = \theta(x) + \sum_{i=1}^{n} \alpha_i f(w_i^{-1}(x)).$$

Here, the $\theta(x)$ acts as a condensation function for the iterated function system with grey level maps.

Under suitable conditions on the w_i and the ϕ_i, the operator T is contractive in $\mathcal{F}(X)$. For example in the Banach space $L_p(X)$ it suffices that

$$(\sum_{i=1}^{n} |s_i \alpha_i^p|)^{1/p} < 1$$

and so it has a unique fixed point $u^* = Tu^*$.

Let $\{q_j(x)\}$ be an orthonormal basis of the Hilbert space of the square integrable functions $L_2(X)$. Given an iterated function system with grey level maps operator T, there exists a corresponding operator M on the sequence Hilbert space $\ell_2(\mathbb{N})$ viewed as sequences of coefficients in expansions of functions with respect to the functions of the orthonormal basis. If the iterated function system with grey level maps operator T is affine or linear, then the operator M will also be affine or linear. If T is contractive in the Hilbert space $L_2(X)$, then M is contractive in the Hilbert space $\ell_2(\mathbb{N})$. When the orthonormal basis is the Haar wavelet basis ψ_{jk}, then the operator M becomes a mapping of wavelet coefficients subtrees to lower subtrees.

9.3 Mandelbrot Set

Let $\mathbb{C}$ be the complex plane. Let $c \in \mathbb{C}$. The *Mandelbrot set* M is defined as follows

$$M := \{\, c \in \mathbb{C} : c,\ c^2 + c,\ (c^2 + c)^2 + c,\ \ldots \not\to \infty \,\}.$$

To find the Mandelbrot set we study the recursion relation

$$z_{t+1} = z_t^2 + c$$

where $t = 0, 1, 2, \ldots$ and the initial value is given by $z_0 = 0$. This is obvious since

$$z_1 = c, \quad z_2 = c^2 + c, \quad z_3 = (c^2 + c)^2 + c, \ldots$$

etc.. Since $z = x + iy$ and $c = c_1 + ic_2$ with $x, y, c_1, c_2 \in \mathbb{R}$ we can rewrite the recursion relation as

$$x_{t+1} = x_t^2 - y_t^2 + c_1, \qquad y_{t+1} = 2x_t y_t + c_2$$

with the initial value $(x_0, y_0) = (0, 0)$. For a given $c \in \mathbb{C}$ (or $(c_1, c_2) \in \mathbb{R}^2$) we can now study whether or not c belongs to M. For example, $(c_1, c_2) = (0, 0)$ and

$$(c_1, c_2) = (1/4, 1/4)$$

belong to M. The point $(c_1, c_2) = (1/2, 0)$ does not belong to M. We find that

$$z_0 = 0, \quad z_1 = \frac{1}{2}, \quad z_2 = \frac{3}{4}, \quad z_3 = \frac{17}{16}, \ldots$$

Thus the set M can be defined using nothing more than complex arithmetic. The polynomial $f_c(z) = z^2 + c$ has a unique critical point $\omega = 0$, i.e., a point ω where the derivative $f_c'(\omega)$ equals 0. The parameter value c is called the critical value $c = f_c(0)$. Since both the variable z and the parameter c fill out a plane, it can cause some confusion; in particular because we jump back and forth between these planes. For fixed c we refer to the z-plane as the dynamical plane for f_c, while we refer to the c-plane as the parameter plane.

```
// Mandelbrot.java

import java.awt.*;
import java.awt.event.*;
import java.awt.Graphics;
import java.awt.image.*;

public class Mandelbrot extends Frame
{
 public Mandelbrot()
 {
 setSize(600,500);
 addWindowListener(new WindowAdapter()
 { public void windowClosing(WindowEvent event) { System.exit(0); }}); }

 public void paint(Graphics g)
 {
 Image img;
 int w = 256, h = 256;
 int[] pix = new int[w*h];
 int index = 0;  int iter;
 double a, b, p, q, psq, qsq, pnew, qnew;
 for(int y=0;y<h;y++)
 { b = ((double)(y-128))/64;
 for(int x=0;x<w;x++)
 { a = ((double)(x-128))/64;
 p = q = 0.0;
 iter = 0;
 while(iter < 32)
 {
 psq = p*p; qsq = q*q;
 if(psq+qsq >= 4.0) break;
 pnew = psq-qsq+a; qnew = 2.0*p*q+b; p = pnew; q = qnew;
 iter++;
 }
 if(iter==32) { pix[index] = 255 << 24 | 255; }
 index++;
 }
 }
 img = createImage(new MemoryImageSource(w,h,pix,0,w));
 g.drawImage(img,0,0,null);
 }

 public static void main(String[] args)
 { Frame f = new Mandelbrot(); f.setVisible(true); }
}
```

9.4 Julia Set

Let f_λ be a mapping in the complex plane, where λ is a complex parameter. The Julia set (Julia [107], Beardon [16]) of f_λ can be defined in two different ways.

Dynamical Definition. $J(f_\lambda)$ is the closure of the set of repelling periodic points. A point $z_0 \in \mathbb{C}$ is periodic if

$$f_\lambda^{(n)}(z_0) = z_0$$

for some n, where $f_\lambda^{(n)}$ denotes the n-fold composition of f_λ with itself. This periodic point is repelling if

$$\left|(f_\lambda^{(n)})'(z_0)\right| > 1$$

where $'$ denotes the derivative.

Complex Analytic Definition. $J(f_\lambda)$ is the set of points z such that the family of functions

$$\left\{ f_\lambda^{(n)} \right\}$$

fails to be a normal family of functions in every neighbourhood of z.

We also find the definition in the form:

z is a stable point of the (rational) map f_λ if there is a neighbourhood U of z on which the iterates

$$f_\lambda^{(n)}$$

form a normal family. Let Ω be the stable set of f. The Julia set is the complement of Ω. It is named after the mathematician Julia who, along with Fatou, originated this subject in the 1920's. The set Ω is open by definition. It must either be empty or dense in $\hat{\mathbb{C}}$ and is often not connected.

Definition. A family of complex analytic maps $\{f_\lambda\}$, defined on a domain D is called a *normal family* if it satisfies the following conditions: every infinite subset of $\{f_\lambda\}$ contains a subsequence which converges uniformly on every compact subset of D.

The Julia set has mainly been studied for the map

$$f_c(z) = z^2 + c$$

where $c \in \mathbb{C}$. In the program `Julia.java` we consider the Julia set for this map, where $c = p + iq$ and $z^2 = x^2 - y^2 + i2xy$ with x, y, p, q are real.

```
// Julia.java

import java.awt.*;
import java.awt.event.*;
```

```
import java.awt.Graphics;

public class Julia extends Frame
{
 public Julia()
 {
 setSize(600,500);
 addWindowListener(new WindowAdapter()
 { public void windowClosing(WindowEvent event) { System.exit(0); }}); }

 public void paint(Graphics g)
 {
 int a = 400, b = 400, kmax = 200, m = 100;
 double x0, y0, dx, dy, x, y, r;
 double p = -0.12256117, q = 0.74486177;
 double xmin = -1.5, xmax = 1.5, ymin = -1.5, ymax = 1.5;
 dx = (xmax-xmin)/((double)a-1); dy = (ymax-ymin)/((double)b-1);
 for(int nx=0;nx<=a;nx++)
 for(int ny=0;ny<=b;ny++)
 { int k = 0;
 x0 = xmin+nx*dx; y0 = ymin+ny*dy; r = 0;
 while((k++ <= kmax) && (r < m))
 { x = x0; y = y0;
 x0 = x*x-y*y+p; y0 = 2.0*x*y+q; r = x0*x0+y0*y0;
 if(r > m) { }
 }
 if(r <= m) { g.drawLine(nx,ny,nx,ny); }
 }
 }

 public static void main(String[] args)
 { Frame f = new Julia(); f.setVisible(true); }
}
```

Exercises. (1) Study the Julia set for the map

$$f(z) = \exp(z) .$$

Find the fixed points of the map and study their stability.

(2) Let $\lambda \in \mathbb{R}$. Study the Julia set for the map

$$f_\lambda(z) = \exp(i\lambda z) .$$

Find the fixed points of the map and study their stability.

9.5 Fractals and Kronecker Product

The *Kronecker product* of matrices (also known as the *tensor product* or *direct matrix product*) has been used in a variety of fields (Steeb [194], Steeb [197]). The Kronecker product can also be used in image processing and related fields (Hardy and Steeb [88]). Here we describe how the Kronecker product can be used in the construction of fractals.

Let A be an $m \times n$ matrix and B be an $r \times s$ matrix. The Kronecker product of A and B is defined as the $(m \cdot r) \times (n \cdot s)$ matrix

$$A \otimes B := \begin{pmatrix} a_{11}B & a_{12}B & \cdots & a_{1n}B \\ a_{21}B & a_{22}B & \cdots & a_{2n}B \\ \vdots & \vdots & \ddots & \vdots \\ a_{m1}B & a_{m2}B & \cdots & a_{mn}B \end{pmatrix}.$$

Example. Let

$$A = \begin{pmatrix} 1 & 1 \\ 0 & 1 \end{pmatrix}.$$

Then

$$A \otimes A = \begin{pmatrix} 1 & 1 & 1 & 1 \\ 0 & 1 & 0 & 1 \\ 0 & 0 & 1 & 1 \\ 0 & 0 & 0 & 1 \end{pmatrix}.$$ ♣

The Kronecker product is associative

$$(A \otimes B) \otimes C = A \otimes (B \otimes C).$$

The Kronecker product is also distributive

$$(A + B) \otimes C = A \otimes C + B \otimes C$$

for A and B both $m \times n$ matrices and C a $p \times q$ matrix. Let $c \in \mathbb{C}$. Then

$$((cA) \otimes B) = c(A \otimes B) = (A \otimes (cB)).$$

Let A be an $m \times n$ matrix, B be a $p \times q$ matrix, C be an $n \times r$ matrix and D be a $q \times s$ matrix then

$$(A \otimes B)(C \otimes D) = (AC) \otimes (BD)$$

where AC, BD denote matrix multiplication.

The multiple Kronecker product is defined in a recursive fashion as

$$\otimes_{i=1}^{k} A_i := (\otimes_{i=1}^{k-1} A_i) \otimes A_k = A_1 \otimes A_2 \otimes \cdots \otimes A_k$$

where $\otimes_{i=1}^{1} A_i = A_1$. We can represent or approximate a given matrix A by the sum of several Kronecker products

$$A \cong \sum_{i=1}^{p} A_{i1} \otimes A_{i2} \otimes \cdots \otimes A_{ik}.$$

This is called the multiple Kronecker product sum approximation of the matrix A. Images and fractals can thus be represented by this multiple Kronecker product.

Many fractals have a self-similar nature, and so can be constructed from a union of self-similar sets. A closed, bounded self-similar set of the Euclidean plane $\mathbb{R}^2$ is a set of the form $A = A_1 \cup A_2 \cup \cdots \cup A_k$ with each of the sets A_i non-overlapping with A and congruent to A.

Example. The simplest case of such a set, is the middle-third *Cantor set.* The Cantor set is obtained by starting with a line segment, and removing the middle third. The middle third of each of the remaining two segments is then also removed, and the process is continued ad infinitum. It is clear that the two remaining segments are self-similar to the original segment, scaled by a factor of $\frac{1}{3}$. We can express the middle-third Cantor set using the Kronecker product. Let $\mathbf{x} = (1, 0, 1)$ which represents the fact that the middle third is removed. The operation can be applied again to the remaining two segments by applying the Kronecker product

$$\mathbf{x} \otimes \mathbf{x} = (1, 0, 1, 0, 0, 0, 1, 0, 1).$$

Thus each 1 has been replaced by a copy of the original vector. The middle-third Cantor set is thus given by

$$\bigotimes_{i=1}^{\infty} \mathbf{x}.$$

An approximation is given by $\bigotimes_{i=1}^{n} \mathbf{x}$. This set can be visualized if each entry 0 is identified with a black pixel and an entry 1 with a white pixel. ♣

Other self-similar fractals can be produced in the same way. For any matrix produced in this fashion, each 0 entry is identified by a black pixel and each 1 entry is identified by a white pixel.

Example. The *Sierpinski carpet* is obtained by taking a square, and dividing the square into 9 equal squares. The middle square is removed, and the process is repeated on the 8 remaining squares. The Sierpinski carpet can thus be generated from the defining matrix

$$X = \begin{pmatrix} 1 & 1 & 1 \\ 1 & 0 & 1 \\ 1 & 1 & 1 \end{pmatrix}.$$

Now the Kronecker product is used to create the Sierpinski carpet through iteration. ♣

Exanple. The *Sierpinski triangle* can be generated in the same way with the matrix

$$X = \begin{pmatrix} 1 & 1 \\ 0 & 1 \end{pmatrix}.$$

♣

Gray scale fractal images can be produced if the elements of the matrix are allowed to take on values between 0 and 1. An example would be the matrix

$$X = \begin{pmatrix} 1.0 & 0.5 & 1.0 \\ 0.5 & 1.0 & 0.5 \\ 1.0 & 0.5 & 1.0 \end{pmatrix}.$$

Obviously nonfractal images can also be generated.

Example. The Kronecker product sum

$$\left(\bigotimes_{j=1}^{n} X \right) \otimes B$$

with matrices

$$X = \begin{pmatrix} 1 & 1 \\ 1 & 1 \end{pmatrix} \quad \text{and} \quad B = \begin{pmatrix} 0 & 1 \\ 1 & 0 \end{pmatrix}$$

generates a checkerboard. ♣

We give a *Metapost program* that generates the Sierpinski triangle up to level 5 and the checkerboard pattern up to level 4.

```
% Kronecker product fractals in metapost
% Only 2x2 matrices are supported since reckronecker (below)
% hardcodes the interpretation of the matrix
% r is the matrix that stores the matrix that is repeatedly
% applied (using the kronecker product) on the left
% The format is
% [ r0 r1 ]
% [ r2 r3 ]
numeric r[];
% p is the base matrix (applied on the right)
numeric p[];

def reckronecker(expr d,x,A)=
    if(d=0):
        %fill unitsquare transformed A withcolor white;
    if(x*r0=0): fill unitsquare scaled 0.5 shifted(0.0,0.0)
                transformed A; fi;
    if(x*r1=0): fill unitsquare scaled 0.5 shifted(0.5,0.0)
                transformed A; fi;
    if(x*r2=0): fill unitsquare scaled 0.5 shifted(0.0,-0.5)
                transformed A; fi;
```

```
    if(x*r3=0): fill unitsquare scaled 0.5 shifted(0.5,-0.5)
                transformed A; fi;
    if(x*r0=1): draw unitsquare scaled 0.5 shifted(0.0,0.0)
                transformed A; fi;
    if(x*r1=1): draw unitsquare scaled 0.5 shifted(0.5,0.0)
                transformed A; fi;
    if(x*r2=1): draw unitsquare scaled 0.5 shifted(0.0,-0.5)
                transformed A; fi;
    if(x*r3=1): draw unitsquare scaled 0.5 shifted(0.5,-0.5)
                transformed A; fi;
    else:
    reckronecker(d-1,x*p0,identity scaled 0.5 shifted(0.0,0.0)
                transformed A);
    reckronecker(d-1,x*p1,identity scaled 0.5 shifted(0.5,0.0)
                transformed A);
    reckronecker(d-1,x*p2,identity scaled 0.5 shifted(0.0,-0.5)
                transformed A);
    reckronecker(d-1,x*p3,identity scaled 0.5 shifted(0.5,-0.5)
                transformed A);
    fi
enddef;

def kronecker(expr d)=
      transform A;
      pickup pencircle scaled 0.5pt;
      A=identity scaled 4cm shifted(0,4cm*(1.0-1.0/(2.0**(d+1))));
      reckronecker(d,1,A);
enddef;

%sierpinski triangle to level 5
beginfig(1);
r0:=1; r1:=1; r2:=0; r3:=1; p0:=1; p1:=1; p2:=0; p3:=1;
kronecker(5);
endfig;
%checkerboard to level 4
beginfig(2);
r0:=0; r1:=1; r2:=1; r3:=0; p0:=1; p1:=1; p2:=1; p3:=1;
kronecker(4);
endfig;
end;
```

Exercises. (1) Consider the set C obtained from the unit interval $[0,1]$ by first removing the middle third of the interval and the removing the middle fifths of the two remaining intervals. Now iterate this process, first removing the middle thirds, then removing middle fifths. Then C is a fractal with the fractal dimension $\ln(4)/\ln(15/2)$. Can this process be implemented using the Kronecker product?

(2) What fractal do we generate starting from the matrix

$$X = \begin{pmatrix} 1 & 1 & 0 \\ 1 & 1 & 1 \\ 0 & 1 & 1 \end{pmatrix}.$$

9.6 Lindenmayer Systems and Fractals

A *Lindenmayer system* (A, R, s) is defined as follows (Lindenmayer [130]). Consider a finite set A of characters (the alphabet), a map $R : A \to A^*$ and a non-empty starting word s (initial string, axiom), an element of A^*. A^* are the words with characters from A. For each $a \in A$ the pair $(a, R(a))$ is called a rule and is written as

$$a \to b_1 b_2 \cdots b_n$$

where $R(a) = b_1 b_2 \cdots b_n \in A^*$. a is the left hand side and $b_1 b_2 \cdots b_n$ the right hand side of the rule. A Lindenmayer system describes a language L, a subset of A^*. The language is defined as follows:

- s is an element of L.
- Let w be an element of L and let $\sim w$ be the word where each character a of w has been replaced by $R(a)$. Then $\sim w$ is in L.

From the starting word $s = s_0$ the word s_1 is created by replacing all characters by their rule image (the right hand side of the corresponding rule). From s_1 the word s_2 is created, from that s_3 and so on. Call s_i the i-th generation of the starting word s. The interpretation of the words s_i of the language L can be done using a turtle. It visualizes the words of the language. A turtle is a drawing device which understands a few simple commands. Given a word of the language L each character of the word is interpreted as a command for the turtle. The word turns into a picture with the help of the turtle. A turtle has a position in the plane, a forward direction and a colour. It understands the following commands: Move forward a given number of units and draw a line, move forward a given number of units without drawing, turn left a given number of degrees, turn right a given number of degrees and change our colour to a given colour. Furthermore, a turtle may remember its current state (position, direction and colour) by pushing it onto a stack and by changing its state to a former one by popping it off from the stack. For each character of the alphabet A, one of these turtle commands may be defined. A character may also correspond to no command at all, causing the turtle to do nothing. A turtle command may be one of the identifiers `Move`, `Line`, `Left`, `Right`, `Push` or `Pop`. These commands cause the turtle to move without drawing, to draw a line, to turn left or right and to push or pop its current state. A colour value must be a list of three numbers [r,g,b] defining new red-, green- and blue colour values for the turtle.

The following default rules for the turtle commands exist:

```
"F" = Line = move forward one unit and draw a line
"f" = Move = move forward one unit
"+" = Left = change forward direction by a left rotation
      of deg degrees
"-" = Right = change forward direction by a right rotation
      of deg degrees
"[" = Push = push current state to stack
"]" = Pop = pop current state from stack
```

These rules are used if no other rules for the turtle commands are defined. It is also necessary to specify which generation of the starting word of the system is to be plotted. We write a Java applet that implements the turtle.

We restrict the stack to one position. We use the start F and the rule

```
F -> F[+F]F[-F]F
```

The angle is $30^\circ = \pi/6$.

```
// Lindenmayer.java

import java.awt.*;
import java.applet.*;

public class Lindenmayer extends Applet
{
 Point a, b;
 int lengthF = 3;           // step length
 double rotation = 30.0;  // rotation in deg
 double direction;
 Graphics g; Graphics2D g2;
 public void init() { setBackground(new Color(255,255,255)); }

 public void paint(Graphics g)
 {
 g2 = (Graphics2D) g;      // Anti-Aliasing
 g2.setRenderingHint(RenderingHints.KEY_ANTIALIASING,
                     RenderingHints.VALUE_ANTIALIAS_ON);
 g2.setColor(new Color(110,170,60)); // Color
 a = new Point(115,495); // starting point
 direction = -80;          // starting direction
 turtle(g2,"F",6);
 }

 public void turtle(Graphics g2,String instruc,int depth)
 {
 if(depth==0) return;
 depth -= 1;
 Point aMark = new Point(0,0);
```

```
double directionMark = 0.0;
char c;
for(int i=0;i<instruc.length();i++)
{
c = instruc.charAt(i);   // step forward
if(c=='F')
{ turtle(g2,"F[+F]F[-F]F",depth); // iteration
if(depth==0) // draw
{
double rad = 2.0*Math.PI*direction/360.0; // deg -> rad
int p = (int)(lengthF*Math.cos(rad));
int q = (int)(lengthF*Math.sin(rad));
b = new Point(a.x+p,a.y+q);
g2.drawLine(a.x,a.y,b.x,b.y);
a = b; // new starting point
}
}
else if(c=='+') direction += rotation; // rotation left
else if(c=='-') direction -= rotation; // rotation right
// store position and direction
else if(c=='[') { aMark = a; directionMark = direction; }
else if(c==']') { a = aMark; direction = directionMark; }
}
}
}
```

Example. A simple case of a Lindenmayer system would be

$$\text{alphabet: } \{a,b\}, \quad \text{rules: } a \to ab, \;\; b \to a, \quad \text{axiom: } b$$

Thus we obtain

$$b \to a \to ab \to aba \to abaab \to abaababa \to \cdots$$ ♣

Example. The *Hilbert curve* is a fractal that can be produced by the L-system productions

$$L \to +RF - LFL - FR+, \qquad R \to -LF + RFR + FL - .$$

The symbol $+$ indicates a clockwise rotation of 90 degrees and $-$ indicates an anticlockwise rotation of 90 degrees. F indicates that a line must be drawn in the current direction, 1 unit in length. R and L are ignored when drawing the Hilbert curve and are only used in productions. The initial string used to produce the curve is simply L. ♣

Example. The *Koch snowflake* can also be produced by an L-System. The productions are given by

$$F \to F + F - -F + F .$$

The initial string describing the snowflake is given by $F - -F - -F$. Once again F indicates that a line should be drawn in the current direction, $+$ indicates a clockwise rotation of 60 degrees and $-$ indicates an anticlockwise rotation of 60 degrees. ♣

9.7 Weierstrass Function

One way to obtain a function whose graph has a fine structure (fractal structure) is to add together a sequence of functions which oscillate increasingly rapidly. Thus if

$$\sum_{k=1}^{\infty} |a_k| < \infty$$

and $\lambda_k \to \infty$ if $k \to \infty$, the function defined by the trigonometric series

$$f(x) = \sum_{j=1}^{\infty} a_j \sin(\lambda_j x)$$

might be expected to have a graph of dimension greater than 1 if a_k and λ_k are chosen suitably. The best-known example of this type is the function

$$f(x) = \sum_{j=1}^{\infty} \lambda^{(s-2)j} \sin(\lambda^j x)$$

where $1 < s < 2$ and $\lambda > 1$, constructed by Weierstrass to be continuous but nowhere differentiable. It is conjectured that the *Weierstrass function* has a graph of dimension s, but this does not appear to have been proven rigorously. The dimension cannot exceed s. There are a number of variants of the Weierstrass function.

There are two ways of measuring the regularity of a continuous non-differentiable function $f : K \to \mathbb{R}$. Here K is a compact set of $\mathbb{R}$. The first is based on the *Hölder properties* of the continuous function f. From a global point of view one looks for the largest $\alpha_g > 0$ such that there exists $C > 0$ so that for all $x, y \in K$

$$|f(x) - f(y)| < C|x - y|^{\alpha_g}.$$

A local approach is to look only at a neighbourhood of x_0 for the largest exponent such that

$$\sup_{x,y \in B(x_0,r)} \frac{|f(x) - f(y)|}{|x - y|^{\alpha_\ell}}$$

is finite and then to make r tend to 0. The pointwise exponent is the largest α_p such that

$$\sup_{x,y \in B(x_0,r)} \frac{|f(x) - f(y)|}{|x - y|^{\alpha_p}}$$

is finite. The second way of evaluating the regularity of a continuous non-differentiable function is to measure the dimension of its graph, for example utilizing the Hausdorff dimension or the box counting dimension.

For the Java-Applet program we select $s = 3/2$ and $\lambda = 4$. Then we draw the function

$$f_N(x) = \sum_{j=1}^{N} \frac{1}{2^j} \sin(4^j x)$$

which is an approximation of the Weierstrass function with $N = 3$. The second and third pieces of the applet tag indicate the width and the height of the applet in pixels. The upper left corner of the applet is always at x-coordinate 0 and y-coordinate 0.

```
// Weierstrass.java

import java.awt.Graphics;

public class Weierstrass extends java.applet.Applet
{
 double f(double x)
 {
 double expr = 1.0/2.0*Math.sin(4.0*x)+1.0/4.0*Math.sin(16.0*x)
               +1.0/8.0*Math.sin(64.0*x);
 return expr*getSize().height/4.0+150.0;
 }

 public void paint(Graphics g)
 {
 for(int n=0;n<getSize().width;n++)
 { g.drawLine(n,(int)f(n),n+1,(int)f(n+1)); }
 }
}
```

The *Weierstrass-Takagi function* (Yamaguchi et al [222]) is an extension of the Weierstrass function and is defined by

$$W(x) := \sum_{k=0}^{\infty} a^{-k} g(f^{(k)}(x))$$

where f is a one-dimensional (chaotic) map $f : [0,1) \to [0,1)$, g is an arbitrary periodic function, and $f^{(k)}$ denotes the k-th iterate of the function f.

The function W satisfies the functional equation

$$W(f(x)) = aW(x) - ag(x)\,.$$

Example. Let $f : [0,1) \to [0,1)$

$$f(x) = 2x \pmod 1$$

and

$$g(x) = \cos(2\pi x)\,.$$

With $a = 2$ we obtain the Weierstrass function. ♣

Example. Let

$$f(x) = \begin{cases} 2x & \text{if } 0 \le x \le 1/2 \\ 2-2x & \text{if } 1/2 \le x < 1 \end{cases}$$

and $g(x) = f(x)/2$. With $a = 2$ we obtain the Takagi function. ♣

Exercises. (1) Find the derivative of the Weierstrass function in the sense of generalized functions.

(2) Find the Fourier transform of the Weierstrass function in the sense of generalized functions.

(3) Study the Weierstrass-Takagi function if f is the bungalow-tent map.

9.8 Lévy-Flight Random Walk

There is a close connection between the Weierstrass function and a *Lévy-flight random walk* (Shlesinger et al [186], Chechkin et al [29]). Lévy flights are a special class of random walks whose step lengths are not constant but rather are chosen from a probability distribution with a power-law tail. We start with the discrete jump probability distribution

$$p(x) = \frac{\lambda - 1}{2\lambda} \sum_{k=0}^{\infty} \lambda^{-k} (\delta(x, +b^k) + \delta(x, -b^k))$$

where $b > \lambda > 1$ are parameters that characterize the distribution, and $\delta(x, y)$ is the Kronecker delta, which is equal to 1 when $x = y$ and zero otherwise. This distribution function allows jumps of length $1, b, b^2, b^3 \ldots$. However, whenever the length of the jump increases by an order of magnitude (in base b), its probability of occurring decreases by an order of magnitude (in base λ). Typically one gets a cluster of λ jumps roughly of length 1 before there is a jump of length b. Approximately λ such clusters, separated by lengths of order b, are formed before one sees a jump of order b^2. This goes on, forming a hierarchy of clusters within clusters. The Fourier transform of $p(x)$ is

$$\widetilde{p}(k) = \frac{\lambda - 1}{2\lambda} \sum_{j=0}^{\infty} \lambda^{-j} \cos(b^j k)$$

which is the self-similar Weierstrass function. The self-similarity of the Weierstrass function appears explicitly through the equation

$$\widetilde{p}(k) = \frac{1}{\lambda} \widetilde{p}(bk) + \frac{\lambda - 1}{\lambda} \cos(k)$$

which has a small-k solution

$$\widetilde{p}(k) \approx \exp(-|k|^{\beta})$$

with $\beta = \log(\lambda)/\log(b)$.

Normal and anomalous diffusion in Hamilton systems has been studied by Kaneko and Konishi [110]. They considered the map

$$q_{n+1}(j) = q_n(j) + p_{n+1}(j)$$
$$p_{n+1}(j) = p_n(j) + \frac{K}{2\pi}(\sin(2\pi(q_n(j+1) - q_n(j))) + \sin(2\pi(q_n(j-1) - q_n(j))))$$

where $j = 0, 1, \ldots, N-1$ with periodic boundary conditions ($N \equiv 0$). The map can be obtained from the Hamilton function

$$H(p, q, t) = \frac{1}{2}\sum_{j=0}^{N-1}(p(j))^2 - \frac{K}{(2\pi)^2}(\sum_{j=0}^{N-1}\cos(2\pi(q(j+1) - q(j)))\sum_{n=-\infty}^{n=\infty}\delta(t-n)).$$

Anomalous diffusion (also known as Lévy flight) exists only up to some crossover time beyond which the diffusion is normal.

Chapter 10

Cellular Automata

10.1 Introduction

Cellular automata (Wolfram [218], Wolfram [219], Hedlund [92], Frisch et al [62]) may be considered as discrete dynamical systems. They are discrete in several aspects. Firstly, they consist of a discrete spatial lattice of sites. Secondly they evolve in discrete time steps, i.e. $t = 0, 1, 2, \ldots$. Thirdly each lattice site (or box, or cell) has only a finite discrete set of possible values. The simplest case is that the dependent variable takes two values, 0 and 1. The simplest case of a lattice is a linear chain with N lattice sites (or boxes). Periodic boundary conditions (also called cyclic boundary conditions) or open end boundary conditions can be imposed. In two dimensions we can consider different lattices (grids), for example rectangular, triangular or hexagonal lattices.

Any cellular automata rule in one dimension can be described by an evolution equation of the form

$$a_i(t+1) = f[a_{\{i\}}(t)], \qquad t = 0, 1, 2, \ldots$$

where $a_i(t)$ is the state of the site i at time t and $f[a_{\{i\}}(t)]$ is a function of the states of sites in a neighbourhood of lattice site i at time t with the initial states given by $a_i(t=0)$. In the one-dimensional case (i.e., one has a linear chain) the equation can also be written as

$$a_i(t+1) = f[a_{i-r}(t), a_{i-r+1}(t), \ldots, a_{i+r}(t)].$$

The local rule f has the range of r sites. In most cases, one considers the case $r = 1$. Thus a 1-dimensional cellular automaton consists of a line of sites with values a_i between 0 and $k-1$. These values are updated in parallel (synchronously) in discrete time steps. Much of the section is concerned with the study of a particular $k = 2$, $r = 1$ cellular automaton.

For example, the rule for evolution could take the value of a site at a particular time step to be the sum modulo two of the values of its two nearest neighbours on the previous time step. The pattern is found to be self-similar, and is characterized by a fractal dimension $log_2 3$. Even with an initial state consisting of a random sequence of 0 and 1 sites, say each with probability $\frac{1}{2}$, the evolution of such a cellular automaton leads to correlations between separated sites and the appearance of structure. This behaviour contradicts the second law of thermodynamics for systems with reversible dynamics, and is made possible by the irreversible nature of the cellular automaton evolution. Starting from a maximum entropy ensemble in which all possible configurations appear with equal probability, the evolution increases the probabilities of some configurations at the expense of others.

Despite the simplicity of their construction, cellular automata are found to be capable of diverse and complex behaviour. Numerical studies suggests that the pattern generated in the time evolution of cellular automata from disordered initial states can be classified as follows (Wolfram [218]):

(i) Evolves to homogeneous state

(ii) Evolves to simple separated periodic structures

(iii) Evolves to chaotic aperiodic patterns

(iv) Evolves to complex pattern of localized structures.

Constants of motion, if any exist, in classical mechanics and conservation laws, if any exist, in field theory play an important rôle in studying the behaviour of the dynamical system. If invariants exist for a cellular automaton, then one has a partition of its state space. Both irreversible and reversible cellular automata can exist. A rule of cellular automata is said to be reversible if it is backwards deterministic.

For mathematical purposes, it is often convenient to consider cellular automata with an infinite number of sites. However practical implementations must contain a finite number of sites N. These are typically arranged in a circular register to obtain periodic boundary conditions. It is also possible to arrange the sites in a feedback shift register.

Cellular automata can be considered as discrete approximations to partial differential equations, and can be used as direct models for a variety of natural systems. They can also be considered as discrete dynamical systems corresponding to continuous mappings on the Cantor set. Finally they can be viewed as computational systems whose evolution processes information contained in their initial configurations.

For one-dimensional cellular automata with $r = 2$ and $k = 1$ we can define rule numbers as follows. Let j be a fixed lattice point (cell). Then consider its two neighbours $j-1$ and $j+1$. For these three lattice sites we can have the following $8=2^3$ configurations at time t

$$111 \quad 110 \quad 101 \quad 100 \quad 011 \quad 010 \quad 001 \quad 000\,.$$

Example. Consider the cellular automata

$$a_j(t+1) = (a_{j-1}(t) + a_{j+1}(t)) \bmod 2\,.$$

Then the eight configurations given above are mapped into $a_j(t+1)$ as follows

$$111 \to 0, \quad 110 \to 1. \quad 101 \to 0, \quad 100 \to 1,$$

$$011 \to 1, \quad 010 \to 0, \quad 001 \to 1, \quad 000 \to 0$$

where we used the mod 2 addition (XOR), i.e.

$$0+0=0, \quad 0+1=1, \quad 1+0=1, \quad 1+1=0\,.$$

Thus the map provides the binary string 01011010. Since

$$0\cdot 2^7 + 1\cdot 2^6 + 1\cdot 2^4 + 1\cdot 2^3 + 1\cdot 2^1 = 90$$

we associate the map given above with the rule 90. ♣

Vice versa, from the rule number we can derive the map. For example consider the rule 56. Since

$$56 = 0\cdot 2^7 + 0\cdot 2^6 + 1\cdot 2^5 + 1\cdot 2^4 + 1\cdot 2^3 + 0\cdot 2^2 + 0\cdot 2^1 + 0\cdot 2^0$$

we obtain the binary string 00111000. Thus we have the maps

$$111 \to 0, \quad 110 \to 0, \quad 101 \to 1, \quad 100 \to 1$$

$$011 \to 1, \quad 010 \to 0, \quad 001 \to 0, \quad 000 \to 0\,.$$

Example. For rule 30 we have

$$30 = 0\cdot 2^7 + 0\cdot 2^6 + 0\cdot 2^5 + 1\cdot 2^4 + 1\cdot 2^3 + 1\cdot 2^2 + 1\cdot 2^1 + 0\cdot 2^0\,.$$

Thus we have the maps

$$111 \to 0, \quad 110 \to 0, \quad 101 \to 0, \quad 100 \to 1$$

$$011 \to 1, \quad 010 \to 1, \quad 001 \to 1, \quad 000 \to 0\,.$$

Exercise. Show that for rule 62 we obtain the bitstring `00111110` and the map

$$\underbrace{111}_{0},\ \underbrace{110}_{0},\ \underbrace{101}_{1},\ \underbrace{100}_{1},\ \underbrace{011}_{1},\ \underbrace{010}_{1},\ \underbrace{001}_{1},\ \underbrace{000}_{0}\,.$$

Show that starting from the bitstring `0001000` we obtain for the next two steps of the map `0011100` and `0110010`.

10.2 A Spin System and Cellular Automata

We consider a one-dimensional ring of N sites labelled sequentially by the index i starting from zero, i.e. $i = 0, 1, \ldots, N-1$. We impose periodic boundary conditions, i.e. $N = 0$. Each site i can take the values $a_i = 0$ or $a_i = 1$. Let a_i evolve as a function $a_i(t)$ of discrete time steps $t = 0, 1, 2, \ldots$ according to the map

$$a_i(t+1) = (a_{i-1}(t) + a_{i+1}(t)) \mod 2\,.$$

Since the map involves the sum $(a_{i-1} + a_{i+1})$ mod 2, it is a nonlinear map. The map can also be expressed using spin variables $S_j \in \{+1, -1\}$, where $S_i = -1$ corresponds to $a_i = 0$ and $S_i = +1$ corresponds to $a_i = 1$. Then the map can be written as

$$S_i(t+1) = -S_{i-1}(t)S_{i+1}(t).$$

This map has the following properties

$$S_1(t)S_2(t)\cdots S_N(t) = (-1)^N$$

for $t = 1, 2, \ldots$ and

$$S_i(t+2^n) = -S_{i-2^n}(t)S_{i+2^n}(t)$$

for $t = 0, 1, \ldots$ and $n = 0, 1, \ldots$.

For $N = 2^k$, $k = 1, 2, \ldots$, one has all $S_i(t) = -1$ for all times $t \geq 2^{k-1}$, irrespective of the initial spin configuration of the ring.

In the program `spin.cpp` we set $N = 5$, i.e. we have five cells (or sites). We apply periodic boundary conditions. The number of time steps is $T = 7$. As expected we find an eventually periodic pattern, i.e., $\mathbf{S}(6) = \mathbf{S}(1)$.

```
// spin.cpp

#include <iostream>
using namespace std;

int main(void)
{
 int N = 5; // number of cells
 int* S = new int[N]; int* W = new int[N]; // memory allocation
 S[0] = -1; S[1] = -1; S[2] = 1; S[3] = -1; S[4] = -1; // initial values
 int T = 7; // number of time steps
 // iterations
 for(int t=1;t<T;t++)
 {
 for(int j=0;j<N;j++) { W[j] = S[j]; }
 for(j=0;j<N;j++) { S[j] = -W[((j-1)+N)%N]*W[(j+1)%N]; }
 // display the result for each time step
 cout << "t = " << t << "  ";
```

```
for(j=0;j<N;j++) { cout << "S[" << j << "] = " << S[j] << "  "; }
cout << endl;
} // end iteration
delete[] S; delete[] W;
return 0;
}
```

Another example of a spin system is given by

$$u_j(t+1) = \text{sgn}(1 - |u_{j-1}(t) - u_j(t) + u_{j+1}(t) - 1|)$$

where $u_j(t) \in \{-1, +1\}$. Here we have

$u_{j-1}(t)$	$u_j(t)$	$u_{j+1}(t)$	$u_j(t+1)$
−1	−1	−1	−1
−1	−1	1	1
−1	1	−1	−1
−1	1	1	−1
1	−1	−1	1
1	−1	1	−1
1	1	−1	−1
1	1	1	1

10.3 Sznajd Model

Sznajd-Weron and Sznajd [202] introduced a one-dimensional Ising spin model with periodic boundary conditions where each spin (or lattice site) $j = 0, 1, \ldots, N-1$ can be found in one of the two states $u_j \in \{-1, +1\}$ or $\{-, +\}$ for short, which in the context of opinion formation, shall refer to two opposite opinions. Owing to the periodic boundary condition we have $u_N \equiv u_0$. For the spin interaction of neighbouring spins, the following two rules are proposed:

Rule 1. If two consecutive lattice sites j and $j+1$ have the same opinion (either $+1$ or -1), i.e. $u_j u_{j+1} = 1$, then the two neighbouring sites $j-1$ and $j+2$ will adopt the opinion of the pair $\{j, j+1\}$, i.e. $u_{j-1} = u_{j+2} = u_j = u_{j+1}$. This rule refers to ferromagnetism.

Rule 2. If two consecutive lattice sites have a different opinion (either $+1$ or -1) i.e., $u_j u_{j+1} = -1$, then the two neighbouring sites $j-1$ and $j+2$ will adopt their opinion from the second nearest neighbours as follows: $u_{j-1} = u_{j+1}$, $u_{j+2} = u_j$. This rule refers to anti-ferromagnetism.

Thus we have a one-dimensional cellular automata. We consider determininstic dynamics, i.e. the rules apply with probability one - which is similar to an Ising system at temperature $T = 0$. However, there is still randomness in the system in the sense that (i) there is an initial random distribution of the opinions with the mean value of the frequencies $f_{+1} = f_{-1} = 0.5$, and (ii) during the computer simulations, the site j for the next step is randomly chosen, i.e., the dynamics is governed by a random sequential update as asynchronous update. Two spins are flipped at a time. With the rules above, we find the following possible transitions in a neighbourhood of $N = 4$.

u_{j-1}	u_j	u_{j+1}	u_{j+2}	$\rightarrow$	u_{j-1}	u_j	u_{j+1}	u_{j+2}	rule
?	+	+	?		+	+	+	+	(1)
?	−	−	?		−	−	−	−	(1)
?	+	−	?		−	+	−	+	(2)
?	−	+	?		+	−	+	−	(2)

Through simulations one finds that the one-dimensional Sznajd model for any random initial configuration asymptotically reaches one of the three possible attractors, two of which refer to ferromagnetism and one to anti-ferromagnetism. These possible attractors are reached with different probability:

attractor $ferro_+$: $\{++++++++++\}$ with probability $p = 0.25$
attractor $ferro_+$: $\{----------\}$ with probability $p = 0.25$
attractor *anti-ferro*: $\{-+-+-+-+-+\}$ with probability $p = 0.5$

In order to verify these probabilities, consider a lattice of size N with periodic boundary conditions and an initial random distribution of $+$ and $-$. Then the number of consecutive pairs is also N. The initial probability of finding either a ferromagnetic or anti-ferromagnetic pair adds up to 0.5, i.e.

$$p_f = p_{++} + p_{--} = 0.5, \qquad p_{af} = p_{-+} + p_{+-} = 0.5, \qquad p_f + p_{af} = 1\,.$$

During the first q steps we may assume that the initial distribution is not changed much by the dynamics, i.e. the equation given above remains valid and the probabilities are given by the bionomial distribution

$$\sum_{k=0}^{q} \binom{q}{k} p_{af}^k p_f^{q-k} = 1\,.$$

If during the first q steps more than $q/2$ anti-ferromagnetic pairs are selected, then p_{af} increases since each selection will lead to two new antiferromagnetic pairs. The case of ferromagnetic pair selection can be treated similarly. If q is an even number, the probability is given by

$$\sum_{k=(q/2)+1}^{q} \binom{q}{k} p_{af}^k p_f^{q-k} + \frac{1}{2}\binom{q}{q/2} p_{af}^{q/2} p_f^{q/2} = 0.5\,.$$

The first term denotes the probability of selecting more than $q/2$ anti-ferromagnetic pairs (favouring anti-ferromagnetism) and the second term denotes the probability of selecting exactly $q/2$ pairs (favouring both ferromagnetism and anti-ferromagnetism with probability 0.5). Thus we can conclude that the probability for the system to reach the antiferromagnetic attractor is given by 0.5. This equation is valid as long as $p_f = p_{af} = 0.5$, i.e., for $t \leq q$. After the initial time lag, the symmetry is broken and the system dynamics goes towards one of the possible anti-ferromagnetic or ferromagnetic attractors with probability one.

For the implementation of the Sznajd model we use the `bitset` class of the Standard Template Library of C++.

```
// sznajd.cpp

#include <bitset>
#include <cstdlib>
#include <iostream>
using namespace std;

template<const size_t n> bitset<n> random_bitstring()
{
 int i, ones, zeros;
 bitset<n> b;
 ones = zeros = n/2;
 for(i=0;i<int(n);i++)
 {
 if(double(rand())/RAND_MAX < double(zeros)/(ones+zeros))
 { b[i] = 0; zeros--; }
 else { b[i] = 1; ones--; }
 }
 return b;
}

template<const size_t n>
bitset<n> sznajd(const bitset<n> &b,int steps)
{
 bitset<n> u = b;
 int j;
 while(steps > 0)
 {
 j = rand()%n;
 if(u[j]==u[(j+1)%n]) u[(j+n-1)%n]=u[(j+2)%n]=u[j];      // rule 1
 else { u[(j+n-1)%n] = u[(j+1)%n]; u[(j+2)%n] = u[j]; } // rule 2
 steps--;
 }
 return u;
}

int main(void)
{
 srand(time(NULL));
```

```
  bitset<8> b = random_bitstring<8>();
  cout << b << endl;
  cout << sznajd(b,1000) << endl;
  return 0;
}
```

10.4 Conservation Laws

A large class of cellular automata is provided by discretization (of space and time coordinates) of partial differential equations. A simple example is the one-dimensional linear diffusion equation

$$\frac{\partial u}{\partial t} = \frac{\partial^2 u}{\partial x^2}$$

with the conserved quantity

$$\int_{\mathbb{R}} u(x,t)dx = C$$

where C is the total amount of the diffusing substance. Here it is assumed that u and its derivative with respect to x go to zero as $|x| \to \infty$. The simplest discretization of the one-dimensional diffusion equation is given by

$$u_j(t+1) - u_j(t) = u_{j+1}(t) - 2u_j(t) + u_{j-1}(t)$$

with unit discretization steps. In modulo 2 integer arithmetic the term $2u_j$ vanishes. Therefore

$$u_j(t+1) = u_{j+1}(t) + u_j(t) + u_{j-1}(t) \mod 2.$$

This equation corresponds to rule 150. Again we impose periodic boundary conditions. As a conservation law we find

$$\sum_{j=0}^{N-1} u_j(t) = C$$

where the constant C is given by

$$C = \sum_{j=0}^{N-1} u_j(0).$$

Note that C can only take the values 0 or 1. The proof is straightforward. We take the sum over j of the left and right hand side of equation for $u_j(t+1)$ and bear in mind that we have modulo 2 integer arithmetic and cyclic boundary conditions. Let $N = 2^5 = 32$ with the initial configuration $u_j(0) = 0$ for $j = 0, 1, \ldots, 14, 16, \ldots, 31$ and $u_{15}(0) = 1$. The number of time steps is 40. The system evolves to a periodic structure. Let $N = 31$ with initial configuration $u_j(0) = 0$ for $j = 0, 1, \ldots, 14, 16, \ldots 30$ and $u_{15}(0) = 1$. Again the system evolves to a periodic structure.

Consider *Burgers equation*

$$\frac{\partial u}{\partial t} = \frac{\partial^2 u}{\partial x^2} - u\frac{\partial u}{\partial x}.$$

The equation can be written as a conservation law

$$\frac{\partial u}{\partial t} = \frac{\partial}{\partial x}\left(-\frac{u^2}{2} + \frac{\partial u}{\partial x}\right)$$

with the conserved quantity

$$\int_{\mathbb{R}} u(x,t)dx = C.$$

Here we have assumed that u and its derivate with respect to x go to zero as $|x| \to \infty$. The simplest discretization yields

$$u_j(t+1) - u_j(t) = u_{j+2}(t) - 2u_{j+1}(t) + u_j(t) - u_j(t)[u_{j+1}(t) - u_j(t)]$$

with unit discretization steps. In modulo 2 integer arithmetic we have $u_j u_{j+1} = -u_j u_{j+1}$ and $u_j = u_j^2$ for all t. Therefore this equation simplifies to

$$u_j(t+1) = u_{j+2}(t) + u_j(t)u_{j+1}(t) + u_j(t) \mod 2.$$

We see that $u_j(t) = 0$ for all $i = 0, 1, \ldots, N-1$ and $u_j(t) = 1$ for all $j = 0, 1, \ldots, N-1$ are invariant states. We see that $\Sigma_{j=0}^{N-1} u_j(t) = C$ is no longer a conservation law. Let $N = 32$ and the initial configuration $u_j(0) = 0$ for $j = 0, 1, \ldots, 14, 16, \ldots, 31$ and $u_{15}(0) = 1$. The number of time steps is 20. The cellular automaton tends to the fixed point

$$u_0^* = u_1^* = \cdots = u_{31}^* = 0.$$

Let $N = 31$ and the initial configuration $u_j(0) = 0$ for $j = 0, 1, \ldots, 14, 16, \ldots, 30$ and $u_{15}(0) = 1$. The pattern reached has a simple periodic structure.

10.5 Two-Dimensional Cellular Automata

The extension to two dimensions is significant for comparisons with many experimental results on pattern formation in physical systems (Wolfram [218]). Applications include dendritic crystal growth, reaction-diffusion systems and turbulent flow patterns. The Navier-Stokes equations for fluid flow appear to admit turbulent solutions only in two or more dimensions. A cellular automaton consists of a regular lattice of sites. Each site takes on k possible values, and is updated in discrete time steps according to a rule f that depends on the value of sites in some neighbourhood around it. There are several possible lattices and neighbourhood structures for two-dimensional cellular automata. In literature mainly square lattices with nearest neighbour interaction are studied. Triangular and hexagonal lattices are also possible, but are not used in the examples given here.

Neighbourhood structures considered for two-dimensional cellular automata. In the cellular automaton evolution, the value of the centre cell is updated according to a rule that depends on the values of the neighbouring cells.

Let `[i][j]` be a lattice site. Then we can consider a four neighbours square cellular automaton, where the neighbours are

`[i][j+1],   [i+1][j],   [i][j-1],    [i-1][j] .`

Thus the time evolution is given by

$$a_{i,j}(t+1) = f(a_{i,j}(t), a_{i,j+1}(t), a_{i+1,j}(t), a_{i,j-1}(t), a_{i-1,j}(t)) .$$

Cellular automata with this neighbourhood are termed "five-neighbour square". Here we often consider the special class of totalistic rules, in which the value of a site depends only on the sum of the values in the neighbourhood

$$a_{i,j}(t+1) = f[a_{i,j}(t) + a_{i,j+1}(t) + a_{i+1,j}(t) + a_{i,j-1}(t) + a_{i-1,j}(t)] .$$

We also can consider the case with eight neighbours given by

```
[i+1][j],   [i+1][j+1],   [i][j+1],   [i-1][j+1]
[i-1][j],   [i-1][j-1],   [i][j-1],  [i+1][j-1] .
```

This neighbourhood is termed "nine-neighbour square." These neighbourhoods are sometimes referred to as the von Neumann and Moore neighbourhoods, respectively.

Totalistic cellular automaton rules assumes the value of the centre site depends only on the sum of the values of the sites in the neighbourhood. With outer totalistic rules, sites are updated according to their previous values and the sum of the values of the other sites in the neighbourhood. The five-neighbour square, triangular, and hexagonal cellular automaton rules may all be considered as special cases of general nine-neighbour square rules.

The map

$$a_{i,j}(t+1) = f[a_{i-1,j}(t) + a_{i,j-1}(t) + a_{i+1,j}(t) + a_{i,j+1}(t)]$$

can be specified by a code

$$C = \sum_n f(n) k^n .$$

One can also consider outer totalistic rules, in which the value of a site depends separately on the sum of the values of sites in a neighbourhood, and on the value of the site itself

$$a_{i,j}(t+1) = g(a_{i,j}(t), a_{i,j+1}(t) + a_{i+1,j}(t) + a_{i,j-1}(t) + a_{i-1,j}(t)) .$$

Such rules are specified by a code

$$\widetilde{C} = \sum_n g[a,n] k^{kn+a} .$$

We consider two-dimensional cellular automata with values 0 or 1 at each site, corresponding to $k = 2$.

An example of an outer totalistic nine-neighbour square cellular automaton is the *Game of Life*, with a rule specified by code $\widetilde{C} = 224$. This cellular automata was introduced by a Cambridge mathematician named John Horton Conway (see Gardner [68]). Game of Life is played by populating a rectangular, two-dimensional grid of cells with critters. Each cell can take the value 0 (dead) or 1 (alive). Then we observe the behaviour of the population over succeeding generations. Each cell $[i][j]$ has eight neighbouring cells at

```
[i+1][j],   [i+1][j+1],   [i][j+1],   [i-1][j+1]
[i-1][j],   [i-1][j-1],   [i][j-1],   [i+1][j-1] .
```

For the C++ implementation we consider $N \times N$ cells, where $N = 16$ and impose a periodic boundary condition in the i and j directions, i.e. $N \equiv 0$. Both i and j start counting from zero. The rules for Conway's Game of Life are

a) A new cell is born (set to 1) when there are 3 live neighbours
b) A cell stays alive when surrounded by 2 or 3 live cells.
c) Each cell with four or more neighbours dies of overpopulation and each cell surrounded by one or no live cell dies of solitude.

```
// twocellular.cpp

#include <iostream>
using namespace std;

int neighb(int a,int b,int c,int d,int e,int f,int g,int h)
{
 int result = a+b+c+d+e+f+g+h;
 return result;
}

int main(void)
{
 int i, j;
 int N = 8; // size of the grid
 int** A = new int* [N];   // memory allocation
 for(i=0;i<N;i++) A[i] = new int[N];
 int** B = new int* [N];   // memory allocation
 for(i=0;i<N;i++) B[i] = new int[N];
 // initial configuration
 for(i=0;i<N;i++)
  for(j=0;j<N;j++) A[i][j] = 0;
 // set the following grid points alive
 A[5][5] = 1; A[5][6] = 1; A[6][5] = 1; A[6][6] = 1;
```

```
int T = 10;   // number of iterations
for(int t=0;t<T;t++)
{
for(i=0;i<N;i++)
{ for(j=0;j<N;j++) { B[i][j] = A[i][j]; } }
for(i=0;i<N;i++)
{
for(j=0;j<N;j++)
{
int temp =
neighb(B[((i+1)%N)][j],B[((i+1)%N)][((j+1)%N)],B[i][((j+1)%N)],
       B[((i-1+N)%N)][((j+1)%N)],B[((i-1+N)%N)][j],
       B[((i-1+N)%N)][((j-1+N)%N)],B[i][((j-1+N)%N)],
       B[((i+1)%N)][((j-1+N)%N)]);
// rule 1
if((B[i][j]==0) && (temp==3)) { A[i][j] = 1; }
// rule 2
if((B[i][j]==1) && ((temp==2) || (temp==3))) { A[i][j] = 1; }
// rule 3
if((B[i][j]==1) && ((temp>=4) || (temp<=1))) { A[i][j] = 0; }
}
}
cout << endl << "t = " << t << endl;
// output for each time step
for(i=0;i<N;i++)
{
for(j=0;j<N;j++)
{
cout << "B[" << i << "]" << "[" << j << "] = " << B[i][j] << " ";
}
}
cout << endl << endl;
} // end for loop t
// free the memory
for(i=0;i<N;i++) delete[] A[i]; delete[] A;
for(i=0;i<N;i++) delete[] B[i]; delete[] B;
return 0;
}
```

The output is t=9 and

```
B[0][0] = 0 B[0][1] = 0 B[0][2] = 0 B[0][3] = 0 B[0][4] = 0
B[0][5] = 0 B[0][6] = 0 B[0][7] = 0 B[1][0] = 0 B[1][1] = 0
B[1][2] = 0 B[1][3] = 0 B[1][4] = 0 B[1][5] = 0 B[1][6] = 0
B[1][7] = 0 B[2][0] = 0 B[2][1] = 0 B[2][2] = 0 B[2][3] = 0
B[2][4] = 0 B[2][5] = 0 B[2][6] = 0 B[2][7] = 0 B[3][0] = 0
B[3][1] = 0 B[3][2] = 0 B[3][3] = 0 B[3][4] = 0 B[3][5] = 0
B[3][6] = 0 B[3][7] = 0 B[4][0] = 0 B[4][1] = 0 B[4][2] = 0
```

```
B[4][3] = 0 B[4][4] = 0 B[4][5] = 0 B[4][6] = 0 B[4][7] = 0
B[5][0] = 0 B[5][1] = 0 B[5][2] = 0 B[5][3] = 0 B[5][4] = 0
B[5][5] = 1 B[5][6] = 1 B[5][7] = 0 B[6][0] = 0 B[6][1] = 0
B[6][2] = 0 B[6][3] = 0 B[6][4] = 0 B[6][5] = 1 B[6][6] = 1
B[6][7] = 0 B[7][0] = 0 B[7][1] = 0 B[7][2] = 0 B[7][3] = 0
B[7][4] = 0 B[7][5] = 0 B[7][6] = 0 B[7][7] = 0
```

10.6 Button Game

The button game described below is closely related to two-dimensional cellular automata. It is also worth inclusion since the Java program `JButtonGame.java` shows how two-dimensional cellular automata can be implemented using the GUI capability (`JPanel` class, `JTextField` class, `JButton` class, etc.) of Java. Given a grid with 4×4 boxes, where one of the boxes is empty. The other 15 boxes contain the numbers 1, 2, ..., 15 so that each number occupies one box. For example

```
 8  4  15   5
 2  1      10
 7  6   9  12
 3 11  13  14
```

Clicking on one of the nearest neighbours of the empty box moves the number of the box into the empty box and the clicked box with the number becomes the empty box. For example, clicking on the box with the number `10` we obtain the pattern

```
 8  4  15   5
 2  1  10
 7  6   9  12
 3 11  13  14
```

Note that there are no cyclic boundary conditions imposed. Nearest neighbours are only on the vertical and horizontal line. Neighbours on the diagonal are not considered as nearest neighbours. The task is now to obtain the pattern

```
 1   2   3  4
 5   6   7  8
 9  10  11 12
13  14  15
```

The class `JPanel` is a container class in which components can be placed. `JPanel` is derived from class `Container`. Components are placed on containers with container method `add`. In Java we can divide a top-level window into panels. Panels act as (smaller) containers for interface elements and can themselves be arranged inside the window. For example we can have one panel for the buttons and another for the text fields. Push buttons are created with the `JButton` class. `JTextfield(String text, int cols)` constructs a new `JTextfield`, where `text` is the text to be displayed and `cols` the number of columns. `Gridlayout` arranges the components into rows and columns.

```
// JButtonGame.java

import java.awt.*;
import java.awt.event.*;
import javax.swing.*;

class JButtonGame extends JFrame implements ActionListener
{
 private int nRows, nCols, nButtons;
 private int blankCol, blankRow, clickedRow, clickedCol;
 private JButton[][] buttons;
 private JTextField textField;

 public JButtonGame()
 {
 setSize (200,200);
 addWindowListener(new WindowAdapter ()
 { public void windowClosing(WindowEvent event)
 { System.exit(0);}});

 nRows = 4; nCols = 4; nButtons = nRows*nCols;
 JPanel panel = new JPanel();
 panel.setLayout(new GridLayout(nRows,nCols));
 buttons = new JButton [nRows][nCols];

 for(int nRow=0;nRow<nRows;nRow++)
 {
 for(int nCol=0;nCol<nCols;nCol++)
 {
 buttons[nRow][nCol] = new JButton("");
 buttons[nRow][nCol].addActionListener(this);
 panel.add(buttons[nRow][nCol]);
 }
 }

 getContentPane().add("Center",panel);
 textField = new JTextField("",80);
 textField.setEditable(false);
 getContentPane().add("South",textField);
 int labelsUsed[] = new int [nButtons];
 for(int i=0;i<nButtons;i++)
 { boolean labelUsed; int label;
 do
 { label = random(nButtons)+1; labelUsed = false;
 for(int j=0;j<i;j++)
  labelUsed = ((labelUsed) || (label==labelsUsed[j]));
 } while (labelUsed);
```

```
labelsUsed[i] = label;
int nRow = i/nCols; int nCol = i-nRow*nCols;
buttons[nRow][nCol].setText((new Integer (label)).toString());
}
getButtonPosition((new Integer (nButtons)).toString());
blankRow = clickedRow; blankCol = clickedCol;
JButton blank = buttons[clickedRow][clickedCol];
blank.setText("");
blank.setBackground(Color.green);
} // end constructor JButtonGame()

private int random(int k)
{ return (int)(k*Math.random()-0.1); } // end method random(int)

private void getButtonPosition(String label)
{
for(int nr=0;nr<nRows;nr++)
{
for(int nc=0;nc<nCols;nc++)
{
if(buttons[nr][nc].getText().equals(label))
{
clickedRow = nr; clickedCol = nc;
textField.setText("[" + nr + ',' + nc + "]" + label);
}
}
}
} // end method getButtonPosition(String)

public void actionPerformed(ActionEvent e)
{
getButtonPosition(e.getActionCommand());
textField.setText("[" + blankRow + "," + blankCol
    + "] = > [" + clickedRow + "," + clickedCol + "]");
if(clickedRow == blankRow)
{
if(Math.abs(clickedCol-blankCol)==1)
moveBlank(blankRow,blankCol,clickedRow,clickedCol);
}
else if(clickedCol==blankCol)
{
if(Math.abs(clickedRow-blankRow)==1)
moveBlank(blankRow,blankCol,clickedRow,clickedCol);
}
} // end method actionPerformed

public void moveBlank(int oldRow,int oldCol,int newRow,int newCol)
{
```

```
JButton oldBlank = buttons[oldRow][oldCol];
JButton newBlank = buttons[newRow][newCol];
String label = newBlank.getText();
newBlank.setText ("");
newBlank.setBackground(Color.green);
oldBlank.setText(label);
oldBlank.setBackground(Color.lightGray);
blankRow = newRow; blankCol = newCol;
} // end method moveBlank

public static void main(String [] args)
{ new JButtonGame().setVisible(true); }  // end method main
}
```

10.7 Langton's Ant

Langton's ant (Langton [126], Gajardo [66], Dorbec [49]) is a cellular automata defined over a two-dimensional rectangular grid, i.e. $\mathbb{Z} \times \mathbb{Z}$. Instead of open ends one could also impose periodic boundary conditions. First one has to intialize all the vertices, for example all black or all white or some mixed pattern. It is assumed at the beginning that the ant enters the vertex $(0,0)$ from $(x_1, x_2) = (1,0)$. The rules are: When the ant enters a vertex it decides the direction in which it moves according to the colour of the vertex. The ant turns to its left if the vertex is black and to its right otherwise and afterwards it flips the colour of the vertex.

For example assume the ant starts at $(0,0)$ coming from $(1,0)$ and all vertices are colored black. Thus the ant moves to the left (i.e. to $(0,-1)$) and then the color of the vertex $(0,0)$ is flipped to white.

This can be summarized in the following definition (Dorbec [49]):

An ant is specified by its position on the grid $\mathbf{x} \in \mathbb{Z}^2$, and its internal state $q \in Q$. The state of the dynamical system is given by the three-tuple: $(c, \mathbf{x}, q)$, where

$$c : \mathbb{Z}^2 \to C$$

is a colorization. The phase space is then

$$X = C^{\mathbb{Z}^2} \times \mathbb{Z}^2 \times Q\,.$$

The dynamics is then given by the global transition function: $F : X \to X$.

Langton's ant is time-reversible. Given any state the dynamical system can be reversed to the initial state. When the ant starts moving on the rectangular grid whose vertices are in the same color (say black) it has an apparently "chaotic" behaviour for about 10000 time steps and then falls into a periodic movement with a drift.

Extensions to other lattice types and higher dimensions have also been studied.

In the C++ code we impose periodic boundary conditions. After some time t the motion of the ant is periodic.

```
// ant.cpp

#include <iostream>
using namespace std;

int main(void)
{
 int N = 5;
 int** G = new int*[N];
 for(int i=0;i<N;i++) G[i] = new int[N];
 for(int j=0;j<N;j++)
  for(int k=0;k<N;k++) { G[j][k] = 1;}

 int mold = 3; int nold = 2; int mnew = 2; int nnew = 2;
 int t = 0; int T = 40;
 while(t < T)
 {
 if((mnew < mold) && (nnew == nold))
 { G[mnew][nnew] = -G[mnew][nnew];
   mold = mnew; nold = nnew; nnew = (nnew+1)%N; goto L; };

 if((mnew == mold) && (nnew < nold))
 { G[mnew][nnew] = -G[mnew][nnew];
   mold = mnew; nold = nnew; mnew = (mnew+1)%N; goto L; };

 if((mnew > mold) && (nnew == nold))
 { G[mnew][nnew] = -G[mnew][nnew];
   mold = mnew; nold = nnew; nnew = (nnew-1+N)%N; goto L; };

 if((mnew == mold) && (nnew > nold))
 { G[mnew][nnew] = -G[mnew][nnew];
   mold = mnew; nold = nnew; mnew = (mnew-1+N)%N; goto L; };
 L:
 cout << mnew << " " << nnew << " " << G[mnew][nnew] << endl;
 t++;
 }
 for(int l=0;l<N;l++) delete[] G[l]; delete[] G;
 return 0;
}
```

Exercise. (1) Extend Langton's ant to $\mathbb{Z} \times \mathbb{Z} \times \mathbb{Z}$.

Chapter 11

Solving Differential Equations

11.1 Introduction

In this chapter we consider numerical methods of integration for nonlinear ordinary differential equations (Scheid [182], Fröberg [64], Hairer et al [79]). Almost all nonlinear differential equations can not be solved in closed form. Numerical methods must be applied. A number of problems arise when we solve differential equations numerically. For example, if the differential equations are derived from a conservative Hamilton system then the numerical scheme should preserve the total energy which is a constant of motion. For a nonlinear differential equation it also can happen that the discretization scheme leads to a chaotic map even though the original nonlinear differential equation did not show chaotic behaviour. Stable fixed points of nonlinear differential equations can become unstable under the discretization scheme. We also find so-called spurious solutions (also called ghost solutions).

We consider autonomous first order

$$\frac{d\mathbf{u}}{dt} = \mathbf{f}(\mathbf{u})$$

and nonautonomous first order systems of ordinary differential equations

$$\frac{d\mathbf{u}}{dt} = \mathbf{g}(\mathbf{u}, t)\,.$$

We assume that we have an initial value problem $\mathbf{u}(t = 0) = \mathbf{u}_0$ and that the functions $\mathbf{f}$ and $\mathbf{g}$ are analytic or at least C^2.

We discuss the Euler method, the Lie series technique (which is closely related to the Taylor series technique), the Runge-Kutta-Fehlberg fifth order method and symplectic integration. The Verlet method important in molecular dynamics is also introduced. We also investigate spurious solutions and invisible chaos. Finally numerical integration of systems of ordinary differential equations which admit first integrals is discussed.

Theorem. (Existence and Uniqueness) Consider the first order ordinary differential equation $d\mathbf{u}/dt = \mathbf{f}(\mathbf{u})$, where both $\mathbf{f}$ and its first partial derivatives with respect to u_j are continuous on an open set of $U \subseteq \mathbb{R}^n$. Then for any real number t_0 and real vector $\mathbf{u}_0$ there is an open interval containing t_0, on which there exists a solution satisfying the initial condition $\mathbf{u}(t_0) = \mathbf{u}_0$, and this solution is unique.

Definition. Let U be an open set in $\mathbb{R}^n$. A function $\mathbf{f}$ is said to be Lipshitz on U if there exists a constant L such that

$$|\mathbf{f}(\mathbf{u}) - \mathbf{f}(\mathbf{v})| \leq L|\mathbf{u} - \mathbf{v}|$$

for all $\mathbf{u}$, $\mathbf{v}$ in U. The constant L is called a *Lipshitz constant* for $\mathbf{f}$.

Example. Consider the nonlinear first order differential equation

$$\frac{du}{dt} = -u^{1/3}, \qquad u(0) = u_0 > 0\,.$$

Integrating yields the solution of the initial value problem

$$u(t) = \left(-\frac{2t}{3} + u_0^{2/3}\right)^{3/2}, \qquad 0 \leq t \leq \frac{3}{2}u_0^{3/2}.$$

The Lipshitz condition is not satisfied. ♣

Two neighbouring solutions to the same system of a first order differential equation can separate from each other at a rate no greater than e^{Lt}, where L is the Lipshitz constant of the system of differential equations. The *Gronwall inequality* is the basis of continuity of the flow as a function of the initial condition (Gronwall [74], Bellman [18]).

Theorem. (Continuous dependence on initial conditions) Let $\mathbf{f}$ be defined on the open set U in $\mathbb{R}^n$. Assume that $\mathbf{f}$ has the Lipshitz constant L in the variable $\mathbf{u}$. Let $\mathbf{v}(t)$ and $\mathbf{w}(t)$ be solutions of $d\mathbf{u}/dt = \mathbf{f}(\mathbf{u})$, and let $[t_0, t_1]$ be a subset of a domain of both solutions. Then

$$|\mathbf{v}(t) - \mathbf{w}(t)| \leq |\mathbf{v}(t_0) - \mathbf{w}(t_0)|e^{L(t-t_0)}$$

for all t in $[t_0, t_1]$.

This means nearby solutions can diverge no faster than an exponential state determined by the Lipshitz constant of the differential equation.

11.2 Euler Method

The *Euler method* for solving numerically systems of first order ordinary differential equations

$$\frac{d\mathbf{u}}{dt} = \mathbf{g}(\mathbf{u}, t)$$

involves computing a discrete set of $\mathbf{u}_k$ values, for arguments t_k, using the difference equation

$$\mathbf{u}_{k+1} = \mathbf{u}_k + h\mathbf{g}(t_k, \mathbf{u}_k), \qquad h := t_{k+1} - t_k .$$

Here h is the step length. By expanding $\mathbf{u}(t+h)$ in a Taylor series with remainder about t, it can be seen that

$$\frac{\mathbf{u}(t+h)-\mathbf{u}(t)}{h} = \frac{\mathbf{u}(t)+h\mathbf{u}'(t)+\frac{1}{2}h^2\mathbf{u}''(\xi)-\mathbf{u}(t)}{h} = \mathbf{u}'(t)+\frac{1}{2}h\mathbf{u}''(\xi)$$

$(t \le \xi \le t+h)$ so that the left-hand side of the Euler method is an $O(h)$ approximation to the derivative it replaced, provided the second derivative of the solution $\mathbf{u}$ is bounded in the time interval of interest. As h goes to zero, the Euler approximation better and better represents the original differential equation; that is, it is consistent with the differential equation. For more details of the error estimates we refer to Hairer et al [79].

Example. Consider the nonlinear first order differential equation

$$\frac{du}{dt} = u(1-u)$$

with the initial condition $u(t=0) = u_0 > 0$. This nonlinear differential equation admits the fixed points $u_1^* = 0$, $u_2^* = 1$. The fixed point $u_1^* = 0$ is unstable, whereas the fixed point $u_2^* = 1$ is stable. The Euler method leads to the difference equation

$$u_{k+1} = u_k + hf(u_k) = u_k + hu_k(1-u_k), \qquad k = 0,1,2,\ldots$$

where h is the step length.

In the Java program `Euler.java` we set $h = 0.005$. The initial value is $u_0 = 0.1$. We find that u_k tends to the stable fixed point $u_2^* = 1$ as $k \to \infty$. ♣

```
// Euler.java

import java.awt.*;
import java.awt.event.*;
import java.awt.Graphics;

public class Euler extends Frame
{
 public Euler()
 {
 setSize(600,500);
 addWindowListener(new WindowAdapter()
 { public void windowClosing(WindowEvent event) { System.exit(0); }}); }

 public double f(double u)  { return u*(1.0-u); }
```

```
 public void paint(Graphics g)
 {
 g.drawLine(10,200,410,200);  g.drawRect(10,40,400,400);
 double h = 0.005; // step length
 double u = 0.1;    // initial value
 double u1; double t = 0.0;
 for(int i=0;i<2000;i++)
 {
 u1 = u;
 int m = (int)(40.0*t+10.5); int n = (int)(200.0-50.0*u1+0.5);
 u = u1+h*f(u1);
 int p = (int)(40.0*(t+h)+10.5); int q = (int)(200.0-50.0*u+0.5);
 g.drawLine(m,n,p,q);
 t += h;
 }
 }

 public static void main(String[] args)
 { Frame f = new Euler(); f.setVisible(true); }
}
```

11.3 Lie Series Technique

The Lie series technique has been studied in particular by Gröbner and co-workers (Gröbner [73], Knapp and Wanner [116]). Let V be the linear differential operator (vector field)

$$V := f_1(\mathbf{z})\frac{\partial}{\partial z_1} + f_2(\mathbf{z})\frac{\partial}{\partial z_2} + \cdots + f_n(\mathbf{z})\frac{\partial}{\partial z_n}$$

where the f_i's are holomorphic functions of the complex variables $z_1, z_2, \ldots, z_n$ in the neighbourhood of one and the same point. If g is a function which is holomorphic in the neighbourhood of the same point, we can apply the linear operator V on g

$$Vg := f_1(\mathbf{z})\frac{\partial g}{\partial z_1} + f_2(\mathbf{z})\frac{\partial g}{\partial z_2} + \cdots + f_n(\mathbf{z})\frac{\partial g}{\partial z_n}.$$

Since the derivatives of a holomorphic function are holomorphic, we obtain a function which is holomorphic at the same point. This holds for all iterated operations. This means that all functions

$$V^2 g := V(Vg), \quad \ldots, \quad V^n g := V(V^{n-1}g), \ldots$$

are holomorphic at the point under investigation and can be expanded in regular convergent power series.

Definition. The series

$$\exp(tV)g(\mathbf{z}) \equiv \sum_{k=0}^{\infty} \frac{t^k}{k!} V^k g(\mathbf{z}) = g(\mathbf{z}) + tVg(\mathbf{z}) + \frac{t^2}{2!}V^2 g(\mathbf{z}) + \cdots$$

is called a *Lie series.*

It is formally explained by the symbols for the series written down, for each term is composed of a factor t^n, $n!$ and a holomorphic function $V^n g$ of the complex variables $z_1, \ldots, z_n$. Consider t to be a new complex variable, which is independent of the variables $z_1, \ldots, z_n$. This series also converges as a power series in t, and is therefore a holomorphic function of the $n+1$ complex variables $z_1, \ldots, z_n, t$. The convergence of the Lie series is proved by the method of Cauchy's majorants (Gröbner [73], Knapp and Wanner [116]).

Theorem. If C is a finite, closed domain of the $\mathbf{z}$-space, in which the differential operator V and the function g are holomorphic, a positive number T exists such that the Lie series converges absolutely and uniformly for $|t| \leq T$ throughout the entire domain G, where it represents a holomorphic function of the $n+1$ complex variables $z_1, \ldots, z_n, t$.

The Lie series may be differentiated according to these variables term by term and any number of times in the interior of G and $|t| \leq T$

$$\frac{\partial}{\partial t}\left(\sum_{k=0}^{\infty} \frac{t^k}{k!} V^k g(\mathbf{z})\right) = \sum_{k=0}^{\infty} \frac{t^k}{k!} V^{k+1} g(\mathbf{z})$$

$$\frac{\partial}{\partial z_j}\left(\sum_{k=0}^{\infty} \frac{t^k}{k!} V^k g(\mathbf{z})\right) = \sum_{k=0}^{\infty} \frac{t^k}{k!} \frac{\partial}{\partial z_j}(V^k g(\mathbf{z})).$$

For the sums and products of Lie series we have

$$e^{tV}(c_1 g_1(\mathbf{z}) + c_2 g_2(\mathbf{z})) = c_1 e^{tV} g_1(\mathbf{z}) + c_2 e^{tV} g_2(\mathbf{z})$$
$$e^{tV}(g_1(\mathbf{z}) g_2(\mathbf{z})) = (e^{tV} g_1(\mathbf{z}))(e^{tV} g_2(\mathbf{z})).$$

If $P(Z_1, Z_2)$ denotes a polynomial in Z_1, Z_2, it holds that

$$e^{tV} P(g_1(\mathbf{z}), g_2(\mathbf{z})) = P(e^{tV} g_1(\mathbf{z}), e^{tV} g_2(\mathbf{z})).$$

Theorem. (Commutation Theorem) If $F(\mathbf{z})$ denotes any function which is holomorphic in a neighbourhood of $\{z_1, \ldots, z_n\}$, the corresponding power series expansion of which still converges at the point $\{Z_1, \ldots, Z_n\}$ (which is certainly the case for sufficiently small values of t), it is true that

$$F(\mathbf{Z}) = \sum_{k=0}^{\infty} \frac{t^k}{k!} V^k F(\mathbf{z})$$

or

$$F(e^{tV}\mathbf{z}) = e^{tV} F(\mathbf{z}).$$

Another important result is obtained by the differentiation of the function Z_j with respect to t. Let

$$Z_j := \exp(tV) z_j, \qquad j = 1, 2, \ldots, n.$$

Differentiating the Lie series term by term gives

$$\frac{\partial Z_j}{\partial t} = e^{tV}(Vz_j), \qquad j = 1,2,\ldots,n.$$

Since $Vz_j = V_j(\mathbf{z})$ we find

$$e^{tV}V_j(\mathbf{z}) = V_j(\mathbf{Z}).$$

It follows that

$$\frac{\partial Z_j}{\partial t} = V_j(\mathbf{Z}), \qquad j = 1,2,\ldots,n.$$

Theorem. If $\{z_1,\ldots,z_n\}$ is a point of the holomorphic domain of V, the functions $\exp(tV)z_j$ in a sufficiently small environment of $t = 0$ satisfy the system of differential equations

$$\frac{\partial Z_j}{\partial t} = V_j(\mathbf{Z})$$

with the initial conditions $Z_j(t = 0) = z_j$. Thus $\exp(tV)z_j$ are the solutions of the system of differential equations which are uniquely determined by these initial conditions.

The uniqueness of the solution can, if uniqueness is to be determined within the domain of all continuous and continuously differentiable functions, be proved by the known methods of real analysis.

The Lie series discussed above can be restricted to the real domain. For our numerical studies we consider the autonomous system of ordinary differential equations

$$\frac{d\mathbf{u}}{dt} = \mathbf{f}(\mathbf{u})$$

where $\mathbf{u} = (u_1,\ldots,u_n)^T$. Let f_j $(j = 1,\ldots,n)$ be analytic functions defined on $\mathbb{R}^n$. Consider the vector field

$$V := \sum_{j=1}^{n} f_j(\mathbf{u})\frac{\partial}{\partial u_j}.$$

Then the solution of the initial value problem of the autonomous system for a sufficiently small t can be given as a Lie series

$$u_j(t) = \exp(tV)u_j|_{\mathbf{u}=\mathbf{u}(0)}$$

where $j = 1,2,\ldots,n$. Expanding the exponential function yields

$$u_j(t) = u_j(0) + tV(u_j)|_{\mathbf{u}=\mathbf{u}(0)} + \frac{t^2}{2!}V(V(u_j))|_{\mathbf{u}=\mathbf{u}(0)} + \cdots$$

where $j = 1,2,\ldots,n$. In most practical cases only a finite number of terms in this expansion can be taken for the numerical integration of the differential equation. This approach has been used in chapter 3 for dissipative systems. Another approximation is to consider

$$\exp(t(V_1 + V_2)) = \prod_{j=1}^{n} \exp(c_j tV_1)\exp(d_j tV_2) + O(t^{n+1})$$

if the vector field V can be written as $V = V_1 + V_2$. Up to second order we have

$$\exp(t(V_1 + V_2)) = \exp(tV_1)\exp(tV_2) + O(t^2)$$

where $k = 1$ with $c_1 = d_1 = 1$. Up to third order we have

$$\exp(t(V_1 + V_2)) = \exp(\frac{1}{2}tV_1)\exp(tV_2)\exp(\frac{1}{2}tV_1) + O(t^3).$$

These are symplectic integrators (Yoshida [224]).

The study of the singularity structure in the complex time-plane of a dynamical system is of interest for systems with chaotic behaviour. The Standard Template Library (STL) of C++ provides a class `complex`. This class is utilized in the following program for the implementation of Lie series.

```
// complexlie.cpp

#include <iostream>
#include <complex>
using namespace std;

int main(void)
{
 complex<double> r(40.0,0.0), s(16.0,0), b(4.0,0.0);
 complex<double> count(0.0,0.0);
 complex<double> eps(0.0,0.01);
 complex<double> half(0.5,0);
 complex<double> us1(0.8,0.0), us2(0.8,0.0), us3(0.8,0.0);
 complex<double> u1, u2, u3;
 complex<double> V1, V2, V3, W1, W2, W3;

 while(abs(count) < 2.0)
 {
 u1 = us1; u2 = us2; u3 = us3;
 V1 = s*(u2-u1); V2 = -u2-u1*u3+r*u1; V3 = u1*u2-b*u3;
 W1 = s*s*(u1-u2) + s*(-u2-u1*u3+r*u1);
 W2 = u2+u1*u3-r*u1-s*(u2-u1)*u3-(u1*u2-b*u3)*u1+r*s*(u2-u1);
 W3 = s*u2*(u2-u1)+(-u2-u1*u3+r*u1)*u1-b*(u1*u2-b*u3);
 us1 = u1+eps*V1+half*eps*eps*W1;
 us2 = u2+eps*V2+half*eps*eps*W2;
 us3 = u3+eps*V3+half*eps*eps*W3;
 count += eps;
 }
 cout << "us1 = " << us1 << endl;
 cout << "us2 = " << us2 << endl;
 cout << "us3 = " << us3 << endl;
 return 0;
}
```

11.4 Runge-Kutta-Fehlberg Technique

Runge-Kutta-Fehlberg methods (Scheid [182], Fröberg [64], Hairer et al [79]) were developed to avoid the computation of high order derivatives which the Taylor method may involve. In place of these derivatives extra values of the given function $\mathbf{g}(\mathbf{u}, t)$ are used in a way which essentially duplicates the accuracy of a Taylor polynomial. We present the formulas on which a *Runge-Kutta-Fehlberg method* of order five is based. Suppose that the initial-value problem

$$\frac{d\mathbf{u}}{dt} = \mathbf{g}(\mathbf{u}, t), \qquad \mathbf{u}(t=0) = \mathbf{u}_0$$

for an autonomous system of ordinary differential equations of first order is to be integrated, where $\mathbf{u}(t_0) = \mathbf{u}_0$. A typical integration step approximates $\mathbf{u}$ at $t = t_0 + h$, where h is the step length. The formulas are

$$\mathbf{u}(t) = \mathbf{u}_0 + h \sum_{k=0}^{5} c_k \mathbf{g}^{(k)}, \qquad t = t_0 + h$$

with

$$\mathbf{g}^{(0)} = \mathbf{g}(\mathbf{u}_0), \qquad \mathbf{g}^{(k)} = \mathbf{g}(\mathbf{u}_0 + h \sum_{j=0}^{k-1} b_{kj} \mathbf{g}^{(j)}).$$

Thus $\mathbf{u}(t)$ approximates the exact solution. The coefficients c_k ($k = 0, 1, \ldots, 5$) are

$$c[0] = \frac{16}{135}, \quad c[1] = 0, \quad c[2] = \frac{6656}{12825},$$

$$c[3] = \frac{28561}{56430}, \quad c[4] = -\frac{9}{50}, \quad c[5] = \frac{2}{55}.$$

The coefficients for b_{jk} ($j = 0, 1, \ldots, 5$, $k = 0, 1, \ldots, 4$) are

$$b[0][0] = b[0][1] = b[0][2] = b[0][3] = b[0][4] = 0$$

$$b[1][0] = \frac{1}{4}, \quad b[1][1] = b[1][2] = b[1][3] = b[1][4] = 0$$

$$b[2][0] = \frac{3}{32}, \quad b[2][1] = \frac{9}{32}, \quad b[2][2] = b[2][3] = b[2][4] = 0$$

$$b[3][0] = \frac{1932}{2197}, \quad b[3][1] = \frac{-7200}{2197}, \quad b[3][2] = \frac{7296}{2197}, \quad b[3][3] = b[3][4] = 0$$

$$b[4][0] = \frac{439}{216}, \quad b[4][1] = -8, \quad b[4][2] = \frac{3680}{513}, \quad b[4][3] = -\frac{845}{4104}, \quad b[4][4] = 0$$

$$b[5][0] = -\frac{8}{27}, \quad b[5][1] = 2, \quad b[5][2] = -\frac{3544}{2565}, \quad b[5][3] = \frac{1859}{4104}, \quad b[5][4] = -\frac{11}{40}.$$

For the error estimation we refer to Hairer et al [79]. The method has been used extensively in chapters 3 to 8.

11.5 Ghost Solutions

When we discretize differential equations it can happen that the resulting difference equation shows chaotic behaviour even if the original differential equations tends to a fixed point. The global behaviour of numerical solutions computed by the difference equation is sensitive to the initial condition, the time-mesh length and the precision of computation employed. The global behaviour of the numerical solutions also depends on the calculators. This phenomenon is caused by roundoff errors.

Example. Consider the nonlinear ordinary differential equation

$$\frac{du}{dt} = u(1-u)$$

with the initial condition $u(0) = u_0 > 0$. The fixed points are given by $u^* = 0$, $u^* = 1$. The fixed point $u^* = 1$ is asymptotically stable. The exact solution of the initial value problem of the differential equation is given by

$$u(t) = \frac{u_0 e^t}{1 - u_0 + u_0 e^t}.$$

The exact solution starting from initial value $u_0 = 0.5$ is monotonically increasing and it converges to 1 as t tends to ∞. For $t = \ln 9999 \approx 9.21024$ we find that

$$u(t) = \frac{e^t}{1+e^t} = \frac{9999}{10000} = 0.9999$$

so that $u(t)$ is already quite near the asymptotically stable fixed point $u^* = 1$. In order to integrate this equation by a *finite difference scheme*, we apply the *central difference scheme*

$$\frac{du}{dt} \to \frac{u_{n+1} - u_{n-1}}{2h}.$$

Thus the differential equation takes the form

$$\frac{u_{n+1} - u_{n-1}}{2h} = u_n(1-u_n)$$

with initial conditions $u_0 = u_0$, $u_1 = u_0 + hu_0(1-u_0)$. We obtain

$$u_{n+1} = u_{n-1} + 2hu_n(1-u_n).$$

Introducing $v_n = u_{n-1}$ we obtain a system of first order difference equations

$$u_{n+1} = v_n + 2hu_n(1-u_n), \qquad v_{n+1} = u_n. \qquad ♣$$

In the Java program we compute a numerical solution by the difference equation starting from initial value $u_0 = 0.5$ and using time-mesh length $h = 0.05$. The difference equation is not stable at the fixed point $u^* = 1$. We find oscillating behaviour. Such a phenomenon is called a *ghost solution* or a *spurious solution*. For $0 < t < 8.5$ the numerical solution gives a good approximation of the true solution.

The solution u_n increases monotonically and tends to 1.000. After $t = 8.6$, the numerical solution is not monotone any more. At $t = 9.7$, the value u_n takes a value slightly greater than 1 for the first time and the solution begins to oscillate. The amplitude of the oscillation grows larger and larger. The growth of this amplitude is geometric and the rate of growth is such that the amplitude is multiplied by about $e = 2.71\ldots$ while t is increased by one, until about $t = 17.0$, when the oscillation loses its symmetry with respect to $u^* = 1$. The repetition of such cycles seems to be nearly periodic. The ghost solutions also appear even if h is quite small. One of the reasons is that the central difference scheme is a second order difference scheme and that the instability enters at the fixed points $u^* = 1$ and $u^* = 0$.

```
// Ghost.java

import java.awt.*;
import java.awt.event.*;
import java.awt.Graphics;

public class Ghost extends Frame
{
 public Ghost()
 {
 setSize(600,500);
 addWindowListener(new WindowAdapter()
 { public void windowClosing(WindowEvent event) { System.exit(0); }}); }

 public void paint(Graphics g)
 {
 double h = 0.05;              // step length
 double u = 0.5;               // initial value
 double v = u+h*u*(1.0-u);     // initial value
 double u1, v1;
 double t0 = 0.0;

 while(t0 <= 30.5)
 {
 u1 = u; v1 = v; u = v1+2.0*h*u1*(1.0-u1); v = u1;
 int tb = (int) Math.floor(20.0*t0+10.0);
 int m = (int) Math.floor(300.0-150.0*u1);
 int te = (int) Math.floor(20.0*(t0+h)+10.0);
 int n = (int) Math.floor(300.0-150.0*u);
 g.drawLine(tb,m,te,n);
 t0 += h;
 }
 }

 public static void main(String[] args)
 { Frame f = new Ghost(); f.setVisible(true); }
}
```

In the C++ program `ghost.cpp` we use the `Rational` and `Verylong` classes of SymbolicC++ to do the iteration of the difference scheme. The result is stored in the file `ghost.dat`. It shows that the oscillating behaviour is not caused by finite precision and rounding errors.

```
// ghost.cpp

#include <fstream>
#include "rational.h"
#include "verylong.h"
using namespace std;

const Rational<Verylong> h("1/10");      // time step
const Rational<Verylong> x0("99/100");

int main(void)
{
 Rational<Verylong> u, v, u1, twoh, t;
 Rational<Verylong> zero("0"), one("1"), two("2"), five("5");
 ofstream sout("ghost.dat");
 u = x0; v = u+h*u*(one-u); // initial values
 t = zero;
 twoh = two*h;
 while(t <= five)
 {
 u1 = u; u = v;
 v = u1+two*u*(one-u);
 t += h;
 sout << t << " " << v << endl;
 }
 sout.close();
 return 0;
}
```

The Lie series technique (C++ program given below) leads to the exact result, i.e., we do not find oscillating behaviour as in the case of the central difference scheme. Thus with the Lie series technique we find $u \to 1$ as $t \to \infty$.

```
// GhostLie.cpp

#include <iostream>
using namespace std;

int main(void)
{
 double h = 0.001;  // step length
 double u = 0.5;    // initial value
 double t = 0.0;
 double u1;
```

```
 while(t <= 5.0)
 {
 t += h;
 u1 = u+h*u*(1.0-u)+h*h*u*(1.0-u)*(1.0-2.0*u)/2.0;
 cout << "t = " << t << " " << "u = " << u1 << endl;
 u = u1;
 }
 return 0;
}
```

11.6 Symplectic Integration

The Hamilton's equations of motion in standard or canonical form (Abraham and Marsden [1], Guillemin and Sternberg [76]) are given by

$$\frac{dq_j}{dt} = \frac{\partial H}{\partial p_j}, \qquad \frac{dp_j}{dt} = -\frac{\partial H}{\partial q_j}, \qquad j = 1, 2, \ldots, N.$$

In the Hamiltonian description of mechanics, the evolution of a system is described in terms of $2N$ first order differential equations. The $2N$ variables $q_1, \ldots, q_N, p_1, \ldots, p_N$ are often referred to as the canonical variables. They define a $2N$-dimensional phase space. The solution to Hamilton's equations

$$q_j(t) = q_j(\mathbf{q}_0, \mathbf{p}_0, t), \qquad p_j(t) = p_j(\mathbf{q}_0, \mathbf{p}_0, t)$$

where $\mathbf{q}_0 = (q_1(0), \ldots, q_N(0))$ and $\mathbf{p}_0 = (p_1(0), \ldots, p_N(0))$ are the set of initial conditions, define the state of the system at time t. In its time-evolution, $(\mathbf{q}(t), \mathbf{p}(t))$ map out a trajectory exploring different regions in the phase-space. The N coordinates and n momenta can be considered as a single set of $2N$ coordinates, z_j where

$$\mathbf{z} = (q_1, \ldots, q_N, p_1, \ldots, p_N).$$

Using this notation, Hamilton's equations can be written in concise form as

$$\frac{d\mathbf{z}}{dt} = J_{2N} \cdot \nabla H(\mathbf{z})$$

where $\nabla := (\partial/\partial z_1, \ldots, \partial/\partial z_{2n})$. The $2n \times 2n$ matrix J_{2N} is called the *symplectic matrix*,

$$J_{2N} := \begin{pmatrix} \mathbf{0} & I_N \\ -I_N & \mathbf{0} \end{pmatrix}$$

where I_N is the $N \times N$ unit matrix. They satisfy the incompressibility condition

$$\sum_{j=1}^{N} \left(\frac{\partial \dot{q}_j}{\partial q_j} + \frac{\partial \dot{p}_j}{\partial p_j} \right) = 0.$$

This means that the divergence of the Hamilton system is zero. Thus a volume element in phase-space is preserved under the Hamilton flow. This result is known as

Liouville's theorem, and is one of the important properties of Hamilton dynamical systems.

If a dynamical system evolves under a Hamilton flow, a number of important quantities are left invariant. The most fundamental of these is a geometric entity the differential two-form

$$\omega_2 := \sum_{j=1}^{N} dp_j \wedge dq_j$$

where $d\omega_2 = 0$ with d denoting the exterior derivative and $\wedge$ is the exterior product, i.e. $dp_j \wedge dq_j = -dq_j \wedge dp_j$ for $j = 1, \ldots, N$. If we denote the Hamilton phase flow by Φ_t, where Φ_t maps the initial conditions to the solution at time t, we have

$$(\Phi_t)^* \omega_2 = \omega_2 \, .$$

Thus we have the definition

Definition. A transformation $(\mathbf{q}, \mathbf{p}) \to (\mathbf{Q}, \mathbf{P})$ is called *symplectic* if it preserves the 2-form ω_2.

Any transformation preserving ω_2 also preserves the form of Hamilton's equations. The converse is not true. A flow having this property is termed *symplectic*. In the case of one degree of freedom ($N = 1$), this simply means preservation of the (oriented) phase-area.

Example. Let $N = 1$ and consider the transformation

$$P(p, q) = -q, \qquad Q(p, q) = p \, .$$

Then $dP \wedge dQ = -dq \wedge dp = dp \wedge dq$. ♣

This preservation of the 2-form is a fundamental property of Hamilton systems. In fact, a Hamilton flow can be characterized solely in terms of the 2-form. If a domain D in $\mathbb{R}^{2N}$ is simply connected (i.e., it has no holes), and

$$\frac{d\mathbf{p}}{dt} = \mathbf{f}(\mathbf{q}, \mathbf{p}), \qquad \frac{d\mathbf{q}}{dt} = \mathbf{g}(\mathbf{q}, \mathbf{p})$$

is a smooth differential system whose flow preserves the 2-form, then this system of differential equations is a Hamilton system for some Hamilton function H.

The preservation of the 2-form ω_2 is one of a whole hierarchy of quantities preserved by the Hamiltonian flow first studied by Poincaré, which he termed the *integral invariants*.

Liouville's Theorem. The Hamilton flow preserves the volume element in phase space

$$\int \prod_{j=1}^{N} dp_j \wedge dq_j = \int \prod_{j=1}^{N} dP_j \wedge dQ_j$$

in which the integral sign represents a $2N$-dimensional integration over a prescribed volume in phase-space.

Another property of conservative Hamilton systems is that H is a constant of motion, i.e., $dH/dt = 0$. Since we are interested in numerical discretizations of a conservative Hamilton systems, one wants the discretization capture as much as possible of the original Hamilton structure. This motivates a study of transformations preserving the differential two-form. A numerical discretization of a continuous system cannot be expected to be exact. We cannot preserve all of the properties of the original flow. We cannot preserve the original form of the Hamilton function and the area-preserving properties of the original flow. Preservation of both these quantities amounts to the exact solution of the system. Since preservation of the differential two-form (symplecticity) is such a fundamental property of Hamilton systems, one attempts to preserve this property in the discretized system.

When a continuous flow is discretized, the discrete flow becomes a succession of transformations from one time-step to another. Preservation of the 2-form can be retained by ensuring that the transformations in the discrete system are symplectic. Another reason why symplectic transformations are useful, is that it is often possible to simplify the integration of the equations of motion of a system by transforming to a different set of coordinates. We have two sets of independent variables $\mathbf{p}$ and $\mathbf{q}$, which are very much on an equal footing. Therefore we can transform from one set of phase-space variables $(\mathbf{q}, \mathbf{p})$ to a new set $(\mathbf{Q}, \mathbf{P})$. This transformation can be written as

$$P_i = P_i(q_1, \ldots, q_N, p_1, \ldots, p_N), \qquad Q_i = Q_i(q_1, \ldots, q_N, p_1, \ldots, p_N).$$

There are various ways of constructing symplectic transformations. The simplest of these is the method of *generating functions*. Consider the one degree of freedom system

$$\frac{dq}{dt} = \frac{\partial H(p,q)}{\partial p}, \qquad \frac{dp}{dt} = -\frac{\partial H(p,q)}{\partial q}.$$

This system can be discretized with Euler's method, yielding

$$Q = q + \tau \frac{\partial H(p,q)}{\partial p}, \qquad P = p - \tau \frac{\partial H(p,q)}{\partial q}.$$

The discrete system defines a map from (q,p) (the approximate solution at time t) to (Q,P) (the approximate solution at $t+\tau$). The Jacobian determinant for this transformation is given by

$$\det\left(\frac{\partial(Q,P)}{\partial(q,p)}\right) = 1 + O(\tau).$$

Thus this transformation does not preserve area. Euler's method is therefore not symplectic. Consider now

$$P = p - \tau \frac{\partial H(q,P)}{\partial q}, \qquad Q = q + \tau \frac{\partial H(q,P)}{\partial p}.$$

Then the Jacobian determinant satisfies

$$\det \left(\frac{\partial(Q,P)}{\partial(q,p)} \right) = 1.$$

The modified method is therefore symplectic.

For a higher dimensional system to be symplectic, the preservation of volume is a necessary but not sufficient property.

Example. Assume a separable planar ($n = 1$) Hamilton function

$$H(q,p) = \frac{1}{2}p^2 + V(q).$$

Choosing a generating function of the second kind

$$F^2(q,P,t) = qP + tH(q,P) = qP + \frac{1}{2}tP^2 + tV(q)$$

the symplectic map $S^t : (q,p) \to (Q,P)$ is obtained as

$$p = \frac{\partial F^2}{\partial q} = P + tV'(q), \qquad Q = \frac{\partial F^2}{\partial P} = q + tP\,.$$

Thus the symplectic scheme is

$$p_{m+1} = p_m - \tau V'(q_m), \qquad q_{m+1} = q_m + \tau p_{m+1}\,.$$

Consequently

$$p_{m+1} = p_m - \tau V'(q_m), \qquad q_{m+1} = q_m + \tau(p_m - \tau V'(q_m))\,. \quad ♣$$

Higher order symplectic methods (Yoshida [224]) are constructed as follows. We assume that the vector field V can be written as $V = V_1 + V_2$. If

$$[V_1, V_2] = 0$$

then for t sufficiently small we have

$$\exp(tV) = \exp(tV_1 + tV_2) = \exp(tV_1)\exp(tV_2)$$

where [,] denotes the commutator. In general, we have

$$[V_1, V_2] \neq 0\,.$$

The problem is as follows. Let V_1 and V_2 be two non-commutative vector fields and t be a sufficiently small real number. For a given positive integer n which is called the *order of integrator*, find a set of real numbers $c_1, c_2, \ldots, c_k$ and $d_1, d_2, \ldots, d_k$ such that the difference of the exponential function $\exp(t(V_1+V_2))$ and the product of the exponential functions

$$\exp(c_1 tV_1)\exp(d_1 tV_2)\exp(c_2 tV_1)\exp(d_2 tV_2)\cdots\exp(c_k tV_1)\exp(d_k tV_2)$$

is of the order t^{n+1}, i.e., the following equality holds

$$\exp(t(V_1+V_2)) = \prod_{j=1}^{k}\exp(c_j tV_1)\exp(d_j tV_2) + O(t^{n+1}).$$

If $n=1$, a trivial solution is $c_1 = d_1 = 1$ $(k=1)$, and we have

$$\exp(t(V_1+V_2)) = \exp(tV_1)\exp(tV_2) + O(t^2).$$

When $n=2$, we find that $c_1 = c_2 = \frac{1}{2}$, $d_1 = 1$, $d_2 = 0$ $(k=2)$. Thus

$$\exp(t(V_1+V_2)) = \exp(\frac{1}{2}tV_1)\exp(tV_2)\exp(\frac{1}{2}tV_1) + O(t^3).$$

For the construction of higher order symplectic integrators we refer to Yoshida [224]. The construction is based on the *Baker-Campbell-Hausdorff formula.*

The formula is as follows. For any non-commutative operators X and Y, the product of the two exponential functions, $\exp(X)\exp(Y)$, can be expressed in the form of a single exponential function as

$$\exp(X)\exp(Y) = \exp(Z)$$

where

$$Z = X + Y + \frac{1}{2}[X,Y] + \frac{1}{12}([X,X,Y] + [Y,Y,X]) + \cdots.$$

Here $[\,,\,]$ denotes the commutator, and higher order commutators like

$$[X,X,Y] := [X,[X,Y]].$$

The feature of the Baker-Campbell-Hausdorff formula is that only commutators of X and Y appear except for the linear terms in the series. Symplectic integrators have been applied in chapter 8.

Another technique which can be used for integration is the *Trotter formula* (Trotter [205]). Let A be the generator of a contractive C_0 - semigroup $\exp(tA)_{t\geq 0}$ on a Banach space E, and let $B \in \mathcal{L}(E)$ be a linear dissipative operator, where $\mathcal{L}(E)$ denotes the vector space of all linear bounded maps $E \to E$. Then $A+B$ generates a C_0 - semigroup which is given by Trotter's formula

$$\exp(t(A+B)) = \lim_{n\to\infty}\left(\exp\left(\frac{t}{n}A\right)\exp\left(\frac{t}{n}B\right)\right)^n$$

where the limit is taken in the strong operator topology. Thus the formula in particular applies if A and B are $n \times n$ matrices.

11.7 Verlet Method

The *Verlet algorithm* (Verlet [212]) is a method for integrating second order ordinary differential equations

$$\frac{d^2\mathbf{x}}{dt^2} = \mathbf{F}(\mathbf{x}(t), t)\,.$$

The Verlet algorithm is used in *molecular dynamics simulations*. It has a fixed time discretization interval h and it needs only one evaluation of the force $\mathbf{F}$ per step. The algorithm is derived by adding the *Taylor expansions* for the coordinates $\mathbf{x}$ at $t = \pm h$ about 0

$$\mathbf{x}(h) = \mathbf{x}(0) + h\frac{d\mathbf{x}(0)}{dt} + \frac{h^2}{2}\mathbf{F}(\mathbf{x}(0), 0) + \frac{h^3}{6}\frac{d^3\mathbf{x}(0)}{dt^3} + O(h^4)$$
$$\mathbf{x}(-h) = \mathbf{x}(0) - h\frac{d\mathbf{x}(0)}{dt} + \frac{h^2}{2}\mathbf{F}(\mathbf{x}(0), 0) - \frac{h^3}{6}\frac{d^3\mathbf{x}(0)}{dt^3} + O(h^4)$$

leading, after addition of the two expressions, to

$$\mathbf{x}(h) = 2\mathbf{x}(0) - \mathbf{x}(-h) + h^2\mathbf{F}(\mathbf{x}(0), 0) + O(h^4)\,.$$

Knowing the values of $\mathbf{x}$ at time 0 and $-h$, this algorithm predicts the value of $\mathbf{x}(h)$. Thus we need the last two values of $\mathbf{x}$ to produce the next one. If we only have the initial position $\mathbf{x}(0)$ and initial velocity $\mathbf{v}(0)$ at our disposal, we approximate $\mathbf{x}(h)$ by

$$\mathbf{x}(h) \approx \mathbf{x}(0) + h\mathbf{v}(0) + \frac{h^2}{2}\mathbf{F}(\mathbf{x}(0), 0)$$

i.e., we set

$$\mathbf{v}(0) = \frac{\mathbf{x}(0) - \mathbf{x}(-h)}{h}\,.$$

Example. Consider the one-dimensional *harmonic oscillator*

$$\frac{d^2x}{dt^2} + \Omega^2 x = 0$$

where Ω is a constant frequency. We find the approximation

$$x(t+h) = 2x(t) - x(t-h) - h^2\Omega^2 x(t)\,.$$

The analytic solution to this difference equation can be written in the form

$$x(t) = \exp(i\omega t)$$

with frequency ω satisfying the condition

$$2 - 2\cos(\omega h) = h^2\Omega^2\,.$$

If $h^2\Omega^2 \equiv (h\Omega)^2 > 4$, the frequency ω becomes imaginary, and the analytical solution becomes unstable. Thus it is useful to introduce a dimensionless time τ via

$$\widetilde{x}(\tau(t)) = x(t), \qquad \tau(t) = \Omega t\,.$$

Then the linear differential equation for the harmonic oscillator takes the form

$$\frac{d^2\widetilde{x}}{d\tau^2} + \widetilde{x} = 0 . \qquad \clubsuit$$

In the C++ program we give an implementation of the Verlet method for the one-dimensional pendulum $d^2\theta/dt^2 = -(g/L)\sin(\theta)$.

```
// Verlet.cpp

#include <iostream>
#include <cmath>
using namespace std;

int main(void)
{
 const double pi = 3.141592654;
 double theta = 1.4;     // initial angle (in radians)
 double omega = 0.0;     // initial velocity
 double g_over_L = 1.0; // the constant g/L
 double time = 0.0;      // initial time
 double time_old;        // time of previous reversal
 double tau = 0.005;     // time step size
 int nStep = 2000;       // number of steps

 // take one backward step to start Verlet
 double accel = -g_over_L*sin(theta); // gravitational acceleration
 double theta_old = theta-omega*tau+0.5*tau*tau*accel;

 for(int i=1;i<=nStep;i++)
 {
 time += tau;
 accel = -g_over_L*sin(theta);
 double theta_new = 2.0*theta-theta_old+tau*tau*accel;
 cout << "time = " << time << " " << "theta = " << theta_new << " "
      << "accel = " << accel << endl;
 theta_old = theta; theta = theta_new;
 }
 return 0;
}
```

11.8 Störmer Method

We consider the system of second-order ordinary differential equations

$$m\frac{d^2\mathbf{x}}{dt^2} = \mathbf{F}(\mathbf{x})$$

where $\mathbf{x}$ is the collective position vector, m is a diagonal matrix of masses, and $\mathbf{F}$ is the collective force vector. The discretization known as the second-order *Störmer method* (Hairer et al [79]) is given by

$$\frac{1}{\Delta t^2} m(\mathbf{X}^{n+1} - 2\mathbf{X}^n + \mathbf{X}^{n-1}) = \mathbf{F}(\mathbf{X}^n)$$

where Δt is the timestep, and $\mathbf{X}^n$ denotes the difference approximation to $\mathbf{x}$ at time $n\Delta t$. This method can be used as an integrator for *molecular dynamics* together with the formula

$$\mathbf{V}^n = \frac{1}{2\Delta t}(\mathbf{X}^{n+1} - \mathbf{X}^{n-1})$$

for calculating the velocity $\mathbf{v} = d\mathbf{x}/dt$. This combination is equivalent to the *leapfrog method*, defined as

$$\mathbf{V}^{n+1/2} = \mathbf{V}^{n-1/2} + \Delta t m^{-1}\mathbf{F}(\mathbf{X}^n)$$
$$\mathbf{X}^{n+1} = \mathbf{X}^n + \Delta t \mathbf{V}^{n+1/2} .$$

The implicit discretization scheme with a different right-hand side

$$\frac{1}{\Delta t^2} m(\mathbf{X}^{n+1} - 2\mathbf{X}^n + \mathbf{X}^{n-1}) = \frac{1}{12}\mathbf{F}(\mathbf{X}^{n-1}) + \frac{5}{6}\mathbf{F}(\mathbf{X}^n) + \frac{1}{12}\mathbf{F}(\mathbf{X}^{n+1})$$

is often called *Cowell's method.* Another method which can be used for integrating the differential equation is

$$\frac{1}{\Delta t^2} m(\mathbf{X}^{n+1} - 2\mathbf{X}^n + \mathbf{X}^{n-1}) = \mathbf{F}\left(\frac{1}{2}(\mathbf{X}^{n-1} + \mathbf{X}^{n+1})\right) .$$

11.9 Invisible Chaos

Discretization of nonlinear differential equations can lead to maps which show chaotic behaviour. We consider the anharmonic system

$$\frac{d^2u}{dt^2} + u^3 = 0 .$$

Using a finite difference scheme this differential equation can be written as

$$\frac{u_{n+1} - 2u_n + u_{n-1}}{(\Delta t)^2} = -u_n^3$$

to obtain approximate solutions of the differential equation. For a fixed time step Δt, $\Delta t > 0$, this scheme is equivalent to the scheme

$$U_{n+1} - 2U_n + U_{n-1} = -U_n^3$$

by the replacement $u_n = U_n/\Delta t$. We define $V_n := U_{n-1}$. Consequently we obtain the two-dimensional map

$$U_{n+1} = -V_n + 2U_n - U_n^3, \qquad V_{n+1} = U_n .$$

This two-dimensional map is invertible and area-preserving and $(0,0)$ is a fixed point. Numerical solutions of the difference scheme for small Δt correspond to orbits near the origin $(0,0)$ of the map via the transformation $u_n = U_n/\Delta t$. Numerical experiments show that there are invariant circles of the map around the origin $(0,0)$.

```
// Invisible.java

import java.awt.*;
import java.awt.event.*;
import java.awt.Graphics;

public class Invisible extends Frame
{
 public Invisible()
 {
 setSize(600,500);
 addWindowListener(new WindowAdapter()
 { public void windowClosing(WindowEvent event) { System.exit(0); }}); }

 public void paint(Graphics g)
 {
 double u = 0.9, v = 0.75;  // initial values
 int T = 3000;              // number of iterations
 double u1, v1;
 for(int t=0;t<T;t++) {
 u1 = u; v1 = v;
 u = -v1+2.0*u1-u1*u1*u1;  v = u1;
 int mx = (int) Math.floor(100.0*u+300.0+0.5);
 int ny = (int) Math.floor(100.0*v+220.0+0.5);
 g.drawLine(mx,ny,mx,ny);
 }
 }

 public static void main(String[] args)
 { Frame f = new Invisible(); f.setVisible(true); }
}
```

11.10 First Integrals and Numerical Integration

Consider the autonomous system of first order ordinary differential equations

$$\frac{d\mathbf{u}}{dt} = \mathbf{f}(\mathbf{u})$$

where the vector field $\mathbf{f} : \mathbb{R}^n \to \mathbb{R}^n$ is analytic. Some of these systems admit a first integral, i.e., there exists a scalar function I such that

$$\frac{dI(\mathbf{u})}{dt} = 0\,.$$

Thus

$$\sum_{j=1}^{n} \frac{\partial I}{\partial u_j} \frac{du_j}{dt} \equiv \sum_{j=1}^{n} \frac{\partial I}{\partial u_j} f_j(\mathbf{u}) = 0\,.$$

Preserving first integrals in numerical integration is important because of their physical relevance, e.g. in mechanics and astronomy, but also because they can ensure long-term stabilising effects. Thus we want to find a discrete approximation to the system of differential equations

$$\frac{\mathbf{u}' - \mathbf{u}}{\tau} = \mathbf{g}(\mathbf{u}, \mathbf{u}', \tau)$$

($\mathbf{u} \equiv \mathbf{u}(n\tau)$, $\mathbf{u}' \equiv \mathbf{u}((n+1)\tau)$) such that the first integral is preserved exactly, i.e. $I(\mathbf{u}') = I(\mathbf{u})$. For example the implicit midpoint rule

$$\frac{\mathbf{u}' - \mathbf{u}}{\tau} = \mathbf{f}\left(\frac{\mathbf{u}' + \mathbf{u}}{2}\right)$$

preserves quadratic integrals.

McLaren and Quispel [144] provide a more general case. The system of ordinary differential equations can be written in the form

$$\frac{d\mathbf{u}}{dt} = S \cdot \nabla I(\mathbf{u})$$

where S is a skew-symmetric $n \times n$ matrix over $\mathbb{R}$, i.e. $S^T = -S$ and ∇ denotes the gradient. An integral-preserving discrete version of this is

$$\frac{\mathbf{u}' - \mathbf{u}}{\tau} = \widetilde{S}(\mathbf{u}, \mathbf{u}', \tau) \overline{\nabla} I(\mathbf{u}, \mathbf{u}')$$

where $\mathbf{u}$, $\mathbf{u}'$ denote $\mathbf{u}_n$ respectively $\mathbf{u}_{n+1}$, and where $\widetilde{S}$ is a skew symmetric matrix satisfying (for consistency)

$$\widetilde{S}(\mathbf{u}, \mathbf{u}', \tau) = S(\mathbf{u}) + O(\tau)\,.$$

The general discrete gradient $\overline{\nabla} I$ is defined as

$$(\mathbf{u}' - \mathbf{u}) \cdot \nabla \overline{I}(\mathbf{u}', \mathbf{u}) := I(\mathbf{u}') - I(\mathbf{u})$$

and may be expanded in the form

$$\overline{\nabla} I(\mathbf{u}, \mathbf{u}') = \nabla I + B(\mathbf{u})(\mathbf{u}' - \mathbf{u}) + (\mathbf{u}' - \mathbf{u})^T M(\mathbf{u})(\mathbf{u}' - \mathbf{u}) + O(\|\mathbf{u}' - \mathbf{u}\|^3)\,.$$

This leads to the conditions

$$B_{ij} + B_{ji} = \frac{\partial^2 I}{\partial u_i \partial u_j}, \qquad M_{ijk} + M_{jki} + M_{kij} = \frac{1}{2} \frac{\partial^3 I}{\partial u_i \partial u_j \partial u_k}\,.$$

The order of accuracy of an integral-preserving integrator based on the discretization given above is determined by $\widetilde{S}$ and by the choice of the discrete gradient $\overline{\nabla} I(\mathbf{u}, \mathbf{u}')$, i.e. by $\widetilde{S}$ and the matrix B, the tensor M, and the higher order parts of $\nabla\overline{I}$.

Exercises. (1) Consider an autonomous system $d\mathbf{u}/dt = \mathbf{f}(\mathbf{u})$ in $\mathbb{R}^3$, where f_j's are analytic function. Then $\mathbf{f}$ can be written as

$$\mathbf{f} = \mathrm{grad}\Phi + \mathrm{curl}\mathbf{W}$$

(*Hodge-Helmholtz decomposition*). Consider the Lorenz model

$$\frac{du_1}{dt} = -\sigma u_1 + \sigma u_2$$
$$\frac{du_2}{dt} = -u_1 u_3 + r u_1 - u_2$$
$$\frac{du_3}{dt} = u_1 u_2 - b u_3 \,.$$

Show that

$$\Phi(\mathbf{u}) = -\frac{\sigma}{2}u_1^2 - \frac{1}{2}u_2^2 - \frac{b}{2}u_3^2 + u_1 u_2 u_3$$

and

$$W_1(\mathbf{u}) = -\frac{1}{2}u_1 u_2^2 - \frac{1}{6}u_1^3, \quad W_2(\mathbf{u}) = -\sigma u_2 u_3 + \frac{1}{6}u_2^3, \quad W_3(\mathbf{u}) = \frac{1}{2}u_1^2 u_3 - \frac{r}{2}u_1^2 + \frac{\sigma}{2}u_3^2 \,.$$

(2) Find the Hodge-Helmholtz decomposition for the Rikitake two disc dynamo

$$\frac{du_1}{dt} = -\mu u_1 + u_2 u_3$$
$$\frac{du_2}{dt} = -\mu u_2 + (u_3 - a) u_1$$
$$\frac{du_3}{dt} = 1 - u_1 u_2 \,.$$

Chapter 12

Optimization

12.1 Lagrange Multiplier Method

In mathematical optimization problems, Lagrange multipliers are a method for finding the minima and maxima of a differentiable function f with equality constraints. The Lagrange multiplier method could also fail even if there is a solution.

The *Lagrange multiplier method* is as follows (Protter [168]). Let M be a manifold and f be a real valued function of class $C^{(2)}$ on some open set containing M. We consider the problem of finding the extrema of the function $f|M$. This is called a problem of constrained extrema. Assume that f has a constrained extremum at $\mathbf{x}^* = (x_1^*, x_2^*, \ldots, x_n^*)$. Let

$$g_1(\mathbf{x}) = 0, \ldots, g_m(\mathbf{x}) = 0$$

be the constraints (manifolds) with $m < n$. We assume that f and g_j $(j = 1, \ldots, m)$ are continuously differentiable in a neighbourhood of $\mathbf{x}^*$. Then there exist real numbers $\lambda_1, \ldots, \lambda_m$ such that $\mathbf{x}^*$ is a critical point of the function (called the *Lagrange function*)

$$L(\mathbf{x}) := f(\mathbf{x}) + \lambda_1 g_1(\mathbf{x}) + \cdots + \lambda_m g_m(\mathbf{x}).$$

The numbers $\lambda_1, \ldots, \lambda_m$ are called Lagrange multipliers. Thus we have to solve

$$\nabla L(\mathbf{x}^*) = \mathbf{0}$$

and

$$g_j(\mathbf{x}^*) = 0, \qquad j = 1, 2, \ldots, m$$

with respect to $x_1^*, \ldots, x_n^*, \lambda_1^*, \ldots, \lambda_m^*$. Here ∇ denotes the gradient and we have to assume that the rank of the matrix $\nabla \mathbf{g}(\mathbf{x}^*)$ is m and we assume that $m < n$. We will see later that the Lagrange multiplier method can fail even if there is a solution.

We have the following theorem.

Theorem. Let $f : \mathbb{R}^n \to \mathbb{R}$ be a twice continuously differentiable function in an open set $\Omega \subseteq \mathbb{R}^n$. Let S be an open set $S \subseteq \mathbb{R}^n$. Let $\mathbf{g} = (g_1, g_2, \ldots, g_m) : S \to \mathbb{R}^m$ be twice continuously differentiable, and assume that $m < n$. Let X_0 be the subset of S where g vanishes, that is

$$X_0 := \{\, \mathbf{x} \in S \,:\, \mathbf{g}(\mathbf{x}) = \mathbf{0} \,\}.$$

Suppose that $\mathbf{x}^* \in X_0$ and assume that there is a neighbourhood N of $\mathbf{x}^*$ such that f achieves maximum or minimum at $\mathbf{x}^*$ in $N \cap X_0$. Assume that the determinant of the $m \times m$ matrix $(\partial g_i(\mathbf{x}^*)/\partial x_j)$ does not vanish. Then there exist m real numbers $\lambda_1, \ldots, \lambda_m$ such that the following n equations are satisfied

$$\frac{\partial}{\partial x_r} f(\mathbf{x}^*) + \sum_{k=1}^{m} \lambda_k \frac{\partial}{\partial x_r} g_k(\mathbf{x}^*) = 0, \qquad r = 1, \ldots, n.$$

The *Hessian matrix* of the Lagrange function $L = f + \sum_{j=1}^{m} \lambda_j g_j$ is given by

$$\begin{pmatrix} L_{x_1^2} & L_{x_1x_2} & \ldots & L_{x_1x_n} & L_{x_1\lambda_1} & \ldots & L_{x_1\lambda_m} \\ L_{x_2x_1} & L_{x_2^2} & \ldots & L_{x_2x_n} & L_{x_2\lambda_1} & \ldots & L_{x_2\lambda_n} \\ \vdots & \vdots & \vdots & \vdots & \vdots & & \\ L_{x_nx_1} & L_{x_nx_2} & \ldots & L_{x_nx_n} & L_{x_n\lambda_1} & \ldots & L_{x_n\lambda_n} \\ L_{\lambda_1x_1} & L_{\lambda_1x_2} & \ldots & L_{\lambda_1x_n} & L_{\lambda_1\lambda_1} & \ldots & L_{\lambda_1\lambda_m} \\ \vdots & \vdots & \vdots & \vdots & \vdots & & \\ L_{\lambda_nx_1} & L_{\lambda_nx_2} & \ldots & L_{\lambda_nx_n} & L_{\lambda_n\lambda_1} & \ldots & L_{\lambda_n\lambda_m} \end{pmatrix}$$

where $L_{x_ix_j} := \partial^2 L/\partial x_i \partial x_j$ etc.. Obviously we have $L_{\lambda_i\lambda_j} = 0$. This Hessian matrix is also called *bordered Hessian matrix*. It is used to determine whether the critical points are local maxima or minima. If for every nonzero (column) vector $\mathbf{v} \in \mathbb{R}^n$ satisfying $\mathbf{v}^T \nabla g_j(\mathbf{x}^*) = 0$ ($j = 1, 2, \ldots, m$) it follows that

$$\mathbf{v}^T \nabla_{\mathbf{x}}^2 L(\mathbf{x}^*, \lambda^*) \mathbf{v} > 0$$

then the function f has a strict local minimum at $\mathbf{x}^*$ subject to $g_j(\mathbf{x}) = 0$ ($j = 1, 2, \ldots, m$).

Example 1. The norm of an $n \times n$ matrix over the real numbers $\mathbb{R}$ is given by

$$\|A\| := \sup_{\|\mathbf{x}\|=1} \|A\mathbf{x}\|.$$

This is a problem with the constraint $\|\mathbf{x}\| = 1$ i.e., the length of the vector $\mathbf{x} \in \mathbb{R}^n$ must be 1. It can be solved with the Lagrange multiplier method. To find the norm of an $n \times n$ matrix one considers the Lagrange function

$$L(\mathbf{x}) := \|A\mathbf{x}\|^2 + \lambda \|\mathbf{x}\|^2$$

where λ is the Lagrange multiplier. Consider the 2×2 matrix

$$A = \begin{pmatrix} 1 & 1 \\ 2 & 2 \end{pmatrix}.$$

Then we have

$$\|A\mathbf{x}\|^2 = (A\mathbf{x})^T A\mathbf{x} = \mathbf{x}^T A^T A\mathbf{x} = 5x_1^2 + 10x_1x_2 + 5x_2^2.$$

From the constraint $\|\mathbf{x}\|^2 = 1$ we obtain $x_1^2 + x_2^2 = 1$. Differentiation the function

$$L(\mathbf{x}) = 5x_1^2 + 10x_1x_2 + 5x_2 + \lambda(x_1^2 + x_2^2)$$

and setting the derivative equal to zero yields

$$\frac{\partial L}{\partial x_1} = 10x_1 + 10x_2 + 2\lambda x_1 = 0, \qquad \frac{\partial L}{\partial x_2} = 10x_1 + 10x_2 + 2\lambda x_2 = 0.$$

Thus we find that $x_1 = x_2$ and

$$(x_1, x_2) = (1/\sqrt{2}, 1/\sqrt{2}), \qquad (x_1, x_2) = (-1/\sqrt{2}, -1/\sqrt{2}).$$

It follows that $\|A\|^2 = 10$. ♣

Example 2. We calculate the shortest Euclidean distance between the curves

$$x^2 + (y-5)^2 = 1, \qquad y = x^2$$

in $\mathbb{R}^2$ applying the Lagrange multiplier method. The square of the distance between two points (x_1, y_1) and (x_2, y_2) on the curves is given by

$$d^2 := (x_1 - x_2)^2 + (y_1 - y_2)^2.$$

We have two constraints. This means we have two Lagrange multipliers. Thus we define

$$L(x_1, y_1, x_2, y_2) := (x_1 - x_2)^2 + (y_1 - y_2)^2 + \lambda_1(x_1^2 + (y_1 - 5)^2 - 1) + \lambda_2(y_2 - x_2^2).$$

Hence we obtain the equations

$$\frac{\partial L}{\partial x_1} = 2(x_1 - x_2) + 2\lambda_1 x_1 = 0$$
$$\frac{\partial L}{\partial x_2} = -2(x_1 - x_2) - 2\lambda_2 x_2 = 0$$
$$\frac{\partial L}{\partial y_1} = 2(y_1 - y_2) + 2\lambda_1(y_1 - 5) = 0$$
$$\frac{\partial L}{\partial y_2} = -2(y_1 - y_2) + \lambda_2 = 0.$$

Adding the first two equations and the last two equations yields

$$\lambda_1 x_1 = \lambda_2 x_2, \qquad 2\lambda_1(y_1 - 5) = -\lambda_2 .$$

Of course we still have $(1+\lambda_1)x_1 = x_2$, $(1+\lambda_1)(y_1-5) = y_2 - 5$ from the first set of equations. If $\lambda_1 = 0$, we have $\lambda_2 = 0$ so that $x_1 = x_2$ and $y_1 = y_2$. Thus we obtain

$$x_1^4 - 9x_1^2 + 24 = 0 \quad \Rightarrow \quad x_1^2 = \frac{9 \pm \sqrt{-15}}{2}.$$

Thus $\lambda_1 = 0$ does not give a valid solution. Suppose $\lambda_2 = 0$. Once again $x_1 = x_2$ and $y_1 = y_2$, thus $\lambda_2 = 0$ does not give a valid solution. Now suppose $x_1 = 0$, thus $x_2 = y_2 = 0$ and $y_1 = 6$ or $y_1 = 4$. Lastly suppose $x_1 \neq 0$. Thus

$$\begin{aligned} \frac{x_2}{x_1} = \frac{\lambda_1}{\lambda_2} &= (1 + \lambda_1) \\ \frac{\lambda_1}{\lambda_2}(y_1 - 5) &= (1 + \lambda_1)(y_1 - 5) = -\frac{1}{2} \\ (1 + \lambda_1)(y_1 - 5) &= y_2 - 5 \\ y_2 &= \frac{9}{2} \end{aligned}$$

From $y_2 = x_2^2$ we obtain $x_2 = \pm 3/\sqrt{2}$. Furthermore we have

$$\frac{1}{1+\lambda_1} = \frac{x_1}{x_2} = \frac{y_1 - 5}{y_2 - 5}.$$

From $x_1^2 + (y_1 - 5)^2 = 1$ we obtain $2x_1^2/9 = 4(y_1 - 5)^2$ and

$$x_1 = \pm \frac{6}{\sqrt{38}}, \qquad y_1 = \pm\sqrt{\frac{2}{38}} + 5 .$$

We tabulate the solutions

x_1	y_1	x_2	y_2	Value
0	4	0	0	4
0	6	0	0	6
$\frac{6}{\sqrt{38}}$	$\sqrt{\frac{2}{38}} + 5$	$-\frac{3}{\sqrt{2}}$	$\frac{9}{2}$	3.179449
$\frac{6}{\sqrt{38}}$	$-\sqrt{\frac{2}{38}} + 5$	$\frac{3}{\sqrt{2}}$	$\frac{9}{2}$	1.179449
$-\frac{6}{\sqrt{38}}$	$\sqrt{\frac{2}{38}} + 5$	$\frac{3}{\sqrt{2}}$	$\frac{9}{2}$	3.179449
$-\frac{6}{\sqrt{38}}$	$-\sqrt{\frac{2}{38}} + 5$	$-\frac{3}{\sqrt{2}}$	$\frac{9}{2}$	1.179449

The minimum distance is approximately 1.179449. ♣

Example 3. A firm uses two inputs to produce one output. Its production function is

$$f(x_1, x_2) = x_1^a x_2^b, \qquad a, b > 1 .$$

The price of the output is p, and the prices of the inputs are w_1 and w_2. The firm is constrained by a law that says it must use exactly the same number of units of both inputs. We use the Lagrange multiplier method to maximize the function

$$g(x_1, x_2) = pf(x_1, x_2) - w_1 x_1 - w_2 x_2$$

subject to $x_2 - x_1 = 0$. The Lagrange function is given by

$$L(x_1, x_2) = px_1^a x_2^b - w_1 x_1 - w_2 x_2 - \lambda(x_2 - x_1)\,.$$

Thus from $\partial L/\partial x_1 = 0$, $\partial L/\partial x_2 = 0$ we find the system of equations

$$apx_1^{a-1} x_2^b - w_1 + \lambda = 0, \qquad bpx_1^a x_2^{b-1} - w_2 - \lambda = 0\,.$$

Furthermore we have the constraint $x_2 = x_1$. These three equations have a single solution

$$x_1^* = x_2^* = \left(\frac{w_1 + w_2}{p(a+b)}\right)^{1/(a+b-1)}, \qquad \lambda^* = \frac{bw_1 - aw_2}{a+b}\,.$$

Thus

$$g(x_1^*, x_2^*) = p(x_1^*)^{a+b} - x_1^*(w_1 + w_2)\,. \qquad \clubsuit$$

Example 4. Given the *Lagrange function*

$$L(\mathbf{x}(t), \dot{\mathbf{x}}(t)) = \frac{1}{2}\sum_{j=1}^{3}(\dot{x}_j^2 - \omega_j^2 x_j^2) - \frac{1}{2}\lambda(t)\left(\sum_{j=1}^{3} x_j^2 - 1\right) \qquad (1)$$

where λ is the time-dependent Lagrange multiplier. The system describes an $n = 3$ dimensional *harmonic oscillator* constrainted to a unit $n - 1 = 2$ sphere, i.e.,

$$\sum_{j=1}^{3} x_j^2 = 1\,. \qquad (2)$$

We find the equations of motion using the *Euler-Lagrange equations*

$$\frac{d}{dt}\frac{\partial L}{\partial \dot{x}_j} - \frac{\partial L}{\partial x_j} = 0, \qquad j = 1, 2, 3\,. \qquad (3)$$

Inserting L into the Euler-Lagrange equation we find the equation of motion

$$\ddot{x}_j + \omega_j^2 x_j + \lambda(t) x_j = 0, \qquad j = 1, 2, 3\,. \qquad (4)$$

To eliminate the time dependent Lagrange multiplier λ we proceed as follows. From the constraint we obtain by differentiating with respect to t

$$\sum_{j=1}^{3} \dot{x}_j x_j = 0$$

and differentiating twice with respect to t yields

$$\sum_{j=1}^{3}(\ddot{x}_j x_j + \dot{x}_j^2) = 0\,. \tag{5}$$

From (4) we obtain $\ddot{x}_j x_j + \omega_j^2 x_j^2 + \lambda(t) x_j^2 = 0$. Summation yields

$$\sum_{j=1}^{3} \ddot{x}_j x_j + \sum_{j=1}^{3} \omega_j^2 x_j^2 + \lambda(t) = 0$$

where we used the constraint. Thus

$$\lambda(t) = -\sum_{j=1}^{3} \ddot{x}_j x_j - \sum_{j=1}^{3} \omega_j^2 x_j^2 = \sum_{j=1}^{3} \dot{x}_j^2 - \sum_{j=1}^{3} \omega_j^2 x_j^2$$

where we used (5). Inserting $\lambda(t)$ into (4) yields the equations of motion

$$\ddot{x}_j + \omega_j^2 x_j + x_j \sum_{j=1}^{3}(\dot{x}_j^2 - \omega_j^2 x_j^2) = 0\,.$$

♣

Example 5. We want to minimize $f(x_1, x_2) = x_1^2 + x_2^2$ subject to the side condition $x_2 = x_1 - 1$. Obviously we find $x_1^* = 1/2$ and $x_2^* = -1/2$. We can also apply the *penalty method* to find this solution. The penalty method, which is an approximation method where the failure of the minimizer to fulfill the side condition is penalized and one seeks the minimum of the function

$$g(x_1, x_2) = x_1^2 + x_2^2 + \frac{\gamma}{2}(x_2 - x_1 + 1)^2\,.$$

Here γ, the penalty parameter, is chosen to a fixed large number and not allowed to vary. Note that the side condition must be squared, otherwise the sign would preclude the existence of a minimum. Minimization of g with respect to x_1 and x_2 yields

$$x_1 = \frac{\gamma}{2(\gamma+1)}, \qquad x_2 = -\frac{\gamma}{2(\gamma+1)}$$

which in the limit $\gamma \to \infty$ returns $x_1 = 1/2$, $x_2 = -1/2$. ♣

12.2 Coordinate Systems

In some cases the constraints can be eliminated using a suitable coordinate system, for example spherical coordinates, cylindrical coordinates, prolate spheroidal coordinates, oblate coordinates, parabolic coordinates etc.

The constraint

$$\|\mathbf{x}\|^2 = 1 \Leftrightarrow x_1^2 + x_2^2 + \cdots + x_n^2 = 1$$

can be eliminated using n-dimensional *spherical coordinates*

$$\begin{aligned} x_1 &= r\cos\theta_1 \\ x_2 &= r\sin\theta_1\cos\theta_2 \\ x_3 &= r\sin\theta_1\sin\theta_2\cos\theta_3 \\ x_4 &= r\sin\theta_1\sin\theta_2\sin\theta_3\cos\theta_4 \\ &\vdots \\ x_{n-1} &= r\sin\theta_1\sin\theta_2\sin\theta_3\cdots\sin\theta_{n-2}\cos\theta_{n-1} \\ x_n &= r\sin\theta_1\sin\theta_2\sin\theta_3\cdots\sin\theta_{n-2}\sin\theta_{n-1} \end{aligned}$$

or

$$\begin{aligned} x_k &= r\cos\theta_k\prod_{\ell=1}^{k-1}\sin\theta_\ell, \quad \text{for} \quad k=1,2,\ldots,n-1 \\ x_n &= r\prod_{\ell=1}^{n-1}\sin\theta_\ell \end{aligned}$$

with $r=1$, $-\pi\le\theta_1\le\pi$, and $0\le\theta_j\le\pi$, $2\le j\le n-1$. The inverse transform is

$$\begin{aligned} \theta_k &= \arccos\left(\frac{x_k}{\sqrt{r^2-\sum_{j=1}^{k-1}x_j^2}}\right), \quad \text{for} \quad k=1,2,\ldots,n-2 \\ \theta_{n-1} &= \arctan\left(\frac{x_n}{x_{n-1}}\right). \end{aligned}$$

The Jacobian determinant J of this transform is

$$J = r^{n-1}\prod_{p=1}^{n-1}(\sin\theta_p)^{n-p-1}.$$

Other usefule coordinate systems for problems in $\mathbb{R}^3$ are: *Prolate spheroidal coordinates* are given by ($a>0$)

$$\begin{aligned} x_1(\eta,\alpha,\phi) &= a\sinh\eta\sin\alpha\cos\phi \\ x_2(\eta,\alpha,\phi) &= a\sinh\eta\sin\alpha\sin\phi \\ x_3(\eta,\alpha,\phi) &= a\cosh\eta\cos\phi. \end{aligned}$$

In three-dimensional Euclidean space *oblate spheroidal coordinates* are given by ($a>0, \phi\in[0,2\pi)$)

$$\begin{aligned} x_1(\eta,\alpha,\phi) &= a\cosh\eta\sin\alpha\cos\phi \\ x_2(\eta,\alpha,\phi) &= a\cosh\eta\sin\alpha\sin\phi \\ x_3(\eta,\alpha,\phi) &= a\sinh\eta\cos\phi. \end{aligned}$$

In three-dimensional Euclidean space *parabolic coordinates* are given by ($\eta > 0, \xi > 0$)

$$\begin{aligned} x_1(\xi,\eta,\phi) &= \xi\eta\cos\phi \\ x_2(\xi,\eta,\phi) &= \xi\eta\sin\phi \\ x_3(\eta,\alpha,\phi) &= \frac{1}{2}(\xi^2-\eta^2). \end{aligned}$$

In three-dimensional Euclidean space *sphero-conical coordinates* are given by (sn, cn, dn are the Jacobi elliptic functions)

$$\begin{aligned} x_1(r,\mu,\nu) &= rk\mathrm{sn}(\mu,k)\mathrm{sn}(\nu,k) \\ x_2(r,\mu,\nu) &= r\frac{ik}{k'}\mathrm{cn}(\mu,k)\mathrm{cn}(\nu,k) \\ x_3(r,\mu,\nu) &= r\frac{i}{k'}\mathrm{dn}(\mu,k)\mathrm{dn}(\nu,k) \end{aligned}$$

where k is the modulus ($0 \leq k \leq 1$) and $k' = \sqrt{1-k^2}$.

12.3 Differential Forms

Using *differential forms* (Cartan [26], Flanders [61], von Westenholz [213], Steeb [198], Zizza [230]) to solve constrained max-min problems has the advantage over the Lagrange multiplier method that the Lagrange multipliers are eliminated. Thus the number of equations to solve are less.

As described above the approach of Lagrange's condition for maximizing or minimizing f, subject to one constraint of the form $g(\mathbf{x}) = c$ utilizes the observation that at a critical point the contours of f and g are tangential. The equation

$$df|_P = \lambda dg|_P$$

between differential one forms is equivalent to the Lagrange condition

$$\nabla f|_P = \lambda \nabla g|_P.$$

However geometrically the differential condition says that the tangent lines to the contour and the constraint curves are identical, while the gradient condition says that the normal vectors of these lines are parallel.

For the case $m = n-1$ we obtain an additional equation (Zizza [230]). Besides the m conditions from the constraint, we have

$$df \wedge dg_1 \wedge \cdots \wedge dg_m = 0$$

where $\wedge$ denotes the exterior product (wedge product) with

$$dx_j \wedge dx_k = -dx_k \wedge dx_j$$

and d denotes the exterior derivative. As a consequence of this equation we have $dx_j \wedge dx_j = 0$.

Example. We find the optimum values for $f(x_1, x_2) = 2x_1^2 + x_2^2$ subject to the contraint $g(x_1, x_2) = x_1 + x_2 - 1 = 0$. For the differentials we have

$$df = 4x_1 dx_1 + 2x_2 dx_2, \qquad dg = dx_1 + dx_2 .$$

Thus with $dx_1 \wedge dx_2 = -dx_2 \wedge dx_1$, $dx_1 \wedge dx_1 = dx_2 \wedge dx_2 = 0$ we arrive at

$$df \wedge dg = (4x_1 - 2x_2) dx_1 \wedge dx_2 .$$

From the condition $df \wedge dg = 0$ it follows that $2x_1 - x_2 = 0$. This equation together with $x_1 + x_2 = 1$ has to be solved. We obtain the solution $x_1^* = 1/3$, $x_2^* = 2/3$. ♣

This can be implemented using SymbolicC++ as follows

```
// lagrange.cpp

#include <iostream>
#include "symbolicc++.h"
using namespace std;

int main(void)
{
 Symbolic x("x"), y("y"), dx("dx"), dy("dy");
 dx = ~dx; dy = ~dy;  //non-commutative
 Symbolic f = 2*x*x+y*y;
 Symbolic g = x+y-1;
 cout << "f = " << f << endl;
 cout << "g = " << g << endl;
 Symbolic d_f = df(f,x)*dx+df(f,y)*dy;
 Symbolic d_g = df(g,x)*dx+df(g,y)*dy;
 cout << "d_f = " << d_f << endl;
 cout << "d_g = " << d_g << endl;
 Symbolic w = (d_f*d_g).subst_all((dx*dx==0,dy*dy==0,dy*dx==-dx*dy));
 cout << (w.coeff(dx*dy) == 0) << endl;
 return 0;
}
```

If $m < n-1$ (Zizza [230]) then we apply a coordinate system $s_1, s_2, \ldots, s_n$ adapted to the problem so that the first m coordinates are defined by $s_j = g_j(x_1, \ldots, x_n)$ ($j = 1, 2, \ldots, m$), while the remaining $n - m$ coordinates are any functions of x_1, $x_2, \ldots, x_n$ as long as the result is indeed a coordinate system. Then the conditions, besides the constraints, are

$$\begin{aligned}
df \wedge dg_1 \wedge dg_2 \wedge \cdots \wedge dg_m \wedge \widehat{ds}_1 \wedge ds_2 \wedge \cdots \wedge ds_{n-m} &= 0 \\
df \wedge dg_1 \wedge dg_2 \wedge \cdots \wedge dg_m \wedge ds_1 \wedge \widehat{ds}_2 \wedge \cdots \wedge ds_{n-m} &= 0 \\
&\vdots \\
df \wedge dg_1 \wedge dg_2 \wedge \cdots \wedge dg_m \wedge ds_1 \wedge ds_2 \wedge \cdots \wedge \widehat{ds}_{n-m} &= 0 .
\end{aligned}$$

and $\widehat{\ }$ indicates omission.

Example. Consider minimizing the function

$$f(x_1, x_2, y_1, y_2) = (x_1 - x_2)^2 + (y_1 - y_2)^2$$

subject to the constraints

$$g_1(x_1, x_2, y_1, y_2) = x_1^2 + (y_1 - 5)^2 - 1 = 0, \quad g_2(x_1, x_2, y_1, y_2) = y_2 - x_2^2 = 0\,.$$

Thus $n = 4$ and $m = 2$. This example has been studied above using the Lagrange multiplier method. Since

$$\begin{aligned} df &= 2(x_1 - x_2)dx_1 - 2(x_1 - x_2)dx_2 + 2(y_1 - y_2)dy_1 - 2(y_1 - y_2)dy_2 \\ dg_1 &= 2x_1 dx_1 + 2(y_1 - 5)dy_1 \\ dg_2 &= dy_2 - 2x_2 dx_2 \end{aligned}$$

we obtain

$$\begin{aligned} df \wedge dg_1 \wedge dg_2 = &(-8x_1x_2(y_1 - y_2) + 8(x_1 - x_2)x_2(y_1 - 5))dx_1 \wedge dx_2 \wedge dy_1 \\ &+(4(x_1 - x_2)x_1 + 8(y_1 - y_2)x_1x_2)dx_1 \wedge dx_2 \wedge dy_2 \\ &+4((x_1 - x_2)(y_1 - 5) - 4(y_1 - y_2)x_1)dx_1 \wedge dy_1 \wedge dy_2 \\ &+(-4(x_1 - x_2)(y_1 - 5) - 8(y_1 - y_2)x_2(y_1 - 5))dx_2 \wedge dy_1 \wedge dy_2\,. \end{aligned}$$

We now set $s_1 = x_1$, $s_2 = x_2$. From the conditions

$$df \wedge dg_1 \wedge dg_2 \wedge dx_1 = 0, \qquad df \wedge dg_1 \wedge dg_2 \wedge dx_2 = 0$$

we obtain the two equations

$$\begin{aligned} (x_1 - x_2)(y_1 - 5) - (y_1 - y_2)x_1 &= 0 \\ (x_1 - x_2)(y_1 - 5) + 2(y_1 - y_2)x_2(y_1 - 5) &= 0 \end{aligned}$$

plus two equations for the constraints instead of the six equations we obtain with the Lagrange multiplier method. ♣

Example. The Lagrange multiplier can fail even if there is a solution. The differential form method still provides a solution. Consider the differentiable function $f : \mathbb{R}^2 \to \mathbb{R}$

$$f(x_1, x_2) = 2x_1^3 - 3x_1^2$$

subject to the constraint

$$g(x_1, x_2) = (3 - x_1)^2 - x_2^2 = 0\,.$$

Then from the Lagrange function

$$L(x_1, x_2, \lambda) = 2x_1^3 - 3x_1^2 + \lambda((3 - x_1)^2 - x_2^2)$$

we obtain the system of equations

$$6x_1^2 - 6x_1 - 3\lambda(3-x_1)^2 = 0, \quad -2\lambda x_2 = 0, \quad (3-x_1)^3 - x_2^2 = 0\,.$$

We have to do a case study. From the second condition we have $\lambda = 0$ or $x_2 = 0$. If $x_2 = 0$, then the third condition implies $x_1 = 3$, but $x_1 = 3$ violates the first condition. The case $\lambda = 0$ implies

$$(x_1, x_2, \lambda) = (0, \sqrt{27}, 0), \quad \text{and} \quad (x_1, x_2, \lambda) = (1, \sqrt{8}, 0)\,.$$

However the unique solution for the global maximum is $(x_1, x_2) = (3, 0)$ with $f(x_1 = 3, x_2 = 0) = 27$. ♣

Using differential forms we obtain the correct result. From f and g we obtain

$$df = 6x_1^2 dx_1 - 6x_1 dx_1, \qquad dg = -3(3-x_1)^2 dx_1 - 2x_2 dx_2\,.$$

It follows that

$$df \wedge dg = -12(x_1 x_2(x_1 - 1))dx_1 \wedge dx_2 = 0\,.$$

Thus we have to solve the two equations

$$x_1 x_2(x_1 - 1) = 0, \qquad (3-x_1)^3 - x_2^2 = 0\,.$$

The case $x_1 = 0$ implies $x_2 = \sqrt{27}$ with $f(x_1 = 0, x_2 = \sqrt{27}) = 0$. The case $x_1 = 1$ implies $x_2 = \sqrt{8}$ with $f(x_1 = 1, x_2 = \sqrt{8}) = -1$. The case $x_2 = 0$ yields $x_1 = 3$ the global maximum.

12.4 Karush-Kuhn-Tucker Conditions

Many optimization problems also include inequalities. Karush-Kuhn-Tucker (Kuhn and Tucker [122], Bazaraa and Shetty [15]) extended the Lagrange multiplier method to include inequality constraints.

Given an optimization problem with convex domain $\Omega \subseteq \mathbb{R}^n$,

$$\begin{aligned} \text{minimize} \quad & f(\mathbf{x}), \quad \mathbf{x} \in \Omega \\ \text{subject to} \quad & g_j(\mathbf{x}) \le 0\,,\ j = 1, \ldots, k \\ & h_j(\mathbf{x}) = 0\,,\ j = 1, \ldots, l\,. \end{aligned}$$

We define the generalized Lagrangian function as

$$\begin{aligned} L(\mathbf{x}, \boldsymbol{\alpha}, \boldsymbol{\beta}) &= f(\mathbf{x}) + \sum_{j=1}^{k} \alpha_j g_j(\mathbf{x}) + \sum_{j=1}^{l} \beta_j h_j(\mathbf{x}) \\ &\equiv f(\mathbf{x}) + \boldsymbol{\alpha}^T \mathbf{g}(\mathbf{x}) + \boldsymbol{\beta}^T \mathbf{h}(\mathbf{x})\,. \end{aligned}$$

We assume that the functions f, g_j, h_j are continuously differentiable functions. Then we have the following

Theorem. Given the above optimization problem with convex domain $\Omega \subseteq \mathbb{R}^n$, necessary and sufficient conditions for a normal point $\mathbf{x}^*$ to be an optimum are the existence of $\boldsymbol{\alpha}^*$ and $\boldsymbol{\beta}^*$ such that

$$\frac{\partial L(\mathbf{x}^*, \boldsymbol{\alpha}^*, \boldsymbol{\beta}^*)}{\partial \mathbf{x}} = 0$$
$$\frac{\partial L(\mathbf{x}^*, \boldsymbol{\alpha}^*, \boldsymbol{\beta}^*)}{\partial \boldsymbol{\beta}} = 0$$
$$\alpha_j^* g_j(\mathbf{x}^*) = 0, \; j = 1, \ldots, k$$
$$g_j(\mathbf{x}^*) \le 0, \; j = 1, \ldots, k$$
$$\alpha_j^* \ge 0, \; j = 1, \ldots, k.$$

The third relation is known as Karush-Kuhn-Tucker complementarity condition. That is, either a constraint is active, meaning $g_i(\mathbf{x}) = 0$, or the corresponding multiplier satisfies $\alpha_i^* = 0$. If in addition f, h_j, g_j are twice continuously differentiable, there holds

$$\mathbf{y}^T \nabla_{\mathbf{xx}} L(\mathbf{x}^*, \boldsymbol{\lambda}^*, \boldsymbol{\mu}^*) \mathbf{y} \ge 0$$

for all column vectors $\mathbf{y} \in \mathbb{R}^n$ such that

$$\nabla h_i(\mathbf{x}^*)^T \mathbf{y} = 0, \quad i = 1, \ldots, m, \qquad (\nabla g_j(\mathbf{x})^*)^T) \mathbf{y} = 0$$

for all $j \in A(\mathbf{x}^*)$ where $A(\mathbf{x}^*)$ is the set of active constraints at $\mathbf{x}^*$.

We can also formulate the *Karush-Kuhn-Tucker condition* as follows. Here we assume that the constraints $g_k(\mathbf{x}) \ge 0$. Consider the constraint nonlinear programming problem with inequality and equality constraints: minimize the scalar function f subject to the inequality constraints $g_k(\mathbf{x}) \ge 0$, $k = 1, 2, \ldots, K$ and the equality constraints $h_m(\mathbf{x}) = 0$, $m = 1, 2, \ldots, M$. For this problem we can construct the *Lagrange function*

$$L(\mathbf{x}, \boldsymbol{\lambda}, \boldsymbol{\mu}) = f(\mathbf{x}) - \sum_{k=1}^{K} \lambda_k g_k(\mathbf{x}) - \sum_{m=1}^{M} \mu_m h_m(\mathbf{x})$$

where λ_k and μ_m are the Lagrange multipliers. If the problem has a solution

$$\mathbf{x}^* = (x_1^*, x_2^*, \ldots, x_n^*)$$

i.e., $\min_{\mathbf{x}} f(\mathbf{x}) = f(\mathbf{x}^*)$ and all constraints are satisfied, then the following Karush-Kuhn-Tucker conditions hold

$$\nabla f(\mathbf{x}^*) - \sum_{k=1}^{K} \lambda_k^* \nabla g_k(\mathbf{x}^*) - \sum_{m=1}^{M} \mu_m^* \nabla h_m(\mathbf{x}^*) = \mathbf{0}$$

and

$$\begin{aligned} g_k(\mathbf{x}^*) &\geq 0 \quad k = 1, 2, \ldots, K \\ h_m(\mathbf{x}^*) &= 0 \quad m = 1, 2, \ldots, M \\ \lambda_k^* g_k(\mathbf{x}^*) &= 0 \quad k = 1, 2, \ldots, K \\ \lambda_k^* &\geq 0 \quad k = 1, 2, \ldots, K. \end{aligned}$$

In convex programming problems, the Karush-Kuhn-Tucker conditions are necessary and sufficient for a global minimum.

Example 1. Find the minimum of the function

$$f(\mathbf{x}) = (x_1 - 2)^2 + (x_2 - 1)^2$$

under the constraints

$$g_1(\mathbf{x}) = x_2 - x_1^2 \geq 0, \quad g_2(\mathbf{x}) = 2 - x_1 - x_2 \geq 0, \quad g_3(\mathbf{x}) = x_1 \geq 0.$$

The Lagrange function is

$$L(\mathbf{x}, \boldsymbol{\lambda}) = (x_1 - 2)^2 + (x_2 - 1)^2 - \lambda_1(x_2 - x_1^2) - \lambda_2(2 - x_1 - x_2) - \lambda_3 x_1.$$

Thus we find the Karush-Kuhn-Tucker conditions

$$\begin{aligned} 2(x_1 - 2) + 2\lambda_1 x_1 + \lambda_2 - \lambda_3 &= 0 \\ 2(x_2 - 1) - \lambda_1 + \lambda_2 &= 0 \\ x_2 - x_1^2 &\geq 0 \\ 2 - x_1 - x_2 &\geq 0 \\ x_1 &\geq 0 \\ \lambda_1(x_2 - x_1^2) &= 0 \\ \lambda_2(2 - x_1 - x_2) &= 0 \\ \lambda_3 x_1 &= 0 \end{aligned}$$

and $\lambda_1 \geq 0$, $\lambda_2 \geq 0$, $\lambda_3 \geq 0$. These equations and inequalities can be solved starting from $\lambda_3 x_1 = 0$ with the cases $\lambda_3^* = 0$ or $x_1^* = 0$. It turns out that the case $x_1^* = 0$ is not admissible. Proceeding with the case study we find

$$x_1^* = 1, \; x_2^* = 1, \; \lambda_1^* = \frac{2}{3}, \; \lambda_2^* = \frac{2}{3}, \; \lambda_3^* = 0$$

for which $f(x_1^*, x_2^*) = 1$. ♣

Example 2. We want to maximize

$$f(x_1, x_2) = 3.6x_1 - 0.4x_1^2 + 1.6x_2 - 0.2x_2^2$$

subject to

$$2x_1 + x_2 \leq 10, \quad x_1 \geq 0, \quad x_2 \geq 0\,.$$

We need to change the problem to find the minimum of

$$h(x_1, x_2) = -3.6x_1 + 0.4x_1^2 - 1.6x_2 + 0.2x_2^2$$

and we have to rewrite the constraints as $g_1(x_1, x_2) = 10 - 2x_1 - x_2 \geq 0$, $g_2(x_1, x_2) = x_1 \geq 0$, and $g_3(x_1, x_2) = x_2 \geq 0$ in order to apply the Karush-Kuhn-Tucker conditions as follows

$$\begin{aligned} -3.6 + 0.8x_1 + 2\lambda_1 - \lambda_2 &= 0 \\ -1.6 + 0.4x_2 + \lambda_1 - \lambda_3 &= 0 \\ 10 - 2x_1 - x_2 &\geq 0 \\ \lambda_1(10 - 2x_1 - x_2) &= 0 \\ \lambda_2 x_1 &= 0 \\ \lambda_3 x_2 &= 0 \end{aligned}$$

and $x_1 \geq 0$, $x_2 \geq 0$, $\lambda_1 \geq 0$, $\lambda_2 \geq 0$, $\lambda_3 \geq 0$. A case study of these system of equations and inequalities yields $x_1 = 3.5$, $x_2 = 3.0$, $\lambda_1 = 0.4$, $\lambda_2 = 0$, $\lambda_3 = 0$. The Hessian for h

$$\begin{pmatrix} \frac{\partial^2 h}{\partial x_1^2} & \frac{\partial^2 h}{\partial x_1 \partial x_2} \\ \frac{\partial^2 h}{\partial x_1 \partial x_2} & \frac{\partial^2 h}{\partial x_2^2} \end{pmatrix} = \begin{pmatrix} 0.8 & 0 \\ 0 & 0.4 \end{pmatrix}$$

is positive definite. Thus, h is convex and the inequality constraints are all linear and thus concave. Thus the solution $(x_1, x_2) = (3.5, 3)$ is optimal. ♣

Example 3. For problems in *linear programming*, normally the Simplex method is used. However the Karush-Kuhn-Tucker condition can also be applied. Consider the function

$$f(x_1, x_2, x_3) = 2x_1 + 3x_2 + 3x_3$$

with has to be maximized subject to

$$\begin{aligned} g_1(x_1, x_2, x_3) &= 60 - 3x_1 - 2x_2 \geq 0 \\ g_2(x_1, x_2, x_3) &= 10 + x_1 - x_2 - 4x_3 \geq 0 \\ g_3(x_1, x_2, x_3) &= 50 - 2x_1 + 2x_2 - 5x_3 \geq 0 \\ g_4(x_1, x_2, x_3) &= x_1 \geq 0 \\ g_5(x_1, x_2, x_3) &= x_2 \geq 0 \\ g_6(x_1, x_2, x_3) &= x_3 \geq 0\,. \end{aligned}$$

The Lagrangian to minimize is

$$\begin{aligned} L(\mathbf{x}, \boldsymbol{\lambda}) = {} & 2x_1 + 3x_2 + 3x_3 + \lambda_1(60 - 3x_1 - 2x_2) + \lambda_2(10 + x_1 - x_2 - 4x_3) \\ & + \lambda_3(50 - 2x_1 + 2x_2 - 5x_3) + \lambda_4 x_1 + \lambda_5 x_2 + \lambda_6 x_3 \end{aligned}$$

The necessary conditions become

$$\frac{\partial}{\partial x_1}L(\mathbf{x},\boldsymbol{\lambda}) = 2 - 3\lambda_1 + \lambda_2 - 2\lambda_3 + \lambda_4 = 0$$
$$\frac{\partial}{\partial x_2}L(\mathbf{x},\boldsymbol{\lambda}) = 3 - 2\lambda_1 - \lambda_2 + 2\lambda_3 + \lambda_5 = 0$$
$$\frac{\partial}{\partial x_3}L(\mathbf{x},\boldsymbol{\lambda}) = 3 - 4\lambda_2 - 5\lambda_3 + \lambda_6 = 0$$

$$60 - 3x_1 - 2x_2 \geq 0, \quad 10 + x_1 - x_2 - 4x_3 \geq 0, \quad 50 - 2x_1 + 2x_2 - 5x_3 \geq 0$$

and $x_1 \geq 0$, $x_2 \geq 0$, $x_3 \geq 0$, $\lambda_1 \geq 0$, $\lambda_2 \geq 0$, $\lambda_3 \geq 0$, $\lambda_4 \geq 0$, $\lambda_5 \geq 0$, $\lambda_6 \geq 0$. Furthermore we have the equalities

$$\lambda_1(60 - 3x_1 - 2x_2) = 0$$
$$\lambda_2(10 + x_1 - x_2 - 4x_3) = 0$$
$$\lambda_3(50 - 2x_1 + 2x_2 - 5x_3) = 0$$

$$\lambda_4 x_1 = 0, \quad \lambda_5 x_2 = 0, \quad \lambda_6 x_3 = 0.$$

A case study of these equations and inequalities gives the global maximum point $(x_1, x_2, x_3) = (8, 18, 0)$. The maximum global value is $f(8, 18, 0) = 70$.

12.5 Support Vector Machine

12.5.1 Introduction

The support vector machine (Vapnik [210], Vapnik [211], Wang [215]) is an algorithm for learning linear classifiers. It is motivated by the idea of maximizing margins. There is an efficient extension to non-linear support vector machines through use of kernels. We have n vectors $\mathbf{v}_0$, $\mathbf{v}_1$, ..., $\mathbf{v}_{n-1}$ from the vector space $\mathbb{R}^m$ which are from two classes. The training data are $\{(\mathbf{v}_j, y_j) : j = 0, 1, \ldots, n-1\}$ where $y_j \in \{+1, -1\}$. The goal is to learn a classification rule from the data which only makes small classification errors on the n-examples but also has good generalizations.

12.5.2 Linear Decision Boundaries

Consider the training set of two separate classes be represented by the set of vectors

$$(\mathbf{v}_0, y_0), \quad (\mathbf{v}_1, y_1), \quad \ldots \quad, (\mathbf{v}_{n-1}, y_{n-1})$$

where $\mathbf{v}_j$ ($j = 0, 1, \ldots, n-1$) is a vector in the m-dimensional real Hilbert space $\mathbb{R}^m$ and $y_j \in \{-1, +1\}$ indicates the class label. Given a weight vector $\mathbf{w}$ and a bias b, it is assumed that these two classes can be separated by two margins parallel to the hyperplane

$$\mathbf{w}^T\mathbf{v}_j + b \geq 1, \quad \text{for} \quad y_j = +1 \qquad (1)$$
$$\mathbf{w}^T\mathbf{v}_j + b \leq -1, \quad \text{for} \quad y_j = -1 \qquad (2)$$

for $j = 0, 1, \ldots, n-1$ and $\mathbf{w} = (w_0, w_1, \ldots, w_{m-1})^T$ is a column vector of m-elements. Inequalities (1) and (2) can be combinded into a single inequality

$$y_j(\mathbf{w}^T\mathbf{v}_j + b) \geq 1 \quad \text{for } j = 0, 1, \ldots, n-1. \tag{3}$$

There exist a number of separate hyperplanes for an identical group of training data. The objective of the *support vector machine* (Vapnik [210], Vapnik [211], Wang [215]) is to determine the optimal weight $\mathbf{w}^*$ and the optimal bias b^* such that the corresponding hyperplane separates the positive and negative training data with maximum margin and it produces the best generation performance. This hyperplane is called an optimal separating hyperplane. The equation for an arbitrary hyperplane is given by

$$\mathbf{w}^T\mathbf{x} + b = 0 \tag{4}$$

and the distance between the two corresponding margins is

$$\gamma(\mathbf{w}, b) = \min_{\{\mathbf{v}\,|\,y=+1\}} \frac{\mathbf{w}^T\mathbf{v}}{\|\mathbf{w}\|} - \max_{\{\mathbf{v}\,|\,y=-1\}} \frac{\mathbf{w}^T\mathbf{v}}{\|\mathbf{w}\|}. \tag{5}$$

The optimal separating hyperplane can be obtained by maximizing the above distance or minimizing the norm of $\|\mathbf{w}\|$ under the inequality constraint (3), and

$$\gamma_{max} = \gamma(\mathbf{w}^*, b^*) = \frac{2}{\|\mathbf{w}\|}. \tag{6}$$

The saddle point of the Lagrange function

$$L_P(\mathbf{w}, b, \boldsymbol{\alpha}) = \frac{1}{2}\mathbf{w}^T\mathbf{w} - \sum_{j=0}^{n-1} \alpha_j(y_j(\mathbf{w}^T\mathbf{v}_j + b) - 1) \tag{7}$$

gives solutions to the minimization problem, where $\alpha_j \geq 0$ are Lagrange multiplier. The solution of this quadratic programming optimization problem requires that the gradient of $L_P(\mathbf{w}, b, \boldsymbol{\alpha})$ with respect to $\mathbf{w}$ and b vanishes, i.e.,

$$\left.\frac{\partial L_P}{\partial \mathbf{w}}\right|_{\mathbf{w}=\mathbf{w}^*} = \mathbf{0}, \qquad \left.\frac{\partial L_P}{\partial b}\right|_{b=b^*} = 0.$$

We obtain for the weight vector

$$\mathbf{w}^* = \sum_{j=0}^{n-1} \alpha_j y_j \mathbf{v}_j$$

and the constraint

$$\sum_{j=0}^{n-1} \alpha_j y_j = 0.$$

Inserting $\mathbf{w}^*$ and the constraint into L_P yields the Lagrange function

$$L_D(\boldsymbol{\alpha}) = \sum_{i=0}^{n-1} \alpha_i - \frac{1}{2}\sum_{i=0}^{n-1}\sum_{j=0}^{n-1} \alpha_i\alpha_j y_i y_j \mathbf{v}_i^T\mathbf{v}_j$$

under the constraints

$$\sum_{j=0}^{n-1} \alpha_j y_j = 0$$

and $\alpha_j \geq 0$, $j = 0, 1, \ldots, n-1$. The function $L_D(\boldsymbol{\alpha})$ has be maximized. Note that L_P and L_D arise from the same objective function but with different constraints; and the solution is found by minimizing L_P or by maximizing L_D. The points located on the two optimal margins will have nonzero coefficients α_j among the solutions of $\max L_D(\boldsymbol{\alpha})$ and the constraints. These vectors with nonzero coefficients α_j are called support vectors. The bias can be calculated as follows

$$b^* = -\frac{1}{2}\left(\min_{\{\mathbf{v}_j \,|\, y_j=+1\}} \mathbf{w}^{*T}\mathbf{v}_j + \max_{\{\mathbf{v}_j \,|\, y_j=-1\}} \mathbf{w}^{*T}\mathbf{v}_j\right).$$

After determination of the support vectors and bias, the decision function that separates the two classes can be written as

$$f(\mathbf{x}) = \operatorname{sgn}\left(\sum_{j=0}^{n-1} \alpha_j y_j \mathbf{v}_j^T \mathbf{x} + b^*\right).$$

Example. We apply this classification technique to the data set (AND-gate)

j	Training set $\mathbf{v}_j$	Target y_j
0	(0,0)	1
1	(0,1)	1
2	(1,0)	1
3	(1,1)	-1

For this data set we find

$$L_D(\boldsymbol{\alpha}) = \sum_{j=0}^{3} \alpha_j - \frac{1}{2}\alpha_1^2 + \alpha_1\alpha_3 - \frac{1}{2}\alpha_2^2 + \alpha_2\alpha_3 - \alpha_3^2$$

since for the scalar products we have

$$\mathbf{v}_0^T\mathbf{v}_j = 0, \qquad j = 0, 1, 2, 3$$

and

$$\mathbf{v}_1^T\mathbf{v}_1 = 1,\ \mathbf{v}_1^T\mathbf{v}_2 = 0,\ \mathbf{v}_1^T\mathbf{v}_3 = 1,\ \mathbf{v}_2^T\mathbf{v}_2 = 1,\ \mathbf{v}_2^T\mathbf{v}_3 = 1,\ \mathbf{v}_3^T\mathbf{v}_3 = 2.$$

The constraints are

$$\alpha_0 \geq 0, \quad \alpha_1 \geq 0, \quad \alpha_2 \geq 0, \quad \alpha_3 \geq 0$$

and

$$\alpha_0 + \alpha_1 + \alpha_2 - \alpha_3 = 0.$$

To apply the Karush-Kuhn-Tucker conditions (which is formulated for a minimum) we have to change $L_D(\boldsymbol{\alpha})$ to $-L_D(\boldsymbol{\alpha})$. Thus we have the Lagrangian

$$\begin{aligned}\widetilde{L}(\boldsymbol{\alpha}) =& -\sum_{j=0}^{3}\alpha_j + \frac{1}{2}\alpha_1^2 - \alpha_1\alpha_3 + \frac{1}{2}\alpha_2^2 - \alpha_2\alpha_3 + \alpha_3^2 \\ & -\mu(\alpha_0+\alpha_1+\alpha_2-\alpha_3) - \lambda_0\alpha_0 - \lambda_1\alpha_1 - \lambda_2\alpha_2 - \lambda_3\alpha_3 .\end{aligned}$$

Thus we have to solve the sytem of equations

$$\begin{aligned}\frac{\partial\widetilde{L}}{\partial\alpha_0} &= 0 \to -1-\mu-\lambda_0 = 0 \\ \frac{\partial\widetilde{L}}{\partial\alpha_1} &= 0 \to -1+\alpha_1-\alpha_3-\mu-\lambda_1 = 0 \\ \frac{\partial\widetilde{L}}{\partial\alpha_2} &= 0 \to -1+\alpha_2-\alpha_3-\mu-\lambda_2 = 0 \\ \frac{\partial\widetilde{L}}{\partial\alpha_3} &= 0 \to -1-\alpha_1-\alpha_2+2\alpha_3+\mu-\lambda_3 = 0\end{aligned}$$

together with

$$\lambda_0\alpha_0 = 0, \quad \lambda_1\alpha_1 = 0, \quad \lambda_2\alpha_2 = 0, \quad \lambda_3\alpha_3 = 0$$

$$\lambda_0 \geq 0, \quad \lambda_1 \geq 0, \quad \lambda_2 \geq 0, \quad \lambda_3 \geq 0$$

and the constraint $\alpha_0+\alpha_1+\alpha_2-\alpha_3 = 0$. We find the solution

$$\alpha_0 = 0, \quad \alpha_1 = 2, \quad \alpha_2 = 2, \quad \alpha_3 = 4$$

$$\lambda_0 = 2, \quad \lambda_1 = 0, \quad \lambda_2 = 0, \quad \lambda_3 = 0, \quad \mu = -3 .$$

Thus the weight vector is

$$\mathbf{w}^* = \sum_{j=0}^{3}\alpha_j y_j \mathbf{v}_j = (-2,-2) .$$

For b^* we obtain $b^* = 3$. The decision function that separates the two classes is given by

$$\begin{aligned}f(\mathbf{x}) &= \mathrm{sgn}(\alpha_1 y_1 \mathbf{v}_1^T\mathbf{x} + \alpha_2 y_2 \mathbf{v}_2^T\mathbf{x} + \alpha_3 y_3 \mathbf{v}_3^T\mathbf{x} + b^*) \\ &= \mathrm{sgn}(2x_2 + 2x_1 - 4(x_1+x_2) + b^*) \\ &= \mathrm{sgn}(-2x_1 - 2x_2 + 3) \\ &= \mathrm{sgn}\left(-x_1 - x_2 + \frac{3}{2}\right) .\end{aligned}$$

Thus we obtain the line in $\mathbb{R}^2$

$$x_1 + x_2 = \frac{3}{2} .$$

This solution can also be seen on inspection of the data set. ♣

12.5.3 Nonlinear Decision Boundaries

In the previous problem we have considered a data set which can be separated by a hyperplane. For nonlinear decision boundaries (Vapnik [210], Vapnik [211], Wang [215]) we can extend the method as follows. The data points, $\mathbf{v}_j$ only appear inside a scalar product. We map the datapoints into an alternative higher dimensional space, called *feature space*, through

$$\mathbf{v}_i^T \mathbf{v}_j \to \langle \phi(\mathbf{v}_i), \phi(\mathbf{v}_j) \rangle$$

where $\langle \, , \, \rangle$ denotes the scalar product in the feature space. The map $\phi(\mathbf{v}_i)$ does not need to be known since it is implicitly defined by the choice of the positive definite kernel

$$K(\mathbf{v}_i, \mathbf{v}_j) = \langle \phi(\mathbf{v}_i), \phi(\mathbf{v}_j) \rangle .$$

It is assumed that $K(\mathbf{v}_i, \mathbf{v}_j) = K(\mathbf{v}_j, \mathbf{v}_i)$. Examples are the radial base function kernel

$$K(\mathbf{v}_i, \mathbf{v}_j) = \exp(-\|\mathbf{v}_i - \mathbf{v}_j\|^2 / (2\sigma^2))$$

and the polynomial kernel

$$K(\mathbf{v}_i, \mathbf{v}_j) = (1 + \mathbf{v}_i^T \mathbf{v}_j)^d .$$

For binary classification with a given choice of kernel the learning task therefore involves maximisation of the Lagrangian

$$L_D(\boldsymbol{\alpha}) = \sum_{i=0}^{n-1} \alpha_i - \frac{1}{2} \sum_{i=0}^{n-1} \sum_{j=0}^{n-1} \alpha_i \alpha_j y_i y_j K(\mathbf{v}_i, \mathbf{v}_j)$$

subject to the constraints

$$\sum_{i=0}^{n-1} \alpha_i y_i = 0, \qquad \alpha_i \geq 0, \qquad i = 0, 1, \ldots, n-1 .$$

After the optimal values α_i^* have been found the *decision function* is given by

$$f(\mathbf{x}) = \text{sign} \left(\sum_{i=0}^{n-1} \alpha_i^* y_i K(\mathbf{x}, \mathbf{v}_i) + b \right) .$$

The bias b is found from the primal constraints

$$b = -\frac{1}{2} \left(\max_{\{i : y_i = -1\}} \left(\sum_{j=0}^{n-1} \alpha_j y_j K(\mathbf{v}_i, \mathbf{v}_j) \right) + \min_{\{i : y_i = +1\}} \left(\sum_{j=0}^{n-1} \alpha_j y_j K(\mathbf{v}_i, \mathbf{v}_j) \right) \right) .$$

For the Karush-Kuhn-Tucker conditions (which are formulated for a minimum) we have to change L_D to $-L_D$. Thus taking into account the constraints we have the Lagrangian

$$\widetilde{L}(\boldsymbol{\alpha}) = -\sum_{j=0}^{n-1} \alpha_j + \frac{1}{2} \sum_{i=0}^{n-1} \sum_{j=0}^{n-1} \alpha_i \alpha_j y_i y_j K(\mathbf{v}_i, \mathbf{v}_j) - \mu \sum_{j=0}^{n-1} \alpha_j y_j - \sum_{j=0}^{n-1} \lambda_j \alpha_j .$$

From $\partial L/\partial \alpha_k = 0$ we find

$$-1 + y_k \sum_{j=0}^{n-1} \alpha_j y_j K(\mathbf{v}_k, \mathbf{v}_j) - \mu y_k - \lambda_k = 0$$

for $k = 0, 1, \ldots, n-1$. The other Karush-Kuhn-Tucker conditions are

$$\sum_{j=0}^{n-1} \alpha_j y_j = 0$$
$$\alpha_j \geq 0, \quad j = 0, 1, \ldots, n-1$$
$$\lambda_j \alpha_j = 0, \quad j = 0, 1, \ldots, n-1$$
$$\lambda_j \geq 0, \quad j = 0, 1, \ldots, n-1 .$$

Note that there is no condition on the Lagrange multiplier μ.

Example. Consider the XOR-problem. Let $\mathbf{v} = (v_1, v_2)^T \in \mathbb{R}^2$ and a feature map ϕ that maps

$$\mathbf{v} \to \phi(\mathbf{v}) = (v_1^2, v_2^2, \sqrt{2}v_1v_2, \sqrt{2}v_1, \sqrt{2}v_2, 1)^T \in \mathbb{R}^6 .$$

Now note that

$$\phi^T(\mathbf{v}_1)\phi(\mathbf{v}_2) = 1 + 2v_{11}v_{21} + 2v_{12}v_{22} + v_{11}^2v_{21}^2 + v_{12}^2v_{22}^2 + 2v_{11}v_{12}v_{21}v_{22} .$$

Thus the kernel function for the feature space is

$$K(\mathbf{v}_i, \mathbf{v}_j) = (v_{i1}v_{j1} + v_{i2}v_{j2} + 1)^2 = (1 + (\mathbf{v}_i^T\mathbf{v}_j))^2 .$$

We find the solutions of the Karush-Kuhn-Tucker conditions for this kernel and the data set

j	Training set $\mathbf{v}_j$	Target y_j
0	$(-1, -1)$	-1
1	$(-1, +1)$	$+1$
2	$(+1, -1)$	$+1$
3	$(+1, +1)$	-1

Inserting the data set into the kernel we obtain a (positive definite) matrix with $K(\mathbf{v}_i, \mathbf{v}_i) = 9$ and $K(\mathbf{v}_i, \mathbf{v}_j) = 1$ for $i \neq j$ with $i, j = 0, 1, 2, 3$. The solution of the Karush-Kuhn-Tucker conditions is given by

$$\alpha_0 = \alpha_1 = \alpha_2 = \alpha_3 = \frac{1}{8}$$

and $\lambda_0 = \lambda_1 = \lambda_2 = \lambda_3 = \mu = 0$. Furthermore we have $b = 0$. Thus

$$f(\mathbf{x}) = \text{sgn}(-x_1x_2) .$$ ♣

One derives a relation between the learning rate η and the kernel using

$$L(\boldsymbol{\alpha}) = \sum_{j=0}^{n-1} \alpha_j - \frac{1}{2}\sum_{i=0}^{n-1}\sum_{j=0}^{n-1} \alpha_i \alpha_j y_i y_j K(\mathbf{v}_i, \mathbf{v}_j) - \mu \sum_{j=0}^{n-1} \alpha_j y_j .$$

and choose the gradient ascent algorithm

$$\delta\alpha_k = \eta \frac{\partial L}{\partial \alpha_k} = \eta \left(1 - y_k \sum_{j=0}^{n-1} \alpha_j y_j K(\mathbf{v}_j, \mathbf{v}_k) - \mu y_k \right).$$

It follows that

$$\begin{aligned}
\Delta L_k &:= L(\alpha_0, \ldots, \alpha_k + \delta\alpha_k, \ldots, \alpha_{n-1}) - L(\alpha_0, \ldots, \alpha_k, \ldots, \alpha_{n-1}) \\
&= \delta\alpha_k \left(1 - y_k \sum_{j=0}^{n-1} \alpha_j y_j K(\mathbf{v}_j, \mathbf{v}_k) - \mu y_k \right) - \frac{1}{2}(\delta\alpha_k)^2 K(\mathbf{v}_k, \mathbf{v_k}) \\
&= \left(\frac{1}{\eta} - \frac{K(\mathbf{v}_k, \mathbf{v}_k)}{2} \right) (\delta\alpha_k)^2 .
\end{aligned}$$

Given that $\Delta L_k > 0$ we find

$$0 < \eta K(\mathbf{v}_k, \mathbf{v}_k) < 2$$

and thus

$$0 < \eta < \frac{2}{K(\mathbf{v}_k, \mathbf{v}_k)} .$$

When L reaches a maximum value and the we have a stable solution, $\delta\alpha_k = 0$. It follows that

$$1 - y_k \sum_{j=0}^{n-1} \alpha_j y_j K(\mathbf{v}_j, \mathbf{v}_k) - \mu y_k = y_k (y_k - \sum_{j=0}^{n-1} \alpha_j y_j K(\mathbf{v}_j, \mathbf{v}_k) - \mu) = 0$$

where we used that $y_k^2 = +1$.

The *Kernel-Adatron algorithm* (Frieß et al [63]) is given by

1. Initialize $\alpha_0 = \alpha_1 = \cdots = \alpha_{n-1} = 1$, $\theta = 0$.
2. For $i = 0, 1, \ldots, n-1$ calculate
$$z_i = \sum_{j=0}^{n-1} \alpha_j y_j K(\mathbf{v}_i, \mathbf{v}_j)$$
3. For $i = 0, 1, \ldots, n-1$ calculate $\gamma_i = y_i(z_i - \theta)$.
4. Let $\delta\alpha_i := \eta(1 - \gamma_i)$ be the proposed change to α_i.

(a) if $\alpha_i + \delta\alpha_i \leq 0$ then $\alpha_i = 0$.

(b) if $\alpha_i + \delta > 0$ then $\alpha_i = \alpha_i + \delta\alpha_i$.

5. Calculate the new threshold

$$\theta := \frac{1}{2}(\min_i(z_i^+) + \max_i(z_i^-))$$

where z_i^+ are those patterns i with class label $+1$ and z_i^- those with class label -1.

6. If a maximum number of presentations of the pattern set has been exceeded or the margin

$$m := \frac{1}{2}(\min_i(z_i^+) - \max_i(z_i^-))$$

has approached 1 then stop, otherwise return to step 2.

The C++ program could be improved by calculating the kernel beforehand.

```
// kerneladatron.cpp

#include <iostream>
#include <cmath>       // for fabs
using namespace std;

double K(double vi[2], double vj[2]) // kernel
{
 double k1 = 1.0+vi[0]*vj[0]+vi[1]*vj[1];
 return k1*k1;
}

int main(void)
{
 const int m = 4;
 double v[m][2] = { {-1,-1}, {-1,+1}, {+1,-1}, {+1,+1} };
 double y[m]    = {       +1,      -1,      -1,      +1 };
 double alpha[m];
 double eta = 0.01, eps = 0.00001, margin = 0.0, theta = 0.0;
 double min, max;
 int i, j, mininit, maxinit;
 for(i=0;i<m;i++) alpha[i] = 1.0;
 while(fabs(margin-1.0) > eps)
 { mininit = maxinit = 1;
 for(i=0;i<m;i++)
 {
 double z = 0.0;
 for(j=0;j<m;j++) z += alpha[j]*y[j]*K(v[i],v[j]);
 double delta=eta*(1.0-y[i]*(z-theta));
```

```
 if(alpha[i]+delta<=0.0) alpha[i] = 0.0;
 else alpha[i] += delta;
 if((mininit || z<min) && y[i]>0) { min=z; mininit=0; }
 if((maxinit || z>max) && y[i]<0) { max=z; maxinit=0; }
 }
 margin=(min-max)/2.0; theta=(min+max)/2.0;
 }
 for(i=0;i<m;i++) cout << "alpha[" << i << "] = " << alpha[i] << endl;
 cout << "theta = " << theta << endl;
 return 0;
}
```

12.5.4 Kernel Fisher Discriminant

The *Linear Fisher discriminant* (Fisher [60], Mika et al [146], Mika et al [147], Saadi et al [178]) is a two-class discriminative technique. It finds the optimal projection direction such that the distance between the two mean values of the projected classes is maximized while each class variance is minimized. Thus the linear Fisher discriminant is capable of performing feature dimensionality reduction for classification, because only one-dimensional features are extracted for the two-class problems. The linear Fisher discriminant can be extended to a kernel version (i.e., nonlinear Fisher discriminant).

Consider first the linear Fisher discriminant. Let

$$\chi := \{ \mathbf{x}_1, \mathbf{x}_2, \ldots, \mathbf{x}_\ell \} \equiv \{ \chi_1 \,,\ \chi_2 \} \subset \mathbb{R}^d$$

and

$$\chi_1 := \{ \mathbf{x}_1^{(1)}, \ldots, \mathbf{x}_{\ell_1}^{(1)} \}, \qquad \chi_2 := \{ \mathbf{x}_1^{(2)}, \ldots, \mathbf{x}_{\ell_2}^{(2)} \}$$

with $\ell = \ell_1 + \ell_2$. Here χ_1, χ_2 are the sets of training samples for two different pattern classes. Each sample here is a (column) vector in $\mathbb{R}^d$. Thus ℓ_1 and ℓ_2 are the number of training samples corresponding to each class. Let $\ell = \ell_1 + \ell_2$ be the total number of training samples of all classes. Fisher's linear discrimant attempts to find a linear combination of input variables, $\mathbf{w}^T\mathbf{x}$, that maximises the average separation of the projections of points belonging to class C_1 and C_2, whilst minimising the within class variance of the projection of those points. The linear Fisher discriminant is given by the column vector $\mathbf{w}$ which maximizes the following Rayleigh coefficients

$$J(\mathbf{w}) = \frac{\mathbf{w}^T S_B \mathbf{w}}{\mathbf{w}^T S_W \mathbf{w}}$$

where

$$S_B := (\mathbf{m}_1 - \mathbf{m}_2)(\mathbf{m}_1 - \mathbf{m}_2)^T, \quad S_W := \sum_{j=1}^{2} \sum_{i=1}^{\ell_j} (\mathbf{x}_i^{(j)} - \mathbf{m}_j)(\mathbf{x}_i^{(j)} - \mathbf{m}_j)^T$$

are between-class and within-class scatter matrices, respectively. The column vector $\mathbf{m}_j$ is defined by

$$\mathbf{m}_j := \frac{1}{\ell_j} \sum_{i=1}^{\ell_j} \mathbf{x}_i^j .$$

The optimal discrimination mask can be computed in a closed form finding

$$\mathbf{w}^* = \arg\max_{\omega} J(\mathbf{w}) = S_W^{-1}(\mathbf{m}_1 - \mathbf{m}_2)$$

where $\mathbf{w}^*$ is an optimal linear feature extractor. In this case, the optimal discriminative texture features can be directly computed using $\mathbf{w}^*$ by the projection.

The linear Fisher discrimininant has a connection to the optimal linear Bayesian classifier. The optimal projection direction corresponds to the optimal Bayesian classifier. Its optimality depends on the assumption that all the classes have equal covariance matrices. Real-world data are usually not linearly separable. To overcome this problem, the *kernel Fisher discriminant* has been introduced to find a nonlinear projection direction for two-class problems (Mika et al [146], Mika et al [147], Saadi et al [178]). Its implementation can be achieved by employing the kernel trick introduced by Vapnik [211]. Accordingly, $\mathbf{w}$ becomes a nonlinear texture discrimination mask. The kernel Fisher discriminant analysis finds the direction in a feature space, defined implicitly by a kernel, onto which the projections of positive and negative classes are well separated in terms of the Fisher discriminant ratio. Its classification performance depends on the choice of the kernel.

Suppose there is a feature mapping ϕ which maps the input data into a higher-dimensional inner-product vector space F, that is,

$$\phi : \chi \to F .$$

Consequently, the linear Fisher discriminant can be applied in F (corresponding to a nonlinear operation in the input space χ). It is equivalent to maximizing with respect to $\mathbf{w}$ the following criterion

$$J(\mathbf{w}) = \frac{\mathbf{w}^T S_B^{\phi} \mathbf{w}}{\mathbf{w}^T S_W^{\phi} \mathbf{w}}$$

where $\mathbf{w} \in F$. S_B^{ϕ} and S_W^{ϕ} are the corresponding between-class and within-class scatter matrices, respectively, formed in F, i.e.,

$$S_B^{\phi} := (\mathbf{m}_1^{\phi} - \mathbf{m}_2^{\phi})(\mathbf{m}_1^{\phi} - \mathbf{m}_2^{\phi})^T$$

$$S_W^{\phi} := \sum_{j=1}^{2} \sum_{i=1}^{\ell_j} (\phi(\mathbf{x}_i^{(j)}) - \mathbf{m}_j^{\phi})(\phi(\mathbf{x}_i^{(j)}) - \mathbf{m}_j^{\phi})^T$$

with

$$\mathbf{m}_j^{\phi} = \frac{1}{\ell_j} \sum_{i=1}^{\ell_j} \phi(\mathbf{x}_i^j) .$$

From the theory of reproducing kernels, the solution of $\mathbf{w} \in F$ must lie in the span of all the training samples in F. Thus, the vector $\mathbf{w}$ can be formed by a linear combination of the mapped training samples in F as follows

$$\mathbf{w} = \sum_{i=1}^{\ell} \alpha_i \phi(\mathbf{x}_i)\,.$$

Using the definition of $\mathbf{m}_j^{\phi}$ we find the projection between the two vectors in F

$$\mathbf{w}^T \mathbf{m}_j^{\phi} = \frac{1}{\ell_j} \sum_{i=1}^{\ell} \sum_{k=1}^{\ell_j} \alpha_i \langle \phi(\mathbf{x}_i), \phi(\mathbf{x}_k^j) \rangle$$

where $\langle \phi(\mathbf{x}_k), \phi(\mathbf{x}_k^j) \rangle$ is the inner product between $\phi(\mathbf{x}_i)$ and $\phi(\mathbf{x}_k^j)$. By introducing a kernel function $k(\mathbf{x}, \mathbf{y})$ to represent the inner product $\langle \mathbf{x}, \mathbf{y} \rangle$ in F, we obtain

$$\mathbf{w}^T \mathbf{m}_j^{\phi} = \frac{1}{\ell_j} \sum_{i=1}^{\ell} \sum_{k=1}^{l_j} \alpha_i k(\mathbf{x}_i, \mathbf{x}_k^j) = \boldsymbol{\alpha}^T \boldsymbol{\mu}_j, \qquad j = 1, 2$$

where

$$\mu_j^i := \frac{1}{\ell_j} \sum_{k=1}^{l_j} k(\mathbf{x}_i, \mathbf{x}_k^j)\,.$$

Applying the definition of S_B^{ϕ} we have

$$\mathbf{w}^T S_B^{\phi} \mathbf{w} = \boldsymbol{\alpha}^T M \boldsymbol{\alpha}$$

where the matrix M is given by

$$M = (\boldsymbol{\mu}_1 - \boldsymbol{\mu}_2)(\boldsymbol{\mu}_1 - \boldsymbol{\mu}_2)^T\,.$$

Analogously we obtain

$$\mathbf{w}^T S_W^{\phi} \mathbf{w} = \boldsymbol{\alpha}^T N \boldsymbol{\alpha}$$

where the matrix N is given by

$$N = \sum_{j=1}^{2} K_j (I - L_j) K_j^T$$

and K_j is an $\ell \times \ell_j$ kernel matrix of class j with

$$(K_j)_{nm} := k(\mathbf{x}_n, \mathbf{x}_m^j)\,.$$

Here I is the identity matrix and L_j is the matrix with all entries l_j^{-1}. Thus, the optimization of $J(\mathbf{w})$ is equivalent to finding the optimal value of $\boldsymbol{\alpha}$ by maximizing

$$J(\boldsymbol{\alpha}) = \frac{\boldsymbol{\alpha}^T M \boldsymbol{\alpha}}{\boldsymbol{\alpha}^T N \boldsymbol{\alpha}}$$

with respect to $\boldsymbol{\alpha}$. The optimal vector $\boldsymbol{\alpha}^*$ can be computed by finding the leading eigenvector of the matrix $N^{-1}M$. Then the projection of a test pattern $\mathbf{x}_t$ onto $\mathbf{w}$ can be computed by

$$\langle \mathbf{w}, \phi(\mathbf{x}_t) \rangle = \sum_{i=1}^{\ell} \alpha_i k(\mathbf{x}_i, \mathbf{x}_t) .$$

Rather than computing the left-hand side of this equation, the right-hand side can be much more easily obtained via a linear combination of the inner products which is independent of the mapping operator ϕ. Thus we only need to define a kernel form of an inner product instead of computing the explicit form of this mapping. Without considering the mapping ϕ explicitly, the kernel Fisher discrimimant can be constructed by selecting the proper kernel. Kernel functions are

- Gaussian kernel (*radial basis function*):

$$k(\mathbf{x}, \mathbf{y}) = \exp\left(\frac{-\|\mathbf{x} - \mathbf{y}\|^2}{\sigma} \right)$$

- Polynomial kernel:

$$k(\mathbf{x}, \mathbf{y}) = (1 + \mathbf{x}^T \mathbf{y})^n$$

- Tangent hyperbolic kernel:

$$k(\mathbf{x}, \mathbf{y}) = \tanh(\mathbf{x}^T \mathbf{y} + \theta)$$

where σ, n, and θ are the parameters of the three kernels, respectively. The dimensionality of F is usually much higher than the number of training samples which could cause the matrix N to be non-positive definite. Consequently, finding the optimal value of the vector $\boldsymbol{\alpha}$ is an ill-posed problem. The commonly-used approach to solve this problem is to employ regularization, which adds a multiple of the identity matrix or the kernel matrix K to N to guarantee that N is positive definite.

Exercise. Consider the vectors in $\mathbb{R}^4$ (standard basis)

$$\begin{pmatrix} 1 \\ 0 \\ 0 \\ 0 \end{pmatrix}, \quad \begin{pmatrix} 0 \\ 1 \\ 0 \\ 0 \end{pmatrix}, \quad \begin{pmatrix} 0 \\ 0 \\ 1 \\ 0 \end{pmatrix}, \quad \begin{pmatrix} 0 \\ 0 \\ 0 \\ 1 \end{pmatrix}$$

which belong to the class $+1$ and the vectors in $\mathbb{R}^4$ (*Bell basis*)

$$\frac{1}{\sqrt{2}} \begin{pmatrix} 1 \\ 0 \\ 0 \\ 1 \end{pmatrix}, \quad \frac{1}{\sqrt{2}} \begin{pmatrix} 1 \\ 0 \\ 0 \\ -1 \end{pmatrix}, \quad \frac{1}{\sqrt{2}} \begin{pmatrix} 0 \\ 1 \\ 1 \\ 0 \end{pmatrix}, \quad \frac{1}{\sqrt{2}} \begin{pmatrix} 0 \\ 1 \\ -1 \\ 0 \end{pmatrix}$$

which belong to the class -1. Apply the kernel Fisher discriminant analysis.

Chapter 13

Neural Networks

13.1 Introduction

Neural networks are models of the brain's cognitive process (Arbib [4]). The brain has a multiprocessor architecture that is highly interconnected. Neural networks have a potential to advance the types of problems that are being solved by computers. The neuron is the basic processor in neural networks. Each neuron has one output, which is generally related to the state of the neuron -its activation - and which may fan out to several other neurons. Each neuron receives several inputs over these connections, called synapses. The inputs are the activations of the incoming neurons multiplied by the weights of the synapses. The activation of the neuron is computed by applying a threshold function to this product. This threshold function is modelled by a nonlinear function.

The basic artificial neuron (Fausett [57], Haykin [90], Cichocki and Unbehauen [35], Hassoun [89], Rojas [173]) can be modelled as a multi-input nonlinear device with weighted interconnections w_{ji}, also called *synaptic weights* or strengths. The cell body (soma) is represented by a nonlinear limiting or threshold function f. The simplest model of an artificial neuron sums the n weighted inputs and passes the result through a nonlinearity given by a monotonically increasing function f

$$y_j = f\left(\sum_{i=1}^{n} w_{ji}x_i - \theta_j\right) .$$

Here f is called a *threshold function* (also called an *activation function*. θ_j ($\theta_j \in \mathbb{R}$) is the external threshold, also called an offset or bias, w_{ji} are the synaptic weights or strengths, x_i are the inputs ($i = 1, 2, \ldots, n$), n is the number of inputs and y_j represents the output. The activation function is also called the nonlinear transfer characteristic or the squashing function. The activation function f is a monotonically increasing function.

A threshold value θ_j may be introduced by employing an additional input x_0 equal to +1 and the corresponding weight w_{j0} equal to minus the threshold value. Thus we can write

$$y_j = f\left(\sum_{i=0}^{n} w_{ji} x_i\right)$$

where

$$w_{j0} = -\theta_j, \qquad x_0 = 1.$$

The basic artificial neuron is characterized by its nonlinearity and the threshold θ_j. The *McCulloch-Pitts model* of the neuron used only the binary (hard-limiting) function (step function or *Heaviside function*), i.e.,

$$H(x) := \begin{cases} 1 \text{ if } x \geq 0 \\ 0 \text{ if } x < 0 \,. \end{cases}$$

In this model a weighted sum of all inputs is compared with a threshold θ_j. If this sum exceeds the threshold, the neuron output is set to 1, otherwise to 0. For bipolar representation we can use the *sign function*

$$\text{sign}(x) := \begin{cases} 1 \text{ if } x > 0 \\ 0 \text{ if } x = 0 \\ -1 \text{ if } x < 0 \,. \end{cases}$$

The threshold (step) function may be replaced by a more general nonlinear function and consequently the output of the neuron y_j can either assume a value of a discrete set (e.g. $\{-1, 1\}$) or vary continuously (e.g. between -1 and 1 or generally between $y_{\min}$ and $y_{\max} > y_{\min}$). The activation level or the state of the neuron is measured by the output signal y_j, e.g. $y_j = 1$ if the neuron is firing (active) and $y_j = 0$ if the neuron is quiescent in the unipolar case and $y_j = -1$ for the bipolar case.

In the basic neural model the output signal is usually determined by a monotonically increasing sigmoid function of a weighted sum of the input signals. Such a *sigmoid function* can be described for example as

$$y_j = \tanh(\lambda u_j) \equiv \frac{1 - e^{-2\lambda u_j}}{1 + e^{-2\lambda u_j}}$$

for a symmetrical (bipolar) representation. For an unsymmetrical unipolar representation we have

$$y_j = \frac{1}{1 + e^{-\lambda u_j}}$$

where λ is a positive constant or variable which controls the steepness (slope) of the sigmoidal function. The quantity u_j is given by

$$u_j := \sum_{i=0}^{n} w_{ji} x_i \,.$$

The function

$$f_\lambda(x) = \frac{1}{1+e^{-\lambda x}}, \qquad \lambda > 0$$

satisfies the nonlinear ordinary differential equation

$$\frac{df_\lambda}{dx} = \lambda f_\lambda(1 - f_\lambda)$$

with $f_\lambda(0) = 1/2$. The function

$$g_\lambda(x) = \tanh(\lambda x), \qquad \lambda > 0$$

satisfies the nonlinear ordinary differential equation

$$\frac{dg_\lambda}{dx} = \lambda(1 - g_\lambda^2)$$

with $g_\lambda(0) = 0$. These differential equations are important for the backpropagation algorithm.

A model of the above sigmoid function can be built using traditional electronic circuit components (Cichocki and Unbehauen [35], Luo and Unbehauen [135]). A voltage amplifier simulates the cell body (soma), the wires replace the input structure (dendrites) and output structure (axon) and the variable resistors model the synaptic weights (synapses). The amplifier output voltage y_j replaces the variable pulse rate of a real neuron. The sigmoid activation function is naturally provided by the saturating characteristic of the amplifier. The input voltage signals x_i supply current into the wire-dendrites in proportion to the sum of the products of the input voltages and the appropriate conductances. By applying *Kirchhoff's Current Law* at the input node of the amplifier we obtain the expression

$$y_j = f(u_j), \qquad u_j := \frac{\sum_{i=0}^{n} G_{ji} x_i}{\sum_{i=0}^{n} G_{ji}}$$

where x_i $(i = 0, 1, \ldots, n)$ are the input voltages, u_j denotes the input voltage of the j-th amplifier, y_j means the output voltage of the j-th neuron, f is the sigmoid activation function of the amplifier, $G_{ji} = R_{ji}^{-1}$ is the conductance of the resistor connecting the amplifier i with the j-th amplifier and I_{ji} is the current flowing through the resistor R_{ji} (from neuron i to neuron j). The above equation can be written in the form

$$y_j = f\left(\sum_{i=0}^{n} w_{ji} x_i\right), \qquad w_{ji} = \frac{G_{ji}}{\sum_{i=0}^{n} G_{ji}}.$$

```
// threshold.cpp

#include <iostream>
#include <cmath>   // for exp, tanh
using namespace std;

int H(double* w,double* x,int n)
{
 double sum = 0.0;
 for(int i=0;i<=n;i++) { sum += w[i]*x[i]; }
 if(sum >= 0.0) return 1;
 else return 0;
}

int sign(double* w,double* x,int n)
{
 double sum = 0.0;
 for(int i=0;i<=n;i++) { sum += w[i]*x[i]; }
 if(sum >= 0.0) return 1;
 else return -1;
}

double unipolar(double* w,double* x,int n)
{
 double lambda = 1.0; double sum = 0.0;
 for(int i=0;i<=n;i++) { sum += w[i]*x[i]; }
 return 1.0/(1.0+exp(-lambda*sum));
}

double bipolar(double* w,double* x,int n)
{
 double lambda = 1.0; double sum = 0.0;
 for(int i=0;i<=n;i++) { sum += w[i]*x[i]; }
 return tanh(lambda*sum);
}

int main(void)
{
 int n = 5;           // length of input vector includes bias
 double theta = 0.5;  // threshold
 // memory allocation for weight vector w
 double* w = new double[n];
 w[0] = -theta; w[1] = 0.7; w[2] = -1.1; w[3] = 4.5; w[4] = 1.5;
 double* x = new double[n]; // memory allocation
 x[0] = 1.0;  // bias
 x[1] = 0.7; x[2] = 1.2; x[3] = 1.5; x[4] = -4.5;
 int r1 = H(w,x,n); cout << "r1 = " << r1 << endl;
 int r2 = sign(w,x,n); cout << "r2 = " << r2 << endl;
```

```
double r3 = unipolar(w,x,n); cout << "r3 = " << r3 << endl;
double r4 = bipolar(w,x,n); cout << "r4 = " << r4 << endl;
delete[] w; delete[] x;
return 0;
}
```

13.2 Hopfield Model

13.2.1 Introduction

The problem is formulated as follows (Hopfield [99], Kamp and Hasler [109], Rojas [173]). Store a set of p patterns

$$\mathbf{x}_k, \qquad k = 0, 1, \ldots, p-1$$

in such a way that when presented with a new pattern $\mathbf{s}$, the network responds by producing whichever one of the stored patterns most closely resembles $\mathbf{s}$. A binary Hopfield net can be used to determine whether an input vector (pattern) is a known vector (pattern) (i.e., one that was stored in the net) or an unknown vector (pattern). The net recognizes a known vector by producing a pattern of activation on the units of the net that is the same as a vector stored in the net. It can happen that an input vector converges to an activation vector that is none of the stored patterns. Such a pattern is called a *spurious stable state*. The *Hopfield network* is a recurrent network that embodies a profound physical principle, namely, that of storing information in a dynamically stable configuration. One locates each pattern to be stored at the bottom of a "valley" of an energy landscape, and then permitting a dynamical procedure to minimize the energy of the network in such a way that the valley becomes a basin of attraction. The standard discrete-time version of the Hopfield network uses the McCulloch-Pitts model for the neurons. Retrieval of information stored in the network uses a dynamical procedure of updating the state of a neuron selected from among those that want to change, with that particular neuron being picked randomly and one at the time. This asynchronous dynamical procedure is repeated until there are no further state changes to report. An extension of the Hopfield network is that the firing mechanism of the neurons (i.e. switching them on or off) follows a probabilistic law. Here one refers to the neurons as stochastic neurons.

For the recurrent network with symmetric coupling we may define an *energy function* (also called a *Liapunov function*). When the network is started in any initial state, it will move in a downhill direction of the energy function E until it reaches a local minimum. Then it stops changing with time. A recurrent network with symmetric coupling cannot oscillate despite the abundant presence of feedback. One refers to the space of all possible states of the network as the phase space. It is also referred to as the state space. The local minima of the energy function E represent the stable points of the phase space. These points are also referred to as attractors in the sense that each attractor exercises a substantial domain of influence (i.e. basin of attraction) around it. Accordingly, symmetric recurrent networks are sometimes referred to as attractor neural networks. The Hopfield network may be viewed as a nonlinear associative memory or content-addressable memory, the primary function of which is to retrieve a pattern (item) stored in memory in response to the presentation of an incomplete or noisy version of that pattern. The essence of a content-addressable memory is to map a fundamental memory $\mathbf{x}_k$ onto a stable fixed point of a dynamical system. The stable fixed points of the phase space of the network are the fundamental memories of the network. We may then represent their particular pattern as a starting point in the phase space. Provided that the starting point is close to the stable fixed point representing the memory being retrieved, the system should evolve with time and finally converge onto the memory state itself. Thus the Hopfield network is a dynamic system whose phase space contains a set of stable fixed points representing the fundamental memories of the system.

The problem could also be solved by storing the patterns and then calculating the *Hamming distance* between the test pattern $\mathbf{s}$ and each of the stored patterns $\mathbf{x}_k$. The Hamming distance between two sequences of binary numbers of the same length is the number of bits that are different in the two numbers.

In the following we have vectors with elements 1 or -1. Then we use the following definition for the Hamming distance.

Definition. Let $\mathbf{x}$ and $\mathbf{y}$ be two vectors of the same length N with $x_i, y_i \in \{1, -1\}$. Then the Hamming distance is defined as

$$d(\mathbf{x}, \mathbf{y}) := \frac{1}{2} \sum_{i=0}^{N-1} |x_i - y_i|.$$

Example. Consider $\mathbf{x} = (1, -1, 1, 1)^T$, $\mathbf{y} = (-1, 1, 1, -1)^T$. Then $d(\mathbf{x}, \mathbf{y}) = 3$. ♣

Our configuration space consists of N cells. Every cell can take two values (in our case $+1$ and -1). Then the number of configurations (or the number of all possible states of the network) is

$$2^N.$$

Within this space the stored pattern $\mathbf{x}_k$ $(k = 0, 1, \ldots, p-1)$ are attractors. The dynamics of the network maps starting points (initial configurations) into one of the attractors. The whole configuration space is thus divided up into basins of attraction for the different attractors.

```
// Hamming.cpp

#include <iostream>
using namespace std;

int distance(int* x,int* y,int n)
{
```

```
 int d = 0;
 for(int i=0;i<n;i++) { if(x[i] != y[i]) d++; }
 return d;
}

int main(void)
{
 int n = 4;
 int* x = new int[n];    // memory allocation
 x[0] = 1; x[1] = -1; x[2] = 1; x[3] = 1;
 int* y = new int[n];    // memory allocation
 y[0] = -1; y[1] = 1; y[2] = 1; y[3] = -1;
 int result = distance(x,y,n);
 cout << "result = " << result << endl;
 delete[] x; delete[] y;
 return 0;
}
```

13.2.2 Synchronous Operations

The *binary Hopfield model* consists of N neurons

$$s_0, \quad s_1, \quad \ldots \quad , s_{N-1}$$

where

$$s_i = \pm 1, \qquad i = 0, 1, \ldots, N-1 .$$

If $s_i = 1$ we say that the neuron at the site i is active (the neuron fires). If $s_i = -1$ we say that the neuron at site i is inactive (i.e. the neuron does not fire). The neuron at site i is connected with the neuron at site j with $i = 0, 1, \ldots, N-1$ and $j = 0, 1, \ldots, N-1$. In the binary Hopfield model we first impose p patterns, $p \geq 1$, of neuron states. We denote these patterns by

$$\mathbf{x}_0 := (x_{0,0}, x_{0,1}, \ldots, x_{0,N-1})^T, \ \ldots, \ \mathbf{x}_{p-1} := (x_{p-1,0}, x_{p-1,1}, \ldots, x_{p-1,N-1})^T$$

where

$$x_{k,i} = \pm 1, \qquad k = 0, 1, \ldots, p-1, \quad i = 0, 1, \ldots, N-1.$$

Next we assign *connection weights* W_{ij} (W is an $N \times N$ symmetric matrix)

$$W_{ij} := \begin{cases} \displaystyle\sum_{k=0}^{p-1} x_{k,i} x_{k,j} & i \neq j \\ 0 & i = j, \ 0 \leq i, j \leq N-1 \end{cases}$$

where W_{ij} is the connection weight between node i and node j and $i = 0, 1, \ldots, N-1$, $j = 0, 1, \ldots, N-1$. Thus the diagonal elements of the matrix W are equal to zero.

The neurons are updated according to the rule

$$s_i(t+1) = \text{sign}(h_i(t)), \quad t = 0, 1, 2, \ldots,$$

where the *sign function* is defined by

$$\text{sign}(z) := \begin{cases} 1 & \text{if } z > 0 \\ -1 & \text{if } z < 0 \end{cases}$$

with the convention that $s_i(t+1) = s_i(t)$ if $h_i(t) = 0$. The local fields h_i are defined as

$$h_i(t) := \sum_{j=0}^{N-1} W_{ij} s_j(t), \quad t = 0, 1, 2, \ldots$$

Thus the dynamics of the neurons s_i is governed by

$$s_i(t+1) = \text{sign}\left(\sum_{j=0}^{N-1} W_{ij} s_j(t) \right), \qquad t = 0, 1, 2, \ldots$$

where $i = 0, 1, \ldots, N-1$ with the convention described above that

$$s_i(t+1) = s_i(t) \qquad \text{if} \qquad \sum_{j=0}^{N-1} W_{ij} s_j(t) = 0\,.$$

Consequently the dynamics of the neuron is a system of nonlinear difference equations. Thus all components of the state vector $\mathbf{s}$ are updated simultaneously. This is the synchronous operation. In compact form the dynamics can be written as

$$\mathbf{s}(t+1) = \text{sign}(W\mathbf{s}(t))$$

where the sign function is applied to each component of its argument. This operation mode is thus similar to a Jacobi iteration for the solution of linear equations.

Definition. Consider a map $f : S \to S$, where S is a non-empty set. A *fixed point* $x^* \in S$ is defined as a solution of $x^* = f(x^*)$.

Thus for the given case the fixed points of the system of nonlinear difference equations are the solutions of the nonlinear equation

$$\mathbf{s}^* = \text{sign}(W\mathbf{s}^*).$$

They are the time-independent solutions, i.e.,

$$\mathbf{s}^*(t+1) = \mathbf{s}^*(t)\,.$$

Thus the *Hopfield algorithm* is as follows.

1) Initialize the input patterns $\mathbf{x}_0, \ldots, \mathbf{x}_{p-1}$. This is the teaching stage of the algorithm.

2) Calculate the connection weights W_{ij} from the input pattern, i.e. the matrix W.

3) Give the start configuration $\mathbf{s}(t=0)$, where

$$\mathbf{s}(t=0) = (s_0(t=0), s_1(t=0), \ldots, s_{N-1}(t=0))^T.$$

4) Iterate

$$s_i(t+1) = \text{sign}\left(\sum_{j=0}^{N-1} W_{ij} s_j(t)\right), \qquad t = 0, 1, 2, \ldots$$

until convergence. The net is allowed to iterate in discrete time steps, until it reaches a stable configuration. This means the output pattern remains unchanged. The net thus converges to a solution.

Hopfield found experimentally that the number of binary patterns p that can be stored and recalled in a net with reasonable accuracy, is given approximately by

$$p \approx 0.15N$$

where N is the number of neurons in the net. Another estimate is

$$p \approx \frac{N}{2\log_2 N}.$$

For $N = 40$ Hopfield's estimation is $p = 6$.

13.2.3 Energy Function

Hopfield proved that the discrete Hopfield net will converge to a stable limit point (pattern of activation of the units) by considering an *energy function* for the system. The energy function is also called a *Liapunov function*. An energy function (Kamp and Hasler [109]) is a function that is bounded below and is a nonincreasing function of the state of the system. For the Hopfield network the energy function E is

$$E(\mathbf{s}(t)) := -\frac{1}{2}\sum_{i=0}^{N-1}\sum_{j=0}^{N-1} W_{ij} s_i(t) s_j(t) \equiv -\frac{1}{2}\mathbf{s}^T(t) W \mathbf{s}(t).$$

Note that $W_{ij} = W_{ji}$ (i.e. W is symmetric) and $W_{ii} = 0$ for $i = 0, 1, \ldots, N-1$. The factor $1/2$ is used because the identical terms $W_{ij}s_i(t)s_j(t)$ and $W_{ji}s_j(t)s_i(t)$ are presented in the double sum. An unknown input pattern represents a particular point in the energy landscape. As the network iterates its way to a solution, the point moves through the landscape towards one of the hollows. These basins of attraction represent the stable states of the network. The solution from the net occurs when the point moves into the lowest region of the basin; from there, everywhere else in the close vicinity is uphill, and so it will stay where it is. This is directly analogous to the three-dimensional case where a ball placed on a landscape of valleys and hillsides will move down towards the nearest hollow, setting into a stable state that does not alter with time when it reaches the bottom. With the assumption given above we can show that

$$E(\mathbf{s}(t+1)) - E(\mathbf{s}(t)) \leq 0.$$

This can be seen as follows. The energy does not depend upon cell numbering, because $W_{ij} = W_{ji}$. Now if cell k is chosen to reevaluate its activation, it will keep the same value (and therefore leave the energy unchanged) unless

1. $s_k(t) = -1$ and $\sum_{j=0}^{N-1} W_{kj} s_j(t) > 0$. Then $s_k(t+1)$ becomes $+1$

2. $s_k(t) = +1$ and $\sum_{j=0}^{N-1} W_{kj} s_j(t) < 0$. Then $s_k(t+1)$ becomes -1

In either case the energy changes by

$$\Delta E := E(\mathbf{s}(t+1)) - E(\mathbf{s}(t)) = -(s_k(t+1) - s_k(t)) \sum_{j \neq k} W_{k,j} s_j(t)\,.$$

Consequently

$$\Delta E = -2 s_k(t+1) \sum_{j=0}^{N-1} W_{k,j} s_j(t) \leq 0\,.$$

Notice that we used the fact that the weights are symmetrical ($W_{i,j} = W_{j,i}$) and that $W_{k,k} = 0$ for $k = 0, 1, \ldots, N-1$. Moreover we used the fact that $s_j(t+1) = s_j(t)$ for $j \neq k$. Thus energy in a Hopfield model can never increase, and whenever a cell changes activation to -1 the energy strictly decreases. We also have the identity

$$\begin{aligned} E(\mathbf{s}(t+1)) - E(\mathbf{s}(t)) &= -(\mathbf{s}^T(t+1) - \mathbf{s}^T(t)) W \mathbf{s}(t) \\ &\quad - \frac{1}{2}(\mathbf{s}^T(t+1) - \mathbf{s}^T(t)) W (\mathbf{s}(t+1) - \mathbf{s}(t)) \end{aligned}$$

where we used

$$\mathbf{s}^T(t+1) W \mathbf{s}(t) \equiv \mathbf{s}^T(t) W \mathbf{s}(t+1)\,.$$

Each component in $W\mathbf{s}(t)$ has the same sign as the corresponding component in $\mathbf{s}(t+1)$. Thus, if $\mathbf{s}(t+1) \neq \mathbf{s}(t)$, the first term gives a strictly negative contribution and this effect is possibly enhanced by the second term.

13.2.4 Basins and Radii of Attraction

In order to measure the dissimilarity of two vectors $\mathbf{x}$ and $\mathbf{y}$ with

$$x_j,\ y_j \in \{\,1, -1\,\}$$

we introduced the Hamming distance. We define by $N_r(\mathbf{y})$ (Kamp and Hasler [109]) the neighbourhood of $\mathbf{y}$ with radius r as the set of vectors $\mathbf{x}$ located at most at distance r from $\mathbf{y}$. Thus

$$N_r(\mathbf{y}) := \{\, \mathbf{x} \in \{-1,1\}^n \, : \, d(\mathbf{x},\mathbf{y}) \leq r \,\}.$$

The basin of direct attraction $B_1(\mathbf{s}^*)$ of a fixed point $\mathbf{s}^*$ is the largest neighbourhood of $\mathbf{s}^*$ such that any vector in the neighbourhood is attracted by the fixed point in a single iteration

$$B_1(\mathbf{s}^*) := \max_r \{\, N_r(\mathbf{s}^*) : \mathbf{s}(0) \in N_r(\mathbf{s}^*) \to \mathbf{s}(1) = \mathbf{s}^* \,\}.$$

One should distinguish between the basin of direct attraction as defined above and the domain of direct attraction $D_1(\mathbf{s}^*)$ which is the set of all points in $\{\,-1,1\,\}^N$ which are attracted by $\mathbf{s}^*$ in a single iteration

$$D_1(\mathbf{s}^*) := \{\, \mathbf{s}(0) \in \{\,-1,1\,\}^N \, : \, \mathbf{s}(1) = \mathbf{s}^* \,\}.$$

The domain of attraction can include points which lie outside the basin $B_1(\mathbf{s}^*)$ and consequently $B_1(\mathbf{s}^*) \subset D_1(\mathbf{s}^*)$. The basin of direct attraction is independent of the particular operation mode, be it synchronous, asynchronous or block-sequential. One defines the basin of attraction of order k, denoted $B_k(\mathbf{s}^*)$, as the largest neighbourhood of $\mathbf{s}^*$ such that any vector in this neighbourhood converges to $\mathbf{s}^*$ in k iterations at most

$$B_k(\mathbf{s}^*) := \max_r \{\, N_r(\mathbf{s}^*) \, : \, \mathbf{s}(0) \in N_r(\mathbf{s}^*) \to \mathbf{s}(k) = \mathbf{s}^* \,\}.$$

The basin of long-term attraction $B(\mathbf{s}^*)$ is defined as the largest neighbourhood of a fixed point in which attraction take place in a finite number of iterations

$$B(\mathbf{s}^*) := \max_r \{\, N_r(\mathbf{s}^*) \, : \, \mathbf{s}(0) \in N_r(\mathbf{s}^*) \to \exists\, k \text{ such that } \mathbf{s}(t) = \mathbf{s}^* \text{ if } t \geq k \,\}.$$

The radius of attraction of order k, $R_k(\mathbf{s}^*)$ and the radius of long-term attraction, $R(\mathbf{s}^*)$ of a fixed point $\mathbf{s}^*$ are the radii of the neighbourhood $B_k(\mathbf{s}^*)$ and $B(\mathbf{s}^*)$, respectively.

13.2.5 Spurious Attractors

When a Hopfield network is used to store p patterns by means of the Hebb prescription for the synaptic weights, the network is usually found to have *spurious attractors*, also referred to as *spurious states*. Spurious states represent stable states of the network that are different from the fundamental memories of the network. Note that the energy function

$$E(\mathbf{s}) = -\frac{1}{2}\mathbf{s}^T W \mathbf{s}$$

is symmetric in the sense that its value remains unchanged if the states of the neurons are reversed (i.e., the state s_i is replaced by $-s_i$ for all i) since W is a symmetric matrix and $W_{ii} = 0$. Accordingly, if the fundamental memory $\mathbf{x}_k$ corresponds to a particular local minimum of the energy landscape, that same local minimum also corresponds to $-\mathbf{x}_k$. This sign reversal need not pose a problem in the retrieval of stored information if it is agreed to reverse all the remaining bits of a retrieval pattern should it be found that a particular bit designated as the "sign" bit is -1 instead of $+1$. Secondly, there is an attractor for every mixture of the stored patterns. A mixture state corresponds to a linear combination of an odd number of patterns. Thirdly, for a large number p of fundamental memories, the energy landscape has local minima that are not correlated with any of these memories embedded in the network. Such spurious states are sometimes referred to as *spin-glass states*, by analogy with spin-glass models in statistical mechanics. If we set $W_{ii} \neq 0$ for all i, additional stable spurious states might be produced in the neighbourhood of a desired attractor.

13.2.6 Hebb's Law

Based on *Hebb's law* (Hebb [91], Herz et al [94]), the most widely accepted hypothesis explaining the learning mechanism achieved by associative memories is that some functional modification takes place in the synaptic links between the neurons. It is assumed that correlated neuron activities increase the strength of the synaptic link. Usually, this hypothesis of synaptic plasticity is quantitatively expressed by stating that the synaptic weight W_{ij} should increase whenever neurons i and j have simultaneously the same activity level and that it should decrease in the opposite case. Consequently, in order to store a prototype $\mathbf{x}$ according to Hebb's hypothesis, the synaptic weight should be modified by an amount

$$\Delta W_{ij} = \Delta W_{ji} = \eta x_i x_j$$

where η is a positive *learning factor*. For the complete synaptic matrix the modification will thus be

$$\Delta W = \eta \mathbf{x}\mathbf{x}^T .$$

Notice that $\mathbf{x}$ is a column vector with N components and therefore $\mathbf{x}^T$ is a row vector with N components. Thus W is an $N \times N$ matrix. If p prototype vectors $\mathbf{x}_0, \mathbf{x}_1, \ldots, \mathbf{x}_{p-1}$ have to be stored, one considers that the resulting synaptic matrix is given by Hebb's law

$$W = \frac{1}{p} \sum_{k=0}^{p-1} \mathbf{x}_k \mathbf{x}_k^T .$$

Since the sign function applies to the vector $W\mathbf{s}$ the factor $1/p$ can also be omitted. Furthermore it is assumed that $W_{ii} = 0$ for $i = 0, 1, \ldots, N-1$. With this assumption we have

$$W = \sum_{k=0}^{p-1} \mathbf{x}_k \mathbf{x}_k^T - pI_N$$

where I_N is the $N \times N$ unit matrix. When the prototypes (stored pattern) are orthogonal, i.e.

$$\mathbf{x}_k^T \mathbf{x}_l = 0 \quad \text{if} \quad k \neq l$$

then the stored patterns are fixed points of the nonlinear map since

$$W\mathbf{x}_k = (N-p)\mathbf{x}_k, \qquad k = 0, 1, \ldots, p-1 .$$

However, if the prototypes are not orthogonal, the correlations between them may prevent exact retrieval and it is to be expected that the number of vectors which can be stored will be reduced if the Hamming distance between these vectors becomes smaller. The problem of correlated prototypes is unavoidable since, for most applications the selection of these prototypes is not free but imposed by the patterns to be stored.

Hebb's rule does not always provide perfect retrieval. A modification has been proposed which allows us to give variable weightings to the prototypes in order to reinforce those which are more difficult to memorize. The weighted Hebbian rule is defined by

$$W := \sum_{k=0}^{p-1} \lambda_k \mathbf{x}_k \mathbf{x}_k^T, \qquad \lambda_k \leq 1 \quad \text{and} \quad \sum_{k=0}^{p-1} \lambda_k = 1 .$$

The construction of W is quite similar to the *spectral theorem* in matrix theory. Let A be an $N \times N$ symmetric matrix over the real numbers $\mathbb{R}$. Then the eigenvalues λ_0, λ_1, $\ldots$, λ_{N-1} of A are real. Assume that the eigenvalues are pairwise different. Then the corresponding normalized eigenvectors $\mathbf{x}_0$, $\mathbf{x}_1$, $\ldots$, $\mathbf{x}_{N-1}$ are pairwise orthonormal. We consider the eigenvectors as column vectors. Then the symmetric matrix A can be reconstructed as

$$A = \sum_{j=0}^{N-1} \lambda_j \mathbf{x}_j \mathbf{x}_j^T .$$

Example. Consider the 2×2 matrix

$$A = \begin{pmatrix} 0 & 1 \\ 1 & 0 \end{pmatrix} .$$

The eigenvalues are given by $\lambda_0 = 1$, $\lambda_1 = -1$ with the corresponding normalized eigenvectors

$$\mathbf{x}_0 = \frac{1}{\sqrt{2}} \begin{pmatrix} 1 \\ 1 \end{pmatrix}, \qquad \mathbf{x}_1 = \frac{1}{\sqrt{2}} \begin{pmatrix} 1 \\ -1 \end{pmatrix} .$$

Hence $A = \mathbf{x}_0 \mathbf{x}_0^T - \mathbf{x}_1 \mathbf{x}_1^T$. Note that $\mathbf{x}_0^T \mathbf{x}_1 = 0$. ♣

13.2.7 Hopfield Example

Consider the case with $N = 5$ neurons. The input vectors (patterns) to be stored are

$$\mathbf{x}_0 = \begin{pmatrix} 1 \\ 1 \\ 1 \\ 1 \\ -1 \end{pmatrix}, \qquad \mathbf{x}_1 = \begin{pmatrix} -1 \\ 1 \\ -1 \\ 1 \\ 1 \end{pmatrix} .$$

We note that

$$\mathbf{x}_0^T \mathbf{x}_1 = -1$$

i.e., the two input patterns are not orthogonal. Obviously, if the length of the vectors is an odd number the vectors cannot be orthogonal. For the entries of the connection matrix W we have

$$W = \mathbf{x}_0 \mathbf{x}_0^T + \mathbf{x}_1 \mathbf{x}_1^T - pI$$

where $p = 2$ (number of pattern stored) and I is the 5×5 unit matrix. The term $-pI$ is necessary so that the diagonal elements W_{jj} cancel out. Thus the symmetric connection matrix W is

$$W = \begin{pmatrix} 0 & 0 & 2 & 0 & -2 \\ 0 & 0 & 0 & 2 & 0 \\ 2 & 0 & 0 & 0 & -2 \\ 0 & 2 & 0 & 0 & 0 \\ -2 & 0 & -2 & 0 & 0 \end{pmatrix}.$$

Next we show that the two stored patterns are fixed points, i.e., we have to prove that

$$\mathbf{s}^* = \text{sign}(W\mathbf{s}^*).$$

For the first stored pattern $\mathbf{x}_0$ we have

$$\text{sign}(W\mathbf{x}_0) = \text{sign} \begin{pmatrix} 4 \\ 2 \\ 4 \\ 2 \\ -4 \end{pmatrix} = \begin{pmatrix} 1 \\ 1 \\ 1 \\ 1 \\ -1 \end{pmatrix} = \mathbf{x}_0 .$$

Analogously we prove that $\mathbf{x}_1$ is a fixed point, i.e.

$$\text{sign}(W\mathbf{x}_1) = \text{sign} \begin{pmatrix} -4 \\ 2 \\ -4 \\ 2 \\ 4 \end{pmatrix} = \begin{pmatrix} -1 \\ 1 \\ -1 \\ 1 \\ 1 \end{pmatrix} = \mathbf{x}_1 .$$

Next we find the energy values from the energy function. From

$$E = -\frac{1}{2}\mathbf{s}^T W \mathbf{s}$$

we obtain for the first stored pattern $\mathbf{x}_0$ the energy $E = -8$. Analogously, for the second stored pattern $\mathbf{x}_1$ we find $E = -8$. Recall that there are $2^5 = 32$ configurations. Next we look at the time evolution of the initial state

$$\mathbf{s}(0) = \begin{pmatrix} 1 \\ 1 \\ 1 \\ 1 \\ 1 \end{pmatrix}.$$

We find

$$W\mathbf{s}(0) = \begin{pmatrix} 0 \\ 2 \\ 0 \\ 2 \\ -4 \end{pmatrix}, \qquad \mathbf{s}(1) = \text{sign}(W\mathbf{s}(0)) = \begin{pmatrix} 1 \\ 1 \\ 1 \\ 1 \\ -1 \end{pmatrix}$$

where we used the rule that sign(0) is assigned to the corresponding value in the vector $\mathbf{s}(0)$. We find that $\mathbf{s}(1)$ is the stored pattern $\mathbf{x}_0$. This means we have reached a stable fixed point. The energy value of the initial state $\mathbf{s}(0)$ is $E = 0$. $E = -8$ for $\mathbf{s}(1)$ is the lowest energy state the system has.

Exercise. Find the energy value for the other configurations. Recall that there are $2^5 = 32$ configurations. Study what happens if the initial state is given by

$$\mathbf{s}(0) = \begin{pmatrix} -1 \\ -1 \\ 1 \\ 1 \\ 1 \end{pmatrix}.$$

Find the energy value E.

13.2.8 Hopfield C++ Program

In the C++ program we consider synchronous operations. We consider the three patterns given by the numbers 1, 2 and 4 in the figure. A black box indicates +1 and a white box −1. Thus in the example we have $N = 40 = 8 \cdot 5$ neurons, i.e., we have 8 rows and 5 columns in the grid.

We consider the three patterns giving the numbers 1, 2 and 4. For the number 1 we have the pattern

$$\mathbf{x}_0 = (-1,-1,1,-1,-1,-1,1,1,-1,-1,-1,-1,1,-1,-1,-1,-1,1,-1,-1,-1,$$
$$-1,-1,1,-1,-1,-1,-1,1,-1,-1,-1,-1,1,-1,-1,-1,-1,-1,-1,-1)^T.$$

For the number 2 we have the pattern

$$\mathbf{x}_1 = (-1,1,1,1,-1,1,-1,-1,-1,1,-1,-1,-1,-1,1,-1,-1,1,1,-1,$$
$$-1,1,-1,-1,-1,1,-1,-1,-1,-1,1,1,1,1,1,-1,-1,-1,-1,-1)^T.$$

For the number 4 we have the pattern

$$\mathbf{x}_2 = (1,-1,-1,1,-1,1,-1,-1,1,-1,1,-1,-1,1,-1,1,1,1,1,1,$$
$$-1,-1,-1,1,-1,-1,-1,-1,1,-1,-1,-1,-1,1,-1,-1,-1,-1,-1,-1)^T.$$

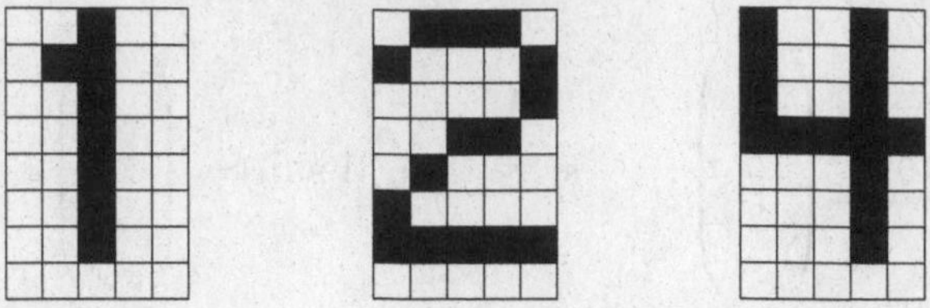

Figure 13.1 Input Patterns for C++ Program

Problem. Calculate the scalar products $\mathbf{x}_0^T\mathbf{x}_1$, $\mathbf{x}_0^T\mathbf{x}_2$ and $\mathbf{x}_1^T\mathbf{x}_2$. Are the patterns, orthogonal to each other? Are the patterns fixed points of the map W?

When we consider the initial configuration

$$\mathbf{s}(t=0) = (1,1,-1,1,-1,-1,1,-1,1,1,1,-1,-1,1,-1,-1,1,1,1,1,$$

$$-1,-1,-1,1,-1,-1,-1,-1,1,-1,-1,-1,-1,1,-1,-1,-1,-1,-1,-1)^T$$

the algorithm tends to the pattern of the number 4. The figure shows the input pattern.

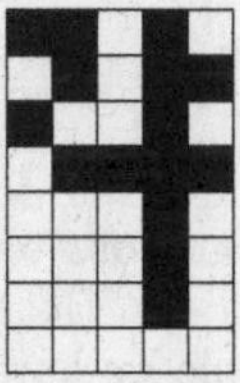

Figure 13.2 Input Pattern

When we consider the initial configuration

$$\mathbf{s}(t=0) = (-1,1,1,-1,-1,-1,-1,1,-1,-1,-1,-1,1,-1,-1,-1,-1,-1,1,-1,$$

$$-1,-1,1,-1,-1,-1,-1,-1,1,-1,-1,-1,1,-1,-1,-1,-1,-1,-1,-1)^T$$

we obtain the pattern of the number 1. The figure shows the input pattern.

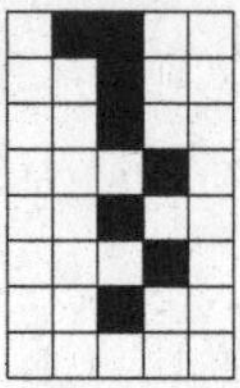

Figure 13.3 Input Pattern

```
// hopfield.cpp

#include <iostream>
using namespace std;

// initialize the weight matrix W with the patterns
void weights(int** W,int* x0,int* x1,int* x2,int N)
{
 for(int i=0;i<N;i++)
  for(int j=0;j<N;j++)
   { W[i][j] = x0[i]*x0[j]+x1[i]*x1[j]+x2[i]*x2[j]; }
 for(int k=0;k<N;k++) W[k][k] = 0;
}

// calculation of sum over j of W[i][j]*s[j]
void mul(int** W,int* s,int* h,int N)
{
 for(int i=0;i<N;i++)
 {
 int sum = 0;
 for(int j=0;j<N;j++) { sum += W[i][j]*s[j]; }
 h[i] = sum;
 }
}

int sign(int y) { if(y > 0) return 1; else return -1; }

int check(int* v1,int* v2,int N)  // checks whether two vectors
{                                 // are the same
 for(int i=0;i<N;i++) { if(v1[i] != v2[i]) return 0; }
 return 1;
}

int energy(int** W,int* s,int N)  // energy of the configuration
{                                 // Ising model
 int E = 0;
 for(int i=0;i<N;i++)
  for(int j=0;j<N;j++) E += W[i][j]*s[i]*s[j];
 return -E;
}

int main(void)
{
 int N = 40;
 int* x0 = new int[N]; int* x1 = new int[N]; int* x2 = new int[N];
 // pattern 0
 x0[0] = -1;  x0[1] = -1; x0[2] = 1;  x0[3] = -1; x0[4] = -1;
 x0[5] = -1;  x0[6] = 1;  x0[7] = 1;  x0[8] = -1; x0[9] = -1;
```

```
x0[10] = -1; x0[11] = -1; x0[12] = 1; x0[13] = -1; x0[14] = -1;
x0[15] = -1; x0[16] = -1; x0[17] = 1; x0[18] = -1; x0[19] = -1;
x0[20] = -1; x0[21] = -1; x0[22] = 1; x0[23] = -1; x0[24] = -1;
x0[25] = -1; x0[26] = -1; x0[27] = 1; x0[28] = -1; x0[29] = -1;
x0[30] = -1; x0[31] = -1; x0[32] = 1; x0[33] = -1; x0[34] = -1;
x0[35] = -1; x0[36] = -1; x0[37] = -1; x0[38] = -1; x0[39] = -1;
// pattern 1
x1[0] = -1;  x1[1] = 1; x1[2] = 1;  x1[3] = 1; x1[4] = -1;
x1[5] = 1;  x1[6] = -1;  x1[7] = -1;  x1[8] = -1; x1[9] = 1;
x1[10] = -1; x1[11] = -1; x1[12] = -1; x1[13] = -1; x1[14] = 1;
x1[15] = -1; x1[16] = -1; x1[17] = 1; x1[18] = 1; x1[19] = -1;
x1[20] = -1; x1[21] = 1; x1[22] = -1; x1[23] = -1; x1[24] = -1;
x1[25] = 1; x1[26] = -1; x1[27] = -1; x1[28] = -1; x1[29] = -1;
x1[30] = 1; x1[31] = 1; x1[32] = 1; x1[33] = 1; x1[34] = 1;
x1[35] = -1; x1[36] = -1; x1[37] = -1; x1[38] = -1; x1[39] = -1;
// pattern 2
x2[0] = 1;  x2[1] = -1; x2[2] = -1; x2[3] = 1; x2[4] = -1;
x2[5] = 1; x2[6] = -1;  x2[7] = -1; x2[8] = 1; x2[9] = -1;
x2[10] = 1; x2[11] = -1; x2[12] = -1; x2[13] = 1; x2[14] = -1;
x2[15] = 1; x2[16] = 1; x2[17] = 1; x2[18] = 1; x2[19] = 1;
x2[20] = -1; x2[21] = -1; x2[22] = -1; x2[23] = 1; x2[24] = -1;
x2[25] = -1; x2[26] = -1; x2[27] = -1; x2[28] = 1; x2[29] = -1;
x2[30] = -1; x2[31] = -1; x2[32] = -1; x2[33] = 1; x2[34] = -1;
x2[35] = -1; x2[36] = -1; x2[37] = -1; x2[38] = -1; x2[39] = -1;

int** W = NULL;    // memory allocation
W = new int *[N];
for(int i=0;i<N;i++) { W[i] = new int [N]; }
for(int i=0;i<N;i++)
 for(int j=0;j<N;j++) W[i][j] = 0;

weights(W,x0,x1,x2,N); // weight function to obtain weight matrix
int* s = new int[N];   // memory allocation
// start configuration
s[0] = 1;  s[1] = 1;   s[2] = -1;   s[3] = 1; s[4] = -1;
s[5] = -1;  s[6] = 1;    s[7] = -1; s[8] = 1; s[9] = 1;
s[10] = 1; s[11] = -1; s[12] = -1;   s[13] = 1;  s[14] =-1;
s[15] = 1;  s[16] = -1;   s[17] = 1;  s[18] = 1; s[19] = 1;
s[20] = -1;  s[21] = -1;   s[22] = -1;   s[23] = -1; s[24] = -1;
s[25] = -1; s[26] = -1;   s[27] = 1; s[28] = 1; s[29] = -1;
s[30] = -1; s[31] = -1;  s[32] = -1;   s[33] = 1;  s[34] = -1;
s[35] = -1; s[36] = -1;   s[37] = -1; s[38] = -1;  s[39] = -1;

int E = energy(W,s,N);
cout << "energy of initial configuration: " << E << endl;

int* h = new int[N]; // memory allocation
for(int p=0;p<N;p++) { h[p] = 0; } // initialising h
```

```
 int* s1 = new int[N]; // memory allocation

 for(i=0;i<N;i++) { s1[i] = s[i]; }
 int result, count;
 count = 0;
 do
 {
 for(int i=0;i<N;i++)  { s1[i] = s[i]; }
 mul(W,s,h,N);
 for(int j=0;j<N;j++)
 {
 if(h[j] != 0) { s[j] = sign(h[j]); }
 if(h[j] == 0) { s[j] = s1[j]; }
 }
 result = check(s,s1,N);
 count++;
 cout << "count = " << count << endl;
 } while((count < 100) && (result != 1));
 cout << "number of iterations: " << count << endl;
 for(int i=0;i<N;i++)
 {
 cout << "s[" << i << "] = " << s[i] << " "; // end configuration
 if(((i+1)%5) == 0) { cout << endl; }
 }
 E = energy(W,s,N);
 cout << "energy of end configuration: " << E << endl;

 delete[] x0; delete[] x1; delete[] x2;
 delete[] s; delete[] s1; delete[] h;
 for(i=0;i<N;i++) { delete[] W[i]; }
 delete[] W;
 return 0;
}
```

13.2.9 Asynchronous Operation

In the asynchronous operation (Kamp and Hasler [109]) each element of the state vector is updated separately, while taking into account the most recent values for the components which have already been updated. Several variants are still possible, but the most usual one is sequential updating where the new element values are computed in the order in which they appear in the state vector. Thus, when updating element s_i, use will be made of the fact that new elements have already been computed for the elements s_0 up to s_{i-1} according to the formula

$$s_i(t+1) = \mathrm{sign}\left(\sum_{j=0}^{i-1} W_{ij}s_j(t+1) + \sum_{j=i}^{N-1} W_{ij}s_j(t)\right).$$

This scheme is close to a *Gauss-Seidel iteration* in linear algebra. The Gauss-Seidel iteration is used to solve systems of linear equations. Starting from the state vector

$$\mathbf{s}(t) = (s_0(t), s_1(t), \ldots, s_{N-1}(t))$$

one complete sequential iteration consists in updating successively each of the N components until one obtains the new state vector $\mathbf{s}(t+1)$. The components of the state vector can be updated in some different order, e.g. according to a fixed permutation of the natural order or at random but in such a way that each element is at least updated once in a given interval of time.

We have assumed that the externally applied threshold vector $\boldsymbol{\theta}$ is the zero vector. If we include this vector then the map, in the synchronous operation, is given by

$$\mathbf{s}(t+1) = \mathrm{sign}(W\mathbf{s}(t) - \boldsymbol{\theta})\,.$$

For the asynchronous operation we have

$$s_i(t+1) = \mathrm{sign}\left(\sum_{j=0}^{i-1} W_{ij}s_j(t+1) + \sum_{j=i}^{N-1} W_{ij}s_j(t) - \theta_i\right)$$

where $i = 0, 1, \ldots, N-1$.

13.2.10 Translation Invariant Pattern Recognition

Those patterns which are similar, except for a translation in the plane, should be identified as the same. This can be done using the two-dimensional discrete Fourier transform of the scanned images. The two-dimensional discrete Fourier transform is invertible. The figure shows an example of two identical patterns positioned with a displacements in x and y directions from each other.

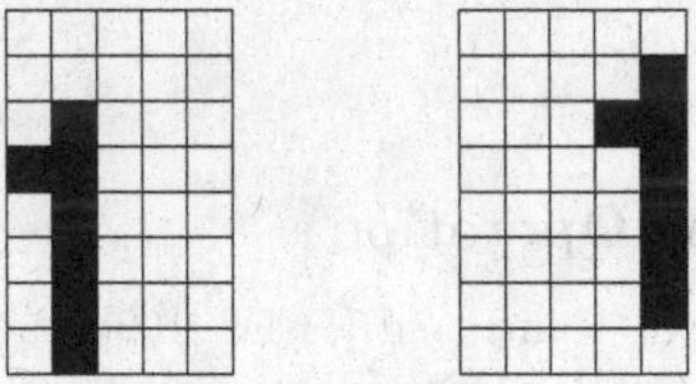

Figure 13.4 Identical Pattern with Displacement

The *two-dimensional discrete Fourier transform* of a two-dimensional array is computed by first performing a one-dimensional Fourier transform of the rows of the array and then a one-dimensional Fourier transform of the columns of the result (or vice versa). The absolute value of the Fourier coefficients does not change under a translation of the two-dimensional pattern. The Fourier transform preserves angles, that is, similar patterns in the original domain are also similar in the Fourier domain.

This preprocessing can be used to implement translation-invariant associative recall. In this case the vectors which are stored in the network are the absolute values of the Fourier coefficients for each pattern.

Definition. Let $x(n_1, n_2)$ denote an array of real values, where n_1, n_2 are integers such that $0 \leq n_1 \leq N_1 - 1$ and $0 \leq n_2 \leq N_2 - 1$. The two-dimensional discrete Fourier transform $\hat{x}(k_1, k_2)$ of $x(n_1, n_2)$ is defined by

$$\hat{x}(k_1, k_2) := \frac{1}{N_1}\frac{1}{N_2}\sum_{n_1=0}^{N_1-1}\sum_{n_2=0}^{N_2-1} x(n_1, n_2) \exp\left(-\frac{2\pi}{N_1} i n_1 k_1 - \frac{2\pi}{N_2} i n_2 k_2\right)$$

where $0 \leq k_1 \leq N_1 - 1$ and $0 \leq k_2 \leq N_2 - 1$.

Consider two arrays $x(n_1, n_2)$ and $y(n_1, n_2)$ of real values. Assume that

$$y(n_1, n_2) = x(n_1 + d_1, n_2 + d_2)$$

where d_1 and d_2 are two given integers. The addition $n_1 + d_1$ is performed modulo N_1 and $n_2 + d_2$ is performed modulo N_2 (torus). We show that

$$\|\hat{y}(k_1, k_2)\| = \|\hat{x}(k_1, k_2)\|.$$

We have

$$\hat{y}(k_1, k_2) = \frac{1}{N_1}\frac{1}{N_2}\sum_{n_1=0}^{N_1-1}\sum_{n_2=0}^{N_2-1} x(n_1 + d_1, n_2 + d_2) \exp\left(-\frac{2\pi}{N_1} i n_1 k_1 - \frac{2\pi}{N_2} i n_2 k_2\right).$$

With the change of indices

$$n_1' = (n_1 + d_1) \bmod N_1, \qquad n_2' = (n_2 + d_2) \bmod N_2$$

we obtain

$$\hat{y}(k_1, k_2) = \frac{1}{N_1}\frac{1}{N_2}\sum_{n_1'=0}^{N_1-1}\sum_{n_2'=0}^{N_2-1} x(n_1', n_2') \exp\left(2\pi i\left(-\frac{n_1' k_1}{N_1} - \frac{n_2' k_2}{N_2} + \frac{d_1 k_1}{N_1} + \frac{d_2 k_2}{N_2}\right)\right).$$

This can be written as

$$\hat{y}(k_1, k_2) = \exp\left(\frac{2\pi}{N_1} i d_1 k_1\right) \exp\left(\frac{2\pi}{N_2} i d_2 k_2\right) \hat{x}(k_1, k_2).$$

This expression tells us that the Fourier coefficients of the array $y(n_1, n_2)$ are the same as the coefficients of the array $x(n_1, n_2)$ except for a phase factor

$$\exp\left(-\frac{2\pi}{N} i d_1 k_1\right) \exp\left(-\frac{2\pi}{N} i d_2 k_2\right).$$

Taking the absolute value results in

$$\|\hat{y}(k_1, k_2)\| = \|\hat{x}(k_1, k_2)\|.$$

Thus the absolute values of the Fourier coefficients for both patterns are identical.

13.3 Similarity Metrics

A key component of most pattern-comparison algorithms is a prescribed measurement of dissimilarity between two feature vectors. Assume we have two feature vectors, $\mathbf{x}$ and $\mathbf{y}$ defined on a vector space V. We define a metric or distance function d on the vector space V as a real-valued function on the Cartesian product $V \times V$ such that

$$0 \leq d(\mathbf{x}, \mathbf{y}) < \infty \text{ for } \mathbf{x}, \mathbf{y} \in V \text{ and } d(\mathbf{x}, \mathbf{y}) = 0 \text{ iff } \mathbf{x} = \mathbf{y}$$

$$d(\mathbf{x}, \mathbf{y}) = d(\mathbf{y}, \mathbf{x}) \text{ for } \mathbf{x}, \mathbf{y} \in V$$

$$d(\mathbf{x}, \mathbf{y}) \leq d(\mathbf{x}, \mathbf{z}) + d(\mathbf{y}, \mathbf{z}), \quad \mathbf{x}, \mathbf{y}, \mathbf{z} \in V .$$

In, addition, a distance funtion is called invariant if

$$d(\mathbf{x} + \mathbf{z}, \mathbf{y} + \mathbf{z}) = d(\mathbf{x}, \mathbf{y}) .$$

Most distance measures, or metrics, are ℓ^p metrics

$$\ell^p(\mathbf{x}, \mathbf{y}) := \left(\sum_{i=1}^{n} |x_i - y_i|^p \right)^{1/p} .$$

If $p = 2$, we have the *Euclidean distance.*

The cosine between two row vectors $\mathbf{x}, \mathbf{y}$ in $\mathbb{R}^n$ defines another metrical similarity measure

$$d(\mathbf{x}, \mathbf{y}) = \frac{\mathbf{x}\mathbf{y}^T}{\|\mathbf{x}\| \, \|\mathbf{y}\|}$$

where ^{T} denotes the transpose.The *cosine distance* is invariant under rotation and dilation.

Most current neural networks require two-dimensional input patterns to be converted into multi-dimensional vectors before training and recognition can be carried out. Learned patterns in these networks are represented as multidimensional vectors of training weights, and the measure of similarity between the presented input and the learned pattern is based on some similarity metric. Three common similarity metrics in use in neural networks today are the Hamming distance, the vector dot product, and the Euclidean distance.

In the case of binary $\{0, 1\}$ vectors of length n, comparable measures of similarity between an input vector $\mathbf{x}$ and a weight vector $\mathbf{y}$ based on each of these metrics can be defined as

$$H := 1 - \frac{d}{n} \qquad \text{Hamming}$$

$$DP := \frac{\mathbf{x}^T \mathbf{y}}{|\mathbf{x}|} \qquad \text{Dot Product}$$

$$E := 1 - \sqrt{\frac{d}{n}} \qquad \text{Euclidean}$$

where d is the number of mismatched elements between $\mathbf{x}$ and $\mathbf{y}$, n is the number of total elements in $\mathbf{x}$ (i.e. the length of the vectors $\mathbf{x}$ and $\mathbf{y}$). $|\mathbf{x}|$ is the number of 1's in the vector $\mathbf{x}$.

Transforming a two-dimensional input pattern into a multi-dimensional vector and then comparing that vector to learned vectors can produce behaviour that is counter-intuitive. Two-dimensional input patterns that appear very similar (to the human eye) to a particular learned pattern can generate poor results when compared to that pattern using these metrics.

Another in many cases more useful metric is the Hausdorff distance (Munkres [152]). The *Hausdorff distance* measures the extent to which each point of an input set lies near some point of a model set. Given two finite sets of vectors in the Euclidean space $\mathbb{R}^n$

$$A := \{\mathbf{x}_0, \mathbf{x}_1, \ldots, \mathbf{x}_{p-1}\}, \qquad B := \{\mathbf{y}_0, \mathbf{y}_1, \ldots, \mathbf{y}_{q-1}\}$$

the (undirected) Hausdorff distance is defined as

$$H(A, B) := \max\{\, h(A, B)\, ,\, h(B, A)\,\}$$

where the function $h(A, B)$ defines the *directed Hausdorff distance* from A to B

$$h(A, B) := \max_{\mathbf{x} \in A} \left\{ \min_{\mathbf{y} \in B} \{\|\mathbf{x} - \mathbf{y}\|\} \right\}.$$

Thus

$$h(B, A) := \max_{\mathbf{y} \in B} \left\{ \min_{\mathbf{x} \in A} \{\|\mathbf{y} - \mathbf{x}\|\} \right\}$$

where $\|\mathbf{x} - \mathbf{y}\|$ is a norm in $\mathbb{R}^n$, for example the Euclidean norm. The directed Hausdorff distance identifies that point in A that is furthest from any point in B and measures the distance from that point to its nearest neighbour in B. If $h(A, B) = d$, all points in A are within distance d of some point in B. The (undirected) Hausdorff distance, then, is the maximum of the two directed distances between two point sets A and B so that if the Hausdorff distance is d, then all points of set A are within distance d of some point in set B and vice versa.

The *pointwise Hausdorff distance* for a vector $\mathbf{x} \in A$ is defined as

$$h(\mathbf{x}, B) := \min_{\mathbf{y} \in B} \{\|\mathbf{x} - \mathbf{y}\|\}.$$

The (undirected) Hausdorff distance has the following properties. It is a metric over the set of all closed, bounded sets. It is everywhere non-negative and it obeys the properties of identity, symmetry, and triangle inequality. In the context of pattern recognition this means that a shape is identical only to itself, that the order of comparison does not matter, and that if two shapes are highly dissimilar they cannot both be similar to some third shape. This final property (triangle inequality) is particularly important for reliable pattern classification.

In the C++ program Hausdorff.cpp we calculate the Hausdorff distance for the two sets

$$X = \{ (0.0, 0.2),\ (0.5, 0.8),\ (0.3, 0.7) \}, \qquad Y = \{ (1.5, 2.1),\ (0.8, 1.5) \}$$

in $\mathbb{R}^2$.

```
// Hausdorff.cpp

#include <iostream>
#include <cmath>        // for sqrt
using namespace std;

// Euclidean distance between to n-dimensional vectors
double distance(double* x,double* y,int n)
{
 double result = 0.0;
 for(int j=0;j<n;j++) { result += (x[j]-y[j])*(x[j]-y[j]); }
 return sqrt(result);
}

double dhausdorff(double** x,double** y,int p,int q,int n)
{
 double max = 0.0;
 double min, temp;
 for(int i=0;i<p;i++) { min = distance(x[i],y[0],n);
 for(int j=0;j<q;j++) { temp = distance(x[i],y[j],n);
 if(temp < min) min = temp;
 }
 if(max < min) max = min;
 }
 return max;
}

int main(void)
{
 int n = 2;   // dimension of Euclidean space
 int p = 3;   // number of vectors in set A
 int q = 2;   // number of vectors in set B
 double** x = NULL; x = new double*[p];
 for(int j=0;j<p;j++) x[j] = new double[n];
 x[0][0] = 0.0; x[0][1] = 0.2;
 x[1][0] = 0.5; x[1][1] = 0.8;
 x[2][0] = 0.3; x[2][1] = 0.7;
 double** y = NULL; y = new double*[q];
 for(int j=0;j<q;j++) y[j] = new double[n];
```

```
   y[0][0] = 1.5; y[0][1] = 2.1; y[1][0] = 0.8; y[1][1] = 1.5;
   double dhd = dhausdorff(x,y,p,q,n);
   cout << "Hausdorff distance = " << dhd << endl;
   for(int j=0;j<p;j++) { delete[] x[j]; } delete[] x;
   for(int j=0;j<q;j++) { delete[] y[j]; } delete[] y;
   return 0;
}
```

As described above, the directed Hausdorff distance $h(A,B)$ identifies the point $a \in A$ that is farthest from any point of B, and measures the distance from a to its nearest neighbour in B. Thus the Hausdorff distance $H(A,B)$ measures the degree of mismatch between to sets, as it reflects the distance of the point of A that is farthest from any point of B and vice versa. The Hausdorff distance is very sensitive to even a single "outlying" point of A and B. For example, consider $B = A \cup \{x\}$, where the point x is some large distance D from any point of A. In this case $H(A,B) = D$. This means it is determined solely by the point x.

Sometimes one uses a generalization of the Hausdorff distance which does not obey the metric properties on A and B, but does obey them on specfic subsets of A and B. This generalized Hausdorff measure is given by taking the k-th ranked distance rather than the maximum, or largest ranked one,

$$h_k(A,B) = k\text{th} \min_{a \in A\, b \in B} \|a - b\|$$

where kth denotes the kth ranked value (or equivalently the quantile of m values). For example, when $k = m$ then kth is max, and thus this measure is the same as $h(\cdot,\cdot)$. When $k = m/2$ then the median of the m individual point distances determines the overall distance. Therefore this measure generalizes the directed Hausdorff measure, by replacing the maximum with a quantile.

Another useful metric for two-dimensional patterns is as follows. Let A, B be two $m \times n$ matrices over $\mathbb{R}$. We can introduce a scalar product

$$\langle A, B \rangle := \text{tr}(AB^T)$$

where ^{T} denotes the transpose. Note that AB^T is an $m \times m$ matrix. The scalar product induces the norm

$$\|A\|^2 = \langle A, A \rangle = \text{tr}(AA^T).$$

Thus we have the metric

$$\|A - B\|^2 = \langle A - B, A - B \rangle = \langle A, A \rangle - \langle A, B \rangle - \langle B, A \rangle + \langle B, B \rangle.$$

Thus

$$\|A - B\|^2 = \text{tr}(AA^T - AB^T - BA^T + BB^T).$$

Since the trace is linear we have

$$\|A - B\|^2 = \mathrm{tr}(AA^T) - 2\mathrm{tr}(AB^T) + \mathrm{tr}(BB^T)$$

where we used that $\mathrm{tr}(AB^T) = \mathrm{tr}(BA^T)$. Finally

$$\|A - B\|^2 = \sum_{i=1}^{m}\sum_{j=1}^{n} a_{ij}^2 - 2\sum_{i=1}^{m}\sum_{j=1}^{n} a_{ij}b_{ij} + \sum_{i=1}^{m}\sum_{j=1}^{n} b_{ij}^2 .$$

13.4 Kohonen Network

13.4.1 Introduction

A feature of neural systems is the ability to produce self-organizing, topology preserving mappings of any given feature space, e.g., the surface of a body. Self-organization means that mapping onto a cortex of neurons is originated only by stochastically presenting the neurons with input features. The mapping is called topology preserving when neighbouring neurons represent neighbouring regions in the feature space. In order to profit best from the given number of neurons, important regions of the feature space (high density of presentation) are represented by more neurons, i.e., the resolution of the mapping is optimized.

The self-organizing neural network (also called Kohonen's feature map or topology-preserving map) (Kohonen [118], Kohonen [119]) assumes a topological structure among the cluster units. This property is observed in the brain, but it is not found in other artificial networks. There are p cluster units, arranged in a one-or two-dimensional array. The input signals $\mathbf{x}_k$ are vectors. The weight vector $\mathbf{w}$ for a cluster unit serves as an exemplar of the input patterns associated with that cluster. During the self-organizing process the cluster unit whose weight vector matches the input pattern most closely (typically the square of the minimum Euclidean distance) is chosen as the winner. The winning unit and its neighbour units (in terms of the topology of the cluster units) update their weights. The weight vectors $\mathbf{w}_j$ of neighbouring units are not, in general, close to the input pattern. For example, for a linear array of cluster units, the neighbourhood of radius r around cluster unit j^* consists of all

$$\max(1, j^* - r) \le j \le \min(j^* + r, p) .$$

When inputs match the node vectors, that area of the map is selectively optimized to represent an average of the training data for that class. From a randomly organized set of nodes the grid settles into a feature map that has local representation and is self-organised. Self-organizing feature maps capture the topology and probability distribution of input data. Mappings from higher to lower dimensions are also possible with a self-organizing feature map and are in general useful for dimensionality of input data.

The topology-preserving maps can be used, for example, for pattern recognition and for finding solutions to the traveling salesman problem.

The Kohonen's feature map allows us to define a criterion function. The Kohonen's rule is a stochastic gradient-descent search that leads, on average and for small learning rates η, to a local minimum of the criterion function.

13.4.2 Kohonen Algorithm

T. Kohonen (Kohonen [118], Kohonen [119]) proposed an algorithm producing the desired mapping. A neuron is identified with a prototype $\mathbf{w}_i$ $(i = 0, 1, \ldots, n-1)$ in the feature space. When all neurons are presented with an input feature, $\mathbf{x}$, a *winner neuron* is determined. To this end, a metric in feature space has to be defined. Taking this metric, the prototype $\mathbf{w}_{j^*}$ of the winner neuron j^* has the smallest distance in the feature space compared to the other neurons i

$$\|\mathbf{w}_{j^*} - \mathbf{x}\| \leq \|\mathbf{w}_i - \mathbf{x}\| \quad \text{for all} \quad i\,.$$

Thus it represents best the input feature $\mathbf{x}$. The main item of Kohonen's algorithm is the learning step, which improves the mapping of the feature space onto the cortex of neurons. The neurons in the cortical neighbourhood of the winner neuron j^* adapt their prototypes in the direction of the input feature $\mathbf{x}$

$$\mathbf{w}_i(t+1) = \mathbf{w}_i(t) + \eta(t)h(d_{ij^*}(t))(\mathbf{x} - \mathbf{w}_i(t)), \qquad t = 0, 1, 2, \ldots$$

where $d_{ij^*}(t)$ denotes the cortical distance of neuron i to the winner neuron j^*. The function $h(d_{ij^*}(t))$ describes the lateral interaction of neurons. Often a *Gaussian function*

$$h(x) = \exp(-x^2/\sigma^2)$$

is used. Another option is $h(x) = \exp(-|x|/\sigma)$. One also often assumes that

$$h(0) = 1 \quad \text{and} \quad h(d < d_{\max}(t)) = 0$$

where $h(d_{ij^*}(t))$ is monotonous in between. With increasing distance from the winner neuron, the relative amount of the learning step decreases. $d_{\max}$ is the maximum distance of interaction. Thus $h(d_{ij^*}(t))$ is the neighbourhood function. This function is 1 for $i = j^*$ and falls off with distance $|\mathbf{r} - \mathbf{r}_j^*|$. Thus units close to the winner, as well as the winner j^* itself, have their weights changed appreciably, while those further away, where d_{ij^*} is small, experience little effect.

The time-dependent parameter η,

$$0 \leq \eta(t) \leq 1$$

controls the absolute amount of a *learning step* (*learning rate*). The learning rate decreases with $t \to \infty$. The time-dependent parameters $\eta(t)$ and $d_{\max}(t)$ denote the plasticity of the neural network. By repeating the whole algorithm (iterating $\mathbf{w}_i(t)$), a mapping with the desired properties will result if the time-dependent parameters $\eta(t)$ and $d_{\max}(t)$ are successively decreasing as $t \to \infty$. Thus the term $\eta(t)$ is a gain term $(0 < \eta(t) < 1)$ that decreases in time, thus slowing the weight adaption. Notice that the neighbourhood $N_{j^*}(t)$ decreases in size as time goes on, thus localising the area of maximum activity. Different neighbourhoods can be chosen, for example a rectangular grid or a hexagonal grid.

If we have more than one input pattern

$$\mathbf{x}_0, \quad \mathbf{x}_1, \quad \dots \quad , \mathbf{x}_{m-1}$$

we proceed as follows. We can write the set of weights $\mathbf{w}_i$ $(i = 0, 1, \dots, n-1)$ as an $n \times N$ matrix

$$W = (\mathbf{w}_0, \mathbf{w}_1, \dots, \mathbf{w}_{n-1})$$

where N is the length of the vectors $\mathbf{x}_i$ and $\mathbf{w}_i$.

1. Initialise network. Define $\mathbf{w}_i(t=0)$ $(0 \le i \le n-1)$ to be the weights. Initialise weights to small random values. Set the initial radius of the neighbourhood around node i, $N_i(0)$, to be large.

2. Present input. Present input patterns $\mathbf{x}_0$, $\mathbf{x}_1$, ..., $\mathbf{x}_{m-1}$ to the network.

3. Calculate distances. For $i = 0, 1, \dots, n-1$ calculate

$$\|\mathbf{w}_i - \mathbf{x}_0\| \,.$$

4. Find the winning neuron j^*, i.e.

$$\|\mathbf{w}_{j^*} - \mathbf{x}_0\| \le \|\mathbf{w}_i - \mathbf{x}_0\|, \qquad i = 0, 1, \dots, n-1$$

5. Update weights. Update the weight according to

$$\mathbf{w}_i(t=1,0) = \mathbf{w}_i(t=0,0) + \eta(t=0)h(d_{ij^*}(t=0))(\mathbf{x}_0 - \mathbf{w}_i(t=0,0))$$

for $i = 0, 1, \dots, n-1$. Thus the biggest change is at the winning neuron $(i = j^*)$ since the function h has the largest value at $i = j^*$. The quantity $\eta(t = 0)$ is the learning rate at $t = 0$.

6. Goto step 3 and repeat the procedure for the input pattern $\mathbf{x}_1$, $\mathbf{x}_2$, ..., $\mathbf{x}_{m-1}$. For each input pattern the matrix W is updated according to the formula given above. This is one epoch. Goto step 3 again and repeat for the next time step, where the learning rate $\eta(t)$ is decreased.

A stopping condition is that no noticeable changes in the feature map are observed.

Let m be the number of input vectors and n be the number of neurons. One can define a *criterion function*

$$J(\mathbf{w}) := \frac{1}{2}\sum_{k=0}^{m-1}\sum_{i=0}^{n-1} h(\mathbf{r}_i - \mathbf{r}_{j^*})\|\mathbf{x}_k - \mathbf{w}_i\|^2$$

where j^* is the label of the winner unit upon presentation of the input vector $\mathbf{x}_k$, and $h(\mathbf{r}_i - \mathbf{r}_{j^*})$ is the neighbourhood function. The criterion function is an extension of the competitive learning criterion function. Performing gradient descent for the criterion function J

$$\Delta\mathbf{w}_i := -\eta\nabla J(\mathbf{w}) \equiv -\eta \begin{pmatrix} \partial J/\partial w_{i0} \\ \vdots \\ \partial J/\partial w_{iN-1} \end{pmatrix}$$

yields

$$\Delta\mathbf{w}_i = \eta\sum_{k=0}^{m-1} h(\mathbf{r}_i - \mathbf{r}_{j^*})(\mathbf{x}_k - \mathbf{w}_i)$$

which is the batch-mode version of Kohonen's self-organizing rule. The Kohonen's rule is a stochastic gradient-descent search that leads, on average and for small learning rate η, to a local minimum of J. These minima are given as solutions to the system of nonlinear equations

$$\Delta\mathbf{w}_i = \sum_{k=0}^{m-1} h(\mathbf{r}_i - \mathbf{r}_{j^*})(\mathbf{x}_k - \mathbf{w}_i) = \mathbf{0}.$$

The solution depends on the choice of the neighbourhood function h. What is desired is the global minimum of the criterion function J. Local minima of J are topological defects like kinks in one-dimensional maps and twists in two-dimensional maps.

13.4.3 Kohonen Example

In the example we follow Fausett [57]. We have four input vectors and two weight vectors each of length 4. In the C++ program we can select different numbers of weight vectors. The counting for all the vectors and matrices and their elements starts from zero. The four input vectors (input patterns) are given by

$$\mathbf{x}_0 = \begin{pmatrix} 1.0 \\ 1.0 \\ 0.0 \\ 0.0 \end{pmatrix}, \quad \mathbf{x}_1 = \begin{pmatrix} 0.0 \\ 0.0 \\ 0.0 \\ 1.0 \end{pmatrix}, \quad \mathbf{x}_2 = \begin{pmatrix} 1.0 \\ 0.0 \\ 0.0 \\ 0.0 \end{pmatrix}, \quad \mathbf{x}_3 = \begin{pmatrix} 0.0 \\ 0.0 \\ 1.0 \\ 1.0 \end{pmatrix}.$$

The initial weight matrix at time $t = 0$ $W(t = 0)$ is given by

$$W(t=0) = \begin{pmatrix} 0.2 & 0.8 \\ 0.6 & 0.4 \\ 0.5 & 0.7 \\ 0.9 & 0.3 \end{pmatrix}.$$

Thus the initial weight matrix consists of two column vectors (two neurons)

$$\mathbf{w}_0(t=0) = \begin{pmatrix} 0.2 \\ 0.6 \\ 0.5 \\ 0.9 \end{pmatrix}, \qquad \mathbf{w}_1(t=0) = \begin{pmatrix} 0.8 \\ 0.4 \\ 0.7 \\ 0.3 \end{pmatrix}.$$

The initial value of the learning rate is $\eta(t=0) = 0.6$. The time evolution of the learning rate is

$$\eta(t+1) = \frac{1}{1.05}\eta(t).$$

With only two clusters (neurons) available only one cluster updates its weights at each step. First we calculate the square of the distance between the first input vector $\mathbf{x}_0$ and $\mathbf{w}_0(t=0)$ and the square of the distance between the first input vector $\mathbf{x}_0$ and $\mathbf{w}_1(t=0)$. We find

$$\begin{aligned} \|\mathbf{w}_0(t=0) - \mathbf{x}_0\|^2 &= (0.2-1)^2 + (0.6-1)^2 + (0.5-0)^2 + (0.9-0)^2 = 1.86 \\ \|\mathbf{w}_1(t=0) - \mathbf{x}_0\|^2 &= (0.8-1)^2 + (0.4-1)^2 + (0.7-0)^2 + (0.3-0)^2 = 0.98. \end{aligned}$$

Thus the vector $\mathbf{w}_1(t=0)$ is closest to the input vector $\mathbf{x}_0$, i.e. $j^* = 1$ (winning neuron). Consequently, we update the column $\mathbf{w}_1(t=0)$ in the weight matrix $W(t=0)$ and leave the column $\mathbf{w}_0(t=0)$ unchanged. Thus we find ($\eta = 0.6$)

$$\begin{aligned} W_{01}(t=1,0) &= W_{01}(t=0) + \eta(t=0)(x_{0,0} - W_{01}(t=0)) = 0.92 \\ W_{11}(t=1,0) &= W_{11}(t=0) + \eta(t=0)(x_{0,1} - W_{11}(t=0)) = 0.76 \\ W_{21}(t=1,0) &= W_{21}(t=0) + \eta(t=0)(x_{0,2} - W_{21}(t=0)) = 0.28 \\ W_{31}(t=1,0) &= W_{31}(t=0) + \eta(t=0)(x_{0,3} - W_{31}(t=0)) = 0.12. \end{aligned}$$

Thus the updated weight matrix $W(t=1,0)$ after taking into account the input vector $\mathbf{x}_0$ is given by

$$W(t=1,0) = \begin{pmatrix} 0.2 & 0.92 \\ 0.6 & 0.76 \\ 0.5 & 0.28 \\ 0.9 & 0.12 \end{pmatrix}.$$

Thus

$$\mathbf{w}_0(t=1,0) = \begin{pmatrix} 0.2 \\ 0.6 \\ 0.5 \\ 0.9 \end{pmatrix}, \qquad \mathbf{w}_1(t=1,0) = \begin{pmatrix} 0.92 \\ 0.76 \\ 0.28 \\ 0.12 \end{pmatrix}.$$

For the second input vector $\mathbf{x}_1$ we find

$$\begin{aligned} \|\mathbf{w}_0(t=1,0) - \mathbf{x}_1\|^2 &= (0.2)^2 + (0.6)^2 + (0.5)^2 + (0.9-1.0)^2 = 0.66 \\ \|\mathbf{w}_1(t=1,0) - \mathbf{x}_1\|^2 &= (0.92)^2 + (0.76)^2 + (0.28)^2 + (0.12-1.0)^2 = 2.2768. \end{aligned}$$

Thus the vector $\mathbf{w}_0(t=1,0)$ is closest to the input vector $\mathbf{x}_1$, i.e., $j^*=0$ (winning neuron). Consequently, we update the column $\mathbf{w}_0(t=1,0)$ in the weight matrix $W(t=1,0)$ and leave the column $\mathbf{w}_1(t=1,0)$ unchanged. Thus we find

$$\begin{aligned}
W_{00}(t=1,1) &= W_{00}(t=1,0)+\eta(t=0)(x_{1,0}-W_{00}(t=1,0)) = 0.08\\
W_{10}(t=1,1) &= W_{10}(t=1,0)+\eta(t=0)(x_{1,1}-W_{10}(t=1,0)) = 0.24\\
W_{20}(t=1,1) &= W_{20}(t=1,0)+\eta(t=0)(x_{1,2}-W_{20}(t=1,0)) = 0.20\\
W_{30}(t=1,1) &= W_{30}(t=1,0)+\eta(t=0)(x_{1,3}-W_{30}(t=1,0)) = 0.96
\end{aligned}$$

where $\eta(t=0)=0.6$. Thus the updated weight matrix after taking into account the first input vector $\mathbf{x}_1$ is given by

$$W(t=1,1)=\begin{pmatrix}0.08 & 0.92\\ 0.24 & 0.76\\ 0.20 & 0.28\\ 0.96 & 0.12\end{pmatrix}.$$

Thus

$$\mathbf{w}_0(t=1,1)=\begin{pmatrix}0.08\\ 0.24\\ 0.20\\ 0.96\end{pmatrix}, \qquad \mathbf{w}_1(t=1,1)=\begin{pmatrix}0.92\\ 0.76\\ 0.28\\ 0.12\end{pmatrix}.$$

For the third input vector $\mathbf{x}_2$ we find

$$\|\mathbf{w}_0(t=1,1)-\mathbf{x}_2\|^2 = 1.8656, \qquad \|\mathbf{w}_1(t=1,1)-\mathbf{x}_2\|^2 = 0.6768.$$

Thus the vector $\mathbf{w}_1(t=1,1)$ is closest to the input vector $\mathbf{x}_2$, i.e. $j^*=1$ (winning neuron). Consequently, we update the column $\mathbf{w}_1(t=1,1)$ in the weight matrix $W(t=1,1)$ and leave the column $\mathbf{w}_0(t=1,1)$ unchanged. It follows that

$$W(t=1,2)=\begin{pmatrix}0.08 & 0.968\\ 0.24 & 0.304\\ 0.20 & 0.112\\ 0.96 & 0.048\end{pmatrix}.$$

Hence

$$\mathbf{w}_0(t=1,2)=\begin{pmatrix}0.08\\ 0.24\\ 0.20\\ 0.96\end{pmatrix}, \qquad \mathbf{w}_1(t=1,2)=\begin{pmatrix}0.968\\ 0.304\\ 0.112\\ 0.048\end{pmatrix}.$$

For the fourth input vector $\mathbf{x}_3$ we find

$$\|\mathbf{w}_0(t=1,2)-\mathbf{x}_3\|^2 = 0.7056, \qquad \|\mathbf{w}_1(t=1,2)-\mathbf{x}_3\|^2 = 2.724.$$

The vector $\mathbf{w}_0(t=1,2)$ is closest to the input vector $\mathbf{x}_3$, i.e. $j^*=0$ (winning neuron). Consequently, we update the column $\mathbf{w}_0(t=1,2)$ in the weight matrix W

and leave the column $\mathbf{w}_1(t = 1, 2)$ unchanged. All four input vectors have now been taken into account. We find

$$W(t=1) = \begin{pmatrix} 0.032 & 0.968 \\ 0.096 & 0.304 \\ 0.680 & 0.112 \\ 0.984 & 0.048 \end{pmatrix}$$

with

$$\mathbf{w}_0(t=1) = \begin{pmatrix} 0.032 \\ 0.096 \\ 0.680 \\ 0.984 \end{pmatrix}, \qquad \mathbf{w}_1(t=1) = \begin{pmatrix} 0.968 \\ 0.304 \\ 0.112 \\ 0.048 \end{pmatrix}.$$

This was an epoch (one iteration, one time step). Now reduce the learning to

$$\eta(t=1) = \frac{1}{1.05}\eta(t=0).$$

We repeat the steps again for each of the input vectors $\mathbf{x}_0$, $\mathbf{x}_1$, $\mathbf{x}_2$ and $\mathbf{x}_3$. After the second epoch (second iteration) the weight matrix is

$$W(t=2) = \begin{pmatrix} 0.016 & 0.980 \\ 0.047 & 0.360 \\ 0.630 & 0.055 \\ 0.999 & 0.024 \end{pmatrix}.$$

Iterating over 100 epochs we find that the weight matrix converges to the matrix

$$W(t=100) \approx \begin{pmatrix} 0.0 & 1.0 \\ 0.0 & 0.5 \\ 0.5 & 0.0 \\ 1.0 & 0.0 \end{pmatrix}.$$

The zeroth column is the arithmetic average of the two vectors $\mathbf{x}_1$ and $\mathbf{x}_3$ and the first column is the arithmetic average of the vectors $\mathbf{x}_0$ and $\mathbf{x}_2$.

In the C++ program we can select different numbers of weight vectors. We consider the case 1, 2, 3, 4 and 5. The weight vectors at $t = 0$ are selected at random using a random number generators. The number of input vectors can also be easily changed. In the program `const int m = 4` is the number of input vectors.

```
// Fausett.cpp

#include <iostream>
#include <cstdlib>     // srand, rand
#include <ctime>
using namespace std;

double euclidean(double *vec1,double *vec2,int n)
```

```
{
 double dist = 0.0;
 for(int i=0;i<n;i++) dist += (vec1[i]-vec2[i])*(vec1[i]-vec2[i]);
 return dist;
}

double distance(int i,int jstar)
{
 return double(i!=jstar);
 // returns 1.0 if i!=jstar, returns 0.0 if i==jstar
}

double h(double d) { return 1.0-d; }

void train(double **W,int n,int cols,double *vec,double rate)
{
 int i,j;
 int win = 0;
 double windist = euclidean(W[0],vec,n),edist;
 for(i=0;i<cols;i++)
 if((edist=euclidean(W[i],vec,n)) < windist) { win = i; windist = edist; }
 for(i=0;i<cols;i++)
  for(j=0;j<n;j++)
   W[i][j] += rate*h(distance(i,win))*(vec[j]-W[i][j]);
}

int main(void)
{
 int i, j;
 int T = 10000;     // number of iterations
 double eta = 0.6;  // learning rate
 const int m = 4;
 int cols;
 // training vectors
 double x0[m] = { 1.0,1.0,0.0,0.0 }; double x1[m] = { 0.0,0.0,0.0,1.0 };
 double x2[m] = { 1.0,0.0,0.0,0.0 }; double x3[m] = { 0.0,0.0,1.0,1.0 };
 cout << "Enter number of columns for weight matrix: ";
 cin >> cols;
 double** W = NULL; W = new double*[cols];
 for(i=0;i<cols;i++) W[i] = new double[m];

 srand(time(NULL));
 for(i=0;i<cols;i++)
  for(j=0;j<m;j++) W[i][j] = rand()/double(RAND_MAX);

 for(i=0;i<T;i++)
 {
 train(W,m,cols,x0,eta); train(W,m,cols,x1,eta);
```

```
 train(W,m,cols,x2,eta); train(W,m,cols,x3,eta);
 eta /= 1.05;    // learning rate decreased
 }
 for(i=0;i<cols;i++)
 {
 cout << "W[" << i << "]= [";
 for(j=0;j<m;j++)  cout << W[i][j] << " ";
 cout << "]" << endl;
 }
 for(i=0;i<cols;i++) delete[] W[i]; delete[] W;
 return 0;
}
```

For the output of the program we must keep in mind that it is sensitive to the initial conditions for the weight vectors which we selected with a random number generator. Furthermore the result is sensitive to the initial value and decrement of the learning rate η. A typical output is

```
Enter number of columns for weight matrix : 1
W[0]= [0.496931  0.245436  0.25464  0.503069]
// average of x0,x1,x2,x3

Enter number of columns for weight matrix: 2
W[0]= [1  0.493825  4.36192e-12  1.17225e-11]
// average of x0,x2
W[1]= [5.91835e-12  7.1785e-15  0.506175  1]
// average of x1,x3

Enter number of columns for weight matrix: 3
W[0]= [7.88199e-12  1.60492e-11  0.506175  1]
W[1]= [0.999996  1.26278e-06  9.35514e-07  8.32493e-07]
W[2]= [0.999999  0.999999  3.48972e-06  2.59697e-06]

Enter number of columns for weight matrix: 4
W[0]= [0.999997  1  5.9738e-07  1.93129e-06]                 // x0
W[1]= [1.74599e-06  2.94386e-06  4.57243e-07  0.999998]  // x1
W[2]= [3.24247e-06  2.76703e-06  0.999999  0.999999]        // x3
W[3]= [0.999999  1.4358e-06  1.08384e-07  4.02981e-06]     // x2

Enter number of columns for weight matrix: 5
W[0]= [1  0.493825  6.02668e-12  1.20011e-12]
// average of x0,x2
W[1]= [0.799493  0.979553  0.88229  0.325358]
W[2]= [2.78495e-06  1.26278e-06  1.37201e-06  1]
// x1
W[3]= [2.09753e-06  1.48308e-06  0.999999  0.999997]
// x3
W[4]= [0.737724  0.589648  0.57268  0.307535]
```

13.4.4 Traveling Salesman Problem

The *traveling salesman problem* is perhaps the most famous problem in all network and combinatorial optimization. Starting from his home base, node 0, a salesman wishes to visit each of several cities, represented by nodes $1, \ldots, n-1$, exactly once and return home, doing so at the lowest possible travel cost. We refer to any feasible solution to this problem as a tour (of the cities). There are $n!$ possible tours if there are n cities, but some of them are the same. The traveling salesman problem is a generic core model that captures the combinatorial essence of most routing problems and, indeed, most other routing problems are extensions of it.

We present the traveling salesman problem as an embedded (directed) network flow structure (Ahuja et al [2]). Let c_{ij} denote the cost of traveling from city i to city j and let y_{ij} be a zero-one variable, indicating whether or not the salesman travels from city i to city j. Moreover, let us define flow variables x_{ij} on each arc (i,j) and assume that the salesman has $n-1$ units available at node 1, which we arbitrarily select as a "source node", and that he must deliver 1 unit to each of the other nodes. Then the model is

$$\text{Minimize} \sum_{(i,j)\in A} c_{ij} y_{ij} \qquad (a)$$

subject to

$$\sum_{0\le j\le n-1} y_{ij} = 1 \qquad \text{for all } i = 0,1,\ldots,n-1 \qquad (b)$$

$$\sum_{0\le i\le n-1} y_{ij} = 1 \qquad \text{for all } j = 0,1,\ldots,n-1 \qquad (c)$$

$$\mathcal{N}x = b \qquad (d)$$

$$x_{ij} \le (n-1)y_{ij} \quad \text{for all } (i,j)\in A \qquad (e)$$

$$x_{ij} \ge 0 \quad \text{for all } (i,j)\in A \qquad (f)$$

$$y_{ij} = 0 \text{ or } 1 \quad \text{for all } (i,j)\in A. \qquad (g)$$

We solve the traveling salesman problem using the Kohonen algorithm. The positions of the cities are given by

$$x[0]=10.1, \quad y[0]=20.2, \qquad x[1]=4.0, \quad y[1]=7.0$$
$$x[2]=0.1, \quad y[2]=2.0, \qquad x[3]=0.5, \quad y[3]=10.7$$
$$x[4]=10.5, \quad y[4]=0.6, \qquad x[5]=8.0, \quad y[5]=10.0$$
$$x[6]=12.0, \quad y[6]=16.0, \qquad x[7]=18.0, \quad y[7]=8.0$$

The distance is given by the Euclidian norm. The output of the program `kohonen.cpp` using the Kohonen algorithm provides the following solution to the traveling salesman problem

$$0 \to 3 \to 5 \to 1 \to 2 \to 4 \to 7 \to 6 \to 0\,.$$

The C++ program for the traveling salesman problem using the Kohonen feature map is as follows.

```
// kohonen.cpp

#include <iostream>
#include <cmath>      // for fabs, exp
#include <cstdlib>
#include <ctime>
#include <iomanip>    // for setprecision
using namespace std;

// find the position in the array d
// which possesses the smallest number
int minimum(double* d,int M)
{
 int m = 0;
 double r = d[0];
 for(int j=1;j<M;j++) { if(d[j] < r) { r = d[j]; m = j; } }
 return m;
}

// lateral interaction of neurons
double map(double a,double b,double c,double d,double s)
{
 double result;
 result = exp(-(fabs(a-c)+fabs(b-d))/(2.0*s*s));
 return result;
}

int main(void)
{
 int N = 8;  // number of cities
 double* x = new double[N]; double* y = new double[N];
 // coordinates of cities
 x[0] = 10.1; x[1] = 4.0;  x[2] = 0.1;  x[3] = 0.5;
 x[4] = 10.5; x[5] = 8.0;  x[6] = 12.0; x[7] = 18.0;
 y[0] = 20.2; y[1] = 7.0;  y[2] = 2.0;  y[3] = 10.7;
 y[4] = 0.6;  y[5] = 10.0; y[6] = 16.0; y[7] = 8.0;

 int M = 3*N;                        // M number of neurons
 double* u = new double[M]; double* v = new double[M];
 const double PI = 3.14159;

 // neuron pattern at t = 0
 for(int j=0;j<M;j++)
 {
 double k = j; double K = M;
 u[j] = 10.0+10.0*sin(2.0*PI*k/K);    // neuron pattern
 v[j] = 10.0+10.0*cos(2.0*PI*k/K);    // neuron pattern
 }
```

```
double* dist = new double[M];
double* delW1 = new double[M];
double* delW2 = new double[M];
double eta0 = 0.8;    // learning rate at t = 0
long T = 15000;       // number of iterations
srand((unsigned long) time(NULL)); // seed for random numbers

// iteration starts here
for(long cnt=0;cnt<T;cnt++)
{
int nrand = rand()%N;
for(j= 0;j<M;j++)
{ dist[j] = fabs(u[j]-x[nrand])+fabs(v[j]-y[nrand]); }

int imin = minimum(dist,M);
double c = cnt;  double it = T;
double s;
s = eta0*(1.0-c/(it+10000.0));

for(j=0;j<M;j++)
{
delW1[j] = 1.0*map(u[j],v[j],u[imin],v[imin],s)*(x[nrand]-u[j]);
delW2[j] = 1.0*map(u[j],v[j],u[imin],v[imin],s)*(y[nrand]-v[j]);
}
for(j=0;j<M;j++) { u[j] += delW1[j]; v[j] += delW2[j]; }
}  // end iteration
cout << endl << endl;
for(j=0;j<M;j++) // display the output
{
cout << "u[" << j << "] = " << setprecision(3) << u[j] << "     ";
cout << "v[" << j << "] = " << setprecision(3) << v[j] << endl;
}
delete[] x; delete[] y; delete[] u; delete[] v;
delete[] delW1; delete[] delW2; delete[] dist;
return 0;
}
```

The output is sensitive to the initial learning rate $\eta(0)$ and the decrease of the learning rate at each time step. Furthermore the output is sensitive to the location of the neurons at time $t = 0$ and the number of neurons. The number of neurons should be chosen between

$$2N \leq M \leq 4N\,.$$

In the program we selected the neurons lying on a circle with centre $(10, 10)$ and radius 10. The center $\mathbf{c}$ could also be selected using the formula

$$\mathbf{c} = \frac{1}{N}\sum_{j=0}^{N-1}\begin{pmatrix} x_j \\ y_j \end{pmatrix}.$$

In chapter 14 we solve the traveling salesman problem using genetic algorithms.

13.5 Perceptron

13.5.1 Introduction

The *perceptron* (Minskey and Papert [148], Fausett [57], Haykin [90], Cichocki and Unbehauen [35], Hassoun [89], Rojas [173]) is the simplest form of a neural network used for the classification of special types of patterns said to be linearly separable (i.e. patterns that lie on opposite sides of a hyperplane). It consists of a single neuron with adjustable synaptic weights w_i and threshold θ.

Definition. A perceptron is a computing unit with threshold θ which, when receiving the n real inputs x_1, x_2, ..., x_n through edges with the associated weights w_1, w_2, ..., w_n, outputs 1 if the inequality

$$\sum_{j=1}^{n} w_j x_j \geq \theta$$

holds otherwise it outputs zero.

The origin of the inputs is not important irrespective of whether they come from other perceptrons or another class of computing units. The geometric interpretation of the processing performed by perceptrons is the same as with McCulloch-Pitts elements. A perceptron separates the input space into two half-spaces. For points belonging to one half-space the result of the computation is 0, for points belonging to the other it is 1. We can also formulate this definition using the Heaviside step function

$$H(x) = \begin{cases} 1 \text{ for } x \geq 0 \\ 0 \text{ for } x < 0 \end{cases} .$$

Thus

$$H(\sum_{j=1}^{n} w_j x_j - \theta) = \begin{cases} 1 \text{ for } (\sum_{j=1}^{n} w_j x_j - \theta) \geq 0 \\ 0 \text{ for } (\sum_{j=1}^{n} w_j x_j - \theta) < 0 \end{cases}$$

With w_1, w_2, ..., w_n and θ given, the equation

$$\sum_{j=1}^{n} w_j x_j = \theta$$

defines a hyperplane which divides the Euclidean space $\mathbb{R}^n$ into two half spaces.

Example. The plane

$$x_1 + 2x_2 - 3x_3 = 4$$

which divides $\mathbb{R}^3$ into two half-spaces. ♣

In many cases it is more convenient to deal with perceptrons of threshold zero only. This corresponds to linear separations which are forced to go through the origin of the input space. The threshold of the perceptron with a threshold has been converted into the weight $-\theta$ of an additional input channel connected to the constant 1. This extra weight connected to a constant is called the *bias* of the element. Thus the input vector $(x_1, x_2, \ldots, x_n)$ must be extended with an additional 1 and the resulting $(n+1)$-dimensional vector

$$(1, x_1, x_2, \ldots, x_n)$$

is called the *extended input vector*, where $x_0 = 1$. The extended weight vector associated with this perceptron is

$$(w_0, w_1, \ldots, w_n)$$

whereby $w_0 = -\theta$.

The threshold computation of a perceptron will be expressed using scalar products. The test computed by the perceptron is thus

$$\mathbf{w}^T\mathbf{x} \geq \theta$$

if $\mathbf{w}$ and $\mathbf{x}$ are the weight and input vectors, and

$$\mathbf{w}^T\mathbf{x} \geq 0$$

if $\mathbf{w}$ and $\mathbf{x}$ are the extended weight and input vectors.

Example. If we are looking for the weights and threshold needed to implement the AND function with a perceptron, the input vectors and their associated outputs are

$$(0,0) \mapsto 0, \qquad (0,1) \mapsto 0, \qquad (1,0) \mapsto 0, \qquad (1,1) \mapsto 1.$$

If a perceptron with threshold zero is used, the input vectors must be extended and the desired mappings are

$$(1,0,0) \mapsto 0, \qquad (1,0,1) \mapsto 0, \qquad (1,1,0) \mapsto 0, \qquad (1,1,1) \mapsto 1.$$

A perceptron with three still unknown weights (w_0, w_1, w_2) can carry out this task.

Example. The *AND gate* can be simulated using the perceptron. The AND gate is given by

0	0	0
0	1	0
1	0	0
1	1	1

Thus the input patterns are

$$\mathbf{x}_0 = \begin{pmatrix} 0 \\ 0 \end{pmatrix}, \qquad \mathbf{x}_1 = \begin{pmatrix} 0 \\ 1 \end{pmatrix}, \qquad \mathbf{x}_2 = \begin{pmatrix} 1 \\ 0 \end{pmatrix}, \qquad \mathbf{x}_3 = \begin{pmatrix} 1 \\ 1 \end{pmatrix}$$

with the corresponding output patter $y_0 = 0$, $y_1 = 0$, $y_2 = 0$, $y_3 = 1$. Let

$$\mathbf{w} = \begin{pmatrix} 1 \\ 1 \end{pmatrix}, \qquad \theta = \frac{3}{2}.$$

Then $\mathbf{w}^T = (1, 1)$ and the evaluation of

$$H(\mathbf{w}^T \mathbf{x}_j - \theta)$$

for $j = 0, 1, 2, 3$ yields

$$\begin{aligned} H(\mathbf{w}^T\mathbf{x}_0 - \theta) &= H(0 - 3/2) = H(-3/2) = 0 \\ H(\mathbf{w}^T\mathbf{x}_1 - \theta) &= H(1 - 3/2) = H(-1/2) = 0 \\ H(\mathbf{w}^T\mathbf{x}_2 - \theta) &= H(1 - 3/2) = H(-1/2) = 0 \\ H(\mathbf{w}^T\mathbf{x}_3 - \theta) &= H(2 - 3/2) = H(1/2) = 1\,. \end{aligned}$$

13.5.2 Boolean Functions

We consider the problem of determining which logical functions can be implemented with a single perceptron. A perceptron network is capable of computing any logical function since perceptrons are even more powerful than unweighted McCulloch-Pitts elements. If we reduce the network to a single element, which functions are still computable? Taking the boolean functions of two variables as an example we can gain some insight into this problem. Let $\wedge$ be the AND operation, $\vee$ the OR operation, $\oplus$ the XOR operation, and $\neg$ the NOT operation.

If one considers logical functions of two variables, there are four possible combinations for the input. The outputs for the four inputs are four bits which uniquely distinguish each logical function. We use the number defined by these four bits as a subindex for the name of the functions. The function $(x_1, x_2) \mapsto 0$, for example, is denoted by f_0 (since 0 corresponds to the bit string 0000). The AND function is denoted by f_8 (since 8 corresponds to the bit string 1000), whereby the output bits are ordered according to the ordering of the inputs: $(1,1)$, $(0,1)$, $(1,0)$, $(0,0)$.

The sixteen possible functions of two variables are thus

$$\begin{aligned} f_0(x_1, x_2) &= f_{0000}(x_1, x_2) = 0 \\ f_1(x_1, x_2) &= f_{0001}(x_1, x_2) = \neg(x_1 \vee x_2) \\ f_2(x_1, x_2) &= f_{0010}(x_1, x_2) = x_1 \wedge \neg x_2 \\ f_3(x_1, x_2) &= f_{0011}(x_1, x_2) = \neg x_2 \\ f_4(x_1, x_2) &= f_{0100}(x_1, x_2) = \neg x_1 \wedge x_2 \end{aligned}$$

$$\begin{aligned}
f_5(x_1,x_2) &= f_{0101}(x_1,x_2) = \neg x_1 \\
f_6(x_1,x_2) &= f_{0110}(x_1,x_2) = x_1 \oplus x_2 \\
f_7(x_1,x_2) &= f_{0111}(x_1,x_2) = \neg(x_1 \wedge x_2) \\
f_8(x_1,x_2) &= f_{1000}(x_1,x_2) = x_1 \wedge x_2 \\
f_9(x_1,x_2) &= f_{1001}(x_1,x_2) = \neg(x_1 \oplus x_2) \\
f_{10}(x_1,x_2) &= f_{1010}(x_1,x_2) = x_1 \\
f_{11}(x_1,x_2) &= f_{1011}(x_1,x_2) = x_1 \vee \neg x_2 \\
f_{12}(x_1,x_2) &= f_{1100}(x_1,x_2) = x_2 \\
f_{13}(x_1,x_2) &= f_{1101}(x_1,x_2) = \neg x_1 \vee x_2 \\
f_{14}(x_1,x_2) &= f_{1110}(x_1,x_2) = x_1 \vee x_2 \\
f_{15}(x_1,x_2) &= f_{1111}(x_1,x_2) = 1 .
\end{aligned}$$

The function f_0, for example, is the zero function whereas f_{14} is the inclusive OR-function. Perceptron-computable functions are those for which the points whose function value is 0 can be separated from the points whose function value is 1 using a line. For the AND function and OR function we can find such a separation. The NAND function which is a universal gate is given by the function f_7 and also can be separated. Two of the functions cannot be computed in this way. These are the function XOR (exclusive OR) (function f_6) and the function XNOR f_9. It is obvious that no line can produce the necessary separation of the input space. This can also be shown analytically.

Let w_1 and w_2 be the weights of a perceptron with two inputs, and θ its threshold. If the perceptron computes the XOR function the following four inequalities must be fulfilled

$$\begin{array}{llll}
x_1 = 0 & x_2 = 0 & w_1x_1 + w_2x_2 = 0 & \Rightarrow \quad 0 < \theta \\
x_1 = 1 & x_2 = 0 & w_1x_1 + w_2x_2 = w_1 & \Rightarrow \quad w_1 \geq \theta \\
x_1 = 0 & x_2 = 1 & w_1x_1 + w_2x_2 = w_2 & \Rightarrow \quad w_2 \geq \theta \\
x_1 = 1 & x_2 = 1 & w_1x_1 + w_2x_2 = w_1 + w_2 & \Rightarrow \quad w_1 + w_2 < \theta .
\end{array}$$

Since the threshold θ is positive, according to the first inequality, w_1 and w_2 are positive too, according to the second and third inequalities. Therefore the inequality

$$w_1 + w_2 < \theta$$

cannot be true. This contradiction implies that no perceptron capable of computing the XOR function exists. An analogous proof holds for the function f_9.

13.5.3 Linearly Separable Sets

Many other logical functions of several arguments must exist which cannot be computed with a threshold element. This fact has to do with the geometry of the n-dimensional hypercube whose vertices represent the combination of logic values of the arguments. Each logical function separates the vertices into two classes. If the points whose function value is 1 cannot be separated with a linear cut from the points whose function value is 0, the function is not perceptron-computable.

Definition. Two sets of points A and B in the n-dimensional space $\mathbb{R}^n$ are called *linearly separable* if $n + 1$ real numbers $w_0, w_1, \ldots, w_n$ exist, such that every point $(x_1, x_2, \ldots, x_n) \in A$ satisfies

$$\sum_{j=1}^{n} w_j x_j \geq w_0$$

and every point $(x_1, x_2, \ldots, x_n) \in B$ satisfies $\Sigma_{i=1}^{n} w_i x_i < w_0$.

Definition. Two sets A and B of points in the n-dimensional space $\mathbb{R}^n$ are called *absolutely linearly separable* if $n+1$ real numbers $w_0, w_1, \ldots, w_n$ exist such that every point $(x_1, x_2, \ldots, x_n) \in A$ satisfies

$$\sum_{i=1}^{n} w_j x_j > w_0$$

and every point $(x_1, x_2, \ldots, x_n) \in B$ satisfies $\Sigma_{i=1}^{n} w_i x_i < w_0$.

Definition. The open (closed) positive half-space associated with the n-dimensional weight vector $\mathbf{w}$ is the set of all points $\mathbf{x} \in \mathbb{R}^n$ for which $\mathbf{w}^T\mathbf{x} > 0$ ($\mathbf{w}^T\mathbf{x} \geq 0$). The open (closed) negative half-space associated with $\mathbf{w}$ is the set of all points $\mathbf{x} \in \mathbb{R}^n$ for which $\mathbf{w}^T\mathbf{x} < 0$) ($\mathbf{w}^T\mathbf{x} \leq 0$).

A perceptron can only compute linearly separable functions. How many linearly separable functions of n binary arguments are there? When $n = 2$, 14 out of the 16 possible Boolean functions are linearly separable. When $n = 3$, 104 out of 256 and when $n = 4$, 1882 out of 65536 possible functions are linearly separable. No formula for expressing the number of linearly separable functions as a function of n has yet been found.

13.5.4 Perceptron Learning

A learning algorithm is an adaptive method by which a network of computing units self-organizes to implement the desired behaviour (Minskey and Papert [148], Fausett [57], Haykin [90], Cichocki and Unbehauen [35], Hassoun [89], Rojas [173]). This is done in some learning algorithms by presenting some examples of the desired input-output mapping to the network. A correction step is executed iteratively until the network learns to produce the desired response. The learning algorithm is a closed loop of presentation of examples and of corrections to the network parameters.

Learning algorithms can be divided into *supervised* and *unsupervised* methods. Supervised learning denotes a method in which some input vectors are collected and presented to the network. The output computed by the network is observed and the deviation from the expected answer is measured. The weights are corrected according to the magnitude of the error in the way defined by the learning algorithm. This kind of learning is called *learning with a teacher*, since a control process knows the correct answer for the set of selected input vectors. The perceptron learning algorithm is an example of supervised learning with reinforcement.

Unsupervised learning is used when, for a given input, the exact numerical output a network should produce is unknown. Assume, for example, that some points in two-dimensional space are to be classified into three clusters. For this task we can use a classifier network with three output lines, one for each class. Each of the three computing units at the output must specialize by firing only for inputs corresponding to elements of each cluster. If one unit fires, the others must keep silent. In this case we do not know a priori which unit is going to specialize on which cluster. We do not know how many well-defined clusters are present. Since no "teacher" is available, the network must organize itself in order to be able to associate clusters with units.

The proof of convergence of the perceptron learning algorithm assumes that each perceptron performs the test $\mathbf{w}^T\mathbf{x} > 0$. So far we have been working with perceptrons which perform the test $\mathbf{w}^T\mathbf{x} \geq 0$.

If a perceptron with threshold zero can linearly separate two finite sets of input vectors, then only a small adjustment to its weights is needed to obtain an absolute linear separation. This is a direct corollary of the proposition.

Proposition. Two finite sets of points, A and B, in n-dimensional space which are linearly separable are also absolutely linearly separable.

A approach for starting the learning algorithm is to initialize the network weights randomly and to improve these initial parameters, looking at each step to see whether a better separation of the training set can be achieved. We identify points $(x_1, x_2, \ldots, x_n)$ in n-dimensional space with the column vector $\mathbf{x} = (x_1, x_2, \ldots, x_n)^T$.

Let P and N be two finite sets of points in the Euclidean space $\mathbb{R}^n$ which we want to separate linearly. A weight vector is sought so that the points in P belong to its associated positive half-space and the points in N to the negative half-space. The error of a perceptron with weight vector $\mathbf{w}$ is the number of incorrectly classified points. The learning algorithm must minimize this error function $E(\mathbf{w})$. The training set consists of two sets, P and N, in n-dimensional extended input space. We have to find a vector $\mathbf{w}$ capable of absolutely separating both sets with all vectors in P belonging to the open positive half-space and all vectors in N to the open negative half-space of the linear separation.

Algorithm. Perceptron learning

start:	The weight vector $\mathbf{w}(t=0)$ is generated randomly
test:	A vector $\mathbf{x} \in P \cup N$ is selected randomly,
	if $\mathbf{x} \in P$ and $\mathbf{w}(t)^T\mathbf{x} > 0$ goto *test*,
	if $\mathbf{x} \in P$ and $\mathbf{w}(t)^T\mathbf{x} \leq 0$ goto *add*,
	if $\mathbf{x} \in N$ and $\mathbf{w}(t)^T\mathbf{x} < 0$ goto *test*,
	if $\mathbf{x} \in N$ and $\mathbf{w}(t)^T\mathbf{x} \geq 0$ goto *subtract*,
add:	set $\mathbf{w}(t+1) = \mathbf{w}(t) + \mathbf{x}$ and $t := t+1$, goto *test*
subtract:	set $\mathbf{w}(t+1) = \mathbf{w}(t) - \mathbf{x}$ and $t := t+1$ goto *test*

This algorithm makes a correction to the weight vector whenever one of the selected vectors in P or N has not been classified correctly. The perceptron convergence theorem guarantees that if the two sets P and N are linearly separable the vector $\mathbf{w}$ is updated only a finite number of times. The routine can be stopped when all vectors are classified correctly.

Example. Consider the sets in the extended space

$$P = \{\,(1.0, 2.0, 2.0),\ (1.0, 1.5, 1.5)\,\}$$

$$N = \{\,(1.0, 0.0, 1.0),\ (1.0, 1.0, 0.0),\ (1.0, 0.0, 0.0)\,\}\,.$$

Thus in $\mathbb{R}^2$ we consider the two sets of points

$$\{\,(2.0, 2.0),\ (1.5, 1.5)\,\}, \qquad \{\,(0.0, 1.0),\ (1.0, 0.0),\ (0.0, 0.0)\,\}\,.$$

These two sets are separable by the line $x_1 + x_2 = \frac{3}{2}$. Thus $\mathbf{w}^T = (-\frac{3}{2}, 1, 1)$. ♣

The C++ program `classify.cpp` implements the algorithm.

```
// classify.cpp

#include <iostream>
#include <cstdlib>
#include <ctime>
using namespace std;

void classify(double **P,double **N,int p,int n,double *w,int d)
{
 int i, j, k, classified = 0;
 double *x, sum;
 srand(time(NULL));
 for(i=0;i<d;i++) w[i] = double(rand())/RAND_MAX;
 k = 0;
 while(!classified)
 {
 i = rand()%(p+n-1);
 if(i<p) x = P[i]; else x = N[i-p];
 for(j=0,sum=0;j<d;j++) sum += w[j]*x[j];
 if((i<p) && (sum<=0))  for(j=0;j<d;j++) w[j] += x[j];
```

```
 if((i>=p) && (sum>=0)) for(j=0;j<d;j++) w[j] -= x[j];
 k++;
 classified = 1;
 // check if the vectors are classified
 if((k%(2*p+2*n)) == 0)
 {
 for(i=0;(i<p) && classified;i++)
 {
 sum = 0.0;
 for(j=0,sum=0;j<d;j++) sum += w[j]*P[i][j];
 if(sum <= 0) classified = 0;
 }
 for(i=0;(i<n) && classified;i++)
 {
 sum = 0.0;
 for(j=0,sum=0;j<d;j++) sum += w[j]*N[i][j];
 if(sum >= 0) classified = 0;
 }
 }
 else classified = 0;
 }
}

int main(void)
{
 double **P = new double*[2];
 P[0] = new double[3]; P[1] = new double[3];
 P[0][0] = 1.0; P[0][1] = 2.0; P[0][2] = 2.0;
 P[1][0] = 1.0; P[1][1] = 1.5; P[1][2] = 1.5;
 double **N = new double*[3];
 N[0] = new double[3]; N[1] = new double[3]; N[2] = new double[3];
 N[0][0] = 1.0; N[0][1] = 0.0; N[0][2] = 1.0;
 N[1][0] = 1.0; N[1][1] = 1.0; N[1][2] = 0.0;
 N[2][0] = 1.0; N[2][1] = 0.0; N[2][2] = 0.0;
 double *w = new double[3];
 classify(P,N,2,3,w,3);
 cout << "w = (" << w[0] << "," << w[1] << "," << w[2] << ") " << endl;
 delete[] P[0]; delete[] P[1];
 delete[] N[0]; delete[] N[1]; delete[] N[2];
 delete[] P; delete[] N; delete w;
 return 0;
}
```

13.5.5 Perceptron Learning Algorithm

A simple perceptron learning algorithm is

1. Initialize the connection weight **w** to small random values.

2. Initialize acceptable error tolerance ϵ_0
3. Set $\epsilon_{\max} = 0$
4. For each of the input patterns $\{ \mathbf{x}_j, j = 0, 1, \dots, m-1 \}$ do

   (a) Calculate the output y_j via
   $$y_j = H(\mathbf{w}^T \mathbf{x}_j - \theta)$$
   where H is the Heaviside function and $j = 0, 1, \dots, m-1$.

   (b) Calculate the difference between the output y_j and the desired output $\widetilde{y}_j$ of the network
   $$d_j := \widetilde{y}_j - y_j \,.$$

   (c) Calculate the changes in the connection strengths
   $$\Delta \mathbf{w}_j := \eta d_j \mathbf{x}_j$$
   where η is the learning rate.

   (d) Update the connection weight $\mathbf{w}$ according to
   $$\mathbf{w} \leftarrow \mathbf{w} + \Delta \mathbf{w}_j$$

   (e) Set
   $$\epsilon_{\max} \leftarrow \max(\epsilon_{\max}, \|d_j\|)$$

5. If $\epsilon_{\max} > \epsilon_0$ return to step 3.

Example. Consider the AND gate, where $m = 4$. Let
$$\mathbf{w}^T = (0.005, 0.006), \quad \theta = \frac{3}{2}, \quad \epsilon_0 = 0.01, \quad \eta = 0.5$$
with the four vectors as input patterns
$$\mathbf{x}_0 = \begin{pmatrix} 0 \\ 0 \end{pmatrix}, \quad \mathbf{x}_1 = \begin{pmatrix} 1 \\ 0 \end{pmatrix}, \quad \mathbf{x}_2 = \begin{pmatrix} 0 \\ 1 \end{pmatrix}, \quad \mathbf{x}_3 = \begin{pmatrix} 1 \\ 1 \end{pmatrix}.$$
The desired output is
$$\widetilde{y}_0 = 0, \quad \widetilde{y}_1 = 0, \quad \widetilde{y}_2 = 0, \quad \widetilde{y}_3 = 1\,.$$
The calculations yield

1) (a) $y_0 = H(\mathbf{w}^T \mathbf{x}_0 - \theta) = 0 \Rightarrow d_0 = \widetilde{y}_0 - y_0 = 0 \Rightarrow \Delta \mathbf{w} = 0$

   (b) $y_1 = H(\mathbf{w}^T \mathbf{x}_1 - \theta) = 0 \Rightarrow d_1 = \widetilde{y}_1 - y_1 = 0 \Rightarrow \Delta \mathbf{w} = 0$

   (c) $y_2 = H(\mathbf{w}^T \mathbf{x}_2 - \theta) = 0 \Rightarrow d_2 = \widetilde{y}_2 - y_2 = 0 \Rightarrow \Delta \mathbf{w} = 0$

(d) $y_3 = H(\mathbf{w}^T\mathbf{x}_3 - \theta) = 0 \Rightarrow d_3 = \widetilde{y}_3 - y_3 = 1 \Rightarrow \Delta\mathbf{w} = \eta d_3(1,1) = (0.5, 0.5)$

$\Rightarrow \mathbf{w} = (0.505, 0.506), \quad \epsilon_{max} = 1$

2) (a) $y_0 = H(\mathbf{w}^T\mathbf{x}_0 - \theta) = 0 \Rightarrow d_0 = 0 \Rightarrow \Delta\mathbf{w} = 0$

(b) $y_1 = H(\mathbf{w}^T\mathbf{x}_1 - \theta) = 0 \Rightarrow d_1 = 0 \Rightarrow \Delta\mathbf{w} = 0$

(c) $y_2 = H(\mathbf{w}^T\mathbf{x}_2 - \theta) = 0 \Rightarrow d_2 = 0 \Rightarrow \Delta\mathbf{w} = 0$

(d) $y_3 = H(\mathbf{w}^T\mathbf{x}_3 - \theta) = 0 \Rightarrow d_3 = 1 \Rightarrow \Delta\mathbf{w} = 0.5(1,1) = (0.5, 0.5)$

$\Rightarrow \mathbf{w} = (1.05, 1.06)$

3) (a) $y_0 = H(\mathbf{w}^T\mathbf{x}_0 - \theta) = 0 \Rightarrow d_0 = 0 \Rightarrow \Delta\mathbf{w} = 0$

(b) $y_1 = H(\mathbf{w}^T\mathbf{x}_1 - \theta) = 0 \Rightarrow d_1 = 0 \Rightarrow \Delta\mathbf{w} = 0$

(c) $y_2 = H(\mathbf{w}^T\mathbf{x}_2 - \theta) = 0 \Rightarrow d_2 = 0 \Rightarrow \Delta\mathbf{w} = 0$

(d) $y_3 = H(\mathbf{w}^T\mathbf{x}_3 - \theta) = 1 \Rightarrow d_3 = 0 \Rightarrow \Delta\mathbf{w} = 0$

Thus with $\mathbf{w}^T = (1.05, 1.06)$, $\theta = \frac{3}{2}$ we can simulate the AND gate. In the extended space we have

$$\mathbf{w}^T = (w_0, w_1, w_2) = (-\theta, w_1, w_2), \qquad \mathbf{x}^T = (1, x_1, x_2)\,. \qquad \clubsuit$$

In the C++ program `perceptronAnd.cpp` we use the notation of the extended space. Furthermore, the threshold is also initialized to a small random value at $t = 0$.

```
// perceptronAnd.cpp

#include <iostream>
#include <cmath>     // for fabs
using namespace std;

double H(double z) { if(z >= 0.0) return 1.0; else return 0.0; }

double scalar(double* u,double* v,int n)
{
 double result = 0.0;
 for(int i=0;i<n;i++) result += u[i]*v[i];
 return result;
}

double distance(double* u,double* v,int n)
{
 double result = 0.0;
 for(int i=0;i<n;i++) result += fabs(u[i]-v[i]);
 return result;
}
```

```
void change(double** x,double* yt,double* w,double eta,int m,int n)
{
 double* d = new double[m];
 for(int j=0;j<m;j++)
 {
 d[j] = yt[j]-H(scalar(w,x[j],n));
 for(int i=0;i<n;i++) { w[i] += eta*d[j]*x[j][i]; }
 }
 delete[] d;
}

int main(void)
{
 // number of input vectors (patterns) is m = 4
 // length of each input vector n = 3
 int m = 4, n = 3;
 double** x = new double*[m];
 for(int k=0;k<m;k++) x[k] = new double[n];
 x[0][0] = 1.0; x[0][1] = 0.0; x[0][2] = 0.0;
 x[1][0] = 1.0; x[1][1] = 0.0; x[1][2] = 1.0;
 x[2][0] = 1.0; x[2][1] = 1.0; x[2][2] = 0.0;
 x[3][0] = 1.0; x[3][1] = 1.0; x[3][2] = 1.0;
 // desired output
 double* yt = new double[m];
 yt[0] = 0.0; yt[1] = 0.0; yt[2] = 0.0; yt[3] = 1.0;

 // weight vector, w[0] = -theta (threshold)
 double* w = new double[n];
 // initialized to small random numbers
 w[0] = 0.01; w[1] = 0.005; w[2] = 0.006;
 // learning rate
 double eta = 0.5;

 double* wt = new double[n];
 for(int i=0;i<n;i++) wt[i] = w[i];
 for(;;)
 {
 change(x,yt,w,eta,m,n);
 double dist = distance(w,wt,n);
 if(dist < 0.0001) break;
 for(i=0;i<n;i++) wt[i] = w[i];
 }
 // display the output of the weight vectors
 for(i=0;i<n;i++) cout << "w[" << i << "] = " << w[i] << " ";

 delete[] w; delete[] wt; delete[] yt;
 for(i=0;i<m;i++) { delete[] x[i]; }
```

```
delete[] x;
return 0;
}
```

The output is given by

```
w[0] = -1.49  w[1] = 1.005  w[2] = 0.506
```

Thus with

$$w_0 = -\theta = -1.49, \qquad w_1 = 1.005, \qquad w_2 = 0.506$$

we can simulate the AND gate.

13.5.6 One and Two Layered Networks

We consider feed-forward networks structured in successive layers of computing units (Fausett [57], Haykin [90], Cichocki and Unbehauen [35], Hassoun [89], Rojas [173]). The networks we want to consider must be defined in terms of their architecture. The atomic elements of any architecture are the computing units and their interconnections. Each computing unit collects the information from n input lines with an *integration function*

$$\Sigma : \mathbb{R}^n \to \mathbb{R}.$$

The total excitation computed in this way is then evaluated using an *activation function*

$$f : \mathbb{R} \to \mathbb{R}.$$

In perceptrons the integration function is the sum of the inputs. The activation, also called output function, compares the sum with a threshold. We can generalize f to produce all values between 0 and 1. In the case of Σ some functions other than addition can also be considered. In this case the networks can compute some difficult functions with fewer computing units. A network architecture is a tuple (I, N, O, E) consisting of a set I of input sites, a set N of computing units, a set O of output sites and a set E of weighted directed edges. A directed edge is a tuple (u, v, w) whereby $u \in I \cup N$, $v \in N \cup O$ and $w \in \mathbb{R}$.

The input sites are entry points for information into the network. The results are transmitted to the output sites. The set N consists of all computing elements in the network. The edges between all computing units are weighted, as are the edges between input and output sites and computing units.

Layered architectures are those in which the set of computing units N is subdivided into ℓ subsets $N_1, N_2, \ldots, N_\ell$ in such a way that only connections from units in N_1 go to units in N_2, from units in N_2 to units in N_3, etc. The input sites are only connected to the units in the subset N_1, and the units in the subset N_ℓ are the only ones connected to the output sites. In the usual terminology, the units in N_ℓ are the output units of the network. The subsets N_i are called the *layers* of the network. The set of input sites is called the *input layer*, the set of output units is called the *output layer*. All other layers with no direct connections from or to the outside are called *hidden layers*. The units in a layer are not connected to each other (although some neural models make use of this kind of architecture). A neural network with a layered architecture does not contain cycles. The input is processed and relayed from the layer to the other, until the final result has been computed.

In layered architectures normally all units from one layer are connected to all other units in the following layer. If there are m units in the first layer and n units in the second one, the total number of weights is mn. The total number of connections can be rather large.

13.5.7 XOR Problem and Two-Layered Networks

The properties of a two-layered network can be discussed using the case of the XOR function as an example. A single perceptron cannot compute this function, but a two-layered network can. The network consists of an input layer, a hidden layer and an output layer and three computing units. One of the units in the hidden layer computes the function $x_1 \wedge \neg x_2$, and the other the function $\neg x_1 \wedge x_2$. The third unit computes the OR function, so that the result of the complete network computation is

$$(x_1 \wedge \neg x_2) \vee (\neg x_1 \wedge x_2)$$

where $\wedge$ is the AND-operation, $\vee$ is the OR-operation and $\neg$ is the NOT-operation. The calculations for the XOR gate are as follows. We work in the extended space. The input vectors are

$$\mathbf{x}_0 = \begin{pmatrix} 1 \\ 0 \\ 0 \end{pmatrix}, \quad \mathbf{x}_1 = \begin{pmatrix} 1 \\ 0 \\ 1 \end{pmatrix}, \quad \mathbf{x}_2 = \begin{pmatrix} 1 \\ 1 \\ 0 \end{pmatrix}, \quad \mathbf{x}_3 = \begin{pmatrix} 1 \\ 1 \\ 1 \end{pmatrix}.$$

1) input layer $\longrightarrow$ hidden layer. The weights are

$$w_{000} = -0.5, \quad w_{001} = 1.0, \quad w_{002} = -1.0$$

$$w_{010} = -0.5, \quad w_{011} = -1.0, \quad w_{012} = 1.0\,.$$

The weight has three indexes. The first index indicates the layer, in this case 0 for the input layer. The second index indicates to which node in the hidden layer it points where the number for the hidden node is incremented by 1 so that we can assign the index 0 to the bias in the hidden layer. The third index indicates the number of the neurons.

Input vector $\mathbf{x}_0$:

a) $H((w_{000}, w_{001}, w_{002}) \begin{pmatrix} 1 \\ 0 \\ 0 \end{pmatrix}) = H(-0.5) = 0 = z_0$

b) $H((w_{010}, w_{011}, w_{012})\begin{pmatrix}1\\0\\0\end{pmatrix}) = H(-0.5) = 0 = z_1$

Input vector $\mathbf{x}_1$:

a) $H((w_{000}, w_{001}, w_{002})\begin{pmatrix}1\\0\\1\end{pmatrix}) = H(-1.5) = 0 = z_0$

b) $H((w_{010}, w_{011}, w_{012})\begin{pmatrix}1\\0\\1\end{pmatrix}) = H(+0.5) = 1 = z_1$

Input vector $\mathbf{x}_2$:

a) $H((w_{000}, w_{001}, w_{002})\begin{pmatrix}1\\1\\0\end{pmatrix}) = H(+0.5) = 1 = z_0$

b) $H((w_{010}, w_{011}, w_{012})\begin{pmatrix}1\\1\\0\end{pmatrix}) = H(-1.5) = 0 = z_1$

Input vector $\mathbf{x}_3$:

a) $H((w_{000}, w_{001}, w_{002})\begin{pmatrix}1\\1\\1\end{pmatrix}) = H(-0.5) = 0 = z_0$

b) $H((w_{000}, w_{021}, w_{022})\begin{pmatrix}1\\1\\1\end{pmatrix}) = H(-0.5) = 0 = z_1$

2) hidden layer $\longrightarrow$ output. From the above calculations for z_0 and z_1 we find that the input pairs from the hidden layer are $(1,0,0)$, $(1,0,1)$, $(1,1,0)$ and $(1,0,0)$. Thus the first and the last patterns are the same. The weights are

$$w_{100} = -0.5, \qquad w_{101} = 1.0, \qquad w_{102} = 1.0\,.$$

Consider input pattern $(1,0,0)$ from hidden layer

a) $H((w_{100}, w_{101}, w_{102})\begin{pmatrix}1\\0\\0\end{pmatrix}) = H(-0.5) = 0\,.$

Input pattern $(1,0,1)$ from hidden layer:

b) $H((w_{100}, w_{101}, w_{102})\begin{pmatrix}1\\0\\1\end{pmatrix}) = H(+0.5) = 1$

Input pattern $(1,1,0)$ from hidden layer:

c) $H((w_{100}, w_{101}, w_{102})\begin{pmatrix}1\\1\\0\end{pmatrix}) = H(+0.5) = 1$

Input pattern $(1,0,0)$ from hidden layer (already considered above)

d) $H((w_{100}, w_{101}, w_{102})\begin{pmatrix}1\\0\\0\end{pmatrix}) = H(-0.5) = 0.$

Thus we have simulated the XOR gate using one hidden layer.

```
// XOR1.cpp

#include <iostream>
using namespace std;

double H(double s) { if(s >= 0.0) return 1.0; else return 0.0; }

double map(double*** w,double* testpattern,int size3)
{
 int k;
 double* z = new double[size3];
 z[0] = 1.0;  z[1] = 0.0;  z[2] = 0.0;
 // input layer to hidden layer
 for(k=0;k<size3;k++)
 { z[1] += w[0][0][k]*testpattern[k]; z[2] += w[0][1][k]*testpattern[k]; }
 z[1] = H(z[1]); z[2] = H(z[2]);
 // hidden layer to output layer
 double y = 0.0;
 for(k=0;k<size3;k++) y += w[1][0][k]*z[k];
 delete[] z;
 y = H(y);
 return y;
}

int main(void)
{
 int size1 = 2, size2 = 2, size3 = 3;
 double*** w = NULL; w = new double** [size1];
 for(int i=0;i<size1;i++)
 {
```

```
 w[i] = new double*[size2];
 for(int j=0;j<size2;j++) { w[i][j] = new double[size3]; }
 }
 w[0][0][0] = -0.5; w[0][0][1] = 1.0; w[0][0][2] = -1.0;
 w[0][1][0] = -0.5; w[0][1][1] = -1.0; w[0][1][2] = 1.0;
 w[1][0][0] = -0.5; w[1][0][1] = 1.0; w[1][0][2] = 1.0;
 w[1][1][0] = 0.0; w[1][1][1] = 0.0; w[1][1][2] = 0.0;
 // input patterns
 int p = 4; // number of input pattern
 int n = 3; // length of each input pattern
 double** x = NULL; x = new double* [p];
 for(int k=0;k<p;k++) { x[k] = new double[n]; }
 x[0][0] = 1.0; x[0][1] = 0.0; x[0][2] = 0.0;
 x[1][0] = 1.0; x[1][1] = 0.0; x[1][2] = 1.0;
 x[2][0] = 1.0; x[2][1] = 1.0; x[2][2] = 0.0;
 x[3][0] = 1.0; x[3][1] = 1.0; x[3][2] = 1.0;
 double result = map(w,x[0],size3);
 cout << "result = " << result << endl;     // => 0
 result = map(w,x[1],size3);
 cout << "result = " << result << endl;     // => 1
 result = map(w,x[2],size3);
 cout << "result = " << result << endl;     // => 1
 result = map(w,x[3],size3);
 cout << "result = " << result << endl;     // => 0
 return 0;
}
```

13.6 Multilayer Perceptrons

13.6.1 Introduction

In a practical application of the back-propagation algorithm, learning results from the many presentations of a prescribed set of training examples to the multilayer perceptron (Fausett [57], Haykin [90], Cichocki and Unbehauen [35], Hassoun [89], Rojas [173]) . One complete presentation of the entire training set during the learning process is called an *epoch.* The learning process is maintained on an epoch-by-epoch basis untilthe synaptic weights and threshold levels of the network stabilize and the average squared error over the entire training set converges to some minimum value. It is good practice to randomize the order of presentation of training examples from one epoch to the next. This randomization tends to make the search in weight space stochastic over the learning cycles, thus avoiding the possibility of limit cycles in the evolution of the synaptic weight vectors. We follow in our notation closely Hassoun [89]. For a given training set, back-propagation learning may thus proceed in one of two basic ways. Let

$$\{\, \mathbf{x}_k,\ \mathbf{d}_k \,\}$$

be the training data, where $k = 0, 1, \ldots, m-1$. Here m is the number of training examples (patterns). The sets $\mathbf{x}_k$ ($k = 0, 1, \ldots, m-1$) are the input pattern and the sets $\mathbf{d}_k$ are the corresponding (desired) output pattern. One complete presentation of the entire training set during the learning process is called an epoch.

1. *Pattern Mode.* In the *pattern mode* of back-propagation learning, weight updating is performed after the presentation of each training example; this is the mode of operation for which the derivation of the back-propagation algorithm presented here applies. To be specific, consider an epoch consisting of m training examples (patterns) arranged in the order

$$\mathbf{x}_0, \mathbf{d}_0, \quad \mathbf{x}_1, \mathbf{d}_1, \quad \ldots, \quad \mathbf{x}_{m-1}, \mathbf{d}_{m-1}\,.$$

The first example $\mathbf{x}_0$, $\mathbf{d}_0$ in the epoch is presented to the network, and the sequence of forward and backward computations described below is performed, resulting in certain adjustments to the synaptic weights and threshold levels of the network. Then, the second example $\mathbf{x}_1$, $\mathbf{d}_1$ in the epoch is presented, and the sequence of forward and backward computations is repeated, resulting in further adjustments to the synaptic weights and threshold levels. This process is continued until the last training pattern $\mathbf{x}_{m-1}$, $\mathbf{d}_{m-1}$ is taken into account.

2. *Batch Mode.* In the *batch mode* of back-propagation learning, weight updating is performed after the presentation of all the training examples that constitute an epoch.

13.6.2 Cybenko's Theorem

Single-hidden-layer neural networks are universal approximators. A rigorous mathematical proof for the universality of feedforward layered neural networks employing continuous sigmoid type activation functions, as well as other more general activation units, was given by Cybenko [43]. Cybenko's proof is based on the Hahn-Banach theorem. The following is the statement of Cybenko's theorem.

Theorem. Let f be any continuous sigmoid-type function, for example

$$f(s) = 1/(1+\exp(-\lambda s)), \qquad \lambda \geq 1\,.$$

Then, given any continuous real-valued function g on $[0,1]^n$ (or any other compact subset of $\mathbb{R}^n$) and $\epsilon > 0$, there exists vectors $\mathbf{w}_1, \mathbf{w}_2, \ldots, \mathbf{w}_N$, $\boldsymbol{\alpha}$, and $\boldsymbol{\theta}$ and a parameterized function

$$G(\cdot, \mathbf{w}, \boldsymbol{\alpha}, \boldsymbol{\theta}) : [0,1]^n \to \mathbb{R}$$

such that

$$|G(\mathbf{x}, \mathbf{w}, \boldsymbol{\alpha}, \boldsymbol{\theta}) - g(\mathbf{x})| < \epsilon \qquad \text{for all} \quad \mathbf{x} \in [0,1]^n$$

where

$$G(\mathbf{x}, \mathbf{w}, \boldsymbol{\alpha}, \boldsymbol{\theta}) = \sum_{j=1}^{N} \alpha_j f(\mathbf{w}_j^T \mathbf{x} + \theta_j)$$

and

$$\mathbf{w}_j \in \mathbb{R}^n, \quad \theta_j \in \mathbb{R}, \quad \mathbf{w} = (\mathbf{w}_1, \mathbf{w}_2, \ldots, \mathbf{w}_N)$$

$$\boldsymbol{\alpha} = (\alpha_1, \alpha_2, \ldots, \alpha_N), \quad \boldsymbol{\theta} = (\theta_1, \theta_2, \ldots, \theta_N).$$

For the proof we refer to Cybenko [43].

Thus a one-hidden layer feedforward neural networks is capable of approximating uniformly any continuous multivariate function to any desired degree of accuracy. This implies that any failure of a function mapping by a multilayer network must arise from inadequate choice of parameters, i.e., poor choices for $\mathbf{w}_1, \mathbf{w}_2, \ldots, \mathbf{w}_N$, $\boldsymbol{\alpha}$, and $\boldsymbol{\theta}$ or an insufficient number of hidden nodes.

Hornik et al. [100] employing the Stone-Weierstrass theorem and Funahashi [65] proved similar theorems stating that a one-hidden-layer feedforward neural network is capable of approximating uniformly any continuous multivariate function to any desired degree of accuracy.

13.6.3 Back-Propagation Algorithm

We consider one hidden layer. The notations we use follow closely Hassoun [89]. A two-layer feedforward architecture is considered. This network receives a set of scalar signals

$$x_0, x_1, x_2, \ldots, x_{n-1}$$

where x_0 is a bias signal set to 1. This set of signals constitutes an input vector $\mathbf{x}_k \in \mathbb{R}^n$. The layer receiving this input signal is called the hidden layer. The hidden layer has J units. The output of the hidden layer is a J dimensional real-valued vector $\mathbf{z}_k = (z_0, z_1, \ldots, z_{J-1})$, where we set $z_0 = 1$ (bias signal). The vector $\mathbf{z}_k$ supplies the input for the output layer of L units. The output layer generates an L-dimensional vector $\mathbf{y}_k$ in response to the input vector $\mathbf{x}_k$ which, when the network is fully trained, should be identical (or suffiently close) to the desired output vector $\mathbf{d}_k$ associated with $\mathbf{x}_k$.

The two activation functions f_h (input layer to hidden layer) and f_o (hidden layer to output layer) are assumed to be differentiable functions. We use the logistic functions

$$f_h(s) = \frac{1}{1+\exp(-\lambda_h s)}, \qquad f_o(s) = \frac{1}{1+\exp(-\lambda_o s)}$$

where $\lambda_h, \lambda_o \geq 1$. The *logistic function*

$$f(s) = \frac{1}{1+\exp(-\lambda s)}$$

satisfies the nonlinear ordinary differential equation

$$\frac{df}{ds} = \lambda f(1-f).$$

The components of the desired output vector $\mathbf{d}_k$ must be chosen within the range of f_o. We denote by w_{ji} the weight of the jth hidden unit associated with the input signal x_i. Thus the index i runs from 0 to $n-1$, where $x_0 = 1$ and j runs from 1 to $J-1$. We set $w_{0i} = 0$. Now we have m input/output pairs of vectors

$$\{ \mathbf{x}_k , \mathbf{d}_k \}$$

where the index k runs from 0 to $m-1$. The aim of the algorithm is to adaptively adjust the $(J-1)n + LJ$ weights of the network such that the underlying function/mapping represented by the training set is approximated or learned. Let w_{lj} be the weights from the hidden layer to the output and let w_{ji} be the weights from the input layer to the hidden layer. An *error function* can be defined since the learning is supervised, i.e. the target outputs are available. We denote by w_{lj} the weight of the lth output unit associated with the input signal z_j from the hidden layer. We derive a supervised learning rule for adjusting the weights w_{ji} and w_{lj} such that the error function

$$E(\mathbf{w}) = \frac{1}{2}\sum_{l=0}^{L-1}(d_l - y_l)^2$$

is minimized (in a local sense) over the training set. Here $\mathbf{w}$ represents the set of all weights in the network.

Since the targets for the output units are given, we can use the delta rule directly for updating the w_{lj} weights. We define

$$\Delta w_{lj} := w_{lj}^{new} - w_{lj}^{c}.$$

Since

$$\Delta w_{lj} = -\eta_o \frac{\partial E}{\partial w_{lj}}$$

we find using the chain rule

$$\Delta w_{lj} = \eta_o (d_l - y_l) f_o'(net_l) z_j$$

where $l = 0, 1, \ldots, L-1$ and $j = 0, 1, \ldots, J-1$. Here

$$net_l := \sum_{j=0}^{J-1} w_{lj} z_j$$

is the weighted sum for the lth output unit, f_o' is the derivative of f_o with respect to net, and w_{lj}^{new} and w_{lj}^{c} are the updated (new) and current weight values, respectively. The z_j values are calculated by propagating the input vector $\mathbf{x}$ through the hidden layer according to

$$z_j = f_h\left(\sum_{i=0}^{n-1} w_{ji} x_i\right) = f_h(net_j)$$

where $j = 1, 2, \ldots, J-1$ and $z_0 = 1$ (bias signal). For the hidden-layer weights w_{ji} we do not have a set of target values (desired outputs) for hidden units. However, we can derive the learning rule for hidden units by attempting to minimize the output-layer error. This amounts to propagating the output errors $(d_l - y_l)$ back through the output layer toward the hidden units in an attempt to estimate dynamic targets for these units. Thus a gradient descent is performed on the criterion function

$$E(\mathbf{w}) = \frac{1}{2}\sum_{l=0}^{L-1}(d_l - y_l)^2$$

where $\mathbf{w}$ represents the set of all weights in the network. The gradient is calculated with respect to the hidden weights

$$\Delta w_{ji} = -\eta_h \frac{\partial E}{\partial w_{ji}}, \qquad j = 1, 2, \ldots, J-1, \qquad i = 0, 1, \ldots, n-1$$

where the partial derivative is to be evaluated at the current weight values. We find

$$\frac{\partial E}{\partial w_{ji}} = \frac{\partial E}{\partial z_j}\frac{\partial z_j}{\partial net_j}\frac{\partial net_j}{\partial w_{ji}}$$

where

$$\frac{\partial net_j}{\partial w_{ji}} = x_i, \qquad \frac{\partial z_j}{\partial net_j} = f_h'(net_j)\,.$$

We used the chain rule in this derivation. Since

$$\frac{\partial E}{\partial z_j} = -\sum_{l=0}^{L-1}(d_l - y_l) f_o'(net_l) w_{lj}$$

we obtain

$$\Delta w_{ji} = \eta_h \left(\sum_{l=0}^{L-1}(d_l - y_l) f_o'(net_l) w_{lj}\right) f_h'(net_j) x_i\,.$$

Now we can define an estimated target d_j for the jth hidden unit implicitly in terms of the backpropagated error signal as follows

$$d_j - z_j := \sum_{l=0}^{L-1}(d_l - y_l) f_o'(net_l) w_{lj}\,.$$

The complete approach for updating weights in a feedforward neural net utilizing these rules can be summarized as follows. We do a pattern-by-pattern updating of the weights.

1. *Initialization.* Initialize all weights to small random values and refer to them as current weights w_{lj}^c and w_{ji}^c.

2. *Learning rate.* Set the learning rates η_o and η_h to small positive values.

3. *Presentation of training example.* Select an input pattern $\mathbf{x}_k$ from the training set (preferably at random) propagate it through the network, thus generating hidden- and output-unit activities based on the current weight settings. Thus find z_j and y_l.

4. *Forward computation.* Select the desired target vector $\mathbf{d}_k$ associated with $\mathbf{x}_k$, and calculate

$$\Delta w_{lj} = \eta_o(d_l - y_l)f'(net_l)z_j \equiv \eta_o(d_l - y_l)\lambda_o f(net_l)(1 - f(net_l))z_j$$

to compute the output layer weight changes Δw_{lj}.

5. *Backward computation.* Use

$$\Delta w_{ji} = \eta_h \left(\sum_{l=0}^{L-1} (d_l - y_l) f_o'(net_l) w_{lj} \right) f_h'(net_j) x_i$$

or

$$\Delta w_{ji} = \eta_h \left(\sum_{l=0}^{L-1} (d_l - y_l)\lambda_o f_o(net_l)(1 - f_o(net_l)) w_{lj} \right) \lambda_h f_h(net_j)(1 - f_h(net_j)) x_i$$

to compute the hidden layer weight changes. The current weights are used in these computations. In general, enhanced error correction may be achieved if one employs the updated output-layer weights

$$w_{lj}^{new} = w_{lj}^c + \Delta w_{lj}.$$

However, this comes at the added cost of recomputing y_l and $f'(net_l)$.

6. *Update weights.* Update all weights according to

$$w_{ji}^{new} = w_{ji}^c + \Delta w_{ji}$$

and

$$w_{lj}^{new} = w_{lj}^c + \Delta w_{lj}$$

for the output and for the hidden layers, respectively.

7. *Test for convergence.* This is done by checking the output error function to see if its magnitude is below some given threshold. Iterate the computation by presenting new epochs of training examples to the network until the free parameters of the network stabilize their values. The order of presentation of training examples should be randomized from epoch to epoch. The learning rate parameter is typically adjusted (and usually decreased) as the number of training iterations increases.

In the C++ program we apply the back-propagation algorithm to the XOR problem, where $m = 4$ is the number of input vectors each of length 3 (includes the bias input). The number of hidden layer units is 3 which includes the bias input $z_0 = 1$. By modifying m, n, J and L the program can easily be adapted to other problems.

```
// backpropagation1.cpp

#include <iostream>
#include <cmath>      // for exp
using namespace std;

// activation function (input layer -> hidden layer)
double fh(double net)
{ double lambdah = 10.0; return 1.0/(1.0+exp(-lambdah*net)); }

// activation function (hidden layer -> output layer)
double fo(double net)
{ double lambdao = 10.0; return 1.0/(1.0+exp(-lambdao*net)); }

double scalar(double* a1,double* a2,int length)
{
 double result = 0.0;
 for(int i=0;i<length;i++) { result += a1[i]*a2[i]; }
 return result;
}

int main()
{
 int k, i, j, l;  // summation index
 // k runs over all input pattern k=0,1,..,m-1
 // l runs over all output units l=0,1,..,L-1
 // j runs over all the hidden layer units j=0,1,..,J-1
 // i runs over the length of the input vector i=0,1,..,n-1
 double etao = 0.05, etah = 0.05;       // learning rate
 double lambdao = 10.0, lambdah = 10.0;
 int m = 4;  // number of input vectors for XOR problem
 int n = 3;  // length of each input vector for XOR problem
 double** x = NULL; //memory allocation
 // input vectors
 x = new double* [m];
 for(k=0;k<m;k++) x[k] = new double [n];
 x[0][0] = 1.0; x[0][1] = 0.0; x[0][2] = 0.0;
 x[1][0] = 1.0; x[1][1] = 0.0; x[1][2] = 1.0;
 x[2][0] = 1.0; x[2][1] = 1.0; x[2][2] = 0.0;
 x[3][0] = 1.0; x[3][1] = 1.0; x[3][2] = 1.0;
 // desired output vectors
 // corresponding to set of input vectors x
 double** d = NULL;
```

```
// number of outputs for XOR problem
int L = 1;
d = new double*[m];
for(k=0;k<m;k++) d[k] = new double [L];
d[0][0] = 0.0; d[1][0] = 1.0; d[2][0] = 1.0; d[3][0] = 0.0;
// error function for each input vector
double* E = new double[m];

double totalE = 0.0; // sum of E[k] k=0,1,...,m

// weight matrix (input layer -> hidden layer);
// number of hidden layers includes 0
// current
int J = 3;
double** Wc = NULL; Wc = new double* [J];
for(j=0;j<J;j++) Wc[j] = new double[n];
Wc[0][0] = 0.0; Wc[0][1] = 0.0; Wc[0][2] = 0.0;
Wc[1][0] = -0.2; Wc[1][1] = 0.5; Wc[1][2] = -0.5;
Wc[2][0] = -0.3; Wc[2][1] = -0.3; Wc[2][2] = 0.7;
// new
double** Wnew = NULL; Wnew = new double*[J];
for(j=0;j<J;j++) Wnew[j] = new double[n];
// weight matrix (hidden layer -> output layer)
// current
double** Whc = NULL; Whc = new double* [L];
for(l=0;l<L;l++) Whc[l] = new double[J];
Whc[0][0] = -0.2; Whc[0][1] = 0.3; Whc[0][2] = 0.5;
// new
double** Whnew = NULL;  Whnew = new double*[L];
for(l=0;l<L;l++) Whnew[l] = new double[J];
// vector in hidden layer
double* z = new double[J];
z[0] = 1.0;
// vector output layer (output layer units)
// for the XOR problem the output layer has only one element
double* y = new double[L];
// increment matrix (input layer -> hidden layer)
double** delW = NULL;   delW = new double*[J];
for(j=0;j<J;j++) delW[j] = new double[n];
// increment matrix (hidden layer -> output layer)
double** delWh = NULL;   delWh = new double*[L];
for(l=0;l<L;l++) delWh[l] = new double[J];
// net vector (input layer -> hidden layer)
double* netj = new double[J];
netj[0] = 0.0;
// net vector (hidden layer -> output layer)
double* netl = new double[L];
// training session
```

```
int T = 10000; // number of iterations
for(int t=0;t<T;t++)
{
// for loop over all input pattern
for(k=0;k<m;k++)
{
for(j=1;j<J;j++)
{ netj[j] = scalar(x[k],Wc[j],n); z[j] = fh(netj[j]); }
for(l=0;l<L;l++)
{ netl[l] = scalar(z,Whc[l],J); y[l] = fo(netl[l]); }
for(l=0;l<L;l++)
 for(j=0;j<J;j++)
  delWh[l][j] =
   etao*(d[k][l]-y[l])*lambdao*fo(netl[l])*(1.0-fo(netl[l]))*z[j];
double* temp = new double [J];
for(j=0;j<J;j++) temp[j] = 0.0;
for(j=0;j<J;j++)
 for(l=0;l<L;l++)
  temp[j] += (d[k][l]-y[l])*fo(netl[l])*(1.0-fo(netl[l]))*Whc[l][j];
for(j=0;j<J;j++)
 for(i=0;i<n;i++)
  delW[j][i] = etah*temp[j]*lambdah*fh(netj[j])*(1.0-fh(netj[j]))*x[k][i];
for(i=0;i<n;i++) delW[0][i] = 0.0;
// updating the weight matrices
for(j=0;j<J;j++)
 for(i=0;i<n;i++) Wnew[j][i] = Wc[j][i]+delW[j][i];
for(l=0;l<L;l++)
 for(j=0;j<J;j++) Whnew[l][j] = Whc[l][j]+delWh[l][j];
// setting new to current
for(j=0;j<J;j++)
 for(i=0;i<n;i++) Wc[j][i] = Wnew[j][i];
for(l=0;l<L;l++)
 for(j=0;j<J;j++) Whc[l][j] = Whnew[l][j];
E[k] = 0.0;
double sum = 0.0;
for(l=0;l<L;l++) sum += (d[k][l]-y[l])*(d[k][l]-y[l]);
E[k] = sum/2.0;
totalE += E[k];
} // end for loop over all input pattern
if(totalE < 0.0005) goto Label;
else totalE = 0.0;
} // end training session
Label:
cout << "number of iterations = " << t << endl;
// output after training
for(j=0;j<J;j++)
 for(i=0;i<n;i++)
  cout << "Wc[" << j << "][" << i << "] = " << Wc[j][i] << endl;
```

```
 cout << endl;
 for(l=0;l<L;l++)
  for(j=0;j<J;j++)
   cout << "Whc[" << l << "][" << j << "] = " << Whc[l][j] << endl;
 // testing the XOR gate
 // input (1,0,0)
 for(j=1;j<J;j++)
 { netj[j] = scalar(x[0],Wc[j],n); z[j] = fh(netj[j]); }
 for(l=0;l<L;l++)
 { netl[l] = scalar(z,Whc[l],J); y[l] = fo(netl[l]);
   cout << "y[" << l << "] = " << y[l] << endl; }
 // input (1,0,1)
 for(j=1;j<J;j++)
 { netj[j] = scalar(x[1],Wc[j],n); z[j] = fh(netj[j]); }
 for(l=0;l<L;l++)
 { netl[l] = scalar(z,Whc[l],J); y[l] = fo(netl[l]);
   cout << "y[" << l << "] = " << y[l] << endl; }
 // input (1,1,0)
 for(j=1;j<J;j++)
 { netj[j] = scalar(x[2],Wc[j],n); z[j] = fh(netj[j]); }
 for(l=0;l<L;l++)
 { netl[l] = scalar(z,Whc[l],J); y[l] = fo(netl[l]);
   cout << "y[" << l << "] = " << y[l] << endl; }
 // input (1,1,1)
 for(j=1;j<J;j++)
 { netj[j] = scalar(x[3],Wc[j],n); z[j] = fh(netj[j]); }
 for(l=0;l<L;l++)
 { netl[l] = scalar(z,Whc[l],J); y[0] = fo(netl[l]);
   cout << "y[" << l << "] = " << y[l] << endl; }
 return 0;
}
```

13.7 Radial Basis Function Networks

Traditionally, radial basis function neural networks (Buhmann [25]) which model functions $y(\mathbf{x})$ mapping $\mathbf{x} \in \mathbb{R}^n$ to $y \in \mathbb{R}$ have a single hidden layer so that the model

$$f(\mathbf{x}) = \sum_{j=1}^{m} w_j h_j(\mathbf{x})$$

is linear in the hidden-to-output weights $\{w_j\}_{j=1}^m$. The characteristic feature of RBF networks is the radial nature of the hidden unit transfer functions, $\{h_j\}_{j=1}^m$, which depend only on the distance between the input $\mathbf{x}$ and the centre $\mathbf{c}_j$ of each hidden unit, scaled by a metric R_j (positive definite $n \times n$ matrix),

$$h_j(\mathbf{x}) = \phi\left((\mathbf{x} - \mathbf{c}_j)^T R_j^{-1} (\mathbf{x} - \mathbf{c}_j)\right)$$

where ϕ is some function which is monotonic for non-negative numbers. Normally one restricts attention to diagonal metrics and Gaussian basis functions so that the transfer functions can be written

$$h_j(\mathbf{x}) = \exp\left(- \sum_{k=1}^{n} \frac{(x_k - c_{jk})^2}{r_{jk}^2} \right)$$

where $\mathbf{r}_j$ is the radius vector of the j-th hidden unit. Sometimes one includes low-order polynomial terms. Here the only non-radial basis function regressor we consider is a single bias unit where, for some particular index j, $h_j(\mathbf{x}) = 1$ for all $\mathbf{x}$. Each method estimates a model of the target function from a training set containing p input-output case $\{(\mathbf{x}_i, y_i)\}_{i=1}^{p}$, i.e. we want to find the weights w_i. From the y_i we form the vector $\mathbf{y} = (y_1\ y_2\ \ldots\ y_p)^T$. The response of the m hidden units of the radial basis function network to the p inputs of the training set can be gathered together in a $p \times m$ matrix, H, called the *design matrix*, whose individual components are obviously

$$H_{ij} = h_j(\mathbf{x}_i)\,.$$

If the network weights are $\mathbf{w} = (w_1\ w_2\ \ldots\ w_m)^T$ then the outputs of the network in response to the p inputs are given by $H\mathbf{w}$. It follows that the sum of the squared errors is

$$E = (\mathbf{y} - H\mathbf{w})^T(\mathbf{y} - H\mathbf{w})\,.$$

This is the quantity which is minimised to find the optimum weights, once the centres and radii on which the matrix H depends have been determined. Now E is quadratic in the weight vector and has a unique minimum at

$$\mathbf{w}^* = (H^T H)^{-1} H^T \mathbf{y}\,.$$

Thus we obtain

$$E^* = \mathbf{y}^T (I_p - H(H^T H)^{-1} H^T)^2 \mathbf{y}\,.$$

The choice of centres is important because it determines the number of free parameters of the model. Too few centres and the network may not be capable of generating a good approximation of the target function, too many centres and it may fit misleading variations due to imprecise or noisy data. One can use unsupervised clustering techniques on the inputs of the training data

$$\{\mathbf{x}_k\}_{k=1}^{p}$$

to generate the centres, but as this ignores the output data

$$\{y_k\}_{k=1}^{p}$$

it provides no control over model complexity. A direct approach to the model complexity issue is to select a subset of centres from a larger set which, if used in its entirety, would overfit the data (produce a model which is too complex).

As a simple example consider the *logistic map*

$$f(x) = 4x(1-x)$$

where we generate a training set of $T = 100$ values, i.e.

$$y_t = f(x_t), \qquad x_{t+1} = y_t$$

with $t = 1, 2, \ldots, T$. For the approximation of the function we select as center the first $n_c = 5$ points, i.e. $c_j = x_j$ with $j = 1, \ldots, 5$. The approximation of the logistic map is

$$\varphi(x, \mathbf{w}) := w_0 + \sum_{j=1}^{n_c} w_j h_j(x)$$

with the hidden unit transfer function

$$h_j(x) = \exp(-\beta(x - c_j)^2), \quad j = 1, \ldots, n_c$$

where β is a fixed positive constant. One considers $h_0(x) = 1$ and w_0 as the bias. The gradient descent method is used to calculate the weights from the traning set

$$w_0(t+1) = w_0(t) + \eta(y_t - \varphi(x_t, \mathbf{w}))$$
$$w_j(t+1) = w_j(t) + \eta h_j(x_t)(y_t - \varphi(x_t, \mathbf{w})), \quad j = 1, \ldots, n_c$$

repeated for $t = 1, 2, \ldots, T$. The constant η is the learning rate.

```
// Radial.cpp

#include <iostream>
#include <cmath>
using namespace std;

int main(void)
{
 int T = 100;
 double* x = new double[T+1];
 double* y = new double[T];
 x[0] = (sqrt(5.0)-1.0)/2.0; y[0] = 0.0;
 for(int t=0;t<T;t++)
 { y[t] = 4.0*x[t]*(1.0-x[t]); x[t+1] = y[t]; }
 double beta = 6.0;
 double eta = 0.3;  // learning rate
 int nc = 6;
 double* w = new double[nc];
 for(int n=0;n<nc;n++) { w[n] = 1.0; }
 double* c = new double[nc];
 c[0] = 0.0; c[1] = 0.1; c[2] = 0.28901376;
 c[3] = 0.36; c[4] = 0.82193922; c[5] = 0.9216;
 double* h = new double[nc]; h[0] = 1.0;
```

```
 int sessions = 1;
 int i, j, k;
 double dx, dy, phi;
 for(k=0;k<sessions;k++)
 {
 for(i=0;i<T;i++)
 {
 for(j=1;j<nc;j++) { dx = x[i]-c[j]; h[j] = exp(-beta*dx*dx); }
 phi = 0.0;
 for(j=0;j<nc;j++) { phi += w[j]*h[j]; }
 dy = y[i]-phi;
 for(j=0;j<nc;j++) { w[j] += eta*h[j]*dy; }
 }
 }

 for(int l=0;l<nc;l++)
 { cout << "w[" << l << "]=" << w[l] << endl; }
 delete[] x; delete[] y; delete[] w; delete[] c; delete[] h;
 return 0;
}
```

Linear models have been studied in statistics for about 200 years and the theory is applicable to radial basis function networks which are just one particular type of linear model. However, the fashion for neural networks, which started in the mid-80's, has given rise to new names for concepts already familiar in statistics. The table gives some examples. Such terms can be used interchangeably.

statistics	*neural networks*
model	network
estimation	learning
regression	supervised learning
interpolation	generalisation
observations	training set
parameters	(synaptic) weights
independent variables	inputs
dependent variables	outputs
ridge regression	weight decay

Table: Equivalent terms in statistics and neural networks

13.8 Recursive Deterministic Perceptron Neural Networks

The recursive deterministic perceptron feed-forward multilayer neural network is a generalization of the single layer perceptron topology (Elizondo et al [54]). This model can solve any two-class classification problem as opposed to the single layer perceptron topology. The construction of a recursive deterministic perceptron feed-forward multilayer neural network is done automatically and convergence is always guaranteed. The basic idea is to map into higher dimensional Euclidean spaces.

We illustrate the approach with the XOR problem. As described above this problem consists of classifying the two classes of vectors in the Euclidean space $\mathbb{R}^2$

$$X := \{ (0,0), (1,1) \}, \qquad Y := \{ (0,1), (1,0) \}$$

which are not linearly separable. To perform the non-linearly separable to linearly separable transform, a subset of patterns of the same class, which is linearly separable from the rest of the patterns is selected. We select the subset

$$\{ (0,0) \} \subset X \cup Y .$$

The sets $\{ (0,0) \}$ and $\{ (0,1), (1,0), (1,1) \}$ are linearly separable by the line

$$P_1 = \{ (x_1, x_2) \in \mathbb{R}^2 \, : \, 2x_1 + 2x_2 - 1 = 0 \} .$$

The intermediate neuron IN1 corresponding to the single layer perceptron topology of weight vector $\mathbf{w} = (2,2)$ and threshold $\theta = -1$ associated with the line P_1 is created. The output of IN1 is added to the (four) input vectors of X and Y. One column is added by assigning the value -1 to the input pattern $(0,0)$ and the value 1 to the remaining three input patterns $(0,1)$, $(1,0)$, $(1,1)$. This single layer perceptron topology provides the following sets of augmented input vectors

$$X' = \{ (0,0,-1), (1,1,1) \}, \qquad Y' = \{ (0,1,1), (1,0,1) \} .$$

These vectors are in the Euclidean space $\mathbb{R}^3$. In $\mathbb{R}^3$ these vectors can be linearly separated by the plane

$$P_2 = \{(x_1, x_2, x_3) \in \mathbb{R}^3 \, : \, -2x_1 - 2x_2 + 4x_3 - 1 = 0 \} .$$

Now a second intermediate neuron IN2 (output neuron) which corresponds to the single layer preceptron topology with the weight vector

$$\mathbf{w} = (-2, -2, 4)$$

and the threshold $\theta = -1$, associated to the plane, is created. Thus we obtain a two layer recursive deterministic perceptron neural networks solving the XOR classification problem since the output value of this neural network is -1 for the vectors $(0,0)$, $(1,1)$ and 1 for the vectors $(0,1)$, $(1,0)$. Its formal description is $[((2,2),-1),((-2,-2,4),-1)]$.

Algorithms for the general case are described by Elizondo et al [54].

Exercise. Apply this approach to classify

$$(0,0,0), \quad (1,1,1) \quad \text{with} \quad +1$$

$$(0,0,1), \ (0,1,0), \ (0,1,1), \ (1,0,0), \ (1,0,1), \ (1,1,0) \quad \text{with} \quad -1 .$$

13.9 Chaotic Neural Networks

In designing a neural network, it is usually of prime importance to guarantee the convergence of dynamics for the corresponding system. On the other hand, richer dynamics provide wider applications. For example, transient chaotic behaviours provide higher searching ability for globally optimal or near optimal solutions, in using neural network models as an approximation method for the combinatorial optimization problems. A chaotic neural network can be constructed with chaotic neurons by considering the spatio-temporal summation of both external inputs and feedback inputs from other chaotic neurons (Chen and Aihara [31]). An example of the dynamics of a transient chaotic neural network with N neurons is given by

$$\begin{aligned} x_i(t) &= \frac{1}{1+\exp(-\lambda y_i(t))} \\ y_i(t+1) &= k y_i(t) + \alpha \left(\sum_{j=1, j\neq i}^{N} W_{ij} x_j(t) + I_i \right) - z_i(t)(x_i(t) - I_0) \\ z_i(t+1) &= (1-\beta) z_i(t) \end{aligned}$$

where $i = 1, 2, \ldots, N$, $t = 0, 1, \ldots$, and

$x_i(t)$ output of neuron i $(0 \leq x_i \leq 1)$
$y_i(t)$ internal state of neuron i
$z_i(t)$ self-feedback connection weight $(z_i > 0)$

with

W_{ij} connection weight between neuron i and j
λ slope parameter of sigmoid function
k decay parameter of y_i $(0 \leq k \leq 1)$
α positive scaling parameter for inputs
β decay parameter of z_i $(0 \leq \beta \leq 1)$
I_i external input to neuron i
I_0 positive constant

We assume that a neuron does not receive a feedback from itself, i.e. $W_{jj} = 0$. The transiently chaotic neural network is characterized as follows. At the initial state, the variable $z_j(0)$ is so large that a network state is chaotic and the network searches a global solution. Then $z_j(t)$ is gradually decreased with time t. The network is also gradually changed from a chaotic state to a steady state. An optimization process of the transiently chaotic neural network is regarded as a chaotic simulated annealing. The network can be used to find solutions of the traveling salesman problem.

Another chaotic neural network considered in the literature (Tan et al [203], Bauer and Martienssen [14]) is given by

$$S_i(t+1) = f\left(\sum_{j=1}^{N} J_{ij} S_j(t)\right), \qquad i = 1, 2, \ldots, N$$

with

$$f(x) := \tanh(\alpha x) e^{-\beta x^2}$$

where α and β are positive constants and $t = 0, 1, \ldots$. Here the N neurons are connected to one another through the synapses J_{ij} ($i, j = 1, 2, \ldots, N$). Depending on the matrix elements J_{ij} and the parameters α and β, the system can be chaotic or nonchaotic. For example, for $N = 4$, $\alpha = 3$, $\beta = 2$, and

$$J = \begin{pmatrix} 0.01 & 1.50 & 0.03 & 0.01 \\ 1.60 & 0.01 & 0.10 & 1.00 \\ 0.03 & 0.02 & 0.01 & 1.50 \\ 0.00 & 1.00 & 1.70 & 0.01 \end{pmatrix}$$

the system is hyperchaotic. This hyperchaotic system with $N = 4$, $\alpha = 3$, $\beta = 2$ has infinitely many unstable periodic points embedded in a finite phase space. This chaotic neural network can be used for pattern recognition using the fixed points of the map as the patterns themselves.

13.10 Neuronal-Oscillator Models

We discuss three different neuronal-oscillator models - the *Stein neuronal model*, *Van der Pol oscillator*, and the *FitzHugh-Nagumo model* (Collins and Stewart [38], Collins and Richmond [39]).

The Stein neuronal model can provide oscillatory output. It is defined by the system of first order differential equations

$$\begin{aligned} \frac{dx_j}{dt} &= a\left(-x_j + \frac{1}{1+\exp(-f_{cj}(t) - by_j + bz_j)}\right) \\ \frac{dy_j}{dt} &= x_j - py_j \\ \frac{dz_j}{dt} &= x_j - qz_j \end{aligned}$$

for $j = 1, 2, 3, 4$ where $x_j(t)$ represents the membrane potential (or the firing rate) of the jth neuronal oscillator, a is a rate constant affecting the frequency of the oscillations, f_{cj} is the driving signal for the oscillator j, b allows the model to adapt to a change in stimulus, and p and q control the rate of adaption. Adaption refers to the time-dependent decline in the firing rate of the model following the application of a step change in the driving stimulus. The driving signal f_{cj} is assumed to have both a steady-state (tonic) component and a periodic component and is given by

$$f_{cj}(t) = f\left(1 + k_1 \sin(\omega t) + \sum_{k=1}^{4} \lambda_{kj} x_k(t)\right)$$

where f is an amplitude parameter, k_1 and ω control the amplitude and frequency, respectively, of the periodic component of the driving signal, λ_{kj} is the coupling term that represents the strength of oscillator k's effect on oscillator j, and $x_k(t)$ is the membrane potential of oscillator k. The value of λ_{kj} is normally set to -0.2 if oscillator k inhibited oscillator j, and its value was set to 0.0 if oscillator k did not affect oscillator j.

The *driven Van der Pol oscillator* as model for a neuronal-oscillator is given by

$$\frac{d^2 x_j}{dt^2} + \mu(x_{aj}^2 - p^2)\frac{dx_j}{dt} + \Omega^2 x_{aj} = q(1 + k_1 \sin(\omega t))$$

for $j = 1, 2, 3, 4$ where $x_j(t)$ is the output signal from oscillator j, x_{aj} is the same signal affected by the coupling, μ controls the degree of nonlinearity of the oscillator and therefore affects the shape of its waveform, p controls the amplitude of the oscillations, Ω influences the frequency of the oscillations, q is an amplitude parameter, and k_1 and ω control the amplitude and frequency, respectively, of the periodic component of the driving signal. A possible coupling between the oscillators could be

$$x_{aj}(t) := x_j(t) + \sum_{k=1}^{4} \lambda_{kj} x_k(t)$$

where λ_{kj} is a coupling term that represents the strength of oscillator k's effect on oscillator j, and $x_k(t)$ is the output signal from oscillator k. The value of λ_{jk} is normally set to -0.2 if oscillator k inhibited oscillator j, and its value is set to 0.0 if oscillator k did not affect oscillator j.

The *FitzHugh-Nagumo model* is the coupled system of first order ordinary differential equations

$$\frac{dx_j}{dt} = c\left(y_j + x_j + \frac{x_j^3}{3} + f_{cj}(t)\right), \quad \frac{dy_j}{dt} = -\frac{1}{c}(x_j - a + by_j)$$

for $j = 1, 2, 3, 4$ where $x_j(t)$ is the membrane potential of the jth neuronal oscillator, $f_{cj}(t)$ is the driving signal for oscillator j, and a, b, and c are constants that do not correspond to any particular physiological parameters. The driving signal $f_{cj}(t)$ is assumed to have a tonic component and a periodic component, i.e.

$$f_{cj}(t) = f_a + f_b\left(k_1 \sin(\omega t) + \sum_{k=1}^{4} \lambda_{kj} x_k(t)\right)$$

where f_a is the steady-state value of the driving signal, f_b is an amplitude parameter that affects the magnitude of the variable component of the driving signal, k_1 and ω control the amplitude and frequency, respectively, of the periodic component of the driving signal, λ_{kj} is a coupling term that represents the strength of oscillator k's effect on oscillator j, and $x_k(t)$ is the membrane potential of oscillator k. The values of λ_{kj} are normally set to -0.2 if oscillator k inhibited oscillator j, and its value is set to 0.0 if oscillator k did not affect oscillator j.

A system of differential equations to model a generic neuron is given by

$$\tau_m \frac{dV_m(t)}{dt} = -V_m(t) + (I + g_K(t)(V_{eK} - V_m(t)))R$$
$$\tau_\theta \frac{d\theta(t)}{dt} = -(\theta(t) - \theta_0) + \theta_{max} V_m(t)$$
$$\tau_{gK} \frac{dg_K(t)}{dt} = -g_K(t) + g_{K,max} H(V_m(t), \theta(t))$$

where

$$H(V_m(t), \theta(t)) = \frac{1}{1 + \exp(-(V_m(t) + \theta(t))/\beta)} .$$

Here $V_m(t)$ is the cellular membrane potential, $H(V_m(t), \theta(t))$ the analogous probability of being in the active state, $\theta(t)$ a threshold, $g_k(t)$ a generic potassium conductance, R the average membrane resistance. The parameters I, V_{eK}, θ_0, θ_{max} and $g_{K,max}$ are described below. The behaviour of this model is as follows: every time the neuron fires (i.e. it generates an action potential), an increment of $g_K(t)$ is released, which subsequently decays with an exponential rate. This ensures, via the bounding between $V_m(t)$ and the dynamic threshold in the second equation, a mechanism of ripolarization, whose temporal modalities are regulated mainly by the parameters θ_{max} (that is the threshold sensitivity, while $g_{K,max}$ holds the same significance for $g_K(t)$) and τ_θ. The neuron's sensitivity to the input current I is determined by θ_0 (the baseline threshold potential) and by V_{eK} (the reversal potential of potassium).

13.11 Neural Network, Matrices and Eigenvalues

Let A be an $n \times n$ over $\mathbb{R}$. We can use a neural network (Liu et al [131]) to find the eigenvalues and eigenvectors of A if A satisfies certain conditions. Let I_n be the $n \times n$ unit matrix and 0_n the $n \times n$ zero matrix. The functional neural network is given by the nonlinear system of first order ordinary differential equations with $2n$ components

$$\frac{d\mathbf{u}}{dt} = (S + B(t))\mathbf{u}$$

where $\mathbf{u}(t) = (u_1(t), u_2(t), \ldots, u_{2n}(t))^T$, S is the $2n \times 2n$ matrix

$$S := \begin{pmatrix} 0_n & A \\ -A & 0_n \end{pmatrix}$$

and

$$B(t) := \begin{pmatrix} -(\mathbf{e}_1\mathbf{u}(t))I_n & (\mathbf{e}_2\mathbf{u}(t))I_n \\ -(\mathbf{e}_2\mathbf{u}(t))I_n & -(\mathbf{e}_1\mathbf{u}(t))I_n \end{pmatrix}$$

with $\mathbf{e}_1$ and $\mathbf{e}_2$ the row vectors

$$\mathbf{e}_1 = (\underbrace{1,\ldots,1}_{n},\underbrace{0,\ldots,0}_{n}), \qquad \mathbf{e}_2 = (\underbrace{0,\ldots,0}_{n},\underbrace{1,\ldots,1}_{n}).$$

Thus $\mathbf{e}_1\mathbf{u}(t)$, $\mathbf{e}_2\mathbf{u}(t)$ are scalar products. Here $\mathbf{u}(t)$ is considered as the states of the neurons, $(S + B(t))$ is viewed as synaptic connection weights, and the activation functions are linear functions. Thus the system of first order differential equations describes a continuous time functional neural network.

Exercise. Let λ_1, λ_2, …, λ_n be the eigenvalues of A. What are the eigenvalues of S? Hint. Show that S can be written as a Kronecker product. Is the matrix S skew-symmetric over $\mathbb{R}$?

We set $\mathbf{u} = (x_1, \ldots, x_n, y_1, \ldots, y_n)^T$. Then the system can be cast into the form

$$\frac{d\mathbf{x}}{dt} = A\mathbf{y}(t) - \sum_{j=1}^{n} x_j(t)\mathbf{x}(t) + \sum_{j=1}^{n} y_j(t)\mathbf{y}(t)$$

$$\frac{d\mathbf{y}}{dt} = -A\mathbf{x}(t) - \sum_{j=1}^{n} y_j(t)\mathbf{x}(t) - \sum_{j=1}^{n} x_j(t)\mathbf{y}(t).$$

Now we define

$$\mathbf{z}(t) := \mathbf{x}(t) + i\mathbf{y}(t).$$

Then the system of differential equations can be written as

$$\frac{d\mathbf{z}(t)}{dt} = -iA\mathbf{z}(t) - \sum_{j=1}^{n} z_j(t)\mathbf{z}(t).$$

This is a *projective Riccati system* (Kowalski and Steeb [121]). The fixed points (equilibrium points) of this system are the solution of the equations

$$-iA\mathbf{z}^* = \left(\sum_{j=1}^{n} z_j^*\right)\mathbf{z}^* \quad \text{or} \quad A\mathbf{z}^* = \left(i\sum_{j=1}^{n} z_j^*\right)\mathbf{z}^*.$$

This is the eigenvalue equation for A (under the assumption $\mathbf{z}^* \neq \mathbf{0}$) with the eigenvalue

$$\lambda = \lambda_r + i\lambda_i = i\sum_{j=1}^{n} z_j^*.$$

If the fixed point is stable, there is a neighbourhood of the fixed point such that

$$\lim_{t\to\infty} \mathbf{z}(t) = \mathbf{z}^*.$$

Let $\lambda_1, \lambda_2, \ldots, \lambda_n$ be the eigenvalues of A and let $\mathbf{v}_1, \mathbf{v}_2, \ldots, \mathbf{v}_n$ be the corresponding eigenvectors. Note that it can happen that the number of eigenvectors of the matrix A can be less than n with the eigenvalues degenerate.

The solution of the initial value problem of the system is given by

$$\mathbf{z}(t) = \frac{\sum_{j=1}^{n} z_j(0)\exp(-i\lambda_j t)\mathbf{v}_j}{1 + \sum_{k=1}^{n} z_k(0)\int_0^t \exp(-i\lambda_k \tau)d\tau}.$$

Depending on the structure of A the solution tends to an eigenvector with the corresponding eigenvalue.

Chapter 14

Genetic Algorithms

14.1 Introduction

The interest in heuristic search algorithms with underpinnings in natural and physical processes began in the 1970s. Simulated annealing is based on thermodynamic considerations, with annealing interpreted as an optimization procedure. Evolutionary methods draw inspiration from the natural search and selection processes leading to the survival of the fittest. Simulated annealing and evolutionary methods use a probabilistic search mechanism to locate the global optimum solution in a multimodal landscape.

Genetic algorithms (Holland [97], Goldberg [69], Michalewicz [145], Mitchell [150]) are self-adapting strategies for searching, based on the random exploration of the solution space coupled with a memory component which enables the algorithms to learn the optimal search path from experience. They are the most prominent, widely used representatives of evolutionary algorithms, a class of probabilistic search algorithms based on the model of organic evolution. The starting point of all evolutionary algorithms is the *population* (also called *farm*) of *individuals* (also called *animals*, *chromosomes*, *strings*). The individuals are composed of genes which may take on a number of values (in most cases 0 and 1) called alleles. This means the value of a gene is called its allelic value, and it ranges on a set that is usually restricted to $\{0,1\}$. Thus these individuals are represented as binary strings of fixed length, for example a bitstring with 16 bits (2 bytes)

```
"1000101110111110"
```

or with 32 bits (4 bytes). The `bitset` class of C++ provides bitstrings of arbitrary length.

If the binary string has length N, then 2^N binary strings can be formed. Each of the individuals represents a search point in the space of potential solutions to a given optimization problem. Then random operators model selection, reproduction, crossover and mutation. The optimization problem gives quality information (fitness function or short fitness) for the individuals and the selection process favours individuals of higher fitness to transfer their information (string) to the next generation. The fitness of each string is the corresponding function value. Genetic algorithms are specifically designed to treat problems involving large search spaces containing multiple local minima. Thus the algorithms can be applied to optimization problems. Examples are solutions of ordinary differential equations, the smooth genetic algorithm, genetic algorithms in coding theory, Markov chain analysis, the DNA molecule.

To find an optimal solution, a *fitness function* (also called *cost function*) is used to represent the quality of the solution. The fitness function must not necessarily be differentiable. The fitness function to be optimized can be viewed as a multi-dimensional surface where the height of a point on the surface gives the value of the function at that point. In case of a minimization problem, the wells represent high-quality solutions while the peaks represent low-quality solutions.

The search techniques can be classified into three basic categories.

(1) *Classical or calculus-based.* This uses a deterministic approach to find the best solution. This method requires the knowledge of the gradient or higher-order derivatives. The technique can be applied to well-behaved problems.

(2) *Enumerative.* With these methods, all possible solutions are generated and tested to find the optimal solution. This requires excessive computation in problems involving a large number of variables.

(3) *Random.* Guided random search methods are enumerative in nature; however, they use additional information to guide the search process. Simulated annealing and evolutionary algorithms are typical examples of this class of search methods.

14.2 Sequential Genetic Algorithm

The genetic algorithm evolves a multiset of elements called a population of individuals or farm of animals. Each individual A_i $(i = 1, \ldots, n)$ of the population $\mathbf{A}$ represents a trial solution of the optimization problem to be solved. Individuals are usually represented by strings of variables, each element of which is called a gene. The value of a gene is called its allelic value, and it ranges on a set that is usually restricted to $\{0, 1\}$.

The population of individuals is also called a farm of animals in the literature. Furthermore an individual or animal is also called a chromosome or string.

A genetic algorithm is capable of maximizing a given fitness function f computed on each individual of the population. If the problem is to minimize a given objective function, then it is required to map increasing objective function values into decreasing f values. This can be achieved by a monotonically decreasing function. The standard genetic algorithm (Holland [97], Goldberg [69], Michalewicz [145], Mitchell [150]) is as follows

Step 1. Randomly generate an initial population $\mathbf{A}(t=0) := (A_1(t=0), \ldots, A_n(t=0))$

Step 2. Compute the values of fitness functions $f(A_i(t))$ of each individual $A_i(t)$ of the current population $\mathbf{A}(t)$

Step 3. Generate an intermediate population $\mathbf{A}_r(t)$ by applying the reproduction operator

Step 4. Generate $\mathbf{A}(t+1)$ by applying some other operators to $\mathbf{A}_r(t)$

Step 5: $t := t+1$ if not *(end_test)* goto Step 2.

The most commonly used operators are the following:

1) Reproduction (selection). This operator produces a new population, $\mathbf{A}_r(t)$, extracting with repetition individuals from the old population, $\mathbf{A}(t)$. The extraction can be carried out in several ways. One of the used method is the roulette wheel selection, where the extraction probability $p_r(A_i(t))$ of each individual $A_i(t)$ is proportional to its value of the fitness function $f(A_i(t))$.

2) Crossover. This operator is applied in probability, where the crossover probability is a system parameter, p_c. To apply the standard crossover operator (several variations have been proposed) the individuals of the population are randomly paired. Each pair is then recombined, choosing one point in accordance with a uniformly distributed probability over the length of the individual strings (parents) and cutting them in two parts accordingly. The new individuals (offspring) are formed by the juxtaposition of the first part of one parent and the last part of the other parent.

3) Mutation. The standard mutation operator modifies each allele of each individual of the population in probability, where the mutation probability is a system parameter, p_m. Usually, the new allelic value is randomly chosen with uniform probability distribution.

4) Local search. The necessity of this operator for optimization problems is still under debate. Local search is usually a simple gradient-descent heuristic search that carries each solution to a local optimum. The idea behind this is that search in the space of local optima is much more effective than search in the whole solution space.

The purpose of *parent selection* (also called setting up the farm of animals) in a genetic algorithm is to give more reproductive chances, on the whole, to those population members that are the most fit. We use a binary string as a chromosome to represent real value of the variable x. The length of the binary string depends on the required precision. A population or farm could look like

```
"1010111001111111"
"0011110101010000"
..................
"1111111010101011" <- individual (chromosome, animal, string)
..................
"1010111001000011"
```

For the *crossover operation* the individuals of the population are randomly paired. Each pair is then recombined, choosing one point in accordance with a uniformly distributed probability over the length of the individual strings (parents) and cutting them in two parts, accordingly. The new individuals (offspring) are formed by the part of one part and the last part of the other. An example is

```
"1011011000100101"  parent 1
"0010110110110111"  parent 2
     |       |
"1011010110110101"  child 1
"0010111000100111"  child 2
```

The *mutation operator* modifies each allele (a bit in the bitstring) of each individual of the population in probability. The new allele value is randomly chosen with uniform probability distribution. An example is

```
"1011011001011001"  parent
    |
"1011111001011001"  child
```

The bit position is randomly selected. Whether the child is selected is decided by the fitness function.

We have to map the binary string into a real number x with a given interval $[a, b]$ ($a < b$). The length of the binary string depends on the required precision. The total length of the interval is $b - a$. The binary string is denoted by

$$s_{N-1}s_{N-2}\cdots s_1 s_0$$

where s_0 is the least significant bit (LSB) and s_{N-1} is the most significant bit (MSB). In the first step we convert from base 2 to base 10

$$m = \sum_{i=0}^{N-1} s_i 2^i \,.$$

In the second step we calculate the corresponding real number on the interval $[a, b]$

$$x = a + m\frac{b-a}{2^N - 1}.$$

Obviously if the bit string is given by "000...00" we obtain $x = a$ and if the bit-string is given by "111...11" we obtain $x = b$ since "111...11" $\rightarrow 2^N - 1$.

Example. In the one-dimensional case consider the binary string 10101101 of length 8 (1 byte) and the interval $[-1, 1]$. Therefore

$$m = 1 \cdot 2^0 + 1 \cdot 2^2 + 1 \cdot 2^3 + 1 \cdot 2^5 + 1 \cdot 2^7 = 173.$$

Thus

$$x = -1 + 173\frac{2}{256-1} = 0.357.$$

♣

In higher dimensions we proceed as follows. We consider the two-dimensional case. The extension to higher dimensions is straightforward. Consider the two-dimensional domain

$$[a, b] \times [c, d], \quad a < b, \quad c < d$$

which is a subset of $\mathbb{R}^2$. The coordinates are x_1 and x_2, i.e. $x_1 \in [a, b]$ and $x_2 \in [c, d]$. Given a bitstring

$$s_{N-1}s_{N-2}\cdots s_{N_1}s_{N_1-1}s_{N_1-2}\cdots s_1 s_0$$

of length $N = N_1 + N_2$. The block

$$s_{N_1-1}s_{N_1-2}\cdots s_1 s_0$$

is identified with m_1, i.e.

$$m_1 = \sum_{i=0}^{N_1-1} s_i 2^i$$

and therefore

$$x_1 = a + m_1\frac{b-a}{2^{N_1} - 1}.$$

The block

$$s_{N-1}s_{N-2}\cdots s_{N_1}$$

is identified with the variable m_2, i.e.

$$m_2 = \sum_{i=N_1}^{N-1} s_i 2^{i-N_1}$$

and therefore

$$x_2 = c + m_2\frac{d-c}{2^{N_2} - 1}$$

where $N_2 = N - N_1$. .

Example. In the two-dimensional case consider the binary string 0000000000000000 with $N_1 = N_2 = 8$ and the domain $[-1, 1] \times [-1, 1]$. Then we find $m_1 = m_2 = 0$, $x_1 = -1$ and $x_2 = -1$.

♣

14.3 Schemata Theorem

A schema (Holland [97],Goldberg [69]) is a similarity template describing a subset of strings with similarities at certain string positions. We consider the binary alphabet $\{0, 1\}$. We introduce a schema by appending a special symbol to this alphabet. We add the $*$ or don't care symbol which matches either 0 or 1 at a particular position. With this extended alphabet we can now create strings (schemata) over the ternary alphabet

```
{ 0, 1, * } .
```

A schema matches a particular string if at every location in the schema `1` matches a `1` in the string, a `0` matches a `0`, or a `*` matches either.

Example. Consider the strings and schemata of length 5. The schema `*101*` describes a subset with four members `01010, 01011, 11010, 11011.` ♣

Consider a population of individuals (strings) A_j, $j = 1, 2, \ldots, n$ contained in the population $\mathbf{A}(t)$ at time (or generation) t ($t = 0, 1, 2, \ldots$) where the boldface is used to denote a population. Besides notation to describe populations, strings, bit positions, and alleles, one needs a notation to describe the schemata contained in individual strings and populations. Consider a schema H taken from the three-letter alphabet

$$V := \{0, 1, *\}.$$

For alphabets of cardinality k, there are $(k+1)^l$ schemata, where l is the length of the string. Furthermore, recall that in a string population with n members there are at most $n \cdot 2^l$ schemata contained in a population because each string is itself a representative of 2^l schemata. These counting arguments provides the magnitude of information being processed by genetic algorithms.

All schemata are not created equal. Some are more specific than others. The schema `011*1**` is a more definite statement about important similarity than the schema `0******`. Certain schemata span more of the total string length than others. The schema `1****1*` spans a larger portion of the string than the schema `1*1****`. Two schema properties are introduced: schema order and defining length.

Definition. The order of a schema H, denoted by $o(H)$, is the number of fixed positions (in a binary alphabet, the number of 1's and 0's) present in the template.

Example. The order of the schema `011*1**` is 4, whereas the order of the schema `0******` is 1. ♣

Definition. The defining length of a schema H, denoted by $\delta(H)$, is the distance between the first and last specific string position.

Example. The schema `011*1**` has defining length $\delta = 4$ because the last specific position is 5 and the first specific position is 1. Thus $\delta(H) = 5 - 1 = 4$. ♣

Schemata provide the basic means for analyzing the net effect of reproduction and genetic operators on building blocks contained within the population. Consider the individual and combined effects of reproduction, crossover, and mutation on schemata contained within a population of strings. Suppose at a given time step t there are $m(H, t)$ examples of a particular schema H contained within the population $\mathbf{A}(t)$. During reproduction, a string is copied according to its fitness, or more precisely a string A_i gets selected with probability

$$p_i = \frac{f_i}{\sum_{j=1}^{n} f_j}.$$

After picking a non-overlapping population of size n with replacement from the population $\mathbf{A}(t)$, we expect to have $m(H, t+1)$ representatives of the schema H in the population at time $t+1$ as given by

$$m(H, t+1) = \frac{m(H, t) n f(H)}{\sum_{j=1}^{n} f_j(t)}$$

where $f(H)$ is the average fitness of the strings representing schema H at time t. The *average fitness* of the entire population is defined as

$$\bar{f} := \frac{1}{n} \sum_{j=1}^{n} f_j .$$

Thus we can write the reproductive schema growth equation as follows

$$m(H, t+1) = m(H, t) \frac{f(H)}{\bar{f}(t)}.$$

Assuming that $f(H)/\bar{f}$ remains relatively constant for $t = 0, 1, \ldots$ the preceding equation is a linear difference equation of first order

$$x(t+1) = kx(t)$$

with constant coefficient k which has the solution of the initial value problem

$$x(t) = k^t x(0) .$$

A particular schema grows as the ratio of the average fitness of the schema to the average fitness of the population. Schemata with fitness values above the population average will receive an increasing number of samples in the next generation, while schemata with fitness values below the population average will receive a decreasing number of samples. This behaviour is carried out with every schema H contained in a particular population $\mathbf{A}$ in parallel. In other words, all the schemata in a population grow or decay according to their schema averages under the operation of reproduction alone. Above-average schemata grow and below-average schemata die off. Suppose we assume that a particular schema H remains an amount $c\bar{f}$ above average with c a constant. Under this assumption we find the difference equation

$$m(H, t+1) = m(H, t)\frac{(\bar{f} + c\bar{f})}{\bar{f}} = (1 + c)m(H, t).$$

Starting at $t = 0$ and assuming a stationary value of c, we obtain the solution

$$m(H, t) = m(H, 0)(1 + c)^t.$$

This is a geometric progression or the discrete analog of an exponential form. Reproduction allocates exponentially increasing (decreasing) numbers of trials to above- (below-) average schemata. The fundamental theorem of genetic algorithms is as follows (Goldberg [69]).

Theorem. By using the selection, crossover, and mutation of the standard genetic algorithm, short, low-order, and above average schemata receive exponentially increasing trials in subsequent populations.

The short, low-order, and above average schemata are called building blocks, and the fundamental theorem indicates that building blocks are expected to dominate the population. Is this good or bad in terms of the original goal of function optimization? The preceding theorem does not answer this question. Rather, the connection between the fundamental theorem and the observed optimizing properties of the genetic algorithm is provided by the following conjecture.

The Building Block Hypothesis. The globally optimal strings in Ω

$$f : \Omega \to \mathbb{R} \quad \text{with} \quad \Omega = \{0, 1\}^n$$

may be partitioned into substrings that are given by the bits of the fixed positions of building blocks.

14.4 Bitwise Operations

14.4.1 Introduction

In genetic algorithms bitwise operations play the central role. In this section we describe these operations. The *truth tables* for the AND-gate, OR-gate, XOR-gate and NOT-gate are

AND		
0	0	0
0	1	0
1	0	0
1	1	1

OR		
0	0	0
0	1	1
1	0	1
1	1	1

XOR		
0	0	0
0	1	1
1	0	1
1	1	0

NOT	
0	1
1	0

The NAND-gate is an AND-gate followed by a NOT-gate. The NOR-gate is an OR-gate followed by a NOT-gate. Both are *universal gates*, i.e. all other gates can be built from these gates.

Let $x, y \in \{0,1\}$. Then the NOT-gate, AND-gate, OR-gate, NAND-gate, NOR-gate and XOR-gate can be expressed using arithmetic operations. We have

$$\begin{aligned} NOT &: 1 - x \\ AND &: xy \\ OR &: x + y - xy \\ NAND &: 1 - xy \\ NOR &: 1 - x - y + xy \\ XOR &: x + y - 2xy\,. \end{aligned}$$

Exercise. Show that the XOR-gate can be built from 4 NAND-gates. Show that the AND-gate can be built from 2 NAND gates. Show that the OR-gate can be built from 3 NAND-gates. Show that the NOR-gate can be built from 4 NAND-gates.

Next we consider bitstrings, such as `"10010010"`. In most cases the length of the bitstring is 8, 16, 32, 64 owing to the length of the data types `char`, `short`, `int`, `float`, `double` in C++. Then the bitwise operation to two bitstrings of the same length is applied to each bit at the same position of the two bitstrings.

Example. Consider the XOR-operation for two bitstrings of length 8. We have

```
       10101100
XOR    00110101
result 10011001
```

since $0 \oplus 0 = 0, 0 \oplus 1 = 1, 1 \oplus 0 = 1, 1 \oplus 1 = 0$, where $\oplus$ denotes the XOR operation. ♣

The bitstrings are counted from right to left starting at 0. In C, C++, and Java the bitwise operators for integer data types are

```
&  bitwise AND
|  bitwise OR   (inclusive OR)
^  bitwise XOR  (exclusive OR)
~  NOT operator (one's complement)
>> right-shift operator
<< left-shift operator
```

The basic bit operations `setbit`, `clearbit`, `swapbit` and `testbit` for bitstrings can be implemented in C++ as follows. The bit position b runs from 0 to 31 from right to left in the bit string since we use the datatype `unsigned long`.

The operation `setbit` sets a bit at a given position b (i.e. the bit at the position b is set to 1)

```
unsigned long b = 3;
unsigned long j = 15;
j |= (1 << b);          // shortcut for j = j | (1 << b);
```

The operation `clearbit` clears a bit at a given position b (i.e. the bit at the position b is set to 0)

```
unsigned long b = 3;
unsigned long j = 15;
j &= ~(1 << b);         // short cut for j = j & ~(1 << b);
```

The operation `swapbit` swaps the bit at the position b, i.e., if the bit is 0 it is set to 1 and if the bit is 1 it is set to 0.

```
unsigned long b = 3;
unsigned long j = 15;
j ^= (1 << b);          // short cut for  j = j^(1 << b);
```

The operation `testbit` returns 1 or 0 depending on whether the bit at the position b is set or not.

```
unsigned long b = 3;
unsigned long j = 15;
unsigned long result = ((j & (1 << b)) != 0);
```

The operations `setbit`, `clearbit`, `swapbit` and `testbit` are written as functions. This leads to the following C++ program. Obviously, for `setbit()`, `clearbit()` and `swapbit()` we pass `x` by reference.

```
// mysetbit.cpp

#include <iostream>
using namespace std;

inline void setbit(unsigned long& j,unsigned long b)
{  j |= (1 << b);  }

inline void clearbit(unsigned long& j,unsigned long b)
{  j &= ~(1 << b);  }

inline void swapbit(unsigned long& j,unsigned long b)
{ j ^= (1 << b);  }

inline unsigned long testbit(unsigned long j,unsigned long b)
{ return ((j & (1 << b)) != 0); }

int main(void)
{
 unsigned long b = 3;
 unsigned long j = 10;  // binary 1010
```

```
 setbit(j,b);
 cout << "j = " << j << endl; // 10 => binary 1010
 clearbit(j,b);
 cout << "j = " << j << endl; // 2  => binary 10
 swapbit(j,b);
 cout << "j = " << j << endl; // 10 => binary 1010
 unsigned long r = testbit(j,b);
 cout << "r = " << r << endl; // 1
 return 0;
}
```

14.4.2 Assembly Language

We could also use assembly language to do bitwise operations. In assembly language (Intel Pentium) the bitwise, shift and rotate commmands are

```
AND   bitwise AND
OR    bitwise inclusive OR
XOR   bitwise exclusive OR
BSF   bit scan forward
BSR   bit scan reverse
BSWAP byte swap
BT    bit test
BTC   bit test and complement
      BTC saves the value of the bit indicated by the base (first operand)
      and the bit offset (second operand) into the carry flag and
      then complements the bit
BTR   bit test and reset
      BTR saves the value of the bit indicated by the base (first operand)
      and the bit offset (second operand) into the carry flag and
      then stores 0 in the bit.
BTS   bit test and set
      BTS saves the value of the bit indicated by the
      base (first operand) and the bit offset (second operand)
      into the carry flag and stores 1 in the bit.
NEG   two's complement negation
NOT   one's complement negation
RCL   rotate through carry left instruction
RCR   rotate through carry right instruction
ROL   rotate left instruction
ROR   rotate right instruction
SAL   shift arithmetic left instruction
SAR   shift arithmetic right instruction
SHL   shift left instruction
SHR   shift right instruction
```

Using inline assembly language in C++ we can use these operations as follows. We want to find the minimum of the function

$$f(n) = 3n - 5n^2 + n^3 + 10$$

where n is of data type **unsigned long**. Then the functional f is of data type **unsigned long** when n is restricted the proper range. In the program the farm consists of four elements, a, b, c and d. The operation MOV is the move data operation. The used 32 bit registers are EAX, EBX, ECX, EDX.

```
// assembler.cpp

#include <iostream>
#include <ctime>
#include <cstdlib>
using namespace std;

unsigned long f(unsigned long n) { return 3*n-5*n*n+n*n*n+10; }

int main(void)
{
 unsigned long a = 1, b = 11, c = 5017, d = 1013;
 unsigned long r1, r2, r3, r4;
 srand((unsigned long) time(NULL));
 int T = 1000;
 for(int j=0;j<T;j++)
 {
 _asm
 {
 MOV EAX, a
 MOV EBX, b
 MOV ECX, c
 MOV EDX, d
 BSWAP EAX
 BSWAP EBX
 MOV r1, EAX
 MOV r2, EBX
 XOR EDX, ECX
 MOV r3, EDX
 AND ECX, EAX
 MOV r4, ECX
 }
 if(f(r1) < f(a)) a = r1; if(f(r2) < f(b)) b = r2;
 if(f(r3) < f(c)) c = r3; if(f(r4) < f(d)) d = r4;
 unsigned long s = rand()%32;

 _asm
 {
 MOV EAX, a
```

```
MOV ECX, s
BTS EAX, ECX
MOV r1, EAX
MOV EAX, b
MOV ECX, s
BTR EAX, ECX
MOV r2, EAX
MOV EAX, c
MOV ECX, s
BTC EAX, ECX
MOV r3, EAX
MOV EAX, d
MOV ECX, s
BTS EAX, ECX
MOV r4, EAX
}
if(f(r1) < f(a)) a = r1; if(f(r2) < f(b)) b = r2;
if(f(r3) < f(c)) c = r3; if(f(r4) < f(d)) d = r4;
} // end for loop
cout << "f(" << a << ") = " << f(a) << endl;
cout << "f(" << b << ") = " << f(b) << endl;
cout << "f(" << c << ") = " << f(c) << endl;
cout << "f(" << d << ") = " << f(d) << endl;
return 0;
}
```

14.4.3 Floating Point Numbers and Bitwise Operations

We can also do bitwise manipulations of floating point numbers, for example on the data type `double` (64 bits) in C++. The value of `double` is stored as

```
sign bit, 11 bit exponent, 52 bit mantissa
```

This means

```
 byte 1   byte 2    byte 3    byte 4          byte 8
SXXX XXXX XXXX MMMM MMMM MMMM MMMM MMMM ... MMMM MMMM
```

Thus the first bit is the sign bit, S, the next eleven bits are the exponent bits, E, the the final 52 bits are the fraction, F. Some of the bit combinations are not a number (NaN) owing to the IEEE Standard 754 (Kahan [108]). For example the values $+\infty$ and $-\infty$ are denoted with an exponent of all 1's and a fraction of all 0's. The sign distinguishes between negative infinity and positive infinity. We also have that -0 and $+0$ are distinct values. However they both compare as equal. If $E = 2047$ and F is nonzero, then V=NaN.

The C++ progran shows how to change a bit in `double`, where `|` is the bitwise OR and `<<` is the shift operation (Hardy et al [85]).

```
// doublebits.cpp

#include <iostream>
using namespace std;

double doublebitstring(int bit,double x)
{
 int* p = (int*) &x;
 if(bit < 32) *p |= (1 << bit);
 if(bit >= 32) *(p+1) |= (1 << (bit-32));
 return x;
}

int main(void)
{
 double x = 3.14159;
 for(int bit=0;bit<64;bit++)
 { cout.precision(20); cout << doublebitstring(bit,x) << endl; }
 return 0;
}
```

14.4.4 Java Bitset Class

Java has a `BitSet` class. The default constructor `BitSet()` creates a new bit set. The constructor `BitSet(int nbits)` creates a bit set whose initial size is the specified number of bits. The methods are

`void and(BitSet set)`

performs a logical AND

`void andNot(BitSet set)`

clears all of the bits in this `BitSet` whose corresponding bit is set in the specified `BitSet`

`void clear(int bitIndex)`

the bit with index `bitIndex` in this `BitSet` is changed to the clear (false) state

`boolean get(int bitIndex)`

returns the value of the bit with the specified index

`void or(Bitset set)`

performs a logical OR of this bit set with the bit set argument

```
void xor(BitSet set)
```

performs a logical XOR of this bit set with the bit set argument.

In the program we find the Hamming distance of two bitstrings of length 8.

```
// hamming.java

import java.util.*;

public class hamming
{
 public static void main(String[] args)
 {
 int length = 8;
 BitSet bs1 = new BitSet(length);
 bs1.set(0); bs1.set(4); bs1.set(7);
 BitSet bs2 = new BitSet(length);
 bs2.set(1); bs2.set(3); bs2.set(7);
 System.out.println("bs1 = " + bs1); System.out.println("bs2 = " + bs2);
 bs1.xor(bs2);    // bs1 contains result of xor operation
 System.out.println("bs1 = " + bs1); System.out.println("bs2 = " + bs2);
 int count = 0;
 for(int i=0;i<bs1.length();i++) { if(bs1.get(i)==true) count++; }
 System.out.println("count = " + count);
 } // end main
}
```

14.4.5 C++ Bitset Class

The Standard Template Library in C++ provides a `bitset` class. We can create bitstrings of any length. It includes all the necessary bitwise operation, for example `set()`, `flip()` and `test()` and the bitwise operation `&`, `|`, `^` as well as the shift operations. An example is

```
// bitset.cpp

#include <iostream>
#include <bitset>
#include <string>
using namespace std;

int main(void)
{
 const unsigned long n = 8;
 bitset<n> s;
 cout << s.set() << endl;    // set all bits to 1
 cout << s.flip(4) << endl; // flip at position 4
```

```
 cout << s.test(5) << endl; // test if bit position i is true
 return 0;
}
```

The next C++ program shows how to convert `unsigned int`, `float`, `double` into a `bitset` string.

```
// tobitset.cpp

#include <bitset>
#include <iostream>
#include <limits>
using namespace std;

template <class T> class bitsetfor
{
 public:
  static const int size
    = sizeof(T)*numeric_limits<unsigned char>::digits;
  bitset<size> bs;
};

template <class T> bitsetfor<T> convert(const T &t)
{
 bitsetfor<T> b;
 int bitsperuchar = numeric_limits<unsigned char>::digits;
 int bitsperT = b.size;
 unsigned char* tp = (unsigned char*)&t;
 for(int i=0;i<bitsperT;tp++)
  for(int j=0;j<bitsperuchar && i<bitsperT;j++,i++)
   b.bs[i] = ((*tp) >> j) & 1;
 return b;
}

int main(void)
{
 unsigned int i = 133;
 float f = 10.3; double d = 10.3;
 bitsetfor<unsigned int> bi = convert(i);
 cout << "unsigned int (" << bi.size << " bits) "
      << i << " -> " << bi.bs << endl << endl;
 bitsetfor<float> bf = convert(f);
 cout << "float (" << bf.size << "bits) "
      << f << " -> " << bf.bs << endl << endl;
 bitsetfor<double> bd = convert(d);
 cout << "double (" << bd.size << "bits) "
      << d << " -> " << bd.bs << endl << endl;
 return 0;
}
```

14.5 Bit Vector Class

In the following we provide a Bit Vector class which can be used for genetic algorithm programs.

```
// Bitvect.h

#include <cstring>
using namespace std;

#ifndef __BITVECTOR
#define __BITVECTOR

const unsigned char _BV_BIT[8] = { 1,2,4,8,16,32,64,128 };

class BitVector
{
 protected:
  unsigned char *bitvec;
  int len;
 public:
  BitVector();                     // default constructor
  BitVector(int nbits);            // constructor
  BitVector(const BitVector&);     // copy constructor
  ~BitVector();                    // destructor
  void SetBit(int bit,int val=1);
  int GetBit(int bit) const;
  void ToggleBit(int bit);
  BitVector operator & (const BitVector&) const;
  BitVector& operator &= (const BitVector&);
  BitVector operator | (const BitVector&) const;
  BitVector& operator |= (const BitVector&);
  BitVector operator ^ (const BitVector&) const;
  BitVector& operator ^= (const BitVector&);
  friend BitVector operator ~ (const BitVector&);
  BitVector& operator = (const BitVector&);
  int operator[](int bit) const;
  void SetLength(int nbits);
};

BitVector::BitVector() { len = 0; bitvec = NULL; }

BitVector::BitVector(int nbits)
{ len = nbits/8+((nbits%8)?1:0); bitvec = new unsigned char[len]; }

BitVector::BitVector(const BitVector& b)
{
 len = b.len;
```

```
 bitvec = new unsigned char[len];
 memcpy(bitvec,b.bitvec,len);
}

BitVector::~BitVector() { if(bitvec != NULL) delete[] bitvec; }

void BitVector::SetBit(int bit,int val)
{
 if(bit < 8*len)
 {
 if(val) bitvec[bit/8] |= _BV_BIT[bit%8];
 else bitvec[bit/8] &= ~_BV_BIT[bit%8];
 }
}

int BitVector::GetBit(int bit) const
{
 if(bit < 8*len) return ((bitvec[bit/8]&_BV_BIT[bit%8])?1:0);
 return -1;
}

void BitVector::ToggleBit(int bit)
{ if(bit<8*len) bitvec[bit/8] ^= _BV_BIT[bit%8]; }

BitVector BitVector::operator & (const BitVector& b) const
{
 int mlen = (len > b.len)?len:b.len;
 BitVector ret(mlen*8);
 for(int i=0;i<mlen;i++) ret.bitvec[i] = bitvec[i] & b.bitvec[i];
 return ret;
}

BitVector& BitVector::operator &= (const BitVector& b)
{
 int mlen = (len>b.len)?len:b.len;
 for(int i=0;i<mlen;i++) bitvec[i] &= b.bitvec[i];
 return *this;
}

BitVector BitVector::operator | (const BitVector& b) const
{
 int mlen = (len>b.len)?len:b.len;
 BitVector ret(mlen*8);
 for(int i=0;i<mlen;i++) ret.bitvec[i] = bitvec[i]|b.bitvec[i];
 return ret;
}

BitVector& BitVector::operator |= (const BitVector& b)
```

```
{
 int mlen = (len>b.len)?len:b.len;
 for(int i=0;i<mlen;i++) bitvec[i] |= b.bitvec[i];
 return *this;
}

BitVector BitVector::operator ^ (const BitVector& b) const
{
 int mlen = (len>b.len)?len:b.len;
 BitVector ret(mlen*8);
 for(int i=0;i<mlen;i++) ret.bitvec[i] = bitvec[i]^b.bitvec[i];
 return ret;
}

BitVector& BitVector::operator ^= (const BitVector& b)
{
 int mlen = (len>b.len)?len:b.len;
 for(int i=0;i<mlen;i++) bitvec[i] ^= b.bitvec[i];
 return *this;
}

BitVector operator ~ (const BitVector& b)
{
 BitVector ret(b.len*8);
 for(int i=0;i<b.len;i++) ret.bitvec[i] = ~b.bitvec[i];
 return ret;
}

BitVector& BitVector::operator = (const BitVector& b)
{
 if(bitvec==b.bitvec) return *this;
 if(bitvec!=NULL) delete[] bitvec;
 len = b.len;
 bitvec = new unsigned char[len];
 memcpy(bitvec,b.bitvec,len);
 return *this;
}

int BitVector::operator[](int bit) const { return GetBit(bit); }

void BitVector::SetLength(int nbits)
{
 if(bitvec!=NULL) delete[] bitvec;
 len = nbits/8+((nbits%8)?1:0);
 bitvec = new unsigned char[len];
}
#endif
```

14.6 Penna Bit-String Model

The Penna bit-string model for biological ageing was introduced by Penna [164] in 1995. In the asexual version of the Penna bit-string model each individual (organism genome) is represented by a single bitstring of 32 bits, for example

```
"00001000 00010100 01000001 01000011"
```

The number of 32 bits is computationally convenient. In the model one starts with an ideal set of zero bits for all individuals, i.e. an initial population of perfect individuals is set up (all bits are 0). For example the initial population could be 1000 individuals. Each individual can live at most for 32 timesteps ("years"). We start counting the bitstring from left to right starting from 0. A bit set to 0 in the bitstring means health, a bit set to 1 means an inherited disease starts to act from the age on which the position of this set bit is in the bitstring. If T (typically, $T = 3$) bits are active (i.e. set to 1 in the bitstring), their combined effect kills the individual. Thus whenever a bit 1 appears in the bitstring it is counted and the total count of 1's until the current individuals age is compared with the given limit T. There is a minimum reproduction age R (typically $R = 8$), from which the individual with probabilty p (typically $p = 0.05$), produces b offsprings every year (typically $b = 1$ or $b = 2$). One could also take into account a pregnancy period: after giving birth, an individual (mother) stays one time steps (year) without reproducing. The offspring genome is a copy of the parent's one, but each 0 in the bitstring can mutate to 1 with probability m. This means the child inherts the mother's genome except for M (typically $M = 1$) mutations of randomly selected bits where a 0 bit is set to 1. The Penna bit-string model does not allow for a positive mutation, i.e. a bit set to 1 cannot mutate into a 0.

For an implementation in C++ one can use the **bitset** class of the Standard Template Library.

```
// penna.cpp

#include <bitset>
#include <cstdlib>
#include <iostream>
#include <list>
using namespace std;

const size_t n = 32;

list<bitset<n> > penna(int size,int steps,int T=3,int R=8,
                       double p=0.1,int b=2,double m=0.1)
{
 int k, t;
 list<bitset<n> > population;
 list<bitset<n> >::iterator i, i1;
```

```
 list<int> ages;
 list<int>::iterator j, j1;

 while(size>0)
 {
 population.push_back(bitset<n>(0));
 ages.push_back(-1);
 size--;
 }

 while(steps>0)
 {
 for(i=population.begin(),j=ages.begin();i!=population.end();)
 {
 // age the individual
 (*j)++;
 // death due to age
 if(*j >= int(n)) { i1 = i; j1 = j; i++; j++;
 population.erase(i1); ages.erase(j1); continue; }

 // death due to disease
 for(t=k=0;k<=*j;k++) if(i->operator[](k)) t++;
 if(t >= T)
 { i1 = i; j1 = j; i++; j++;
 population.erase(i1); ages.erase(j1); continue; }

 // reproduction
 if(*j >= R && double(rand())/RAND_MAX < p)
 for(k=0;k<b;k++)
 {
 population.push_back(*i); ages.push_back(-1);
 // mutation
 for(t=0;t<int(n);t++)
 if(double(rand())/RAND_MAX < m) population.back().set(t);
 }
 i++; j++;
 }
 steps--;
 }
 return population;
}

int main(void)
{
 srand(time(NULL));
 list<bitset<n> > population = penna(100,100);
 list<bitset<n> >::iterator i = population.begin();
 cout << "Population size : " << population.size() << endl;
```

```
 cout << "Population : " << endl;
 for(;i!=population.end();i++) cout << *i << endl;
 return 0;
}
```

14.7 Maximum of One-Dimensional Maps

As an example we consider the fitness function

$$f(x) = \cos(x) - \sin(2x)$$

on the interval $[0 : 2\pi]$. In this interval the differentiable function f has three maxima. The global maximum is at 5.64891 and the two local maxima are at 0 and 2.13862.

A simple C++ program would include the following functions

```
// fitness function of individual
double f(double)

// fitness function value of individual
double f_value(double (*func)(double),int* arr,int& N,
                   double a,double b)

// x_value
double x_value(int* arr,int& N,double a,double b)

// setup of farm
void setup(int** farm,int M,int N)

// crossing two individuals
void crossings(int** farm,int M,int N)

// mutate an individual
void mutate(int** farm,int M,int N)
```

Here N is the length of the binary string and M is the size of the population, which is kept constant at each time step. For the given problem we select $N = 10$ and $M = 12$. The binary string "$s_{N-1}s_{N-2}...s_0$" is mapped into the integer number m and then into the real number x in the interval $[0 : 2\pi]$ as described above.

The farm is set up using a random number generator. The crossing function selects the two fittest strings from the two parents and the two children. The parents are selected by a random number generator. With a population of 12 strings in the farm we find after 1000 iterations the global maximum at 5.64891. For 100 iterations we find as a typical output the global maximum and the second highest maximum.

```
// genetic.cpp

#include <iostream>
#include <cstdlib>
#include <ctime>       // for srand, rand
#include <cmath>       // for cos, sin, pow
using namespace std;

// fitness function where maximum to be found
double f(double x) { return cos(x)-sin(2.0*x); }

// fitness function value for individual
double f_value(double (*func)(double),int* arr,int& N,
       double a,double b)
{
 double res; double m = 0.0;
 for(int j=0;j<N;j++) { double k = j; m += arr[N-j-1]*pow(2.0,k); }
 double x = a+m*(b-a)/(pow(2.0,N)-1.0);
 res = func(x);
 return res;
}

// x_value at global maximum
double x_value(int* arr,int& N,double a,double b)
{
 double m = 0.0;
 for(int j=0;j<N;j++) { double k = j; m += arr[N-j-1]*pow(2.0,k); }
 double x = a+m*(b-a)/(pow(2.0,N)-1.0);
 return x;
}

// setup the population (farm)
void setup(int** farm,int M,int N)
{
  srand((unsigned long) time(NULL));
  for(int j=0;j<M;j++)
   for(int k=0;k<N;k++) farm[j][k] = rand()%2;
}

// cross two individuals
void crossings(int** farm,int& M,int& N,double& a,double& b)
{
 int K = 2;
 int** temp = new int* [K];
 for(int i=0;i<K;i++) temp[i] = new int[N];
 double res[4];
 int r1 = rand()%M; int r2 = rand()%M;
 // random returns a value between 0 and one less than its parameter
```

```
 while(r2==r1) r2 = rand()%M;
 res[0] = f_value(f,farm[r1],N,a,b);
 res[1] = f_value(f,farm[r2],N,a,b);
 for(int j=0;j<N;j++)
 { temp[0][j] = farm[r1][j]; temp[1][j] = farm[r2][j]; }
 int r3 = rand()%(N-2) + 1;
 for(j=r3;j<N;j++)
 { temp[0][j] = farm[r2][j]; temp[1][j] = farm[r1][j]; }
 res[2] = f_value(f,temp[0],N,a,b); res[3] = f_value(f,temp[1],N,a,b);
 if(res[2] > res[0])
 { for(j=0;j<N;j++) farm[r1][j] = temp[0][j]; res[0] = res[2]; }
 if(res[3] > res[1])
 { for(j=0;j<N;j++) farm[r2][j] = temp[1][j]; res[1] = res[3]; }
 for(j=0;j<K;j++) delete[] temp[j]; delete[] temp;
}

// mutate an individual
void mutate(int** farm,int& M,int& N,double& a,double& b)
{
 double res[2];
 int r4 = rand()%N;  int r1 = rand()%M;
 res[0] = f_value(f,farm[r1],N,a,b);
 int v1 = farm[r1][r4];
 if(v1==0) farm[r1][r4] = 1; if(v1==1) farm[r1][r4] = 0;
 double a1 = f_value(f,farm[r1],N,a,b);
 if(a1 < res[0]) farm[r1][r4] = v1;
 int r5 = rand()%N;  int r2 = rand()%M;
 res[1] = f_value(f,farm[r2],N,a,b);
 int v2 = farm[r2][r5];
 if(v2==0) farm[r2][r5] = 1; if(v2==1) farm[r2][r5] = 0;
 double a2 = f_value(f,farm[r2],N,a,b);
 if(a2 < res[1]) farm[r2][r5] = v2;
}

int main(void)
{
 int M = 12;    // population (farm) has 12 individuals (animals)
 int N = 10;    // length of binary string
 int** farm = new int*[M];   // allocate memory for population
 for(int i=0;i<M;i++) { farm[i] = new int[N]; }
 setup(farm,M,N);
 double a = 0.0, b = 6.28318;  // interval [a,b]
 for(int k=0;k<1000;k++)
 { crossings(farm,M,N,a,b); mutate(farm,M,N,a,b); } // end for loop
 for(int j=0;j<N;j++)
 cout << "farm[1][" << j << "] = " << farm[1][j] << endl;
 cout << endl;
 for(j=0;j<M;j++)
```

```
 cout << "fitness f_value[" << j << "] = " << f_value(f,farm[j],N,a,b)
      << "  " << "x_value[" << j << "] = " << x_value(farm[j],N,a,b) << endl;
 for(j=0;j<M;j++) delete[] farm[j];
 delete[] farm;
 return 0;
}
```

In the C++ program given above we store a bit as `int`. This wastes a lot of memory space. A more optimal use of memory is to use a string, for example `"1000111101"`. Then we use 1 byte for 1 or 0. An even more optimal use is to manipulate the bits itself. In the following we use the bit Vector class described above to manipulate the bits. The bit Vector class is included in the header file `bitVect.h`.

```
// findmax.cpp

#include <iostream>
#include <cmath>        // for pow
#include <cstdlib>
#include <ctime>
#include "bitvect.h"
using namespace std;

double f(double x) { return cos(x)-sin(2*x);}

double f_value(double (*func)(double),const BitVector &arr,
               int &N,double a,double b)
{
 double res, m = 0.0;
 for(int j=0;j<N;j++) { double k = j; m += arr[N-j-1]*pow(2.0,k); }
 double x = a+m*(b-a)/(pow(2.0,N)-1.0);
 res = func(x);
 return res;
}

double x_value(const BitVector& arr,int& N,double a,double b)
{
 double m = 0.0;
 for(int j=0;j<N;j++) { double k=j; m += arr[N-j-1]*pow(2.0,k); }
 double x = a+m*(b-a)/(pow(2.0,N)-1.0);
 return x;
}

void setup(BitVector *farm,int M,int N)
{
 srand((unsigned)time(NULL));
 for(int j=0;j<M;j++)
  for(int k=0;k<N;k++) farm[j].SetBit(k,rand()%2);
}
```

```
void crossings(BitVector *farm,int &M,int &N,double &a,double &b)
{
 int K = 2, j;
 BitVector *temp = new BitVector[K];
 for(int i=0;i<K;i++) temp[i].SetLength(N);
 double res[4];
 int r1 = rand()%M; int r2 = rand()%M;
 while(r2==r1) r2 = rand()%M;
 res[0] = f_value(f,farm[r1],N,a,b); res[1] = f_value(f,farm[r2],N,a,b);
 for(j=0;j<N;j++)
 { temp[0].SetBit(j,farm[r1][j]); temp[1].SetBit(j,farm[r2][j]); }
 int r3 = rand()%(N-2)+1;
 for(j=r3;j<N;j++)
 { temp[0].SetBit(j,farm[r2][j]); temp[1].SetBit(j,farm[r1][j]); }
 res[2] = f_value(f,temp[0],N,a,b); res[3] = f_value(f,temp[1],N,a,b);
 if(res[2]>res[0]) { farm[r1] = temp[0]; res[0] = res[2]; }
 if(res[3] > res[1]) { farm[r2] = temp[1]; res[1] = res[3]; }
 delete[] temp;
}

void mutate(BitVector *farm,int &M,int &N,double &a,double &b)
{
 double res[2];
 int r4 = rand()%N;  int r1 = rand()%M;
 res[0] = f_value(f,farm[r1],N,a,b);
 int v1 = farm[r1][r4];
 farm[r1].ToggleBit(r4);
 double a1 = f_value(f,farm[r1],N,a,b);
 if(a1 < res[0]) farm[r1].ToggleBit(r4);
 int r5 = rand()%N;  int r2 = rand()%M;
 res[1] = f_value(f,farm[r2],N,a,b);
 int v2 = farm[r2][r5];
 farm[r2].ToggleBit(r5);
 double a2 = f_value(f,farm[r2],N,a,b);
 if(a2 < res[1]) farm[r2].ToggleBit(r5);
}

int main(void)
{
 int M = 12, N = 10;
 int j;
 BitVector* farm = new BitVector[M];
 for(int i=0;i<M;i++) farm[i].SetLength(N);
 setup(farm,M,N);
 double a = 0.0, b = 6.28318; // interval
 for(int k=0;k<1000;k++)
 { crossings(farm,M,N,a,b); mutate(farm,M,N,a,b); }
 for(j=0;j<N;j++) cout << "farm[1]["<<j<<"]=" << farm[1][j] << endl;
```

```
 cout << endl;
 for(j=0;j<M;j++)
  cout << "fitness f_value["<<j<<"]=" << f_value(f,farm[j],N,a,b)
       <<"  x_value["<<j<<"]=" << x_value(farm[j],N,a,b) << endl;
 delete[] farm;
 return 0;
}
```

A typical output is

```
farm[1][0]=1
farm[1][1]=1
farm[1][2]=1
farm[1][3]=0
farm[1][4]=1
farm[1][5]=0
farm[1][6]=0
farm[1][7]=0
farm[1][8]=0
farm[1][9]=0
fitness f_value[0]=1.75411  x_value[0]=5.6997
fitness f_value[1]=1.75411  x_value[1]=5.6997
fitness f_value[2]=1.75411  x_value[2]=5.6997
fitness f_value[3]=1.75411  x_value[3]=5.6997
fitness f_value[4]=1.75411  x_value[4]=5.6997
fitness f_value[5]=1.75411  x_value[5]=5.6997
fitness f_value[6]=1  x_value[6]=0
fitness f_value[7]=0.59771  x_value[7]=0.196541
fitness f_value[8]=1.75411  x_value[8]=5.6997
fitness f_value[9]=1  x_value[9]=0
fitness f_value[10]=1.75411  x_value[10]=5.6997
fitness f_value[11]=1.75411  x_value[11]=5.6997
```

14.8 Maximum of Two-Dimensional Maps

Here we consider the problem how to find the maximum of a two-dimensional bounded function $f : [a,b] \times [c,d] \to \mathbb{R}$, where $a,b,c,d \in \mathbb{R}$, $a < b$ and $c < d$. We follow in our presentation closely Michalewicz [145]. Michalewicz also gives a detailed example.

We use the following notation. N is the length of the chromosome (binary string). The chromosome includes both the contributions from the x variable and y variable. The size of N depends on the required precision. M denotes the size of the farm (population) which is kept constant at each time step. First we have to decide about the precision. We assume further that the required precision is four decimal places for each variable. First we find the domain of the variable x, i.e. $b-a$. The precision requirement implies that the range $[a, b]$ should be divided into at least $(b-a)\cdot 10000$ equal size ranges. Thus we have to find an integer number N_1 such that

$$2^{N_1-1} < (b-a)\cdot 1000 \leq 2^{N_1}.$$

The domain of variable y has length $d-c$. The same precision requirement implies that we have to find an integer N_2 such that

$$2^{N_2-1} < (d-c)\cdot 1000 \leq 2^{N_2}.$$

The total length of a chromosome (solution vector) is then $N = N_1 + N_2$. The first N_1 bits code x and the remaining N_2 bits code y.

Next we generate the farm. To optimize the function f using a genetic algorithm, one creates a population of `size` $= M$ chromosomes. All N bits in all chromosomes are initialized randomly using a random number generator.

Denote the chromosomes by $v_0, v_1, \ldots, v_{M-1}$. During the evaluation phase we decode each chromosome and calculate the fitness function values $f(x, y)$ from (x, y) values.

Now the system constructs a roulette wheel for the selection process. First we calculate the *total fitness* F of the population

$$F := \sum_{j=0}^{M-1} f(v_j).$$

Next we calculate the probability of a selection p_j and the cumulative probability q_j for each chromosome v_j

$$p_j := \frac{f(v_j)}{F}, \qquad q_j := \sum_{k=0}^{j} p_k, \qquad j = 0, 1, \ldots, M-1.$$

Obviously, $q_{M-1} = 1$. Now we spin the roulette wheel M times. First we generate a (random) sequence of M numbers for the range $[0..1]$. Each time we select a single chromosome for a new population as follows. Let r_0 be the first random number. Then $q_k < r_0 < q_{k+1}$ for a certain k. We selected chromosome $k+1$ for the new population. We do the same selection process for all the other $M-1$ random numbers. This leads to a new farm of chromosomes. Some of the chromosomes can now occur twice.

Next one applies the recombination operator, crossover, to the individuals in the new population. For the probability of crossover we choose $p_c = 0.25$. We proceed in the following way: for each chromosome in the (new) population we generate a random number r from the range $[0..1]$. Thus we generate again a sequence of M random numbers in the interval $[0, 1]$. If $r < 0.25$, we select a given chromosome for crossover. If the number of selected chromosomes is even, so we can pair them. If the number of selected chromosomes were odd, we would either add one extra chromosome or remove one selected chromosome. Now we mate selected chromosomes randomly. For each of these two pairs, we generate a random integer number `pos` from the range $[0..N-2]$. The number `pos` indicates the position of the crossing point. We do now the same process for the second pair of chromosomes and so on. This leads to a new farm of chromosomes.

The next operator, mutation, is performed on a bit-by-bit basis. Let $p_m = 0.01$ be the probability of mutation. Thus one expects that (on average) 1% of bits would undergo mutation. There are $M \times M$ bits in the whole population; we expect (on average) $0.01 \cdot N \cdot M$ mutations per generation. Every bit has an equal chance to be mutated, so, for every bit in the population, we generate a random number r from the range $[0..1]$. If $r < 0.01$, we mutate the bit. Thus we have to generate $N \cdot M$ random numbers. Then we translates the bit position into chromosome number and the bit number within the chromosome. Then we swap the bit. This leads to a new population of the same size M.

Thus we have completed one iteration (i.e., one generation) of the `while` loop in the genetic procedure. Next we find the fitness function for the new population and the total fitness of the new population, which should be higher compared to the old population. The fitness value of the fittest chromosome of the new population should also be higher than the fitness value of the fittest chromosome in the old population. Now we are ready to run the selection process again and apply the genetic operators, evaluate the next generation and so on. A stopping condition could be that the total fitness does not change anymore.

```
// twodimensional.cpp

#include <iostream>
#include <cmath>     // for exp, pow
#include <cstdlib>
#include <ctime>
using namespace std;

// function to optimize
double f(double x,double y) { return exp(-(x-1.0)*(x-1.0)*y*y/2.0); }

// determines the chromosone length required
// to obtain the desired precision
int cLength(int precision,double rangeStart,double rangeEnd)
{
 int length = 0;
 double total = (rangeEnd-rangeStart)*pow(10.0,precision);
 while(total > pow(2.0,length)) length++;
 return length;
}
```

```
void setup(int** farm,int size,int length)
{
 srand((unsigned long) time(NULL));
 for(int i=0;i<size;i++)
  for(int j=0;j<length;j++) farm[i][j] = rand()%2;
}

void printFarm(int** farm,int length,int size)
{
 for(int i=0;i<size;i++)
 {
 cout << "\n";
 for(int j=0;j<length;j++) { cout << farm[i][j]; }
 }
}

double xValue(int* chromosome,int xLength,double* domain)
{
 double m = 0.0;
 for(int i=0;i<xLength;i++)
 { m += chromosome[xLength-i-1]*pow(2.0,i); }
 double x =
  domain[0]+m*(domain[1]-domain[0])/(pow(2.0,xLength)-1.0);
 return x;
}

double yValue(int* chromosome,int yLength,int length,double* domain)
{
 double m = 0.0;
 for(int i=0;i<yLength;i++)
 { m += chromosome[length-i-1]*pow(2.0,i); }
 double y = domain[2]+m*(domain[3]-domain[2])/(pow(2.0,yLength)-1.0);
  return y;
}

double fitnessValue(double (*f)(double,double),int* chromosome,
       int length,double* domain,int xLength,int yLength)
{
 double x = xValue(chromosome,xLength,domain);
 double y = yValue(chromosome,yLength,length,domain);
 double result = f(x,y);
 return result;
}

// New farm is set up by using a roulette wheel parent selection
void roulette(int** farm,int length,int size,double* domain,
              int xLength,int yLength)
```

```
{
 int i, j;
 // fitness matrix contains the fitness of each individual chromosome on farm
 double* fitnessVector = new double[size];
 for(i=0;i<size;i++)
 {
 fitnessVector[i] =
 fitnessValue(f,farm[i],length,domain,xLength,yLength);
 }
 // fitness vector contains the fitness of
 // each individual chromosome of the farm
 double totalFitness = 0.0;
 for(i=0;i<size;i++) { totalFitness += fitnessVector[i]; }
 // calculate probability vector
 double* probabilityVector = NULL;
 probabilityVector = new double[size];
 for(i=0;i<size;i++)
 { probabilityVector[i] = fitnessVector[i]/totalFitness; }
 // calculate cumulative probability vector
 double cumulativeProb = 0.0;
 double* cum_prob_Vector = new double[size];
 for(i=0;i<size;i++)
 {
 cumulativeProb += probabilityVector[i];
 cum_prob_Vector[i] = cumulativeProb;
 }

 // setup random vector
 double* randomVector = new double[size];
 srand((unsigned long) time(NULL));
 for(i=0;i<size;i++)
  randomVector[i] = rand()/double(RAND_MAX);
 // create new population
 int count;
 int** newFarm = new int* [size];
 for(i=0;i<size;i++) newFarm[i] = new int[length];
 for(i=0;i<size;i++)
 {
 count = 0;
 while(randomVector[i] > cum_prob_Vector[count]) count++;
 for(j=0;j<length;j++) { newFarm[i][j] = farm[count][j]; }
 }

 for(i=0;i<size;i++)
  for(j=0;j<length;j++) farm[i][j] = newFarm[i][j];
 delete[] fitnessVector;   delete[] probabilityVector;
 delete[] cum_prob_Vector; delete[] randomVector;
 for(i=0;i<size;i++) delete[] newFarm[i]; delete[] newFarm;
```

```
} // end function roulette

 void crossing(int** farm,int size,int length)
 {
 int i, j, k, m;
 int count = 0;
 int* chosen = new int[size];
 double* randomVector = new double[size];
 srand((unsigned long) time(NULL));
 for(i=0;i<size;i++) randomVector[i] = rand()/double(RAND_MAX);
 // fill chosen with indexes of all random values < 0.25
 for(i=0;i<size;i++)
 { if(randomVector[i] < 0.25) { chosen[count] = i; count++; } }

 // if chosen contains an odd number of chromosomes
 // one more chromosome is to be selected
 if((count%2 != 0) || (count == 1))
 {
 int index = 0;
 while(randomVector[index] < 0.25) index++;
 count++;
 chosen[count-1] = index;
 }

 // cross chromosomes with index given in chosen
 int** temp = new int* [2];
 for(i=0;i<2;i++) temp[i] = new int[length];
 for(i=0;i<count;i=i+2)
 {
 for(j=0;j<length;j++)
 { temp[0][j] = farm[chosen[i]][j]; temp[1][j] = farm[chosen[i+1]][j]; }
 int position = rand()%length;

 for(k=position;k<length;k++)
 { temp[0][k] = farm[chosen[i+1]][k]; temp[1][k] = farm[chosen[i]][k]; }

 for(m=0;m<length;m++)
 { farm[chosen[i]][m] = temp[0][m]; farm[chosen[i+1]][m] = temp[1][m]; }
 }
 delete[] chosen; delete[] randomVector;
 for(i=0;i<2;i++) delete[] temp[i]; delete[] temp;
} // end function crossing

void mutate(int** farm,int size,int length)
{
 int totalbits = size*length;
 double* randomVector = new double[totalbits];
 srand((unsigned long) time(NULL));
```

```
 for(int i=0;i<totalbits;i++)
  randomVector[i] = rand()/double(RAND_MAX);

 int a, b;
 for(int i=0;i<totalbits;i++)
 {
 if(randomVector[i] < 0.01)
 {
 if(i >= length) { a = i/length; b = i%length; }
 else { a = 0; b = i; }
 if(farm[a][b]==0) farm[a][b] = 1; else farm[a][b] = 0;
 }
 }
 delete[] randomVector;
}

void printResult(int** farm,int length,int size,double* domain,
                 int xLength,int yLength,int iterations)
{
 int i;
 double* fitnessVector = new double[size];
 for(i=0;i<size;i++)
  fitnessVector[i] =
   fitnessValue(f,farm[i],length,domain,xLength,yLength);
 // search for chromosome with maximum fitness
 double x, y;
 int pos = 0;
 double max = fitnessVector[0];

 for(i=1;i<size;i++)
 { if(fitnessVector[i] > max) { max = fitnessVector[i]; pos = i; } }
 x = xValue(farm[pos],xLength,domain);
 y = yValue(farm[pos],yLength,length,domain);
 // displaying the result
 cout << "\n After " << iterations << " iterations the fitnesses are:\n";
 for(i=0;i<size;i++)
 {
 cout << "\n fitness of chromosome " << i << ": " << fitnessVector[i];
 }
 cout << "\n\n Maximum fitness: f(" << x << "," << y << ") = " << max;
 delete[] fitnessVector;
}

int main(void)
{
 int size = 32;        // population size
 int precision = 6;    // precision
```

```
int iter = 10000;      // number of iterations
double domain[4];      // variables specifying domain
double x1, x2, y1, y2;
x1 = -2.0; x2 = 2.0; y1 = -2.0; y2 = 2.0;
domain[0] = x1; domain[1] = x2; domain[2] = y1; domain[3] = y2;
int xLength = cLength(precision,domain[0],domain[1]);
cout << "\n\n the xLength is: " << xLength;
int yLength = cLength(precision,domain[2],domain[3]);
cout << "\n the yLength is: " << yLength;
// total length
int length = xLength + yLength; // total length
cout << "\n the chromosone length is: " << length;
int** farm = new int*[size]; // memory allocation
for(int i=0;i<size;i++) { farm[i] = new int[length]; }
setup(farm,size,length);
cout << "\n\n The inital farm: \n";
printFarm(farm,length,size); cout << endl;
// iteration loop
for(int t=0;t<iter;t++)
{
roulette(farm,length,size,domain,xLength,yLength);
crossing(farm,size,length);
roulette(farm,length,size,domain,xLength,yLength);
mutate(farm,size,length);
}
printResult(farm,length,size,domain,xLength,yLength,iter);
for(int k=0;k<size;k++)  { delete[] farm[k]; }
delete[] farm;
return 0;
}
```

14.9 Finding a Fitness Function

Genetic algorithms can also be applied to some problems where no fitness function is given (Steeb [195]). This means we first have to construct a fitness function for the problem. A number of examples are given below. We discuss finding the (real) roots of a polynomial. As a second example we consider finding the solution of the boundary value problem of an ordinary differential equation. Then two examples study solutions to systems of linear equations. Finally we consider the four-colour problem.

Example 1. Consider the polynomial, for example,

$$p(x) = x^4 - 7x^3 + 8x^2 + 2x - 1\,.$$

The *zeros* are given by the solution of the equation

$$p(x^*) = 0\,.$$

As fitness function f we can use

$$f(x) = -p(x) \cdot p(x)$$

which we have to maximize, i.e., the zeros of the polynomial p are found where f takes a global maximum. Obviously the global maximum of f is 0. Another possible fitness function would be

$$f(x) = |p(x)|.$$

This fitness function has to be minimized. For faster calculation of f we use Horner's scheme. We could use the program in section 12.6. The output provides the four zeros 5.47947, 1.65434, 0.271902, −0.405717. ♣

Example 2. Let

$$f_1(x_1, x_2) = x_1^2 + x_2^2 - 1 = 0, \qquad f_2(x_1, x_2) = x_1 - x_2 = 0.$$

To find the roots we can consider the fitness function $|f_1| + |f_2|$. ♣

Example 3. Consider the nonlinear second-order ordinary differential equation

$$(u - x)\frac{d^2u}{dx^2} + \sin^2(x) = 0$$

with the boundary values

$$u(0) = 0, \qquad u(1) = 1 + \sin(1), \qquad x \in [0, 1].$$

As an ansatz for the solution we use the polynomial

$$u(x) = c_0 + c_1 x + c_2 x^2 + c_3 x^3 + c_4 x^4.$$

Obviously we can set $c_0 = 0$. We use the second boundary condition to eliminate one more coefficient, for example c_1. Then we can define a fitness function as

$$f(c_2, c_3, c_4) = -\sum_{j=0}^{1/h} \left((u(j \cdot h) - j \cdot h)\frac{d^2u(j \cdot h)}{dx^2} + \sin^2(j \cdot h) \right)^2$$

where h is the step length (for example $h = 0.1$). For the implementation of the second order derivative we first differentiate u, i.e.,

$$\frac{d^2u}{dx^2} = 2c_2 + 6c_3 x + 12c_4 x^2$$

and then replace x by $j \cdot h$. Inserting $u(j \cdot h)$ for $u(x)$, $j \cdot h$ for x and d^2u/dx^2 into the differential equation yields the fitness function f. ♣

Example 4. Consider the symmetric 10×10 matrix

$$A = \begin{pmatrix} 2 & 1 & 0 & 0 & 0 & 0 & 0 & 0 & 0 & 0 \\ 1 & 2 & 1 & 0 & 0 & 0 & 0 & 0 & 0 & 0 \\ 0 & 1 & 2 & 1 & 0 & 0 & 0 & 0 & 0 & 0 \\ 0 & 0 & 1 & 2 & 1 & 0 & 0 & 0 & 0 & 0 \\ 0 & 0 & 0 & 1 & 2 & 1 & 0 & 0 & 0 & 0 \\ 0 & 0 & 0 & 0 & 1 & 2 & 1 & 0 & 0 & 0 \\ 0 & 0 & 0 & 0 & 0 & 1 & 2 & 1 & 0 & 0 \\ 0 & 0 & 0 & 0 & 0 & 0 & 1 & 2 & 1 & 0 \\ 0 & 0 & 0 & 0 & 0 & 0 & 0 & 1 & 2 & 1 \\ 0 & 0 & 0 & 0 & 0 & 0 & 0 & 0 & 1 & 2 \end{pmatrix}$$

and the system of linear equations

$$A\mathbf{x} = \mathbf{b}$$

where

$$\mathbf{b} = (1\,1\,1\,1\,1\,1\,1\,1\,1\,1)^T .$$

Here ^{T} denotes transpose. A possible fitness function is

$$f(\mathbf{x}) = -\left(\frac{1}{10}\sum_{i=0}^{9}\left|\sum_{j=0}^{9} a_{ij}x_j - b_i\right|\right)$$

where a_{ij} are the matrix elements of A. This function has to be maximized. ♣

Example 5. Let A be a given $m \times m$ symmetric positive-semidefinite matrix over $\mathbb{R}$. This means all eigenvalues are real and nonnegative. Let $\mathbf{b} \in \mathbb{R}^m$, where $\mathbf{b}$ is a given column vector. Consider the quadratic functional

$$E(\mathbf{x}) = \frac{1}{2}\mathbf{x}^T A\mathbf{x} - \mathbf{x}^T\mathbf{b}$$

where ^{T} denotes transpose. The minimum $\mathbf{x}^*$ of $E(\mathbf{x})$ over $\mathbb{R}^m$ is unique and occurs where the gradient of $E(\mathbf{x})$ vanishes, i.e.,

$$\nabla E(\mathbf{x} = \mathbf{x}^*) = A\mathbf{x} - \mathbf{b} = \mathbf{0} .$$

The quadratic minimization problem is thus equivalent to solving the system of linear equations $A\mathbf{x} = \mathbf{b}$. Let

$$A = \begin{pmatrix} 2 & 1 \\ 1 & 1 \end{pmatrix}, \qquad \mathbf{b} = \begin{pmatrix} 1 \\ 1 \end{pmatrix} .$$

We use E as a fitness function to solve the system of linear equations applying genetic algorithms. We have to find the minima of the function

$$\begin{aligned} E(\mathbf{x}) &= \frac{1}{2}(x_1, x_2)\begin{pmatrix} 2 & 1 \\ 1 & 1 \end{pmatrix}\begin{pmatrix} x_1 \\ x_2 \end{pmatrix} - (x_1, x_2)\begin{pmatrix} 1 \\ 1 \end{pmatrix} \\ &= x_1^2 + x_1 x_2 + \frac{1}{2}x_2^2 - x_1 - x_2 . \end{aligned}$$

The solution is $x_1 = 0$, $x_2 = 1$. ♣

Example 6. Let A be an invertible $n \times n$ matrix over $\mathbb{R}$. Consider the functions

$$E_j = \frac{1}{2}(A\mathbf{c}_j - \mathbf{e}_j)^T(A\mathbf{c}_j - \mathbf{e}_j)$$

where $j = 1, \ldots, n$, $\mathbf{c}_j$ is the j-th column of the inverse matrix of A, $\mathbf{e}_j$ is the j-th column of the $n \times n$ identity matrix. This means $\mathbf{e}_1, \ldots, \mathbf{e}_n$ is the standard basis (as column vectors) in $\mathbb{R}^n$. The $\mathbf{c}_j$ are determined by minimizing the E_j with respect to the $\mathbf{c}_j$. Thus a fitness functions would be $|E_1| + \cdots + |E_n|$ which have to be minimized. ♣

Example 7. A map is called n-colourable if each region of the map can be assigned a colour from n different colours such that no two adjacent regions have the same colour. The four colour conjecture is that every map is 4-colourable. In 1976 Appel and Haken proved the four colour conjecture with extensive use of computer calculations. We can describe the m regions of a map using a $m \times m$ *adjacency matrix* A where $A_{ij} = 1$ if region i is adjacent to region j and $A_{ij} = 0$ otherwise. We set $A_{ii} = 0$. For the fitness function we can determine the number of adjacent regions which have the same colour. The lower the number, the fitter the individual. Individuals are represented as strings of characters, where each character represents the colour for the region corresponding to the characters position in the string. We write a Java program that uses genetic algorithms to find a solution of the four colour problem given the adjacency matrix.

The data member `population` is the number of individuals in the population, and `mu` is the probability that an individual is mutated. The method `fitness()` evaluates the fitness of a string using the adjacency matrix to determine when adjacent regions have the same colour. If the fitness is equal to 0 we have found a solution. The adjacency matrix can be modified to solve for any map. The method `mutate()` determines for each individual in the population whether the individual is mutated, and mutates a component of the individual by randomly changing the colour. The method `crossing()` performs the crossing operation. The genetic algorithm is implemented in the method `GA()`. The arguments are an adjacency matrix, a string specifying which colours to use and the number of regions on the map. It returns a string specifying a solution to the problem. One such solution is `YBRBYGYRYB`, where `R` stands for red, `G` for green, `B` for blue and `Y` for yellow.

```
// FourColor.java

public class FourColor
{
 static int population = 1000;
 static double mu = 0.01;

 public static void main(String[] args)
 {
```

```
int[][] adjM = {{0,1,0,1,0,0,0,0,0,0},{1,0,1,0,0,1,0,0,0,0},
                {0,1,0,0,0,0,1,0,0,0},{1,0,0,0,1,1,0,0,0,0},
                {0,0,0,1,0,1,0,1,0,0},{0,1,0,1,1,0,1,0,1,1},
                {0,0,1,0,0,1,0,0,0,1},{0,0,0,0,1,0,0,0,1,0},
                {0,0,0,0,0,1,0,1,0,1},{0,0,0,0,0,1,1,0,1,0}};
System.out.println(GA(adjM,"RGBY",10));
}

static int fitness(int[][] adjM,String s,int N)
{
int count = 0;
for(int i=0;i<N-1;i++)
{
for(int j=i+1;j<N;j++)
{ if((s.charAt(i)==s.charAt(j)) && (adjM[i][j]==1)) count++; }
}
return count;
}

static void mutate(String[] p,String colors)
{
int j;
for(int i=0;i<p.length;i++)
{ if(Math.random()<mu)
{ int pos = (int)(Math.random()*(p[i].length()-1));
int mut = (int)(Math.random()*(colors.length()-2));
char[] ca1 = p[i].toCharArray(); char[] ca2 = colors.toCharArray();
for(j=0;ca1[pos]!=ca2[j];j++) {};
ca1[pos] = ca2[(j+mut)%colors.length()];
p[i] = new String(ca1); }
}
}

static void crossing(String[] p,int[][] adjM)
{
int p1 = (int)(Math.random()*(p.length-1)); int p2 = p1;
int c1 = (int)(Math.random()*(p[0].length()-1)); int c2 = c1;
while(p2==p1) p2 = (int)(Math.random()*(p.length-1));
while(c2==c1) c2 = (int)(Math.random()*(p[0].length()-1));
if(c2<c1) { int temp = c2; c2 = c1; c1 = temp;}
String[] temp = new String[4];
temp[0]=p[p1]; temp[1]=p[p2];
temp[2]=p[p1].substring(0,c1)+p[p2].substring(c1+1,c2)
        +p[p1].substring(c2+1,p[p1].length()-1);
temp[3]=p[p2].substring(0,c1)+p[p1].substring(c1+1,c2)
        +p[p2].substring(c2+1,p[p2].length()-1);
int i, f;
for(i=0,f=0;i<4;i++)
```

```
{
if(fitness(adjM,temp[i],temp[i].length())
  > fitness(adjM,temp[f],temp[f].length())) f = i;
}
{ String tmp=temp[f]; temp[f]=temp[0]; temp[0]=tmp; }
for(i=1,f=1;i<4;i++)
{
if(fitness(adjM,temp[i],temp[i].length())
   > fitness(adjM,temp[f],temp[f].length())) f = i;
}
{ String tmp = temp[f]; temp[f] = temp[1]; temp[1] = tmp; }
p[p1] = temp[2]; p[p2] = temp[3];
}

static String GA(int[][] adjM,String colors,int N)
{
int maxfitness, mfi = 0;
String[] p = new String[population];
char[] temp = new char[N];
for(int i=0;i<population;i++)
{
for(int j=0;j<N;j++)
{ temp[j] = colors.charAt((int)((Math.random()*colors.length()))); }
p[i] = new String(temp);
}
maxfitness=fitness(adjM,p[0],p[0].length());
while(maxfitness!=0)
{
mutate(p,colors); crossing(p,adjM);
for(int i=0;i<p.length;i++)
{ if(fitness(adjM,p[i],p[i].length())<maxfitness)
{ maxfitness = fitness(adjM,p[i],p[i].length()); mfi = i; } }
}
return p[mfi];
}
}
```

Example 8. Consider a set of overdetermined linear equations $A\mathbf{x} = \mathbf{b}$, where A is a given $m \times n$ matrix with $m > n$, $\mathbf{x}$ is a column vector with n rows and $\mathbf{b}$ is a given column vector with m rows. For example

$$\begin{pmatrix} 1 & -1 & 1 \\ 1 & -0.5 & 0.25 \\ 1 & 0 & 0 \\ 1 & 0.5 & 0.25 \\ 1 & 1 & 1 \end{pmatrix} \begin{pmatrix} x_1 \\ x_2 \\ x_3 \end{pmatrix} = \begin{pmatrix} 1 \\ 0.5 \\ 0 \\ 0.5 \\ 2.0 \end{pmatrix}.$$

The Chebyshev or minmax solution to the set of overdetermined linear equations

$A\mathbf{x} = \mathbf{b}$, i.e. the vector $\mathbf{x}$ which minimizes

$$f(\mathbf{x}) = \max_{1 \le i \le m} c_i = \max_{1 \le i \le m} \left| b_i - \sum_{j=1}^{n} a_{ij} x_j \right| .$$

The chromosomes are evaluated as

$$c^{(k)} = \max_{1 \le i \le m} \left| b_i - \sum_{j=1}^{n} a_{ij} x_j^{(k)} \right| , \qquad k = 1, 2, \ldots, P$$

where P is the population size. Then the best chromosome is evaluated as the minimum value of the evaluated $c^{(k)}$

$$c_{min} = \min_{1 \le k \le P} \{ c^{(k)} \} .$$

Then the fitness values are calculated as $e_k := c^k - c_{min}$, where $k = 1, \ldots, P$. Thus all the values e_k are non-negative. The population is sorted according to the evaluation values and the elitism group is formed to include all the chromosomes $\mathbf{x}^{(k)}$ with $e_k < \epsilon$. ♣

Example 9. The training of *radial basis function networks* can also be achieved using genetic algorithms. Let $\mathbf{x}(0)$, $\mathbf{x}(1)$, ..., $\mathbf{x}(T-1)$ be the input vectors with $t = 0, 1, \ldots, T-1$ and

$$\mathbf{x}(t) = (x_0(t), x_1(t), \ldots, x_{N-1}(t))^T .$$

Thus $\mathbf{x}(t)$ is a vector in $\mathbb{R}^N$. Let

$$\hat{\mathbf{y}}(t) = (\hat{y}_0(t), \hat{y}_1(t), \ldots, \hat{y}_{L-1}(t))^T$$

be the (desired) corresponding output vector at time t with $t = 0, 1, \ldots, T-1$. Thus $\hat{\mathbf{y}}(t)$ is a vector in $\mathbb{R}^L$. Let the centre vectors (in $\mathbb{R}^N$) of each hidden neuron be denoted by $\mathbf{C}_j$ with $j = 0, 1, \ldots, H-1$, where H must be provided. The output of the i-th neuron in the output layer at time t is given by

$$y_i(t) = \sum_{j=0}^{H-1} w_{ij} f_j(\|\mathbf{x}(t) - \mathbf{C}_j\|) \qquad (1)$$

where $i = 0, 1, \ldots, L-1$ and L is the neuron number of the output layer. The functions f_j are the transfer functions ($j = 0, 1, \ldots, H-1$). In many cases one uses the Gaussian functions $\exp(-r^2/\sigma^2)$ for all f_j. Now $W = (w_{ij})$ is an $L \times H$ matrix. Let

$$\mathbf{y}(t) = (y_0(t), y_1(t), \ldots, y_{L-1}(t))$$

be the output vector from equation (1). Then the cost function (fitness function) is

$$E(W, \mathbf{C}_0, \mathbf{C}_1, \ldots, \mathbf{C}_{H-1}) = \sum_{t=0}^{T-1} \|\mathbf{y}(t) - \hat{\mathbf{y}}(t)\|$$

which has to be minimized. In many cases the centre vectors are also provided so that the cost function only depends on W. ♣

14.10 Problems with Constraints

14.10.1 Introduction

Thus far, we have discussed genetic algorithms for searching unconstrained objective functions. Many practical problems contain one or more constraints that must also be satisfied. A typical example is the traveling salesman problem, where all cities must be visited exactly once. The traveling salesman problem is stated as follows. For a given $n \times n$ distance matrix $C = (c_{ij})$, find a cyclic permutation π of the set $\{1, 2, \ldots, n\}$ that minimizes the function

$$c(\pi) = \sum_{i=1}^{n} c_{i\pi(i)} \,.$$

The value $c(\pi)$ is referred to as the length (or cost or weight) of the permutation π. The traveling salesman problem is one of the standard problems in combinatorial optimization and has many important applications like routing or production scheduling with job-dependent set-up times. Another example is the knapsack problem, where the weight which can be carried is the constraint. The norm of an $n \times n$ matrix over the real numbers $\mathbb{R}$ is given by

$$\|A\| := \sup_{\|\mathbf{x}\|=1} \|A\mathbf{x}\| \,.$$

This is a problem with the constraint $\|\mathbf{x}\| = 1$, i.e. the length of the vector $\mathbf{x} \in \mathbb{R}^n$ must be 1. This problem can be solved with the *Lagrange multiplier method*, since the functions in this problem are differentiable (see chapter 12).

A difficult problem in genetic algorithms is the inclusion of constraints (Michalewicz [145]). Constraints are usually classified as equality or inequality relations. Equality constraints may be included into the system. A genetic algorithm generates a sequence of parameters to be tested using the system model, objective function, and the constraints. We run the model, evaluate the fitness function, and check to see if any constraints are violated. If not, the parameter set is assigned the fitness value corresponding to the objective function evaluation. If constraints are violated, the solution is infeasible and thus does not have a fitness. This procedure is fine except that many practical problems are highly constrained; finding a feasible point is as difficult as finding the best. One wants to get some information out of infeasible solutions, perhaps by degrading their fitness ranking in relation to the degree of constraint violation. This is what is done in a *penalty method.* In a penalty method, a constrained problem in optimization is transformed to an unconstrained problem by associating a cost or penalty with all constraint violations. This cost is included in the objective function evaluation.

Consider, for example, the original constrained problem in minimization form minimize $g(\mathbf{x})$ subject to

$$h_i(\mathbf{x}) \geq 0, \qquad i = 1, 2, \ldots, n \ .$$

where $\mathbf{x}$ is an m vector. We transform this to the unconstrained form

$$\text{minimize} \quad g(\mathbf{x}) + r \sum_{i=1}^{n} \Phi[h_i(\mathbf{x})]$$

where Φ is the penalty function and r is the penalty coefficient. Other approaches use decoders or repair algorithms.

A detailed discussion of problems with constraints is given by Michalewicz [145]. He proposes that appropriate data structures and specialized genetic operators should do the job of taking care of constraints. He then introduces an approach to handle problems with linear constraints (domain constraints, equalities, and inequalities). We consider here the knapsack problem and traveling salesman problem applying genetic algorithms.

14.10.2 Knapsack Problem

The *knapsack problem* can be stated as follows.

Problem. Given M, the capacity of the knapsack,

$$\{\, w_i \,|\, w_i > 0, \ i = 0, 1, \ldots, n-1 \,\}$$

the weights of the n objects, and

$$\{\, v_i \,|\, v_i > 0, \ i = 0, 1, \ldots, n-1 \,\}$$

their corresponding values,

$$\text{maximize} \ \sum_{i=0}^{n-1} v_i x_i \quad \text{subject to} \quad \sum_{i=0}^{n-1} w_i x_i \leq M$$

where $x_i \in \{0, 1\}$. Here $x_i = 0$ means that item i should not be included in the knapsack, and $x_i = 1$ means that it should be included.

Example. A hiker planning a backpacking trip feels that he can carry at most 20 kilograms. After laying out the items that he wants to take he discovers that their total weight exceeds 20 kilograms. He assigns to each item a "value" rating, as shown in the Table. Which items should he take to maximize the value of what he can carry without exceeding 20 kilograms?

Table. Example for the knapsack problem

Item	tent	canteen (filled)	change of clothes	camp stoves	sleeping bag	dried food
weight	11	7	5	4	3	3
value	20	10	11	5	25	50

Item	first-aid kit	mosquito repellent	flashlight	novel	rain gear	water purifier
weight	3	2	2	2	2	1
value	15	12	6	4	5	30

Although we do not know yet how to obtain the solution, the way to fill the knapsack to carry the most value is to take the

sleeping bag, food, mosquito repellent, first-aid kit, flashlight, water purifier, and change of clothes,

for a total value of 149 with a total weight of 19 kilograms. An interesting aspect of the solution is that it is not directly limited by the weight restriction. The knapsack can be filled with exactly 20 kilograms. For example substituting for the change of clothes the camp stove and rain gear. However this decreases the total value.

The following C++ program uses a genetic algorithm to solve the problem. We use the header file `bitvect.h` given above.

```
// knapsack.cpp

#include <iostream>
#include <fstream>
#include <ctime>
#include <cstdlib>
#include "bitvect.h"
using namespace std;

struct item { char name[50]; double weight, value; };

void readitems(char *file,item *&list,int &n,double &max)
{
 ifstream data(file);
 data >> n;
 list = new item[n];
 for(int i=0;i<n;i++)
 { data>>list[i].name; data>>list[i].weight; data>>list[i].value; }
 data >> max;
}

void destroyitems(item *list) { delete[] list; }
```

```
double value(const BitVector &b,int n,double max,item *list)
{
 double tweight = 0.0, tvalue = 0.0;
 for(int i=0;i<n;i++)
 {
 if(b.GetBit(i))
 { tweight += list[i].weight; tvalue += list[i].value; }
 if(tweight > max) { tvalue = -1.0; i = n; }
 }
 return tvalue;
}

void mutate(BitVector *farm,int m,int n,item *list,double max)
{
 const int tries = 1000;
 int animal = rand()%m;
 int i = 0, pos, pos2;
 BitVector* newanim = new BitVector(farm[animal]);
 pos2 = pos = rand()%n;
 newanim -> ToggleBit(pos);
 while(i<tries)
 {
 while(pos2==pos) pos2 = rand()%n;
 newanim -> ToggleBit(pos2);
 if(value(*newanim,n,max,list) > 0) i=tries;
 else { newanim -> ToggleBit(pos2);i++;pos2=pos; }
 }
 if(value(*newanim,n,max,list)>value(farm[animal],n,max,list))
 farm[animal] = *newanim;
 delete newanim;
}

void crossing(BitVector *farm,int m,int n,item *list,double max)
{
 const int tries = 1000;
 int animal1 = rand()%m; int animal2 = rand()%m;
 int pos;
 while(animal2==animal1) animal2 = rand()%m;
 BitVector *newanim1 = new BitVector(farm[animal1]);
 BitVector *newanim2 = new BitVector(farm[animal2]);
 pos = rand()%n;
 for(int i=pos;i<n;i++)
 {
 newanim1 -> SetBit(i,farm[animal2][i]);
 newanim2 -> SetBit(i,farm[animal1][i]);
 }
 if(value(*newanim1,n,max,list) > value(farm[animal1],n,max,list))
 farm[animal1] = *newanim1;
```

```
 if(value(*newanim2,n,max,list) > value(farm[animal2],n,max,list))
 farm[animal2] = *newanim1;
 delete newanim1; delete newanim2;
}

void setupfarm(BitVector *farm,int m,int n,item *list,double max)
{
 const int tries = 2000;
 double temp;
 int i, j, k;
 srand((unsigned long) time(NULL));
 for(i=0;i<m;i++)
 {
 for(j=0;j<n;j++) farm[i].SetBit(j,0);
 temp = 0.0; k = 0;
 while((temp < max) && (k < tries))
 {
 j = rand()%n;
 if(!farm[i].GetBit(j)) temp+=list[j].weight;
 if(temp < max) farm[i].SetBit(j);
 k++;
 }
 }
}

int main(void)
{
 item* list = NULL;
 int n, m = 100, i, iterations = 500, besti = 0;
 double max, bestv = 0.0, bestw = 0.0, temp;
 BitVector *farm = new BitVector[m];
 readitems("knapsack.dat",list,n,max);
 for(i=0;i<m;i++) farm[i].SetLength(n);
 setupfarm(farm,m,n,list,max);
 for(i=0;i<iterations;i++)
 { crossing(farm,m,n,list,max); mutate(farm,m,n,list,max); }

 for(i=0;i<m;i++)
 if((temp=value(farm[i],n,max,list)) > bestv) { bestv=temp; besti=i; }
 cout << "Items to take:" << endl;
 for(i=0;i<n;i++)
 {
 if(farm[besti].GetBit(i))
 { cout << list[i].name << "," << endl; bestw += list[i].weight; }
 }
 cout << endl;
 cout << "for a weight of " << bestw
      << "kg and value of " << bestv << endl;
```

```
 delete[] farm;
 destroyitems(list);
 return 0;
}
```

The input file knapsack.dat is

```
12
tent                           11 20
canteen_(filled)                7 10
change_of_clothes               5 11
camp_stoves                     4  5
sleeping_bag                    3 25
dried_food                      3 50
first-aid_kit                   3 15
mosquito_repellent              2 12
flashlight                      2  6
novel                           2  4
rain_gear                       2  5
water_purifier                  1 30
20
```

The output is

```
Items to take:
change_of_clothes,
sleeping_bag,
dried_food,
first-aid_kit,
mosquito_repellent,
flashlight,
water_purifier,

for a weight of 19kg and value of 149
```

The knapsack problem can also be solved using *dynamic programming.* Consider the following problem

item (j)	weight (w_j)	value (v_j)
1	2	65
2	3	80
3	1	30

and suppose the capacity of the knapsack is 5. We can add an item more than once. Let the stages be indexed by w, the weight carried. The decision is to determine the last item added to bring the weight to $w = 5$. There is one state per stage. Let $f(w)$ be the maximum benefit that can be gained from a w kg knapsack. For item j the following relates $f(w)$ to previously calculated f values

$$f(w) = \max_{1 \le j \le 3} \{ v_j + f(w - w_j) \} .$$

To fill a w kg knapsack, we must end off by adding some item. If we add item j, we end up with a knapsack of size $w - w_j$ to fill. Thus we have $f(w=0) = 0$ and $f(w=1) = 30$. It follows that

$$\begin{aligned} f(w=2) &= \max\{\, 65 + f(w=0) = 65,\ 30 + f(w=1) = 60 \,\} \\ &= 65 \\ f(w=3) &= \max\{\, 65 + f(w=1) = 95,\ 80 + f(w=0) = 80,\ 30 + f(w=2) = 95 \,\} \\ &= 95 \\ f(w=4) &= \max\{\, 65 + f(w=2) = 130,\ 80 + f(w=1) = 110,\ 30 + f(w=3) = 125 \,\} \\ &= 130 \\ f(w=5) &= \max\{\, 65 + f(w=3) = 160,\ 80 + f(w=2) = 145,\ 30 + f(w=4) = 160 \,\} \\ &= 160\,. \end{aligned}$$

This gives a maximum for the value of 160, which is gained by adding 2 of the item 1 with weight 2 and value 65 and 1 of the item 3 with weight 1 and value 30.

14.10.3 Traveling Salesman Problem

The traveling salesman problem is a combinatorial optimization problem. Many combinatorial optimization problems like the traveling salesman problem can be formulated as follows. Let

$$\pi = \{\, i_1, i_2, \ldots, i_n \,\}$$

be some permutation from the set $\{1, 2, \ldots, n\}$. The number of permutations is $n!$. Let Ω be a space of feasible solutions (states) and $f(\pi)$ the optimality function (criterion). It is necessary to find π^* such that

$$\pi^* = \{\, i_1^*, i_2^*, \ldots, i_n^* \,\} = \arg\{\, f(\pi) \to \min_{\pi \subset \Omega} \,\}\,.$$

The structure of Ω and $f(\pi)$ depends on the problems considered. A typical problem is the traveling salesman problem. Given the distances separating a certain number of cities the aim is to find the shortest tour that visits each city once and ends at the city it started from. There are several problems equivalent to a traveling salesman problem. The problem is of practical importance. The number of all possible tours is finite, therefore the problem is solvable. However, the brute force strategy only works for a small number of cities n, since the number of possible tours grows factorially with n. The traveling salesman problem is the best-known example of the whole class of problems called NP-complete (or NP-hard), which makes the problem especially interesting theoretically. The NP-complete problems are transformable into each other, and the computation time required to solve any of them grows faster than any power of the size of the problem. There are strong arguments that a polynomial time algorithm may not exist at all. The aim of the calculations is usually to find near-optimum solutions.

The following C++ program finds all permutations of the numbers $1, 2, \ldots, n$. We choose $n = 3$. The array element p[0] takes the value 0 at the beginning of the program. The end of the evaluation is indicated by p[0]=1.

```
// permutation.cpp

#include <iostream>
using namespace std;

int main(void)
{
 int i, j, k, t;
 unsigned long n = 3;
 int* p = new int[n+1];
 // starting permutation identity 1,2,...,n -> 1,2,...,n
 for(i=0;i<=n;i++)
 { p[i] = i; cout << "p[" << i << "] = " << p[i] << "  "; }
 cout << endl;
 int test = 1;
 do
 {
 i = n-1;
 while(p[i] > p[i+1]) i = i-1;
 if(i > 0) test = 1; else test = 0;
 j = n;
 while(p[j] <= p[i]) j=j-1;
 t = p[i]; p[i] = p[j]; p[j] = t; i = i+1;  j = n;
 while(i < j)
 { t = p[i]; p[i] = p[j]; p[j] = t; i = i+1; j = j-1; }
 // display result
 for(int tau=0;tau<=n;tau++)
 cout << "p[" << tau << "] = " << p[tau] << "  "; cout << endl;
 } while(test==1);
 delete[] p;
 return 0;
}
```

A C++ program that implements the *Durstenfeld algorithm* for randomly shuffling a one-dimensional array of integers is given by

```
// Durstenfeld.cpp

#include <iostream>
#include <cstdlib>
#include <ctime>
using namespace std;

int main(void)
{
```

```
 srand((unsigned long) time(NULL)); // seed
 unsigned long N = 4;
 int* a = new int[N];
 a[0] = 2; a[1] = 3; a[2] = 7; a[3] = 11;
 unsigned long j;
 for(unsigned long i=N-1;i>=1;i--)
 { j = rand()%(i+1); int temp = a[i]; a[i] = a[j]; a[j] = temp; }
 for(unsigned long k=0;k<N;k++)
 cout << "a[" << k << "] = " << a[k] << endl;
 delete[] a;
 return 0;
}
```

Goldberg and Lingle [70] suggested a crossover operator, the so-called partially mapped crossover. They believe it will lead to an efficient solution of the traveling salesman problem. We explain with an example how the partially mapped operator works. Assume that we have 12 cities and assume that the parents are

```
(1 2 3 4 5 6 7 8 9 10 11 12)   : a1
(7 3 6 11 4 12 5 2 10 9 1 8)   : a2
```

a1 and a2 are integer arrays. Positions count from 0 to $n-1$, where $n=12$. We select two random numbers $r1$ and $r2$

$$0 \leq r1 \leq (n-1), \quad 0 \leq r2 \leq (n-1), \quad r1 \leq r2$$

Let $r1 = 3,\ r2 = 6$. Truncate parents using $r1$ and $r2$.

```
(1 2 3 | 4 5 6 7 | 8 9 10 11 12)
(7 3 6 | 11 4 12 5 | 2 10 9 1 8)
```

We obtain the subarrays `s1 = (4 5 6 7)` and `s2 = (11 4 12 5)`. Next we do the crossing

```
(1 2 3 | 11 4 12 5 | 8 9 10 11 12)
(7 3 6 | 4 5 6 7 | 2 10 9 1 8)
```

Now some cities occur twice while others are missing in the new array. The crossing defines the mappings

```
11 -> 4    4 -> 5    12 -> 6    5 -> 7          (*)
4 -> 11    5 -> 4    6 -> 12    7 -> 5         (**)
```

Positions which must be fixed are indicated by `x`

```
(1 2 3| 11 4 12 5 | 8 9 10 x x)
(x 3 x| 4 5 6 7 | 2 10 9 1 8)
```

We fix the first array using the mapping `(*)`.

a) number 11 at position 10 must be fixed

i) map 11 $\mapsto$ 4 but 4 is in array $s2$

ii) map 4 $\mapsto$ 5 but 5 is in array $s2$

iii) map 5 $\mapsto$ 7 o.k. 7 is not in array $s2$

Thus replace number 11 at position 10 by number 7.

a) number 12 at position 11 must be fixed

i) map 12 $\mapsto$ 6 o.k. 6 is not in array s2

Thus replace number 12 at position 11 by number 6.

a) number 7 at position 0 must be fixed

i) map 7 $\mapsto$ 5 but 5 is in array $s1$

ii) map 5 $\mapsto$ 4 but 4 is in array $s1$

iii) map 4 $\mapsto$ 11 o.k. 11 is not in array $s1$

Thus replace number 7 at position 0 by number 11

b) number 6 at position 2 must be fixed

i) map 6 $\mapsto$ 12 o.k. 12 is not in array $s1$

Thus replace number 6 at position 2 by number 12. Consequently, the children are

```
(1 2 3 11 4 12 5 8 9 10 7 6)
(11 3 12 4 5 6 7 2 10 9 1 8)
```

Bac and Perov [8] proposed another operator of crossings using the *permutation group*. We illustrate the operator with an example and a C++ program. Assume that we have ten cities. Let the parents be given by

```
(0 1 2 3 4 5 6 7 8 9) -> (8 7 3 4 5 6 0 2 1 9)  parent 1
(0 1 2 3 4 5 6 7 8 9) -> (7 6 0 1 2 9 8 4 3 5)  parent 2
```

The permutation map yields

```
0 -> 8 -> 3
1 -> 7 -> 4
2 -> 3 -> 1
```

etc.. Thus the children are given by

```
(0 1 2 3 4 5 6 7 8 9) -> (3 4 1 2 9 8 7 0 6 5)
(0 1 2 3 4 5 6 7 8 9) -> (2 0 8 7 3 9 1 5 4 6)
```

The implementation of this permutation is straightforward.

```
// tsppermutation.cpp

#include <iostream>
using namespace std;

void crossing(int* a1,int* a2,int* a3,int* a4,int n)
{
 for(int i=0;i<n;i++) { int p = a1[i];  a3[i] = a2[p]; }
 for(int j=0;j<n;j++) { int q = a2[j];  a4[j] = a1[q]; }
}

int main(void)
{
 int n = 10;
 int* a1 = new int[n]; int* a2 = new int[n];
 int* a3 = new int[n]; int* a4 = new int[n];
 a1[0] = 8; a1[1] = 7; a1[2] = 3; a1[3] = 4; a1[4] = 5;
 a1[5] = 6; a1[6] = 0; a1[7] = 2; a1[8] = 1; a1[9] = 9;
 a2[0] = 7; a2[1] = 6; a2[2] = 0; a2[3] = 1; a2[4] = 2;
 a2[5] = 9; a2[6] = 8; a2[7] = 4; a2[8] = 3; a2[9] = 5;
 crossing(a1,a2,a3,a4,n);
 for(int i=0;i<n;i++)
 {
 cout << "a3[" << i << "] = " << a3[i] << "  ";
 if(((i+1)%2)==0) { cout << endl; }
 }
 for(int j=0;j<n;j++)
 {
 cout << "a4[" << j << "] = " << a4[j] << "  ";
 if(((j+1)%2)==0) { cout << endl; }
 }
 delete[] a1; delete[] a2; delete[] a3; delete[] a4;
 return 0;
}
```

In the following program `tsp.cpp` we use these operators to find solutions to the traveling salesman problem.

```
// tsp.cpp

#include <iostream>
#include <fstream>
#include <cstdlib>
#include <ctime>
#include "bitvect.h"
using namespace std;
```

```
void readdist(char* filename,double**& dist,int& cities)
{
 int i, j;
 ifstream d(filename);
 d >> cities;
 dist = new double*[cities];
 for(i=0;i<cities;i++) dist[i] = new double[cities];
 for(i=0;i<cities;i++)
  for(j=i+1;j<cities;j++) { d >> dist[i][j]; dist[j][i] = dist[i][j]; }
 for(i=0;i<cities;i++) dist[i][i]=0;
  cout << "d[0][0] = " << dist[0][0] << endl;
 d.close();
}

void destroydist(double **dist,int cities)
{
 for(int i=0;i<cities;i++) delete[] dist[i];
 delete[] dist;
}

double distance(int *seq,int cities,double **dist)
{
 double sumdist = 0.0;
 for(int i=1;i<cities;i++) sumdist += dist[seq[i]][seq[i-1]];
 sumdist += dist[seq[0]][seq[cities-1]];
 return sumdist;
}

void setupfarm(int **farm,int n,int cities)
{
 BitVector used(cities);
 int city,i,j;
 srand(time(NULL));
 for(i=0;i<n;i++)
 {
 for(j=0;j<cities;j++) used.SetBit(j,0);
 for(j=0;j<cities;j++)
 {
 city = rand()%cities;
 if(!used.GetBit(city)) { farm[i][j] = city;used.SetBit(city); }
 else j--;
 }
 }
}

void mutate(int **farm,int n,int cities,double **dist)
{
 int seq = rand()%n;
```

```
 int pos1 = rand()%cities; int pos2 = rand()%cities;
 while(pos2==pos1) pos2 = rand()%cities;
 int *mutated = new int[cities];
 for(int i=0;i<cities;i++) mutated[i] = farm[seq][i];
 mutated[pos1] = farm[seq][pos2];
 mutated[pos2] = farm[seq][pos1];
 if(distance(farm[seq],cities,dist) > distance(mutated,cities,dist))
 { delete farm[seq]; farm[seq] = mutated; }
 else delete mutated;
}

void permutate(int** farm,int n,int cities,double** dist)
{
 int seq1 = rand()%n; int seq2 = rand()%n;
 int *result1, *result2, *result3, *result4;
 while(seq2==seq1) seq2 = rand()%n;
 int *child1 = new int[cities]; int *child2 = new int[cities];
 for(int i=0;i<cities;i++)
 {
 child1[i] = farm[seq2][farm[seq1][i]];
 child2[i] = farm[seq1][farm[seq2][i]];
 }
 if(distance(farm[seq1],cities,dist) > distance(child1,cities,dist))
 result1 = child1;
 else result1 = farm[seq1];
 if(distance(farm[seq2],cities,dist) > distance(child2,cities,dist))
 result2 = child2;
 else result2 = farm[seq2];
 result3 = ((result1==farm[seq1])?child1:farm[seq1]);
 result4 = ((result2==farm[seq2])?child2:farm[seq2]);
 farm[seq1] = result1; farm[seq2] = result2;
 delete[] result3; delete[] result4;
}

int insequence(int el,int *seq,int p1,int p2)
{
 for(int i=p1;i<p2;i++) if(seq[i]==el) return i;
 return -1;
}

void pmx(int **farm,int n,int cities,double **dist)
{
 int i, pos;
 int seq1 = rand()%n; int seq2 = rand()%n;
 int *result1, *result2, *result3, *result4;
 while(seq2==seq1) seq2 = rand()%n;
 int pos1 = rand()%cities; int pos2 = rand()%cities;
 while(pos2==pos1) pos2=rand()%cities;
```

```
 if(pos2<pos1) { i = pos2; pos2 = pos1; pos1 = i; }
 int *child1 = new int[cities]; int *child2 = new int[cities];
 for(i=0;i<cities;i++)
 {
 if((i<pos2) && (i>=pos1))
  { child1[i] = farm[seq2][i]; child2[i] = farm[seq1][i]; }
 else { child1[i] = farm[seq1][i]; child2[i] = farm[seq2][i]; }
 }

 for(i=0;i<cities;i++)
 {
 if((i<pos1) || (i>=pos2))
 while((pos = insequence(child1[i],child1,pos1,pos2)) >= 0)
 child1[i] = child2[pos];
 if((i<pos1) || (i>=pos2))
 while((pos = insequence(child2[i],child2,pos1,pos2)) >= 0)
 child2[i] = child1[pos];
 }

 if(distance(farm[seq1],cities,dist)>distance(child1,cities,dist))
 result1 = child1; else result1 = farm[seq1];
 if(distance(farm[seq2],cities,dist)>distance(child2,cities,dist))
 result2 = child2; else result2 = farm[seq2];
 result3=((result1==farm[seq1])?child1:farm[seq1]);
 result4=((result2==farm[seq2])?child2:farm[seq2]);
 farm[seq1] = result1; farm[seq2] = result2;
 delete[] result3; delete[] result4;
}

int main(void)
{
 int N = 16;   // number of animals/chromosomes
 int i, j;
 int iterations = 300;
 cout << N << endl;
 double** dist = NULL;  // array of distances
 int cities;            // number of cities
 readdist("tsp.dat",dist,cities);
 cout << "cities: " << cities << endl;
 int** farm = new int*[N];
 for(i=0;i<N;i++) farm[i] = new int[cities];
 setupfarm(farm,N,cities);
 for(i=0;i<iterations;i++)
 {
 mutate(farm,N,cities,dist);
 permutate(farm,N,cities,dist);
 pmx(farm,N,cities,dist);
 }
```

```
 for(i=0;i<N;i++)
 {
 for(j=0;j<cities;j++) cout << farm[i][j] << " ";
 cout << " distance:" << distance(farm[i],cities,dist) << endl;
 }
 destroydist(dist,cities);
 return 0;
}
```

The input file for the 28 distances of the eight cities `tsp.dat` is:

```
8
14.5413  20.7663  13.5059  19.6041  10.4139   4.60977  14.5344
 6.34114  5.09313  9.12195  5.0     12.0416  14.0357   8.70919
10.4938  11.2432  18.3742  18.8788  14.213    7.5326  12.6625
17.7071   9.72677 15.4729  10.5361   7.2111  10.198   10.0
```

A typical output is

```
cities: 8
7 4 2 3 1 5 0 6   distance:64.8559
7 4 2 3 1 5 0 6   distance:64.8559
7 4 2 3 1 5 0 6   distance:64.8559
0 6 5 1 3 2 4 7   distance:66.1875
7 4 2 3 1 5 0 6   distance:64.8559
7 4 2 3 1 5 0 6   distance:64.8559
7 4 2 3 1 5 0 6   distance:64.8559
0 6 5 1 3 2 4 7   distance:66.1875
7 4 2 3 1 5 0 6   distance:64.8559
4 2 1 3 0 6 5 7   distance:67.9889
7 4 2 3 1 5 0 6   distance:64.8559
7 4 2 3 1 5 0 6   distance:64.8559
0 6 5 1 3 2 4 7   distance:66.1875
0 6 5 1 3 2 4 7   distance:66.1875
7 4 2 3 1 5 0 6   distance:64.8559
7 4 2 3 1 5 0 6   distance:64.8559
```

The brute force method to look at all possible permutations of the cities also provides the solution `7 4 2 3 1 5 0 6` with the shortest distance 64.8559.

14.11 Simulated Annealing

Annealing is the process of cooling a molten substance with the objective of condensing matter into a crystalline solid. Annealing can be regarded as an optimization process (Michalewicz [145]). The configuration of the system during annealing is defined by the set of atomic positions r_i. A configuration of the system is weighted by its Boltzmann probability factor,

$$\exp(-E(r_i)/kT)$$

where $E(r_i)$ is the energy of the configuration, k is the Boltzmann constant, and T is the temperature. When a substance is subjected to annealing, it is maintained at each temperature for a time long enough to reach thermal equilibrium.

The iterative improvement technique for combinatorial optimization has been compared to rapid quenching of molten metals. During rapid quenching of a molten substance, energy is rapidly extracted from the system by contact with a massive cold substrate. Rapid cooling results in metastable system states; in metallurgy, a glassy substance rather than a crystalline solid is obtained as a result of rapid cooling. The analogy between iterative improvement and rapid cooling of metals stems from the fact that iterative improvement and rapid cooling of metals accepts only those system configurations which decrease the fitness function. In an annealing (slow cooling) process, a new system configuration that does not improve the cost function is accepted based on the Boltzmann probability factor of the configuration. This criterion for accepting a new system state is called the *Metropolis criterion*. If the initial temperature is too low, the process gets quenched very soon and only a local optima is found. If the initial temperature is too high, the process is very slow. Only a single solution is used for the search and this increases the chance of the solution becoming stuck at a local optimum. The changing of the temperature is based on an external procedure which is unrelated to the current quality of the solution, that is, the rate of change of temperature is independent of the solution quality. These problems can be solved by using a population instead of a single solution. The annealing mechanism can also be coupled with the quality of the current solution by making the rate of change of temperature sensitive to the solution quality. The simulated annealing procedure is given below. Simulated annealing consists of repeating the Metropolis procedure for different temperatures. The temperature is gradually decreased at each iteration of the simulated annealing algorithm.

```
procedure simulated annealing
  begin
     t <- 0
     initialize temperature T
     select a current string v_c at random
     evaluate the fitness of v_c
     repeat
     repeat
     select a new string v_n
      in the neighbourhood of v_c
      by flipping a single bit of v_c
      if f(v_c) < f(v_n) then v_c <- v_n
      else if random[0,1] < exp((f(v_n)-f(v_c))/T) then v_c <- v_n
      until (termination condition)
      T <- g(T,t)
      t <- t+1
      until (stop-criterion)
   end
```

In the following C++ program we use simulated annealing to find the minimum of the differentiable function

$$f(x) = x^2 \sin(x) \exp(-x/2.0)$$

in the range $[0, 100]$.

```
// annealing.cpp

#include <iostream>
#include <cmath>    // for sin, exp, fmod
#include <cstdlib>
using namespace std;

inline double f(double& x) { return x*x*sin(x)*exp(-x/2.0); }

inline void randval(double* s)
{
 static const double pi = 3.14159265;
 *s = fmod((*s+pi)*(*s+pi)*(*s+pi)*(*s+pi)*(*s+pi),1.0);
}

inline int accept(double& Ecurrent,double& Enew,double& T,double& s)
{
 double dE = Enew-Ecurrent;
 if(dE < 0.0) return 1;
 if(s < exp(-dE/T)) return 1; else return 0;
}

int main(void)
{
 cout << "Finding the minimum via simulated annealing:" << endl;
 double xlow = 0.0, xhigh = 100.0;
 double Tmax = 500.0, Tmin = 1.0;
 double Tstep = 0.1;
 double s = 0.118;   // seed
 randval(&s);
 double xcurrent = s*(xhigh-xlow);
 double Ecurrent = f(xcurrent);
 for(int T=Tmax;T>Tmin;T=T-Tstep)
 {
 randval(&s);
 double xnew = s*(xhigh-xlow);
 double Enew = f(xnew);
 if(accept(Ecurrent,Enew,T,s)!=0) { xcurrent = xnew; Ecurrent = Enew; }
 } // end for loop
 cout << "The minimum found is " << Ecurrent << " at x = " << xcurrent;
 return 0;
}
```

Exercises. (1) We define the number π as the number where the function

$$f(x) = \sin(x)$$

crosses the x-axis in the interval $[2,4]$. How would we calculate π (of course an approximation of it) using this definition and genetic algorithms?

(2) We define the number $\ln 2$ as the number where the function

$$f(x) = e^{-x} - 1/2$$

crosses the x-axis in the interval $[0,1]$. How would we calculate $\ln 2$ (of course an approximation of it) using this definition and genetic algorithms?

(3) Consider the polynomial

$$p(x) = x^2 - x - 1\,.$$

We define the golden mean number g as the number where the polynomial p crosses the x-axis in the interval $[1,2]$. How would we calculate g (of course an approximation of it) using this definition and genetic algorithms?

(4) (i) Describe a map which maps binary strings of length n into the interval $[2,3]$ uniformly, including 2 and 3. Determine the smallest n if the mapping must map a binary string within 0.001 of a given value in the interval.
(ii) Which binary string represents e? How accurate is this representation?
(iii) Give a fitness function for a genetic algorithm that approximates e, without explicitly using the constant e.

(5) Consider the traveling salesman problem. Given eight cities with the space coordinates

$$(0,0,0),\quad (0,0,1),\quad (0,1,0),\quad (0,1,1),\quad (1,0,0),\quad (1,0,1),\quad (1,1,0),\quad (1,1,1)\,.$$

Consider the coordinates as a bitstring. We identify the bitstrings with base 10 numbers, i.e. $000 \to 0$, $001 \to 1$, $010 \to 2$, $011 \to 3$, $100 \to 4$, $101 \to 5$, $110 \to 6$, $111 \to 7$. Thus we set

$$x_0 = (0,0,0), \quad x_1 = (0,0,1), \quad x_2 = (0,1,0), \quad x_3 = (0,1,1)$$

$$x_4 = (1,0,0), \quad x_5 = (1,0,1), \quad x_6 = (1,1,0), \quad x_7 = (1,1,1)\,.$$

Consider the traveling salesman problem and the eight cities. Find the shortest route starting at $(0,0,0)$ and returning to $(0,0,0)$ after visiting each city once. What is the connection with the *Gray code*? In Gray code two successive bitstrings differ only in one bit. For example `"1010"` `"1011"`. Is there more than one solution? Extend the problem to the n-cube.

Chapter 15

Gene Expression Programming

15.1 Introduction

Gene expression programming has been introduced by C. Ferreira (Ferreira [59], Ferreira [58], Hardy and Steeb [84]). It is a genome/phenome genetic algorithm which combines the simplicity of genetic algorithms and the abilities of genetic programming. In a sense gene expression programming is a generalization of genetic algorithms and genetic programming. Gene expression programming is different from genetic programming because the expense of managing a tree structure and ensuring correctness of programs is eliminated. We provide an introduction in the following.

A *gene* is a symbolic string with a head and a tail. Each symbol represents an operation. The operation + takes two arguments and adds them. For example, +x2 is the operation + with arguments x and 2 giving x+2. The operation * also takes two arguments and multiplies them. The operation x would evaluate to the value of the variable x. The tail consists only of operations which take no arguments. The string represents expressions in prefix notation, i.e. 5-3 would be stored as - 5 3. The reason for the tail is to ensure that the expression is always complete. Suppose the string has h symbols in the head which is specified as an input to the algorithm, and t symbols in the tail which is determined from h. Thus if n is the maximum number of arguments for an operation we must have

$$h + t - 1 = hn\,.$$

The left hand side is the total number of symbols except for the very first symbol. The right hand side is the total number of arguments required for all operations. We assume, of course, that each operation requires the maximum number of arguments so that any string of this length is a valid string for the expression. Thus the equation states that there must be enough symbols to serve as arguments for all operations. Now we can determine the required length for the tail $t = h(n-1)+1$.

Example 1. Consider the symbols x, y and the operations $*$, $+$ which take two arguments. Let $h = 5$. Then $t = 6$ since $n = 2$. Thus a gene would be

`++*xy2|xyy32y`

The vertical line indicates the beginning of the tail. The expression is `(x*y)+2`. The tree structure would be

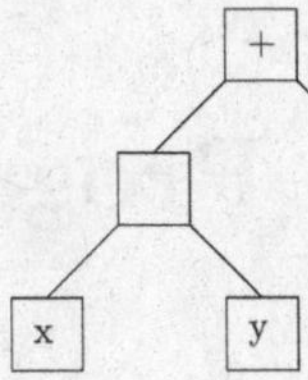

Example 2. Suppose we use $h = 8$, and $n = 2$ for arithmetic operations. Thus the tail length must be $t = 9$. So the total gene length is 17. We could then represent the expression

$$\cos(x^2 + 2) - \sin(x)$$

with the string

`-c+*xx2s|x1x226x31`

The vertical | is used to indicate the beginning of the tail. Here `c` represents `cos` and `s` represents `sin`. We can represent the expressions with trees. For the example above, the root of the tree would be '-' with branches for the parameters. Thus we could represent the expression as follows

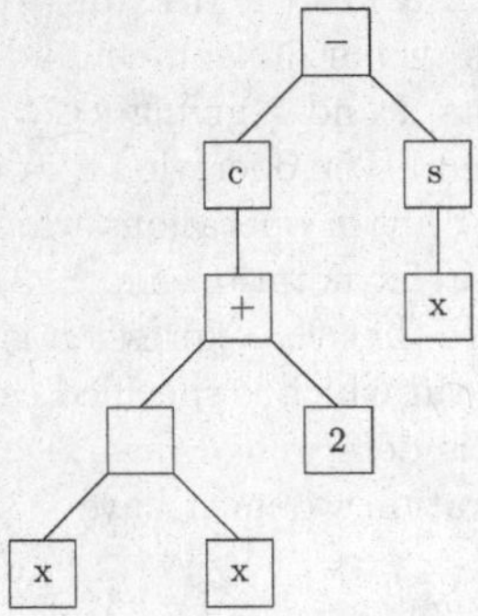

Note that not all of the symbols from the tail are used. Only one symbol from the tail is used in the present example.

A *chromosome* is a collection of genes. The genes combine to form an expression using some operation with the same number of arguments as genes in the chromosome. For example the expressions of genes of a chromosome may be added together. For operations applied to chromosomes we often concatenate the genes to obtain a single string of symbols. For example, suppose we have the following genes forming a chromosome

```
-c+*xx2s|x1x226x31
-c+*xx2s|x1x226x31
-c++x2*x|x1x226x31
```

we form the chromosome by concatenation

```
-c+*xx2s|x1x226x31|-c+*xx2s|x1x226x31|-c++x2*x|x1x226x31
```

where | indicates the beginning of the head and tail portions of the genes. We apply operations to chromosomes in a collection which is called the *population* of chromosomes.

A number of operations are applied to chromosomes.

- **Replication**. The chromosome is unchanged. The roulette wheel selection technique can be used to select chromosomes for replication.

- **Mutation**. Randomly change symbols in a chromosome. Symbols in the tail of a gene may not operate on any arguments. Typically 2 point mutations per chromosome is used. For example, let `++xy2|xyy32y`. Then a mutation could yield `++yy2|xyy32y` or `++xy2|xyy32y` or `++xy-|xyy32y`.

- **Insertion**. A portion of a chromosome is chosen to be inserted in the head of a gene. The tail of the gene is unaffected. Thus symbols are removed from the end of the head to make room for the inserted string. Typically a probability of 0.1 of insertion is used. As an example suppose `+x2` is to be inserted into

  ```
  -c+*xx2s|x1x226x31
  ```

  at the fourth position in the head. We obtain

  ```
  -c++x2*x|x1x226x31
  ```

  which represents the expression $\cos((x+2)+x^2)-1$.

- **Gene transposition**. One gene in a chromosome is randomly chosen to be the first gene. All other genes in the chromosome are shifted downwards in the chromosome to make place for the first gene.

- **Recombination**. The crossover operation. This can be one point (the chromosomes are split in two and corresponding sections are swapped), two point (chromosomes are split in three and the middle portion is swapped) or gene (one entire gene is swapped between chromosomes) recombination. Typically the sum of the probabilities of recombination is 0.7.

Another operation could be swapping. For example, consider again `++xy2|xyy32y`. Then swapping the second and seventh position (counting from left and starting from zero) we obtain `++yy2|xyx32y`.

In the following example we implement these operations. For simplicity we use only one gene in a chromosome and only one point recombination.

15.2 Example

Consider smooth one-dimensional maps $f : \mathbb{R} \to \mathbb{R}$ having the properties

$$f(0) = 0, \quad f(1) = 0, \quad f\left(\frac{1}{2}\right) = 1$$

$$\frac{1}{2} < f\left(\frac{1}{4}\right) < 1, \quad \frac{1}{2} < f\left(\frac{3}{4}\right) < 1.$$

For example the logistic map $g(x) := 4x(1-x)$ satisfies this set of properties. We use Gene Expression Programming to perform symbolic regression in order to obtain maps which satisfy the above properties. We expect that the logistic map should be found since we limit the functions in the implementation to polynomials. We generalize the set of properties as follows. The points of evaluation are specified by a subset of $X \times Y$ where X and Y are given by $f : X \to Y$. Denote by $F_= \subset X \times Y$ the subset of all $(x, y) \in X \times Y$ such that we require of f that $f(x) = y$, by $F_> \subset X \times Y$ the subset of all $(x, y) \in X \times Y$ such that we require of f that $f(x) > y$, by $F_< \subset X \times Y$ the subset of all $(x, y) \in X \times Y$ such that we require of f that $f(x) < y$. Thus we can define the fitness function of a function g (where a smaller value indicates a higher fitness)

$$\begin{aligned}\text{fitness}(g) := &\sum_{(x,y)\in F_=} |g(x) - y| + \sum_{(x,y)\in F_>} |g(x) - y| H(y - g(x)) \\ &+ \sum_{(x,y)\in F_<} |g(x) - y| H(g(x) - y)\end{aligned}$$

where

$$H(x) := \begin{cases} 1 & x > 0 \\ 0 & \text{otherwise} \end{cases}$$

is the step function. We apply gene expression programming until we find a function g such that

$$\text{fitness}(g) < \epsilon$$

for given $\epsilon > 0$. A typical value of ϵ is 0.001.

In the C++ program the function `evalr()` takes a character string and evaluates (i.e., returns type `double`) the corresponding function at the given point (the parameter `x` of type `double`) using recursion. The function `eval()` uses `evalr()` for evaluation without modifying the pointer argument. Similarly `printr()` and `print()` are responsible for output of the symbolic expressions in readable form. A character array and an array of type `double` are passed to the `fitness()` function. The array of `double` consists of triplets: the point of evaluation, the expected value and the comparison type. If the comparison type is zero then the goal is equality (i.e. we can use the absolute value of the difference between the expected value and the actual value), less than zero then the actual value should be less than the expected value, greater than zero then the actual value should be greater than the expected value. Finally, the function gep() implements the gene expression programming algorithm. As arguments gep() takes the data points (point of evaluation, and comparison with a given value), the number of data points N, the size of the population P and the desired fitness eps. The function strncpy of the string header file cstring is used to copy specific regions of the character strings representing the chromosomes. We need to specify the length since the chromosomes are not null terminated.

We use 10 symbols ($h = 10$) for the head portion of the representations, thus the total gene length is 21 (we use only addition, subtraction and multiplication). Since we only use x as a terminal symbol we obtain polynomials of order up to 11, i.e. the highest order polynomial supported by the representation is x^{11}. We randomly choose terminals (symbols which do not take arguments, such as x) and non-terminals (symbols which do take arguments, for example $+$) for the head and then terminals for the tail. We use 0.1 for the probability of mutation, 0.4 for the probability of insertion and 0.7 for the probability of recombination. At each iteration of the algorithm we eliminate the worst half of the population.

```
// gepchaos.cpp

#include <iostream>
#include <cstdlib>
#include <ctime>
#include <cmath>    // for sin, cos
#include <cstring>
using namespace std;

const double pi = 3.1415927;
const int nsymbols = 5;
// 2 terminal symbols (no arguments) x and 1
const int terminals = 2;
// terminal symbols first
const char symbols[nsymbols] = {'1','x','+','-','*'};
const int n = 2;   // for +,- and * which take 2 arguments
int h = 10;

double evalr(char *&e,double x)
{
 switch(*(e++))
 {
 case '1': return 1.0;
 case 'x': return x;
 case 'y': return pi*x;
 case 'c': return cos(evalr(e,x));
 case 's': return sin(evalr(e,x));
 case '+': return evalr(e,x)+evalr(e,x);
 case '-': return evalr(e,x)-evalr(e,x);
 case '*': return evalr(e,x)*evalr(e,x);
```

```
  default : return 0.0;
  }
}

double eval(char *e,double x) { char *c = e; return evalr(c,x); }

void printr(char *&e)
{
 switch(*(e++))
 {
 case '1': cout << '1'; break;
 case 'x': cout << 'x'; break;
 case 'y': cout << "pi*x"; break;
 case 'c': cout << "cos("; printr(e); cout << ")"; break;
 case 's': cout << "sin("; printr(e); cout << ")"; break;
 case '+': cout << '(';  printr(e); cout << '+';  printr(e);
           cout << ')';  break;
 case '-': cout << '(';  printr(e); cout << '-';  printr(e);
           cout << ')';  break;
 case '*': cout << '(';  printr(e); cout << '*';  printr(e);
           cout<<')'; break;
  }
}

void print(char *e) { char *c = e; printr(c); }

double fitness(char *c,double *data,int N)
{
 double sum = 0.0;
 double d;
 for(int j=0;j<N;j++)
 {
 d = eval(c,data[3*j])-data[3*j+1];
 if(data[3*j+2]==0) sum += fabs(d);
 else if(data[3*j+2] > 0) sum -= (d > 0.0)?0.0:d;
 else if(data[3*j+2] < 0) sum += (d < 0.0)?0.0:d;
 }
 return sum;
}

// N number of data points, P population size, eps = accuracy required
void gep(double *data,int N,int P,double eps)
{
 int i,j,k,replace,replace2,rlen,rp;
 int t = h*(n-1)+1;
 int gene_len = h+t;
 int pop_len = P*gene_len;
 int iterations = 0;
```

```
char *population = new char[pop_len];
char *elim = new char[P];
int toelim = P/2;
double bestf,f;        // best fitness, fitness value
double sumf = 0.0;     // sum of fitness values
double pmutate = 0.1; // probability of mutation
double pinsert = 0.4; // probability of insertion
double precomb = 0.7; // probability of recombination
double r,lastf;       // random numbers & roulette wheel selection
char* best = (char*)NULL; // best gene
char* iter;              // iteration variable
// initialize the population
for(i=0;i<pop_len;i++)
if(i%gene_len < h) population[i] = symbols[rand()%nsymbols];
else population[i] = symbols[rand()%terminals];
// initial calculations
bestf = fitness(population,data,N);
best = population;
for(i=0,sumf=0.0,iter=population;i<P;i++,iter+=gene_len)
{
f = fitness(iter,data,N);
sumf += f;
if(f<bestf) { bestf = f; best = population+i*gene_len; }
}
while(bestf >= eps)
{
// reproduction, roulette wheel selection
for(i=0;i<P;i++) elim[i] = 0;
for(i=0;i<toelim;i++)
{
r = sumf*(double(rand())/RAND_MAX);
lastf = 0.0;
for(j=0;j<P;j++)
{
f = fitness(population+j*gene_len,data,N);
if((lastf<=r) && (r<f+lastf)) { elim[j] = 1; j = P; }
lastf += f;
}
}

for(i=0;i<pop_len;)
{
if(population+i == best)
i += gene_len; // never modify/replace best gene
else for(j=0;j<gene_len;j++,i++)
{
// mutation or elimination due to failure in selection
// for reproduction
```

```
if((double(rand())/RAND_MAX < pmutate) || elim[i/gene_len])
if(i%gene_len < h)
population[i] = symbols[rand()%nsymbols];
else population[i] = symbols[rand()%terminals];
}
// insertion
if(double(rand())/RAND_MAX < pinsert)
{
// find a position in the head of this gene for insertion
// -gene_len for the gene since we have already moved
// to the next gene
replace = i-gene_len;
rp = rand()%h;
// a random position for insertion source
replace2 = rand()%pop_len;
// a random length for insertion from the gene
rlen = rand()%(h-rp);
// create the new gene
char *c = new char[gene_len];
// copy the shifted portion of the head
strncpy(c+rp+rlen,population+replace+rp,h-rp-rlen);
// copy the tail
strncpy(c+h,population+replace+h,t);
// copy the segment to be inserted
strncpy(c+rp,population+replace2,rlen);
// if the gene is fitter use it
if(fitness(c,data,N) < fitness(population+replace,data,N))
strncpy(population+replace,c,h);
delete[] c;
}
// recombination
if(double(rand())/RAND_MAX < precomb)
{
// find a random position in the gene for one point recombination
replace = i-gene_len;
rlen = rand()%gene_len;
// a random gene for recombination
replace2 = (rand()%P)*gene_len;
// create the new genes
char *c[5];
c[0] = population+replace;
c[1] = population+replace2;
c[2] = new char[gene_len];
c[3] = new char[gene_len];
c[4] = new char[gene_len];
strncpy(c[2],c[0],rlen);
strncpy(c[2]+rlen,c[1]+rlen,gene_len-rlen);
strncpy(c[3],c[1],rlen);
```

```
 strncpy(c[3]+rlen,c[0]+rlen,gene_len-rlen);
 // take the fittest genes
 for(j=0;j<4;j++)
  for(k=j+1;j<4;j++)
   if(fitness(c[k],data,N) < fitness(c[j],data,N))
   {
   strncpy(c[4],c[j],gene_len);
   strncpy(c[j],c[k],gene_len);
   strncpy(c[k],c[4],gene_len);
   }
 delete[] c[2]; delete[] c[3]; delete[] c[4];
 }
 }
 // fitness
 for(i=0,sumf=0.0,iter=population;i<P;i++,iter+=gene_len)
 {
 f = fitness(iter,data,N);
 sumf += f;
 if(f < bestf) { bestf=f; best=population+i*gene_len; }
 }
 iterations++;
 }
 print(best);
 cout << endl;
 cout << "Fitness of " << bestf << " after "
      << iterations << " iterations" << endl;
 delete[] population; delete[] elim;
}

int main(void)
{
 srand(time(NULL)); // seed for the random number generator
 double data[] = { 0,0,0,1,0,0,0.5,1,0,0.25,0.5,1,0.75,0.5,1,
                   0.25,1,-1,0.75,1,-1 };
 gep(data,7,30,0.001);
 cout << endl;
 return 0;
}
```

We list some typical results below, all with fitness 0.

```
((1-x)*((x+(((x-x)+x)+x))+x))=4x(1-x)
((x+((((((x-x)-x)-x)*x)-x)*x)+x))+x)=x(1-x)(2x+3)
(x-(((x+x)*((x-1)+x))-x))=4x(1-x)
((((1-x)*1)*(((1+1)+1)+1))*x)=4x(1-x)
((((1+1)-(((x+x)*x)+x))*x)+x)=x(1-x)(2x+3)
((1*((x-((((x+x)*x)+x)*x))+x))+x)=x(1-x)(2x+3)
((1-x)*(((((x+x)*x)+x)*1)+x)+x))=x(1-x)(2x+3)
```

```
((((1-(1*(((x+x)*x)-1)))-x)+1)*x)=x(1-x)(2x+3)
```

We find that most of the time either $4x(1-x)$ or $x(1-x)(2x+3)$ is the fittest map. Of course we find the desired functions with less iterations if the poulation size is increased. The map $g : \mathbb{R} \to \mathbb{R}$

$$g(x) = x(1-x)(2x+3)$$

satisfies the conditions given above, but $g(x)$ has values greater than 1 on the interval

$$\left(\frac{1}{2}, \frac{\sqrt{5}}{2} - \frac{1}{2} \right) .$$

Thus we find that almost all initial values $x_0 \in [0,1]$ escape the interval $[0,1]$ under iterations of the map. The set of all points whose iterate stay in $[0,1]$ are of measure zero and form a Cantor set. We can also change the `symbols` array to include the symbols `c` for cosine, `s` for sine and `y` for πx. In this case we find the function $\sin(\pi x)$ nearly every time as the fittest map.

Another application of gene expression programming is to find boolean expressions. In boolean expressions we have the AND, OR, XOR and NOT operation. The NAND-gate is an AND-gate followed by a NOT-gate. The NAND gate is universal gate, where all other gates can be built from. As an example consider the truth table

a	b	O
0	0	1
0	1	1
1	0	0
1	1	1

where a and b are the inputs and O is the output. Assume that $\cdot$ (in C++ the bitwise `&` operator) is the AND-operation, `+` (in C++ the bitwise | operator) is the OR operation and $\overline{a}$ (in C++ the bitwise `~`) is the NOT operation. Using gene expression programming we can form the expressions

```
&a|1~b|aab10bb
|b~a|ababa
```

The first boolean expression would be $a \cdot (1 + \bar{b})$. The second boolean expression would be $b \cdot \bar{a}$. In the first case the tree structure would be given by

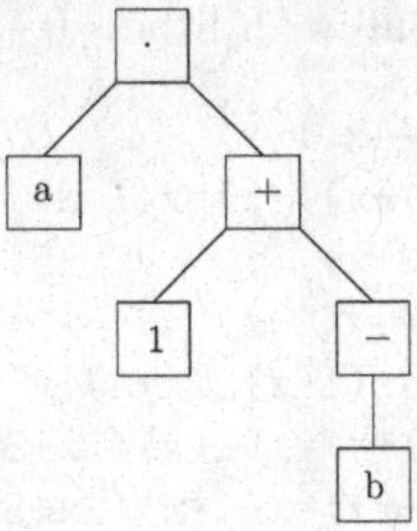

For the fitness function we compare the output for each entry from the given truth table with the output from the gene. For example, consider the expression written in gene expression programming `E(a,b) = ~&a|b0`. The expression could also be written in standard notation

$$E(a,b) = \overline{a \cdot (b+0)}.$$

Thus we find $E(0,0) = 1$, $E(0,1) = 1$, $E(1,0) = 1$, $E(1,1) = 0$. Therefore the fitness is 2, since at two entries the values differ. Of course the fitness we have to achieve is 0.

```
// gepbool.cpp

#include <cstdlib>
#include <ctime>
#include <cmath>
#include <iostream>
#include <string>
using namespace std;

const int nsymbols = 7;
// 4 terminal symbols (no arguments) a, b, 0 and 1
const int terminals = 4;
// terminal symbols first
const char symbols[nsymbols] = {'0','1','a','b','|','~','&'};
const int n = 2; // for +,- and * which take 2 arguments
int h = 10;

int evalr(char *&e,int a,int b)
{
 switch(*(e++))
 {
 case '0': return 0;
 case '1': return 1;
 case 'a': return a;
 case 'b': return b;
 case '|': return evalr(e,a,b) | evalr(e,a,b);
 case '~': return (~evalr(e,a,b)) & 1;
 case '&': return evalr(e,a,b) & evalr(e,a,b);
 default : return 0;
 }
}

int eval(char *e,int a,int b) { char *c = e; return evalr(c,a,b); }

void printr(char *&e)
{
 switch(*(e++))
```

```
 {
 case '0': cout << '0'; break;
 case '1': cout << '1'; break;
 case 'a': cout << 'a'; break;
 case 'b': cout << 'b'; break;
 case '|': cout << '('; printr(e); cout << '|'; printr(e); cout << ')';
           break;
 case '~': cout << "~("; printr(e); cout << ')'; break;
 case '&': cout << '('; printr(e); cout << '&'; printr(e); cout << ')';
           break;
 }
}

void print(char *e) { char *c = e; printr(c); }

double fitness(char *c,int *data,int N)
{
 int d, sum = 0;
 for(int j=0;j<N;j++)
  { d=eval(c,data[3*j],data[3*j+1])^data[3*j+2]; sum += d; }
 return sum;
}

// N number of data points, P population size, eps = accuracy required
void gep(int *data,int N,int P,double eps)
{
 int i,j,k,replace,replace2,rlen,rp;
 int t = h*(n-1)+1;
 int gene_len = h+t;
 int pop_len = P*gene_len;
 int iterations = 0;
 char *population = new char[pop_len];
 char *elim = new char[P];
 int toelim = P/2;
 double bestf,f;          // best fitness, fitness value
 double sumf = 0.0;       // sum of fitness values
 double pmutate = 0.1;    // probability of mutation
 double pinsert = 0.4;    // probability of insertion
 double precomb = 0.7;    // probability of recombination
 double r,lastf;          // random numbers and roulette wheel selection
 char *best = (char*)NULL; //best gene
 char *iter;              // iteration variable

 // initialize the population
 for(i=0;i<pop_len;i++)
  if(i%gene_len < h) population[i] = symbols[rand()%nsymbols];
  else population[i] = symbols[rand()%terminals];
```

```
 // initial calculations
 bestf = fitness(population,data,N);
 best = population;
 for(i=0,sumf=0.0,iter=population;i<P;i++,iter+=gene_len)
 {
  f = fitness(iter,data,N);
  sumf += f;
  if(f<bestf) { bestf = f; best = population+i*gene_len; }
 }

 while(bestf >= eps)
 {
 // reproduction, roulette wheel selection
 for(i=0;i<P;i++) elim[i] = 0;
 for(i=0;i<toelim;i++)
 {
 r = sumf*(double(rand())/RAND_MAX);
 lastf = 0.0;
 for(j=0;j<P;j++)
 {
 f = fitness(population+j*gene_len,data,N);
 if((lastf<=r) && (r<f+lastf)) { elim[j] = 1; j = P; }
 lastf += f;
 }
 }
 for(i=0;i<pop_len;)
 {
 if(population+i==best) i += gene_len; //never modify/replace best gene
 else for(j=0;j<gene_len;j++,i++)
 {
 // mutation or elimination due to failure in selection
 // for reproduction
 if((double(rand())/RAND_MAX < pmutate) || elim[i/gene_len])
 if(i%gene_len < h) population[i] = symbols[rand()%nsymbols];
 else population[i] = symbols[rand()%terminals];
 }
 // insertion
 if(double(rand())/RAND_MAX < pinsert)
 {
 // find a position in the head of this gene for insertion
 // -gene_len for the gene since we have already moved
 // to the next gene
 replace = i-gene_len;
 rp = rand()%h;
 // a random position for insertion source
 replace2 = rand()%pop_len;
 // a random length for insertion from the gene
 rlen = rand()%(h-rp);
```

```
// create the new gene
char *c = new char[gene_len];
// copy the shifted portion of the head
strncpy(c+rp+rlen,population+replace+rp,h-rp-rlen);
// copy the tail
strncpy(c+h,population+replace+h,t);
// copy the segment to be inserted
strncpy(c+rp,population+replace2,rlen);
// if the gene is fitter use it
if(fitness(c,data,N) < fitness(population+replace,data,N))
strncpy(population+replace,c,h);
delete[] c;
}
// recombination
if(double(rand())/RAND_MAX < precomb)
{
// find a random position in the gene for one point recombination
replace = i-gene_len;
rlen = rand()%gene_len;
// a random gene for recombination
replace2 = (rand()%P)*gene_len;
// create the new genes
char *c[5];
c[0] = population+replace; c[1] = population+replace2;
c[2] = new char[gene_len]; c[3] = new char[gene_len];
c[4] = new char[gene_len];
strncpy(c[2],c[0],rlen);
strncpy(c[2]+rlen,c[1]+rlen,gene_len-rlen);
strncpy(c[3],c[1],rlen);
strncpy(c[3]+rlen,c[0]+rlen,gene_len-rlen);
// take the fittest genes
for(j=0;j<4;j++)
 for(k=j+1;j<4;j++)
  if(fitness(c[k],data,N) < fitness(c[j],data,N))
  {
  strncpy(c[4],c[j],gene_len);
  strncpy(c[j],c[k],gene_len);
  strncpy(c[k],c[4],gene_len);
  }
delete[] c[2]; delete[] c[3]; delete[] c[4];
}
}
// fitness
for(i=0,sumf=0.0,iter=population;i<P;i++,iter+=gene_len)
{
f = fitness(iter,data,N);
sumf += f;
if(f < bestf) { bestf = f; best = population+i*gene_len; }
```

```
  }
  iterations++;
  }
  print(best);
  cout << endl;
  cout << "Fitness of " << bestf << " after "
       << iterations << " iterations" << endl;
  delete[] population; delete[] elim;
}

int main(void)
{
  srand(time(NULL)); // seed for the random number generator
  int data[] = {0,0,1, 0,1,1, 1,0,0, 1,1,1};
  gep(data,4,30,0.5);
  return 0;
}
```

15.3 Numerical-Symbolic Manipulation

We consider evolving integrals of limited expressions involving polynomials and exponentials. Suppose we wish to determine the integral

$$\int g(x)dx$$

in terms of elementary functions. The answer should be symbolic. However, we need a fitness function which provides a real valued fitness for any symbolic expression approximating this integral. We know that

$$\frac{d}{dx}\int g(x)dx - g(x) = 0.$$

To evaluate the approximation error amounts to an integration problem again. If we select a representative set of points for example

$$\{-1, -0.9, -0.8, -0.7, \ldots, 0.9, 1\}$$

we can estimate the approximation error over $[-1, 1]$:

$$\sum_{j=0}^{20} \left| \frac{d}{dx}\int g(x)dx - g \right|_{x=-1+0.1j}.$$

Thus a possible fitness function for the integral is given by

$$fitness(f) = \sum_{j=0}^{20} \left| \frac{df}{dx}(x = -1 + 0.1j) - g(x = -1 + 0.1j) \right|$$

i.e. we combine symbolic and numeric calculations. We search for a symbolic expression for which the derivative nearest approximates the integrand over the given set of points. Fitter individuals have a fitness closer to zero.

Another possibility would be to use the length of the simplified expression for

$$\frac{d}{dx}\int g(x)dx - f(x)$$

where a length of 0 is the fittest value (i.e. we found the integral). We use the first measure of fitness in the following program.

```
// gepint.cpp

#include <cstdlib>
#include <ctime>
#include <cmath>
#include <iostream>
#include <string>
#include "symbolicc++.h"
using namespace std;

const int nsymbols = 8;
// 3 terminal symbols (no arguments) x, 0 and 1
const int terminals = 3;
// terminal symbols first
const char symbols[nsymbols] = {'0','1','x','+','-','*','/','e'};
const int n = 2; // for +,-,* and / which take 2 arguments
int h = 5;

Symbolic evalr(char *&e)
{
 switch(*(e++))
 {
 case '0': return 0.0;
 case '1': return 1.0;
 case 'x': return Symbolic("x");
 case '+': return evalr(e)+evalr(e);
 case '-': return evalr(e)-evalr(e);
 case '*': return evalr(e)*evalr(e);
 case '/': return evalr(e)/evalr(e);
 case 'e': return exp(evalr(e));
 default : return 0;
 }
}

Symbolic eval(char *e) { char *c = e; return evalr(c); }
```

```
void printr(char *&e)
{
 switch(*(e++))
 {
 case '0': cout << '0'; break;
 case '1': cout << '1'; break;
 case 'x': cout << 'x'; break;
 case '+': cout << '('; printr(e); cout << '+'; printr(e); cout << ')';
           break;
 case '-': cout << '('; printr(e); cout << '-'; printr(e); cout << ')';
           break;
 case '*': cout << '('; printr(e); cout << '*'; printr(e); cout << ')';
           break;
 case '/': cout << '('; printr(e); cout << '/'; printr(e); cout << ')';
           break;
 case 'e': cout << "exp("; printr(e); cout << ')'; break;
 }
}

void print(char *e) { cout << eval(e) << endl; }

double fitness(char *c,const Symbolic &integrand)
{
 double j;
 double sum = 0;
 Symbolic x("x");
 Symbolic f = eval(c);
 Symbolic d_f = df(f,x);
 for(j=-1;j<=1;j+=0.1)
  sum +=
   fabs(double((d_f-integrand)[x==j,SymbolicConstant::e==exp(1.0)]));
 return sum;
}

// P population size, eps = accuracy required
void gep(const Symbolic &integrand,int P,double eps)
{
 int i,j,k,replace,replace2,rlen,rp;
 int t = h*(n-1)+1;
 int gene_len = h+t;
 int pop_len = P*gene_len;
 int iterations = 0;
 char *population = new char[pop_len];
 char *elim = new char[P];
 int toelim = P/2;
 double bestf,f;        // best fitness, fitness value
 double sumf = 0.0;     // sum of fitness values
 double pmutate = 0.1;  // probability of mutation
```

```
double pinsert = 0.4;  // probability of insertion
double precomb = 0.7;  // probability of recombination
double r,lastf;        // random numbers & roulette wheel selection
char *best = (char*)NULL; // best gene
char *iter;            // iteration variable

// initialize the population
for(i=0;i<pop_len;i++)
 if(i%gene_len < h) population[i] = symbols[rand()%nsymbols];
 else population[i] = symbols[rand()%terminals];

// initial calculations
bestf = fitness(population,integrand);
best = population;
for(i=0,sumf=0.0,iter=population;i<P;i++,iter+=gene_len)
{
f = fitness(iter,integrand);
sumf += f;
if(f<bestf) { bestf = f; best = population+i*gene_len; }
}

while(bestf >= eps)
{
// reproduction, roulette wheel selection
for(i=0;i<P;i++) elim[i] = 0;
for(i=0;i<toelim;i++)
{
r = sumf*(double(rand())/RAND_MAX);
lastf = 0.0;
for(j=0;j<P;j++)
{
f = fitness(population+j*gene_len,integrand);
if((lastf<=r) && (r<f+lastf)) { elim[j] = 1; j = P; }
lastf += f;
}
}

for(i=0;i<pop_len;) // never modify/replace best gene
{
if(population+i == best) i += gene_len;
else for(j=0;j<gene_len;j++,i++)
{
// mutation or elimination due to failure in selection
// for reproduction
if((double(rand())/RAND_MAX < pmutate) || elim[i/gene_len])
if(i%gene_len < h) population[i] = symbols[rand()%nsymbols];
else population[i] = symbols[rand()%terminals];
}
```

```
// insertion
if(double(rand())/RAND_MAX < pinsert)
{
// find a position in the head of this gene for insertion
// -gene_len for the gene since we have already moved to the next gene
replace = i-gene_len;
rp = rand()%h;
// a random position for insertion source
replace2 = rand()%pop_len;
// a random length for insertion from the gene
rlen = rand()%(h-rp);
// create the new gene
char *c = new char[gene_len];
// copy the shifted portion of the head
strncpy(c+rp+rlen,population+replace+rp,h-rp-rlen);
// copy the tail
strncpy(c+h,population+replace+h,t);
// copy the segment to be inserted
strncpy(c+rp,population+replace2,rlen);
// if the gene is fitter use it
if(fitness(c,integrand) < fitness(population+replace,integrand))
strncpy(population+replace,c,h);
delete[] c;
}
// recombination
if(double(rand())/RAND_MAX < precomb)
{
// find a random position in the gene for one point recombination
replace = i-gene_len;
rlen = rand()%gene_len;
// a random gene for recombination
replace2 = (rand()%P)*gene_len;
// create the new genes
char *c[5];
c[0] = population+replace; c[1] = population+replace2;
c[2] = new char[gene_len]; c[3] = new char[gene_len];
c[4] = new char[gene_len];
strncpy(c[2],c[0],rlen); strncpy(c[2]+rlen,c[1]+rlen,gene_len-rlen);
strncpy(c[3],c[1],rlen); strncpy(c[3]+rlen,c[0]+rlen,gene_len-rlen);
// take the fittest genes
for(j=0;j<4;j++)
 for(k=j+1;j<4;j++)
  if(fitness(c[k],integrand) < fitness(c[j],integrand))
  {
  strncpy(c[4],c[j],gene_len);
  strncpy(c[j],c[k],gene_len);
  strncpy(c[k],c[4],gene_len);
  }
```

```
 delete[] c[2]; delete[] c[3]; delete[] c[4];
 }
 }
 // fitness
 for(i=0,sumf=0.0,iter=population;i<P;i++,iter+=gene_len)
 {
 f = fitness(iter,integrand);
 sumf += f;
 if(f < bestf) { bestf = f; best = population+i*gene_len; }
 }
 iterations++;
 print(best);
 }
 print(best);
 cout << endl;
 cout << "Fitness of " << bestf << " after "
      << iterations << " iterations" << endl;
 delete[] population; delete[] elim;
}

int main(void)
{
 Symbolic x("x");
 srand(time(NULL)); // set the seed for the random number generator
 gep(x,10,0.1); cout << endl;
 gep(x*ln(x),10,0.1);
 return 0;
}
```

15.4 Multi Expression Programming

Multi Expression Programming (Oltean [154], Oltean [155]) is a genetic programming variant which encodes multiple expressions in a single chromosome. The expressions are represented in a similar way to gene expression programming, except that expressions may refer to (contain) expressions that appear earlier in the chromosome. The fitness of a chromosome is the best fitness over each of the expressions encoded in the chromosome. Crossing generally occurs at the level of expressions.

Example. The expression

$$\sin b + a * b + c * d + a - b$$

must be represented as a chromosome in multi expression programming which can be constructed as follows

0:	a	The variable a
1:	b	The variable b
2:	c	The variable c
3:	d	The variable d
4:	$*, \mathbf{0}, \mathbf{1}$	The expression $a \times b$
5:	$*, \mathbf{2}, \mathbf{3}$	The expression $c \times d$
6:	$+, \mathbf{4}, \mathbf{5}$	The expression $(a \times b) + (c \times d)$
7:	$+, \mathbf{6}, \mathbf{0}$	The expression $(a \times b) + (c \times d) + a$
8:	$-, \mathbf{7}, \mathbf{1}$	The expression $((a \times b) + (c \times d) + a) - b$
9:	$\sin, \mathbf{1}$	The expression $\sin b$
10:	$+, \mathbf{9}, \mathbf{8}$	The expression $(\sin b) + (((a \times b) + (c \times d) + a) - b)$

The operation $-, x, y$ means $x-y$. The expression must be evaluated for given values of a, b, c and d. The numbers in bold are the expressions which can be referenced (backwards) from within other expressions. A number in bold appearing in an expression must be less than the number of the current expression. The following program constructs this chromosome and evaluates the expression for $a = 3$, $b = 4$, $c = 5$ and $d = 6$.

```
// mep.cpp

#include <cctype>
#include <cstdlib>
#include <iostream>
#include <map>
#include <sstream>
#include <string>
#include <vector>
#include <cmath>
using namespace std;

typedef double (*op)(vector<double>);

// assume appropriate number of arguments
double add(vector<double> v) { return v[0]+v[1]; }
double sub(vector<double> v) { return v[0]-v[1]; }
double mul(vector<double> v) { return v[0]*v[1]; }
double sine(vector<double> v) { return sin(v[0]); }

string show(vector<string> e,string s = "")
{
 size_t i = 0, j = 0;
 string r;
 vector<string> operands;

 if(s=="") return show(e,e.back());
 while(i < s.size() && j != string::npos)
 {
 j = s.find(',',i);
```

```
  operands.push_back(s.substr(i,j-i));
  if(j != string::npos) i=j+1;
  }
  if(operands.size()==1)
  if(isdigit(operands[0][0]))
  {
  int in;
  istringstream is(operands[0]);
  is >> in;
  return show(e,e[in]);
  }
  else return operands[0];
  r = show(e,operands[0]); r += "(";
  for(i=1;i<operands.size()-1;i++)
  { r += show(e,operands[i]); r += ","; }
  r += show(e,operands[i]); r += ")";
  return r;
}

double evaluate(vector<string> e,map<string,double> values,
                map<string,op> ops,string s="")
{
  size_t i = 0, j = 0;
  string r;
  vector<string> operands;
  vector<double> operand_values;
  if(s=="") return evaluate(e,values,ops,e.back());
  while(i < s.size() && j != string::npos)
  {
  j = s.find(',',i);
  operands.push_back(s.substr(i,j-i));
  if(j != string::npos) i=j+1;
  }
  if(operands.size()==1)
  if(isdigit(operands[0][0]))
  {
  int i;
  istringstream is(operands[0]); is >> i;
  return evaluate(e,values,ops,e[i]);
  }
  else return values[operands[0]];
  r = show(e,operands[0]);
  for(i=1;i<operands.size();i++)
  {
  operand_values.push_back(evaluate(e,values,ops,operands[i]));
  }
  return ops[operands[0]](operand_values);
}
```

```
int main(void)
{
 map<string,op>     ops;
 map<string,double> values;
 vector<string>     expression(11);
 // operators used and corresponding arity
 ops["+"] = add;
 ops["-"] = sub;
 ops["*"] = mul;
 ops["s"] = sine;
 // values for evaluation
 values["a"] = 3.0; values["b"] = 4.0;
 values["c"] = 5.0; values["d"] = 6.0;
 // (a*b)+(c*d)+a-b+sin(b)
 expression[0] = "a"; expression[1] = "b";
 expression[2] = "c"; expression[3] = "d";
 expression[4] = "*,0,1";
 expression[5] = "*,2,3";
 expression[6] = "+,4,5";
 expression[7] = "+,6,0";
 expression[8] = "-,7,1";
 expression[9] = "s,1";
 expression[10] = "+,9,8";
 cout << show(expression,"") << endl;
 cout << evaluate(expression, values, ops) << endl;
 return 0;
}
```

Consider the chromosome for the expression

$$\sin b + a * b + c * d + a - b$$

from above

0:	a
1:	b
2:	c
3:	d
4:	$*, \mathbf{0}, \mathbf{1}$
5:	$*, \mathbf{2}, \mathbf{3}$
6:	$+, \mathbf{4}, \mathbf{5}$
7:	$+, \mathbf{6}, \mathbf{0}$
8:	$-, \mathbf{7}, \mathbf{1}$
9:	$\sin, \mathbf{1}$
10:	$+, \mathbf{9}, \mathbf{8}$

Mutation randomly changes a subexpression in the chromosome, for example mutating the first chromosome at position 6 might yield

0:	a
1:	b
2:	c
3:	d
4:	$*, \mathbf{0}, \mathbf{1}$
5:	$*, \mathbf{2}, \mathbf{3}$
6:	$\sin \mathbf{4}$
7:	$+, \mathbf{6}, \mathbf{0}$
8:	$-, \mathbf{7}, \mathbf{1}$
9:	$\sin, \mathbf{1}$
10:	$+, \mathbf{9}, \mathbf{8}$

In other words the entire subexpression $+, \mathbf{4}, \mathbf{5}$ is replaced by the new subexpression $\sin \mathbf{4}$.

Crossing is implemented at the subexpression level in the same way as for genetic algorithms and gene expression programming. Consider for example the one point crossing, at position 4, between the two chromosomes

0:	a	0:	a
1:	b	1:	b
2:	c	2:	$*, \mathbf{1}, \mathbf{0}$
3:	d	3:	$\cos, \mathbf{0}$
4:	$*, \mathbf{0}, \mathbf{1}$	4:	$+, \mathbf{0}, \mathbf{3}$
5:	$*, \mathbf{2}, \mathbf{3}$	5:	$*, \mathbf{3}, \mathbf{3}$
6:	$+, \mathbf{4}, \mathbf{5}$	6:	$+, \mathbf{5}, \mathbf{2}$
7:	$+, \mathbf{6}, \mathbf{0}$	7:	$\sin, \mathbf{2}$
8:	$-, \mathbf{7}, \mathbf{1}$	8:	$-, \mathbf{7}, \mathbf{1}$
9:	$\sin, \mathbf{1}$	9:	$*, \mathbf{1}, \mathbf{1}$
10:	$+, \mathbf{9}, \mathbf{8}$	10:	$+, \mathbf{9}, \mathbf{8}$

which yields the two children

0:	a	0:	a
1:	b	1:	b
2:	c	2:	$*, \mathbf{1}, \mathbf{0}$
3:	d	3:	$\cos, \mathbf{0}$
4:	$+, \mathbf{0}, \mathbf{3}$	4:	$*, \mathbf{0}, \mathbf{1}$
5:	$*, \mathbf{3}, \mathbf{3}$	5:	$*, \mathbf{2}, \mathbf{3}$
6:	$+, \mathbf{5}, \mathbf{2}$	6:	$+, \mathbf{4}, \mathbf{5}$
7:	$\sin, \mathbf{2}$	7:	$+, \mathbf{6}, \mathbf{0}$
8:	$-, \mathbf{7}, \mathbf{1}$	8:	$-, \mathbf{7}, \mathbf{1}$
9:	$*, \mathbf{1}, \mathbf{1}$	9:	$\sin, \mathbf{1}$
10:	$+, \mathbf{9}, \mathbf{8}$	10:	$+, \mathbf{9}, \mathbf{8}$

and the two point crossing between position 3 and 6 which yields the children

0:	a	**0**:	a
1:	b	**1**:	b
2:	c	**2**:	$*, \mathbf{1}, \mathbf{0}$
3:	$\cos, \mathbf{0}$	**3**:	d
4:	$+, \mathbf{0}, \mathbf{3}$	**4**:	$*, \mathbf{0}, \mathbf{1}$
5:	$*, \mathbf{3}, \mathbf{3}$	**5**:	$*, \mathbf{2}, \mathbf{3}$
6:	$+, \mathbf{5}, \mathbf{2}$	**6**:	$+, \mathbf{4}, \mathbf{5}$
7:	$+, \mathbf{6}, \mathbf{0}$	**7**:	$\sin, \mathbf{2}$
8:	$-, \mathbf{7}, \mathbf{1}$	**8**:	$-, \mathbf{7}, \mathbf{1}$
9:	$\sin, \mathbf{1}$	**9**:	$*, \mathbf{1}, \mathbf{1}$
10:	$+, \mathbf{9}, \mathbf{8}$	**10**:	$+, \mathbf{9}, \mathbf{8}$

Mutation and crossing could be refined to change the subexpressions, although this would be a little more involved.

Multiexpression programming can also be applied to boolean expressions. Consider for example the *full adder*

a	b	c_{in}	s	c_{out}
0	0	0	0	0
0	0	1	1	0
0	1	0	1	0
0	1	1	0	1
1	0	0	1	0
1	0	1	0	1
1	1	0	0	1
1	1	1	1	1

Using the NAND-gate the expression in multiexpression programming is

0:	a
1:	b
2:	c
3:	$N, \mathbf{0}, \mathbf{1}$
4:	$N, \mathbf{0}, \mathbf{3}$
5:	$N, \mathbf{1}, \mathbf{3}$
6:	$N, \mathbf{4}, \mathbf{5}$
7:	$N, \mathbf{2}, \mathbf{6}$
8:	$N, \mathbf{6}, \mathbf{7}$
9:	$N, \mathbf{2}, \mathbf{7}$
10:	$N, \mathbf{8}, \mathbf{9}$
11:	$N, \mathbf{3}, \mathbf{7}$

where **10** : provides the sum and **11** : provides the carry out. Thus using nine NAND-gates we can build the full adder.

Exercise. A mathematical expression can be written in a binary tree structure. Thus we can express it using XML. Consider for example the expression

$$b * b - 4 * a * c \,.$$

Then we have

```
<expression>
 <subtract>
  <multiply>
   <symbol> b </symbol>
   <symbol> b </symbol>
  </multiply>
  <multiply>
   <number> 4 </number>
  <multiply>
   <symbol> a </symbol>
   <symbol> c </symbol>
  </multiply>
  </multiply>
 </subtract>
</expression>
```

Write a Java program that can parse through the XML file and evaluates it.

Chapter 16

Wavelets

16.1 Introduction

We discuss one-dimensional discrete wavelet transform (Benedetto and Frazier [19], Erlebacher et al [55], Jensen and La Cour-Harbo [105]) and continuous wavelet transforms (Daubechies [44], Hernández and Weiss [93], Chui [34]). Within the discrete wavelet transform we distinguish between redundant discrete systems (frames) and orthonormal, semi-orthogonal, and biorthogonal bases of wavelets. In most case the discrete wavelet transform (or DWT) is an orthogonal function which can be applied to a finite group of data. Functionally, it is very much like the discrete Fourier transform, in that the transforming function is orthogonal, a signal passed twice (i.e. a forward and a backward transform) through the transformation is unchanged, and the input signal is assumed to be a set of discrete-time samples. Both transforms are convolutions. Whereas the basis function of the Fourier transform is sinusoidal, the wavelet basis is a set of functions which are defined by a recursive difference equation for the *scaling function* ϕ (also called *father wavelet*)

$$\phi(x) = \sum_{k=0}^{M-1} c_k \phi(2x - k)$$

where the range of the summation is determined by the specified number of nonzero coefficients M. Here k is the translation parameter. The number of the coefficients is not arbitrary and is determined by constraints of orthogonality and normalization. Owing to the periodic boundary condition we have

$$c_k \equiv c_{k+nM}$$

where $n \in \mathbb{N}$. We notice that periodic wavelets are only one possibility to deal with signals defined on an interval. Generally, the area under the scaling function over all space should be unity, i.e.

$$\int_{\mathbb{R}} \phi(x) dx = 1 .$$

It follows that

$$\sum_{k=0}^{M-1} c_k = 2\,.$$

In the Hilbert space $L_2(\mathbb{R})$ with the *scalar product*

$$\langle f, g \rangle := \int_{\mathbb{R}} f(x)\bar{g}(x)dx$$

the function ϕ is orthogonal to its translations; i.e.,

$$\int_{\mathbb{R}} \phi(x)\phi(x-k)dx = 0, \qquad k \in \mathbb{Z} \setminus \{0\}\,.$$

What is desired is a function ψ which is also orthogonal to its dilations, or scales, i.e.,

$$\int_{\mathbb{R}} \psi(x)\psi(2x-k)dx = 0\,.$$

Such a function ψ (called the *mother wavelet*) does exist and is given by (the so-called associated *wavelet function*)

$$\psi(x) = \sum_{k=1}^{M} (-1)^k c_{1-k}\phi(2x-k)$$

which is dependent on the solution of ϕ. The following equation follows from the orthonormality of scaling functions

$$\sum_k c_k c_{k-2m} = 2\delta_{0m}$$

which means that the above sum is zero for all m not equal to zero, and that the sum of the squares of all coefficients is two. Another equation which can be derived from $\psi(x)\perp\phi(x-m)$ is

$$\sum_k (-1)^k c_{1-k}c_{k-2m} = 0\,.$$

A way to solve for ϕ is to construct a matrix of coefficient values. This is a square $M \times M$ matrix where M is the number of nonzero coefficients. The matrix is designated L with entries $L_{ij} = c_{2i-j}$. This matrix has an eigenvalue equal to 1, and its corresponding (normalized) eigenvector contains, as its components, the value of the function ϕ at integer values of x. Once these values are known, all other values of the function ϕ can be generated by applying the recursion equation to get values at half-integer x, quarter-integer x, and so on down to the desired dilation. This determines the accuracy of the function approximation.

Example. The *Haar function* ($\psi \in L_2(\mathbb{R})$)

$$\psi(x) := \begin{cases} 1 & 0 \le x < \frac{1}{2} \\ -1 & \frac{1}{2} \le x < 1 \\ 0 & \text{otherwise} \end{cases}$$

is an example for a mother wavelet. The scaling function (*father wavelet*) $\phi \in L_2(\mathbb{R})$ is given by

$$\phi(x) := \begin{cases} 1 & 0 \leq x < 1 \\ 0 & \text{otherwise} \end{cases}.$$

We have

$$\int_{\mathbb{R}} \psi(x)\phi(x)dx = 0$$

and $\psi(x) = \phi(2x) - \phi(2x-1)$. The functions

$$\psi_{m,n}(x) := 2^{\frac{m}{2}} \psi(2^m x - n), \quad m, n \in \mathbb{Z}$$

form a basis in the Hilbert space $L_2(\mathbb{R})$. This means that every function $f \in L_2(\mathbb{R})$ can be expanded with respect to this basis. If we restrict m to $m = 0, 1, 2, \ldots$ and

$$n = 0, 1, 2, \ldots, 2^m - 1$$

we obtain an orthonormal basis in the Hilbert space $L_2[0,1]$. ♣

This class of wavelet functions is constrained, by definition, to be zero outside of a small interval. This is what makes the wavelet transform able to operate on a finite set of data, a property which is formally called compact support. The recursion relation ensures that a scaling function ϕ is non-differentiable everywhere. Of course this is not valid for Haar wavelets. The following table lists coefficients for two wavelet transforms. The sum of the coefficients is normalized to 2.

Wavelet	c_0	c_1	c_2	c_3
Haar	1.0	1.0		
Daubechies-4	$\frac{1}{4}(1+\sqrt{3})$	$\frac{1}{4}(3+\sqrt{3})$	$\frac{1}{4}(3-\sqrt{3})$	$\frac{1}{4}(1-\sqrt{3})$

Table 16.1: Coefficients for Two Wavelet Functions

More generally we consider the Hilbert space $L_2(\mathbb{R})$ of the square integrable functions. Let $\psi \in L_2(\mathbb{R})$ be the mother wavelet. We define

$$\psi_{m,n}(x) := a^{-m/2}\psi(a^{-m}x - nb)$$

for $m, n \in \mathbb{Z}$ and $a > 1$, $b > 0$. The constant $a^{-m/2}$ is a normalization constant so that

$$\int_{\mathbb{R}} |\psi_{m,n}(x)|^2 dx = \int_{\mathbb{R}} |\psi(x)|^2 dx\,.$$

If the support of ψ is $[c,d]$, then the support of $\psi_{m,n}$ is $[a^m c + nb, a^m d + nb]$. However whether ψ has compact support or not the shape of the graph of $\psi_{m,n}$ is a scaled dilated version of the graph of ψ and translated by amount nb. This is one of the main differences between the discrete windowed Fourier transform functions and wavelets.

An important mother wavelet is the *Mexican hat wavelet* given by

$$\psi_\sigma(x) = \frac{1}{\sqrt{2\pi}\sigma^3}\left(1 - \frac{x^2}{\sigma^3}\right)\exp(-x^2/(2\sigma^2)).$$

It is the normalized second derivative of a Gaussian function $\exp(-x^2/(2\sigma^2))$. The extension to more than one dimension is straightforward. It is also part of the *Hermitian wavelets.* The Mexican hat wavelet is a useful tool for point source detection. It has an analytical form that is convenient when making calculations and allows to implement fast algorithm. It is suited for the detection of Gaussian structures since it is obtained by applying the Laplacian

Another mother wavelet with an imaginary part is the *Morlet wavelet.* It consists of a plane wave modulated by a Gaussian function

$$\psi_\Omega(x) = c_\Omega \pi^{-1/4} e^{-x^2/2}(e^{i\Omega t} - \kappa_\Omega)$$

where $\kappa_\Omega = e^{-\Omega^2/2}$ is defined by the admissibility criterion. The normalization constant c_Ω is

$$c_\Omega = (1 + e^{-\Omega^2} - 2e^{-3\Omega^2/4})^{-1/2}.$$

For the *Littlewood-Paley-Stein wavelet* we have

$$\psi \in L_2(\mathbb{R}^n) \quad \text{with} \quad \sum_{j\in\mathbb{Z}} |\hat{\psi}(2^j\xi)|^2 = 1 \text{ a.e.}$$

where $\hat{\psi}$ denotes the Fourier transform of ψ.

Exercises. (1) Show that the Fourier transformation $\hat{\psi}(\omega)$ of the Morlet wavelet is given by

$$\hat{\psi}_\Omega(\omega) = c_\Omega \pi^{-1/4}(e^{-(\Omega-\omega)^2/2} - \kappa_\Omega e^{-\omega/2}).$$

(2) Can the Gaussian function $\exp(-x^2/2)$ be considered as a mother wavelet?

(3) Consider the Hilbert space $L_2(\mathbb{R})$ and the Haar wavelet. Find the expansion of $f \in L_2(\mathbb{R})$

$$f(x) = \exp(-|x|).$$

Is the Haar wavelet a good choice for the expansion?

16.2 Multiresolution Analysis

A multiresolution analysis (Daubechies [44]) of a signal $f(t)$ consists of a collection of nested subspaces $\{ V_j \}_{j\in\mathbb{Z}}$ of the Hilbert space $L_2(\mathbb{R})$ of square integrable functions, satisfying the following set of properties:

(i) $\cap_{j\in\mathbb{Z}} V_j = \{0\}$, $\cup_{j\in\mathbb{Z}} V_j$ is dense in $L_2(\mathbb{R})$

(ii) $V_j \subset V_{j-1}$

(iii) $f(t) \in V_j \longleftrightarrow f(2^j t) \in V_0$

(iv) There exists a function $\phi_0(t)$ in V_0, called the scaling function, such that the collection $\{\phi_0(t-k),\ k \in \mathbb{Z}\}$ is an unconditional Riesz basis for the subspace V_0.

Similarly, the scaled and shifted functions,

$$\{\phi_{j,k}(t) = 2^{-j/2}\phi_0(2^{-j}t - k),\ k \in \mathbb{Z}\}$$

constitute a Riesz basis for the space V_j. Performing a multiresolution analysis of the signal f means successively projecting it into each of the approximation subspaces V_j

$$\text{approx}_j(t) = (\text{Proj}_{V_j} f)(t) = \sum_{k\in\mathbb{Z}} a_f(j,k)\phi_{j,k}(t)\,.$$

Now $V_j \subset V_{j-1}$, approx_j is a coarser approximation of f than is approx_{j-1}. Thus the idea of the multiresolution analysis consists in examining the loss of information, that is the detail, when going from one approximation to the next, coarser one

$$\text{detail}_j(t) = \text{approx}_{j-1}(t) - \text{approx}_j(t)\,.$$

The multiresolution analysis shows that the detail signal detail_j can be obtained from projections of the signal f onto a collection of subspaces, the W_j, called the wavelet subspaces. Let $\oplus$ be the orthogonal sum of two linear spaces. We have

$$V_{j+1} = V_j \oplus W_j, \qquad W_j \subset V_{j+1}\,.$$

Thus we have two mutually orthogonal subspaces, V_j and W_j. Since

$$\text{clos}_{L_2(\mathbb{R})}\left\{\bigcup_{j=-\infty}^{\infty} V_j\right\} = L_2(\mathbb{R})$$

we obtain the expansion

$$L_2(\mathbb{R}) = \bigoplus_{j=-\infty}^{\infty} W_j\,.$$

Any function in the Hilbert space $L_2(\mathbb{R})$ may be expanded in a generalized Fourier series

$$f(x) = \sum_{j=-\infty}^{\infty} w_j \quad \text{with } w_j \in W_j \text{ for all } j \in \mathbb{Z}\,.$$

There exists a function ψ_0, called the mother wavelet, to be derived from ϕ_0, such that its templates $\{\psi_{j,k}(t) = 2^{-j/2}\psi_0(2^{-j}t-k),\ k \in \mathbb{Z}\}$ constitute a Riesz basis for the subspace W_j

$$\text{detail}_j(t) = (\text{Proj}_{W_j} f)(t) = \sum_{k\in\mathbb{Z}} d_f(j,k)\psi_{j,k}(t)\,.$$

Let $\mathcal{H}$ be a separable Hilbert space (here $L_2(\mathbb{R})$) and $\{ h_n \}_{n\in\mathbb{Z}}$ a countable subset of $\mathcal{H}$. We say that $\{ h_n \}$ is a Riesz basis for $\mathcal{H}$ if $\{ h_n \}$ is complete and if there are constants $0 < C_1 \le C_2 < \infty$, for an arbitrary sequence $c := \{c_n\}_{n\in\mathbb{Z}} \in \ell_2(\mathbb{Z})$ such that

$$C_1\|c\| \le \|\sum_{n\in\mathbb{Z}} c_n h_n\| \le C_2\|c\| .$$

The multiresolution analysis consists in rewriting the information in f as a collection of details at different resolutions and a low-resolution approximation

$$\begin{aligned} f(t) &= \text{approx}_J(t) + \sum_{j=1}^{j=J} \text{detail}_j(t) \\ &= \sum_{k\in\mathbb{Z}} a_f(J,k)\phi_{J,k}(t) + \sum_{j=1}^{J}\sum_{k\in\mathbb{Z}} d_f(j,k)\psi_{j,k}(t) . \end{aligned}$$

The approx_j essentially being coarser and coarser approximations of f means that ϕ_0 needs to be a low-pass function. The details_j, being an information 'differential', indicates rather that ψ_0 is a bandpass function, and therefore a small wave, a wavelet. The multiresolution analysis shows that the mother wavelet ψ_0 must satisfy

$$\int_{\mathbb{R}} \psi_0(t)dt = 0$$

and that its Fourier transform obeys

$$|\hat{\psi}_0(\omega)| \sim \omega^N, \quad \omega \to 0$$

where N is a positive integer called the number of vanishing moments of the wavelet.

Given a scaling function ϕ_0 and a mother-wavelet ψ_0, the discrete (or non-redundant) wavelet transform is a mapping $L_2(\mathbb{R}) \to \ell_2(\mathbb{Z})$ between the two Hilbert spaces $L_2(\mathbb{R})$ and $\ell_2(\mathbb{Z})$ given by

$$f(t) \to \{\{a_f(J,k),\, k\in\mathbb{Z}\},\, \{ d_f(j,k),\, j=1,\dots,J,\; k\in\mathbb{Z}\}\} .$$

These coefficients are defined through the inner products of the signal f with two sets of functions

$$a_f(j,k) = \langle f(t), \mathring{\phi}_{j,k}(t)\rangle, \qquad d_f(j,k) = \langle f(t), \mathring{\psi}_{j,k}(t)\rangle$$

where $\mathring{\psi}_{jk}$ (respectively $\mathring{\phi}_{jk}$) are shifted and dilated templates of $\mathring{\psi}_0$ (respectively $\mathring{\phi}_0$), called the dual mother wavelet (respectively the dual scaling function), and whose definition depends on whether one chooses to use an orthogonal, semi-orthogonal or bi-orthogonal discrete wavelet transform. They can pratically be computed by a fast recursive filter-bank-based pyramidal algorithm whith low computational cost.

16.3 Pyramid Algorithm and Discrete Wavelets

For discrete wavelets we have the pyramid algorithm. The *pyramid algorithm* operates on a finite set on N input data $x_0, x_1, \ldots, x_{N-1}$, where N is a power of two; this value will be referred to as the input block size. These data are passed through two *convolution functions*, each of which creates an output stream that is half the length of the original input. These convolution functions are filters, one half of the output is produced by the "low-pass filter"

$$a_i = \frac{1}{2}\sum_{j=0}^{N-1} c_{2i-j+1}x_j, \qquad i = 0, 1, \ldots, \frac{N}{2} - 1$$

and the other half is produced by the "high-pass filter" function

$$b_i = \frac{1}{2}\sum_{j=0}^{N-1}(-1)^j c_{j-2i}x_j, \qquad i = 0, 1, \ldots, \frac{N}{2} - 1$$

where N is the input block size, c_j are the coefficients,

$$\mathbf{x} = (x_0, x_1, \ldots, x_{N-1})$$

is the input sequence, and

$$\mathbf{a} = (a_0, a_1, \ldots, a_{N/2-1}), \qquad \mathbf{b} = (b_0, b_1, \ldots, b_{N/2-1})$$

are the output sequences. In the case of the lattice filter, the low- and high-pass outputs are usually referred to as the odd and even outputs, respectively. In many situations, the odd or low-pass output contains most of the information content of the original input signal. The even, or high-pass output contains the difference between the true input and the value of the reconstructed input if it were to be reconstructed from only the information given in the odd output. Higher order wavelets (i.e. those with more nonzero coefficients) tend to put more information into the odd output, and less into the even output. If the average amplitude of the even output is low enough, then the even half of the signal may be discarded without greatly affecting the quality of the reconstructed signal. An important step in wavelet-based data compression is finding wavelet functions which cause the even terms to be nearly zero. Details can only be neglected for very smooth time series and smooth wavelet filters, a situation which is not satisfied for chaotic time signals.

The Haar wavelet represents a simple interpolation scheme. After passing these data through the filter functions, the output of the low-pass filter consists of the average of every two samples, and the output of the high-pass filter consists of the difference of every two samples. The high-pass filter contains less information than the low pass output. If the signal is reconstructed by an inverse low-pass filter of the form

$$x_j^L = \sum_{i=0}^{N/2-1} c_{2i-j+1}a_i, \qquad j = 0, 1, \ldots, N-1$$

then the result is a duplication of each entry from the low-pass filter output. This is a wavelet reconstruction with 2× data compression. Since the perfect reconstruction is a sum of the inverse low-pass and inverse high-pass filters, the output of the inverse high-pass filter can be calculated. This is the result of the inverse high-pass filter function

$$x_j^H = \sum_{i=0}^{N/2-1} (-1)^j c_{j-1-2i} b_i, \qquad j = 0, 1, \ldots, N-1.$$

The perfectly reconstructed signal is

$$x = x^L + x^H$$

where x is the vector with elements x_j. Using other coefficients and other orders of wavelets yields similar results, except that the outputs are not exactly averages and differences, as in the case using the Haar coefficients.

Let us consider an example. Most discrete signals are filtered using linear difference equations, for example

$$y[t] = ay[t-1] + (1-a)x[t], \qquad t = 0, 1, \ldots$$

where $x[t]$ is the input signal, $y[t]$ the output, $y[-1] = 0$, $x[0] = 0$, and a $(0 < a < 1)$ is the filter parameter. Notice that $y[t]$ depends upon previously computed values of the filter output. Thus the chosen filter is an infinite response filter.

It is well-known that filtered chaotic signals can exhibit increases in observed fractal dimensions, for example the Liapunov dimension. We consider the Liapunov exponent of filtered chaotic one-dimensional maps using wavelets. The input signal is given as the solution of the logistic map

$$x[t+1] = 4x[t](1 - x[t])$$

where $x[0] \in [0, 1]$ and $t = 0, 1, \ldots$. The properties of the logistic map are well-known (see chapter 1). For almost all initial values we find that the Liapunov exponent of the logistic map is given by $\ln(2)$. We apply the Haar wavelet to filter the time series $(x_0, x_1, \ldots, x_{N-1})$ resulting from the logistic map in the chaotic regime. The filter decomposes the signal into the moving average coefficients $\{ a_i \}$ and details $\{ b_i \}$ with $i = 0, 1, \ldots, N/2 - 1$. Owing to Broomhead et al [23] the complete set of coefficients $\{ a_i, b_i \}$ have the same Liapunov exponent as the original time series. The reason is that Haar (and Daubechies) wavelets are finite impulse response filters and that the convolution is performed only once. Therefore we deal with a non-recursive finite impulse filter. Using the time series given by the logistic map we evaluate x_j^L, x_j^H using equations given above. We have

$$x_j^L = x_{j+1}^L$$

for j even and

$$x_j^H = -x_{j+1}^H$$

for j even. We use the series x_j^L for the calculation of the Liapunov exponent. Notice that in the case of the logistic equation the even output resulting from Haar wavelets, i.e. the details, contain nearly as much information as the odd output, i.e. the averages. We remove the duplication of each entry of x_j^L and calculate the Liapunov exponent of this time series. Thus the size of the time series is $N/2$. An algorithm to find the Liapunov exponent from time series is described in chapter 2. The Liapunov exponent is much larger than $\ln(2)$ for this time series, i.e. the time series becomes more chaotic.

Using the coefficients for the Daubehies-4 wavelet we find similar results, i.e., the Liapunov exponent is much larger than $\ln(2)$. Furthermore, for other one-dimensional chaotic maps we also find similar results.

```
// waveletliapunov.cpp

#include <iostream>
#include <cmath>        // for fabs, log
using namespace std;

void find(double* a,int length,int point,double& min,int& position)
{
 int i = 0;
 if(point==0) { min = fabs(a[i]-a[1]); }
 if(point!=0) { min = fabs(a[i]-a[point]); }
 position = i;
 double distance;
 for(i=1;i<(length-1);i++)
 {
 if(i!=point)
 {
 distance = fabs(a[i]-a[point]);
 if(distance < min) { min = distance; position = i; } // end if
 } // end if
 } // end for
}

int main(void)
{
 const double pi = 3.14159;
 // generate time series
 int n = 16384; // length of time series
 int m = n/2;
 int k;
 double* x = new double[n];
 int t;
```

```
x[0] = 1.0/3.0;  // initial value of time series
for(t=0;t<(n-1);t++) { x[t+1] = 4.0*x[t]*(1.0-x[t]); }
// Haar wavelet
double* c = new double[n];
for(k=0;k<n;k++) c[k] = 0.0;
c[0] = 1.0; c[1] = 1.0;

double* a = new double[m];
for(k=0;k<m;k++) a[k] = 0.0;
double* b = new double[m];
for(k=0;k<m;k++) b[k] = 0.0;

int i, j;
for(i=0;i<m;i++) {
 for(j=0;j<n;j++) {
  if(2*i-j+1 < 0) a[i] += c[2*i-j+1+n]*x[j];
  else a[i] += c[2*i-j+1]*x[j];
 }
 a[i] = 0.5*a[i];
 }
for(i=0;i<m;i++) {
 for(j=0;j<n;j++) {
  if(j-2*i < 0) b[i] += pow(-1.0,j)*c[j-2*i+n]*x[j];
  else b[i] += pow(-1.0,j)*c[j-2*i]*x[j];
 }
 b[i] = 0.5*b[i];
 }

// inverse transform
double* xL = new double[n];
double* xH = new double[n];
for(j=0;j<n;j++) xL[j] = 0.0;
for(j=0;j<n;j++) xH[j] = 0.0;

for(j=0;j<n;j++) {
 for(i=0;i<m;i++) {
  if(2*i-j+1 < 0) xL[j] += c[2*i-j+1+n]*a[i];
  else xL[j] += c[2*i-j+1]*a[i];
 }
 }

for(j=0;j<n;j++) {
 for(i=0;i<m;i++) {
  if(j-2*i < 0) xH[j] += pow(-1.0,j)*c[j-2*i+n]*b[i];
  else xH[j] += pow(-1.0,j)*c[j-2*i]*b[i];
```

```
}
}
// input signal reconstructed
double* y = new double[n];
for(k=0;k<n;k++) y[k] = xL[k] + xH[k];
// end wavelet block
// timeseries
int count = 0;
t = 0;
double* series = new double[m];
while(t < n) { series[count] = xL[t]; count++; t += 2; }
// Liapunov block
double* d = new double[m-1];
double* d1 = new double[m-1];
int point = 1; int position;
double min;
for(t=0;t<(m-1);t++)
{
find(series,m,t,min,position);
d[t] = fabs(series[t]-series[position]);
d1[t] = fabs(series[t+1]-series[position+1]);
}
double sum = 0.0;
for(t=0;t<(m-1);t++) { sum += log(d1[t]/d[t]); }
double lambda = sum/((double)(m-1));
cout << "lambda = " << lambda << endl;
delete[] x; delete[] d;  delete[] d1;
return 0;
}
```

16.4 Biorthogonal Wavelets

Let $\langle\, , \rangle$ be the scalar product in the Hilbert space $L_2(\mathbb{R})$. Dilations and translations of the scaling function ϕ leading to $\{\phi_{jk}\}$ constitute a basis for the subspaces V_j and, similarly $\{\psi_{jk}(t)\}$ for the subspaces W_j (Daubechies [44]). We define a dual multiresolution analysis with dual subspaces $\{\widetilde{V}_j\}$ and $\{\widetilde{W}_j\}$ generated from a dual scaling function $\widetilde{\phi}$ and a dual mother wavelet $\widetilde{\psi}$, respectively. This is done so that instead of

$$\langle\phi_{jk},\phi_{jl}\rangle=\delta_{kl},\qquad \langle\psi_{jk},\psi_{j'l}\rangle=\delta_{jj'}\delta_{kl},\qquad \langle\phi_{jk},\psi_{jl}\rangle=0$$

as in an orthogonal case, we have the scalar products

$$\langle\widetilde{\phi}_{jk},\phi_{jl}\rangle=\delta_{kl},\qquad \langle\widetilde{\psi}_{jk},\psi_{j'l}\rangle=\delta_{jj'}\delta_{kl},$$

and

$$\langle \widetilde{\psi}_{jk}, \phi_{jl} \rangle = 0, \qquad \langle \widetilde{\phi}_{jk}, \psi_{jl} \rangle = 0$$

where

$$\widetilde{\phi}_{jk}(t) := 2^{j/2}\widetilde{\phi}(2^j t - k), \qquad \widetilde{\psi}_{jk}(t) := 2^{j/2}\widetilde{\psi}(2^j t - k).$$

These sets constitute bases for the dual subspaces $\{\widetilde{V}_j\}$ and $\{\widetilde{W}_j\}$. The conditions displayed above are referred to as biorthogonality conditions and ψ and $\widetilde{\psi}$ biorthogonal wavelets. In terms of subspaces, the above biorthogonality conditions can be expressed as

$$V_j \perp \widetilde{W}_j, \quad \widetilde{V}_j \perp W_j, \quad \text{and} \quad W_j \perp \widetilde{W}_{j'} \text{ for } j \neq j'.$$

By definition, a scaling function and a mother wavelet satisfy the dilation equation and the wavelet equation. Consequently

$$\phi(t) = \sqrt{2}\sum_{k\in\mathbb{Z}} h_k \phi(2t-k), \qquad \psi(t) = \sqrt{2}\sum_{k\in\mathbb{Z}} g_k \phi(2t-k)$$

and

$$\widetilde{\phi}(t) = \sqrt{2}\sum_{k\in\mathbb{Z}} \widetilde{h}_k \widetilde{\phi}(2t-k), \qquad \widetilde{\psi}(t) = \sqrt{2}\sum_{k\in\mathbb{Z}} \widetilde{g}_k \widetilde{\phi}(2t-k).$$

The coefficients in the above equations can be obtained by taking the scalar product with the appropiate dual function. For example,

$$h_k = \langle \widetilde{\phi}_{1,k}, \phi \rangle = \sqrt{2}\int_{\mathbb{R}} \overline{\widetilde{\phi}(2t-k)}\phi(t)dt$$

$$g_k = \langle \widetilde{\phi}_{1,k}, \psi \rangle = \sqrt{2}\int_{\mathbb{R}} \overline{\widetilde{\phi}(2t-k)}\psi(t)dt.$$

The roles of the two functions ϕ and $\widetilde{\phi}$ or ψ and $\widetilde{\psi}$ can be interchanged. Or if we take the dual of the above equations we obtain the following relations

$$\widetilde{h}_k = \langle \phi_{1,k}, \widetilde{\phi} \rangle = \sqrt{2}\int_{\mathbb{R}} \overline{\phi(2t-k)}\widetilde{\phi}(t)dt$$

$$\widetilde{g}_k = \langle \phi_{1,k}, \widetilde{\psi} \rangle = \sqrt{2}\int_{\mathbb{R}} \overline{\phi(2t-k)}\widetilde{\psi}(t)dt.$$

The Cohen-Daubechies-Feauveau biorthognal linear spline wavelets (Cohen et al [36], Deslauriers et al [46]) have compact support and are known to have optimal time localization for a given number of null moments (frequency localization). An important property of the spline wavelet is that they have zero moments up to order m, i.e.

$$\int_{\mathbb{R}} t^{\ell}\psi(t)dt = 0, \quad \text{for } \ell = 0, 1, \ldots, m.$$

Spline wavelets are extremely regular and unlike other wavelets, they are symmetric in time. A special case of the Cohen-Daubechies-Feauveau wavelet is implemented in the following Java program. The time series is generated from the logistic map.

```
// CDF1.java

public class CDF1
{
 public static void main(String[] arg)
 {
 double[] d = new double[65]; // length for data 2^n+1 : n integer
 d[0] = 1.0/3.0;  // logistic map
 for(int k=1;k<d.length;++k) { d[k] = 4.0*d[k-1]*(1.0-d[k-1]); }

 CDF1 cdf = new CDF1();
 cdf.transform(d);
 // print transformed data
 System.out.println("transformed data:");
 for(int k=0;k<d.length;++k) { System.out.println("d["+k+"] = " + d[k]) }
 cdf.invTransform(d);
 System.out.println("reconstructed data:");
 for(int k=0;k<d.length;++k) { System.out.println("d["+k+"] = "+d[k]); }
 } // end main

 public CDF1() { }  // default constructor

 final static double[] s =
  { 0.03314563036811941, -0.06629126073623882, -0.17677669529663687,
    0.41984465132951254,  0.9943689110435824, 0.41984465132951254,
   -0.17677669529663687, -0.06629126073623882, 0.03314563036811941 };

 final static double[] w = { -0.5, 1.0, -0.5 };

 final static double[][] sLeft =
  { { 1.0275145414117017, 0.7733980419227863, -0.22097086912079608,
      -0.3314563036811941, 0.16572815184059705 },
    { -0.22189158107546605, 0.4437831621509321, 0.902297715576584,
       0.5800485314420897, -0.25687863535292543, -0.06629126073623882,
       0.03314563036811941 },
    { 0.07549838028293866, -0.15099676056587732, -0.0957540432856783,
      0.34250484713723395, 1.0330388131397217, 0.41984465132951254,
      -0.17677669529663687, -0.06629126073623882, 0.03314563036811941 },
    { -0.013810679320049755, 0.02762135864009951, 0.011048543456039804,
      -0.04971844555217912, -0.18506310288866673, 0.41984465132951254,
      0.9943689110435824, 0.41984465132951254, -0.17677669529663687,
      -0.06629126073623882, 0.03314563036811941 } };

final static double[][] sRight =
 {{ 0.03314563036811941, -0.06629126073623882, -0.17677669529663687,
    0.41984465132951254, 0.9943689110435824, 0.41984465132951254,
    -0.18506310288866673, -0.04971844555217912, 0.011048543456039804,
    0.02762135864009951, -0.013810679320049755 },
```

```
      { 0.03314563036811941, -0.06629126073623882, -0.17677669529663687,
          0.41984465132951254, 1.0330388131397217, 0.34250484713723395,
         -0.0957540432856783, -0.15099676056587732, 0.07549838028293866 },
      { 0.03314563036811941, -0.06629126073623882, -0.25687863535292543,
          0.5800485314420897, 0.902297715576584, 0.4437831621509321,
         -0.22189158107546605 },
      { 0.16572815184059705, -0.3314563036811941, -0.22097086912079608,
          0.7733980419227863, 1.0275145414117017 } };

final static double[] sPrimary =
   { 0.35355339059327373, 0.7071067811865475, 0.35355339059327373 };

final static double[] wPrimary =
   { 0.0234375, 0.046875, -0.125, -0.296875,
      0.703125, -0.296875, -0.125, 0.046875, 0.0234375 };

final static double[][] sPrimaryLeft =
   { { 0.7071067811865475, 0.35355339059327373 } };

final static double[][] sPrimaryRight =
   { { 0.35355339059327373, 0.7071067811865475 } };

final static double[][] wPrimaryLeft =
  { { -0.546875, 0.5696614583333334, -0.3138020833333333,
        -0.103515625, 0.10677083333333333, 0.043619791666666664,
        -0.01953125, -0.009765625 },
     { 0.234375, -0.087890625, -0.41015625, 0.673828125,
        -0.2421875, -0.103515625, 0.03515625, 0.017578125 } };

final static double[][] wPrimaryRight =
   { { 0.017578125, 0.03515625, -0.103515625, -0.2421875,
        0.673828125, -0.41015625, -0.087890625, 0.234375 },
      { -0.009765625, -0.01953125,
        0.043619791666666664, 0.10677083333333333,
        -0.103515625, -0.3138020833333333,
        0.5696614583333334, -0.546875 } };

public void transform(double[] v)
{
for(int last=v.length;last>=15;last=(last+1)/2) { transform(v,last); }
} // end method transform

public void invTransform(double[] v)
{
int last;
for(last=v.length;last>=15;last=last/2+1) { ; }
for(;2*last-1<=v.length;last=2*last-1) { invTransform(v,last); }
} // end method invTransform
```

```
public static void transform(double[] v,int last)
{
double[] ans = new double[last];
int half = (last+1)/2;
if(2*half-1 != last) {
  throw new IllegalArgumentException("illegal subband: " + last
        + " within array of length " + v.length);
}
// lowpass
for(int k=0;k<sLeft.length;k++) {
 for(int l=0;l<sLeft[k].length;l++) { ans[k] += sLeft[k][l]*v[l]; }
}
for(int k=sLeft.length;k<half-sRight.length;k++) {
 for(int l=0;l<s.length;l++) { ans[k] += s[l]*v[2*k+l-sLeft.length]; } }
for(int k=0;k<sRight.length;k++) {
 for(int l=0;l<sRight[k].length;l++) {
  ans[k+half-sRight.length] +=
   sRight[k][l]*v[last-sRight[k].length+l];
}
}
// highpass
for(int k=0;k<half-1;k++) {
 for(int l=0;l<w.length;l++) { ans[k+half] += w[l]*v[2*k+l]; }
}
System.arraycopy(ans,0,v,0,last);
} // end method transform

public static void invTransform(double[] v,int last)
{
double[] ans = new double[2*last-1];
// scale coefficients
for(int k=0;k<sPrimaryLeft.length;k++) {
 for(int l=0;l<sPrimaryLeft[k].length;l++)
  { ans[l] += sPrimaryLeft[k][l]*v[k]; }
}
for(int k=sPrimaryLeft.length;k<last-sPrimaryRight.length;k++)
{
for(int l=0;l<sPrimary.length;l++)
 { ans[2*k-1+l] += sPrimary[l]*v[k]; }
}
for(int k=0;k<sPrimaryRight.length;k++) {
 for(int l=0;l<sPrimaryRight[k].length;l++) {
  ans[l-sPrimaryRight[k].length+ans.length] +=
   sPrimaryRight[k][l]*v[k+last-sPrimaryRight.length];
}
}
// wavelet coefficients
```

```
for(int k=0;k<wPrimaryLeft.length;k++) {
 for(int l=0;l<wPrimaryLeft[k].length;l++)
 { ans[l] += wPrimaryLeft[k][l]*v[k+last]; }
}
for(int k=wPrimaryLeft.length;k<last-1-wPrimaryRight.length;k++)
{
for(int l=0;l<wPrimary.length;l++)
{ ans[2*(k-1)-1+l] += wPrimary[l]*v[k+last]; }
}
for(int k=0;k<wPrimaryRight.length;k++) {
 for(int l=0;l<wPrimaryRight[k].length;l++) {
  ans[l-wPrimaryRight[k].length+ans.length] +=
   wPrimaryRight[k][l]*v[k+2*last-1-wPrimaryRight.length];
}
}
System.arraycopy(ans,0,v,0,ans.length);
} // end method invTransform
} // end class
```

16.5 Two-Dimensional Wavelets

There are several methods to construct bases on functional spaces in dimension greater than 1 (Daubechies [44], Antoine et al [3]). The simplest one uses separable wavelets. Consider $d = 2$. The simplest way to build two-dimensional wavelet bases is to use separable products (tensor products) of a one-dimensional wavelet ψ and scaling function ϕ. This provides the following scaling function

$$\Phi(x_1, x_2) = \phi(x_1)\phi(x_2)$$

and there are three wavelets

$$\psi^{(1)}(x_1, x_2) = \phi(x_1)\psi(x_2)$$
$$\psi^{(2)}(x_1, x_2) = \psi(x_1)\phi(x_2)$$
$$\psi^{(3)}(x_1, x_2) = \psi(x_1)\psi(x_2) .$$

There are also non-separable two-dimensional wavelets.

Assume that we have a scaling function Φ and wavelets

$$\Psi := \{ \psi^{(i)} \, : \, i = 1, 2, 3 \}$$

constructed by tensor products described above of a one-dimensional orthogonal wavelet system. If we define

$$\psi^{(i)}_{\mathbf{j},k}(\mathbf{x}) := 2^k \psi^{(i)}(2^k \mathbf{x} - \mathbf{j}), \qquad i = 1, 2, 3$$

where $k \in \mathbb{Z}$, and $\mathbf{j} \in \mathbb{Z}^2$, then any function f in the Hilbert space $L_2(\mathbb{R}^2)$ can be written as the expansion

$$f(\mathbf{x}) = \sum_{\mathbf{j},k,\psi} \langle f, \psi_{\mathbf{j},k} \rangle \psi_{\mathbf{j},k}(\mathbf{x})$$

where the sums range over all $\mathbf{j} \in \mathbb{Z}^2$, all $k \in \mathbb{Z}$, and all $\psi \in \Psi$. Here $\langle \, , \rangle$ denotes the scalar product in the Hilbert space $L_2(\mathbb{R}^2)$, i.e.

$$\langle f, \psi_{\mathbf{j},k}^{(i)} \rangle = c_{\mathbf{j},k}^{(i)} = \int_{\mathbb{R}^2} f(\mathbf{x}) \psi_{\mathbf{j},k}^{(i)}(\mathbf{x}) d\mathbf{x} .$$

Thus we have the expansion

$$f(\mathbf{x}) = \sum_{\mathbf{j},k} \langle f, \psi_{\mathbf{j},k}^{(1)} \rangle \psi_{\mathbf{j},k}^{(1)}(\mathbf{x}) + \sum_{\mathbf{j},k} \langle f, \psi_{\mathbf{j},k}^{(2)} \rangle \psi_{\mathbf{j},k}^{(2)}(\mathbf{x}) + \sum_{\mathbf{j},k} \langle f, \psi_{\mathbf{j},k}^{(3)} \rangle \psi_{\mathbf{j},k}^{(3)}(\mathbf{x}) .$$

Instead of considering the sum over all dyadic levels k, one can sum over $k \geq K$ for a fixed integer K. For this case we have the expansion

$$f(\mathbf{x}) = \sum_{\mathbf{j} \in \mathbb{Z}^2, k \geq K, \psi \in \Psi} c_{\mathbf{j},k,\psi} \psi_{\mathbf{j},k}(\mathbf{x}) + \sum_{\mathbf{j} \in \mathbb{Z}^2} d_{\mathbf{j},K} \Phi_{\mathbf{j},K}(\mathbf{x})$$

where the expansion coefficients are given by

$$c_{\mathbf{j},k,\psi} = \int_{\mathbb{R}^2} f(\mathbf{x}) \psi_{\mathbf{j},k}(\mathbf{x}) d\mathbf{x}$$

and

$$d_{\mathbf{j},K} = \int_{\mathbb{R}^2} f(\mathbf{x}) \Phi_{\mathbf{j},K}(\mathbf{x}) d\mathbf{x} .$$

When we study finite domains, e.g., the unit square I, then two changes must be made to this basis for all of $L_2(\mathbb{R}^2)$ to obtain an orthonormal basis for $L_2(I)$. First, one considers only nonnegative scales $k \geq 0$, and not all shifts $\mathbf{j} \in \mathbb{Z}^2$, but only those shifts for which $\psi_{j,k}$ intersects I nontrivially. Second one must adapt the wavelets that overlap the boundary of I in order to preserve orthogonality on the domain. Consider a function f defined on the unit square $I := [0,1)^2$, which is extended periodically to all of $\mathbb{R}^2$ by

$$f(\mathbf{x} + \mathbf{j}) = f(\mathbf{x})$$

$\mathbf{x} \in I$, $\mathbf{j} \in \mathbb{Z}^2$, where

$$\mathbb{Z}^2 := \{ (j_1, j_2) \, : \, j_1, j_2 \in \mathbb{Z} \} .$$

One can construct periodic wavelets on $L_2(I)$ that can be used to decompose periodic fuctions f on $L_2(I)$. For example, for $\psi \in \Psi$, $k \geq 0$ and $\mathbf{j} \in \{0, 1, \ldots, 2^k - 1\}^2$ one sets

$$\psi_{\mathbf{j},k}^{(p)}(\mathbf{x}) := \sum_{\mathbf{n} \in \mathbb{Z}^2} \psi_{\mathbf{j},k}(\mathbf{x} + \mathbf{n}), \qquad \mathbf{x} \in I .$$

One constructs $\Phi^{(p)}$ in the same way; we have $\Phi^{(p)}(\mathbf{x}) = 1$ for all $\mathbf{x}$, since

$$\{\,\Phi(\cdot - \mathbf{j}) \,:\, \mathbf{j} \in \mathbb{Z}^2\,\}$$

forms a partition of unity. One now has a periodic orthogonal wavelet system on $L_2(I)$ such that

$$f(\mathbf{x}) = \langle f, \Phi^{(p)}\rangle + \sum_{\mathbf{j},k,\psi} \langle f, \psi^{(p)}_{\mathbf{j},k}\rangle \psi^{(p)}_{\mathbf{j},k}(\mathbf{x})\,.$$

In practice we are given only a finite amount of data. Thus one cannot calculate the expansion formula for all $k \geq 0$ and all translations $h \in I$. Assume that we have 2^m rows of 2^m pixels, each of which is the average of f on a square of size $2^{-m} \times 2^{-m}$. Then using the orthogonal wavelets constructed by Daubechies we can calculate these formulae for $k < m$ and average over 2^{2m} different pixel translations $j/2^m$, $\mathbf{j} = (j_1, j_2)$, $0 \leq j_1, j_2 < 2^m$, instead of averaging over $h \in I$. We obtain

$$f(\mathbf{x}) = \int_I f(\mathbf{y})d\mathbf{y} + \sum_{0 \leq k < m, \mathbf{j}, \psi} 2^{2(k-m)} \int_I f(\mathbf{y})\psi_k\left(\mathbf{y} - \frac{\mathbf{j}}{2^m}\right) d\mathbf{y}\psi_k\left(\mathbf{x} - \frac{\mathbf{j}}{2^m}\right).$$

For each dyadic level k we need to compute 3×2^{2m} terms, one for each pixel and one for each $\psi \in \Psi$.

Exercise. Consider the Hilbert space $L_2(\mathbb{R}^2)$ and the function $f \in L_2(\mathbb{R}^2)$

$$f(x_1, x_2) = \exp(-|x_1 + x_2|)\,.$$

Find the expansion using separable products as desribed above and the Haar wavelet. Is the Haar wavelet a good choice for the expansion?

Chapter 17

Discrete Hidden Markov Processes

17.1 Introduction

A *Markov chain* is a finite state machine with probabilities for each transition, that is, a probability that the next state is s_j given that the current state is s_i.

Equivalently, a Markov chain is described by a weighted, directed graph in which the weights correspond to the probability of that transition. In other words, the weights are nonnegative and the total weight of the outgoing edges is positive. If the weights are normalized, the total weight, including self-loops is 1.

The *hidden Markov model* (Rabiner [171], Rabiner and Juang [172], Charniak [28], Jelinek [104]) is a finite state machine with probabilities for each transition, that is, a probability that the next state is s_j given that the current state is s_i. The states are not directly observable. Instead, each state produces one of the observable outputs with certain probability.

Computing a model given sets of sequences of observed outputs is difficult, since the states are not directly observable and transitions are probabilistic. Although the states cannot, by definition, be directly observed, the most likely sequence of sets for a given sequence of observed outputs can be computed in $O(nT)$, where n is the number of states and T is the length of the sequence. Thus a hidden Markov model is a Markov chain, where each state generates an observation. We only see the observations, and the goal is to infer the hidden state sequence.

Hidden Markov models are useful for time-series modelling, since the discrete state-space can be used to approximate many non-linear, non-Gaussian systems.

In a statistical framework, an inventory of elementary probabilistic models of basic linguistic units (e.g., phonemes) is used to build word representations. A sequence of acoustic parameters, extracted from a spoken utterance, is seen as a realization of a concatenation of elementary processes described by hidden Markov models.

A hidden Markov model is a composition of two stochastic processes, a hidden Markov chain, which accounts for temporal variability, and an observable process, which accounts for spectral variability. This combination can cope with the most important sources of speech ambiguity, and flexible enough to allow the realization of recognition systems with dictionaries of hundred of thousands of words.

Applications of the hidden Markov model are

in mobile robots, where

$$\text{states} = \text{location}, \qquad \text{observations} = \text{sensor input}$$

in biological sequencing

$$\text{states} = \text{protein structure}, \qquad \text{observations} = \text{amino acids}$$

In biological sequencing the objective of the algorithm is: Given the structure of a protein, such as insulin, find the amino acids that make up that protein. There are 20 amino acids.

in speech recognition

$$\text{states} = \text{phonemes}, \qquad \text{observations} = \text{acoustic signal}$$

Given a speech signal, find the most probable sequence of words

$$\text{words} = \text{argmax} P(\text{words—speech})$$

Two formal assumptions characterize hidden Markov models as used in speech recognition. The first-order Markov hypothesis states that history has no influence on the chain's future evolution if the present is specified, and the output independence hypothesis states that neither chain evolution nor past observation influences the present oberservation if the last chain transition is specified.

In general, there are many different possible sequences

```
an ice cream
and nice cream
and nice scream
```

A statistical language model can be used to choose the most probable interpretation:

assign probabilities to all possible sequences of words

select most probable sequence from those proposed by the speech analyzer

The hidden Markov model can be used to model a speech utterance. The utterance to be modeled may be a phone, a syllable, a word, or, in principle, an intact sentence or entire paragraph. In small vocabulary systems, the hidden Markov model tends to be used to model words, whereas in a large vocabulary conversational speech recognition system usually the hidden Markov model is usually used to model sub-word units such as phone or syllable. There are two major tasks involved in a typical automatic speech recognition system. First, given a series of training obervations and their associated transcriptions, how do we estimate the parameters of the hidden Markov models which represent the words or phones covered in the transcriptions? Second, given a set of trained hidden Markov models and an input speech observation sequence, how do we find the maximum likelihood of this observation sequence and the corresponding set of hidden Markov models which produce this maximum value? This is the speech recognition problem.

We first introduce Markov chains and then, as an extension, hidden Markov processes.

17.2 Markov Chains

We introduce Markov chains (Doob [48]) and give a number of examples.

First we introduce some definitions.

Definition. A (row) vector $\mathbf{p} = (p_0, p_1, \ldots, p_{N-1})$ is called a *probability vector* if the components are nonnegative and their sum is 1, i.e.,

$$\sum_{j=0}^{N-1} p_j = 1$$

Example. The vector in $\mathbb{R}^5$

$$\mathbf{p} = (1/4, 1/8, 0, 5/8, 0)$$

is a probability vector. ♣

Example. The nonzero vector $(2, 3, 5, 0, 1)$ in $\mathbb{R}^5$ is not a probability vector. However since all numbers are nonnegative it can be normalized to get a probability vector. Since the sum of the components is equal to 11 we obtain the probability vector

$$\mathbf{p} = (2/11, 3/11, 5/11, 0, 1/11)\,.$$ ♣

Definition. An $N \times N$ square matrix $A = (a_{ij})$ is called a *stochastic matrix* if each of its rows is a probability vector, i.e. if each entry of A is nonnegative and the sum of the entries in each row is 1. ♣

Example. The following matrix is a stochastic matrix

$$A = \begin{pmatrix} 0 & 1 & 0 & 0 \\ 1/2 & 1/6 & 1/3 & 0 \\ 0 & 1/3 & 0 & 2/3 \\ 1/4 & 1/4 & 1/4 & 1/4 \end{pmatrix}.$$ ♣

We can easily prove that if A_1 and A_2 are stochastic $N \times N$ matrices, then the matrix product $A_1 A_2$ is also a stochastic $N \times N$ matrix.

Example. Let

$$A_1 = \begin{pmatrix} 0 & 1 \\ 1/2 & 1/2 \end{pmatrix}, \qquad A_2 = \begin{pmatrix} 1/3 & 2/3 \\ 1/4 & 3/4 \end{pmatrix}.$$

Then

$$A_1 A_2 = \begin{pmatrix} 1/4 & 2/3 \\ 7/24 & 17/24 \end{pmatrix}.$$ ♣

If A_1 is a stochastic $N \times N$ matrix and A_2 is a stochastic $M \times M$ matrix, the $A_1 \otimes A_2$ is a stochastic $(NM) \times (NM)$ matrix, where $\otimes$ denotes the Kronecker product.

Example. Let

$$A_1 = \begin{pmatrix} 0 & 1 \\ 1/2 & 1/2 \end{pmatrix}, \qquad A_2 = \begin{pmatrix} 1/3 & 2/3 \\ 1/4 & 3/4 \end{pmatrix}.$$

Then we obtain the 4×4 stochastic matrix

$$A_1 \otimes A_2 = \begin{pmatrix} 0 & 0 & 1/3 & 2/3 \\ 0 & 0 & 1/4 & 3/4 \\ 1/6 & 1/3 & 1/6 & 1/3 \\ 1/8 & 3/8 & 1/8 & 3/8 \end{pmatrix}.$$ ♣

Note that the direct sum of two stochastic matrices is also a stochastic matrix.

Definition. A stochastic matrix A is called *regular* if all the entries of some power of A, A^n, are positive.

Example. The stochastic matrix

$$A = \begin{pmatrix} 0 & 1 \\ 1/2 & 1/2 \end{pmatrix}$$

is regular since

$$A^2 = \begin{pmatrix} 1/2 & 1/2 \\ 1/4 & 3/4 \end{pmatrix}$$

is positive in every entry. ♣

Example. The stochastic matrix

$$A = \begin{pmatrix} 1 & 0 \\ 1/2 & 1/2 \end{pmatrix}$$

is not regular since

$$A^2 = \begin{pmatrix} 1 & 0 \\ 3/4 & 1/4 \end{pmatrix}.$$

♣

Definition. A *fixed point* $\mathbf{p}^*$ of a regular stochastic matrix A is defined as the solution of the system of linear equation

$$\mathbf{p}^* A = \mathbf{p}^* .$$

Example. Consider the regular stochastic matrix

$$A = \begin{pmatrix} 0 & 1 & 0 \\ 0 & 0 & 1 \\ 1/2 & 1/2 & 0 \end{pmatrix}.$$

The vector $\mathbf{p}^*$ can be written as $\mathbf{p}^* = (x, y, 1-x-y)$ with the constraints $x \in [0,1]$ and $y \in [0,1]$. Then the solution to the fixed point equation $\mathbf{p}^* A = \mathbf{p}^*$ is given by

$$\mathbf{p}^* = (1/5, 2/5, 2/5) .$$

♣

The fundamental property of regular stochastic matrices is contained in the following theorem.

Theorem. Let A be a regular stochastic matrix. Then

(i) A has a unique fixed probability vector $\mathbf{p}$, and the components of $\mathbf{q}$ are all positive.

(ii) the sequence of matrices A, A^2, $A^3, \ldots$ of powers of A approaches the matrix B whose rows are each the fixed point $\mathbf{p}^*$.

(iii) if $\mathbf{p}$ is any probability vector, then the sequence of vectors $\mathbf{p}A$, $\mathbf{p}A^2$, $\mathbf{p}A^3, \ldots$ approaches the fixed point $\mathbf{p}^*$.

Next we consider *Markov chains.* We consider a sequence of trials whose outcome, say

$$o_0,\ o_1,\ o_2,\ \ldots,\ o_{T-1}$$

satisfy the following two properties:

(i) Each outcome belongs to a finite set of outcomes

$$\{\, q_0,\ q_1,\ q_2,\ \ldots, q_{N-1} \,\}$$

called the *state space* of the system. If the outcome on the t-th trial is q_i, then we say that the system is in state q_i at time t or at the t-th step.

(ii) The outcome of any trial depends at most upon the outcome of the immediately preceding trial and not upon any other previous outcomes; with each pair of states (q_i, q_j) there is given the probability a_{ij} that q_j occurs immediately after q_i occurs.

Such a stochastic process is called a finite Markov chain. The numbers a_{ij}, called the *transition probabilities* can be arranged in the square $N \times N$ matrix

$$A = \begin{pmatrix} a_{00} & a_{01} & \cdots & a_{0N-1} \\ a_{10} & a_{11} & \cdots & a_{1N-1} \\ \cdots & \cdots & \ddots & \cdots \\ a_{N-10} & a_{N-11} & \cdots & a_{N-1N-1} \end{pmatrix}$$

called the *transition matrix*. The transition matrix A of a Markov chain is a stochastic matrix.

Example. A typical example of a Markov chain is a random walk. Given the set (state space)

$$\{0, 1, 2, 3, 4, 5\}$$

where 0 is the origin and 5 the end point. A woman is at any of these points. She takes a unit step to the right with probability p or to the left with probability $q = 1 - p$, unless she is at the origin where she takes a step to the right to 1 or the point 5 where she takes a step to the left to 4. Let o_t denote her position after t steps. This is a Markov chain with the state space given by the set above. Thus 2 means that the woman is at the point 2. The transition matrix A is

$$A = \begin{pmatrix} 0 & 1 & 0 & 0 & 0 & 0 \\ q & 0 & p & 0 & 0 & 0 \\ 0 & q & 0 & p & 0 & 0 \\ 0 & 0 & q & 0 & p & 0 \\ 0 & 0 & 0 & q & 0 & p \\ 0 & 0 & 0 & 0 & 1 & 0 \end{pmatrix}.$$

Next we discuss the question: What is the probability, denoted by $a_{ij}^{(t)}$, that the system changes from the state q_i to the state q_j in exactly t steps? Let A be the transition matrix of a Markov chain process. Then the t-step transition matrix is equal to the t-th power of A; that is $A^{(t)} = A^t$. The $a_{ij}^{(t)}$ are the elements of the matrix $A^{(t)}$.

Example. Consider again the random walk problem discussed above. Suppose the woman starts at the point 2. To find the probability distribution after three steps we do the following calculation. Since

$$\mathbf{p}^{(0)} = (0, 0, 1, 0, 0, 0)$$

we find

$$\begin{aligned}\mathbf{p}^{(1)} &= \mathbf{p}^{(0)}A = (0,q,0,p,0,0)\\ \mathbf{p}^{(2)} &= \mathbf{p}^{(1)}A = (q^2,0,2pq,0,p^2,0)\\ \mathbf{p}^{(3)} &= \mathbf{p}^{(2)}A = (0,q^2+2pq^2,0,3p^2q,0,p^3)\,.\end{aligned}$$

Thus the probability after three steps to be at the point 1 is q^2+2pq^2. If $p=q=1/2$ we obtain $p^{(3)}=1/2$. Notice that

$$\mathbf{p}^{(3)}A \equiv \mathbf{p}^{(0)}A^3\,.$$

♣

Definition. A state $\mathbf{q}_i$ of a Markov chain is called *absorbing* if the system remains in the state q_i once it enters there. Thus a state $\mathbf{q}_i$ is absorbing if and only if the i-th row of the transition matrix A has a 1 on the main diagonal and obviously zeros everywhere else in the row.

Example. Consider the transition matrix

$$A=\begin{pmatrix}1/4 & 0 & 1/4 & 1/4 & 1/4\\ 0 & 1 & 0 & 0 & 0\\ 1/2 & 0 & 1/4 & 1/4 & 0\\ 0 & 1 & 0 & 0 & 0\\ 0 & 0 & 0 & 0 & 1\end{pmatrix}.$$

The states q_1 and q_4 are each absorbing (notice that we count from 0), since each of the second and fifth rows has a 1 on the main diagonal. ♣

The transition probabilities of a Markov chain can be represented by a diagram, called the *transition diagram*, where a positive probability a_{ij} is denoted by an arrow from the state q_i to the state q_j.

17.3 Discrete Hidden Markov Processes

The following notation is used (Rabiner [171], Rabiner and Juang [172], Charniak [28], Jelinek [104]).

N is the number of states in the model.

M is the total number of distinct observation symbols in the alphabet. If the observations are continuous then M is infinite. We only consider the case for M finite.

T is the length of the sequence of observations (training set), where

$$t=0,1,\ldots,T-1\,.$$

Thus there exist N^T possible state sequences.

Let

$$\Omega_q := \{ q_0, q_1, \ldots, q_{N-1} \}$$

be the finite set of possible states. Let

$$V := \{ v_0, v_1, \ldots, v_{M-1} \}$$

be the finite set of possible observation symbols.

q_t is the random variable denoting the state at time t (state variable).

o_t is the random variable denoting the observation at time t (output variable).

$O = (o_0, o_1, \ldots, o_{T-1})$ is the sequence of actual observations.

The set of state transition probabilities is $A = (a_{ij})$, where $i, j = 0, 1, \ldots, N-1$

$$a_{ij} = p(q_{t+1} = j | q_t = i)$$

where p is the state-transition probability, i.e. the probability of being in state j at time $t+1$ given that we were in state i at time t. We assume that the a_{ij}'s are independent of time t. Obviously we have the conditions

$$a_{ij} \geq 0 \quad \text{for} \quad i, j = 0, 1, \ldots, N-1$$

$$\sum_{j=0}^{N-1} a_{ij} = 1 \quad \text{for} \quad i = 0, 1, \ldots, N-1.$$

Thus A is an $N \times N$ matrix.

Example. If we have six states the transition matrix in speech recognition could look as follows

$$A = \begin{pmatrix} 0.3 & 0.5 & 0.2 & 0 & 0 & 0 \\ 0 & 0.4 & 0.3 & 0.3 & 0 & 0 \\ 0 & 0 & 0.4 & 0.2 & 0.4 & 0 \\ 0 & 0 & 0 & 0.7 & 0.2 & 0.1 \\ 0 & 0 & 0 & 0 & 0.5 & 0.5 \\ 0 & 0 & 0 & 0 & 0 & 1.0 \end{pmatrix}. \qquad \clubsuit$$

The conditional probability distribution of the observation at the t, o_t, given the state j is

$$b_j(k) = p(o_t = v_k | q_t = j)$$

where $j \in \{0, 1, \ldots, N-1\}$ and $k \in \{0, 1, \ldots, M-1\}$, i.e. p is the probability of observing the symbol v_k given that we are in state j. Let $B := \{ b_j(k) \}$. We have

$$b_j(k) \geq 0$$

for $j \in \{0,1,\ldots,N-1\}$ and $k \in \{0,1,\ldots,M-1\}$ and

$$\sum_{k=0}^{M-1} b_j(k) = 1$$

for $j = 0,1,\ldots,N-1$.

The initial state distribution $\pi = \{\pi_i\}$

$$\pi_i = p(q_0 = i)$$

where $i = 0,1,\ldots,N-1$, i.e. π_i is the probability of being in state i at the beginning of the experiment ($t = 0$).

Thus we arrive at the definition (Rabiner [171], Rabiner and Juang [172], Charniak [28], Jelinek [104]):

A hidden Markov model (hidden Markov model) is a five-tuple (Ω_q, V, A, B, π).

Let $\lambda := (A, B, \pi)$ denote the parameters for a given hidden Markov model with fixed Ω_q and V.

The three problems for hidden Markov models are:

1) Given the observation sequence

$$O = (o_0, o_1, \ldots, o_{T-1})$$

and a model $\lambda = (A, B, \pi)$. Find $P(O|\lambda)$: the probability of the observations given the model.

2) Given the observation sequence

$$O = (o_0, o_1, \ldots, o_{T-1})$$

and a model $\lambda = (A, B, \pi)$. Find the most likely state sequence given the model and observations. In other words, given the model $\lambda = (A, B, \pi)$ how do we choose a state sequence $\mathbf{q} = (q_0, q_1, \ldots, q_{T-1})$ so that $P(O, \mathbf{q}|\lambda)$, the joint probability of the observation sequence $O = (o_0, o_1, \ldots, o_{T-1})$ and the state sequence given the model is maximized.

3) Adjust λ to maximize $P(O|\lambda)$. In other words how do we adjust the hidden Markov model parameters $\lambda = (A, B, \pi)$ so that $P(O|\lambda)$ (or $P(O, \mathbf{v}|\lambda)$ is maximized.

Example. Consider the Urn-and-Ball model. We assume that there are N urns (number of states). Within each urn is a number of coloured balls, for example

$$\{\text{red, blue, green, yellow, white, black}\}.$$

Thus the number M of possible observations is 6. The physical process for obtaining observations is as follows. One chooses randomly an initial urn. From this urn, a ball is chosen at random, and its colour is recorded as the observation. The ball is then replaced in the urn from which is was selected. A new urn is then selected according to the random selection procedure associated with the current urn (thus there is a probability that the same urn is selected again), and the ball selection is repeated. This process generates a finite observation sequence of colour. This can be modelled as the observable output of a discrete hidden Markov model. For example

$$\mathrm{O} = (\text{blue, green, red, red, white, black, blue, yellow})\,.$$

Thus the number of observation is $T = 8$. The simplest hidden Markov model that corresponds to the urn-and-ball process is one in which each state corresponds to a specific urn, and for which a ball colour probability is defined for each state. The choice of urns is dictated by the state-transition matrix of the discrete hidden Markov model. The ball colours in each urn may be the same, and the distinction among various urns is in the way the collection of coloured balls is composed. Therefore, an isolated observation of a particular colour ball does not tell which urn it is drawn from. ♣

Example. Consider a discrete hidden Markov model representation of a coin-tossing experiment. Thus we have $M = 2$. We assume that the number of states is $N = 3$ corresponding to three different coins. The probabilities are

```
          state 0   state 1   state 2
=========================================
P(head)   0.5       0.75      0.25
P(tail)   0.5       0.25      0.75
=========================================
```

We set `head=0` and `tail=1`. Thus

$$b_0(0) = 0.5, \qquad b_1(0) = 0.75, \qquad b_2(0) = 0.25$$

$$b_0(1) = 0.5, \qquad b_1(1) = 0.25, \qquad b_2(1) = 0.75\,.$$

Assume that all state-transition probabilities are 1/3 and assume the initial state probability of 1/3. Suppose one observes the sequence ($T = 10$)

$$O = (H, H, H, H, T, H, T, T, T, T)\,.$$

Since all state transitions are equiprobable, the most likely state sequence is the one for which the probability of each individual observation is a maximum. Thus for each `head`, the most likely state is 1 and for each `tail` the most likely state is 2. Consequently the most likely state sequence is

$$\mathbf{q} = (1, 1, 1, 1, 2, 1, 2, 2, 2, 2)\,.$$

The probability of O and $\mathbf{q}$ given the model is

$$P(O, \mathbf{q}|\lambda) = (0.75)^{10} \left(\frac{1}{3}\right)^{10}.$$

Next we calculate the probability that the observation came completely from state 0, i.e.

$$\hat{\mathbf{q}} = (0,0,0,0,0,0,0,0,0,0).$$

Then

$$P(O, \hat{\mathbf{q}}|\lambda) = (0.50)^{10} \left(\frac{1}{3}\right)^{10}.$$

The ratio R of $P(O, \mathbf{q}|\lambda)$ to $P(O, \hat{\mathbf{q}}|\lambda)$ is given by

$$R = \frac{P(O, \mathbf{q}|\lambda)}{P(O, \hat{\mathbf{q}}|\lambda)} = \left(\frac{3}{2}\right)^{10} = 57.67.$$

Thus the state sequence $\mathbf{q}$ is much more likely than the state sequence $\hat{\mathbf{q}}$. ♣

17.4 Forward-Backward Algorithm

Given the model $\lambda = (A, B, \pi)$. Then one has to calculate $P(O|\lambda)$, the probability of occurrence of the observation sequence

$$O = (o_0, o_1, \ldots, o_{T-1}).$$

A straightforward way to find $P(O|\lambda)$ is to find $P(O|\mathbf{q}, \lambda)$ for a fixed state sequence. Then we multiply it by $P(\mathbf{q}|\lambda)$ and then sum up over all possible state sequences. We have

$$P(O|\mathbf{q}, \lambda) = \prod_{t=0}^{T-1} P(o_t|q_t, \lambda)$$

where we have assumed statistical independence of observations. Thus we find

$$P(O|\mathbf{q}, \lambda) = b_{q_0}(o_0) b_{q_1}(o_1) \ldots b_{q_{T-1}}(o_{T-1}).$$

The probability of such a state sequence $\mathbf{q}$ can be written as

$$P(\mathbf{q}|\lambda) = \pi_{q_0} a_{q_0 q_1} a_{q_1 q_2} \ldots a_{q_{T-2} q_{T-1}}.$$

The joint probability of O and $\mathbf{q}$, i.e., the probability that O and $\mathbf{q}$ occur simultaneously, is simply the product of the above two expressions, i.e.

$$P(O, \mathbf{q}|\lambda) = P(O|\mathbf{q}, \lambda) P(\mathbf{q}|\lambda).$$

The probability of O (given the model) is obtained by summing this joint probability over all possible state sequences $\mathbf{q}$. Thus

$$P(O|\lambda) = \sum_{\text{all } \mathbf{q}} P(O|\mathbf{q}, \lambda) P(\mathbf{q}|\lambda).$$

Thus we find

$$P(O|\lambda) = \sum_{q_0,q_1,\ldots,q_{T-1}} \pi_{q_0} b_{q_0}(o_0) a_{q_0 q_1} b_{q_1}(o_1) \ldots a_{q_{T-2} q_{T-1}} b_{q_{T-1}}(o_{T-1}) .$$

The interpretation of the computation is as follows. At time $t = 0$ we are in state q_0 with probability π_{q_0} and generate the symbol o_0 (in this state) with probability $b_{q_0}(o_0)$. The time is incremented by 1 (i.e. $t = 1$) and we make a transition to state q_1 from state q_0 with probability $a_{q_0 q_1}$ and generate symbol o_1 with probability $b_{q_1}(o_1)$. This process continues in this manner until we make the last transition (at time $T-1$) from state q_{T-2} to state q_{T-1} with probability $a_{q_{T-2}q_{T-1}}$ and generate symbol o_{T-1} with probability $b_{q_{T-1}}(o_{T-1})$.

The summand of this equation involves $2T-1$ multiplications and there exists N^T distinct possible state sequences. A direct computation of this equation will involve of the order of $2TN^T$ multiplications. Even for small numbers, for example $N = 5$ and $T = 100$, this means approximately 10^{72} multiplications.

Fortunately a much more efficient technique exists to solve problem 1. It is called *forward-backward procedure*.

The forward procedure is as follows. Consider the forward variable $\alpha_t(i)$ defined as

$$\alpha_t(i) := P(o_0 o_1 \ldots o_t, q_t = i|\lambda)$$

where $i = 0, 1, \ldots, N-1$. Thus $\alpha_i(t)$ is the probability of the partial observation sequence up to time t and the state i at time t, given the model. It can be shown that $\alpha_t(i)$ can be computed as follows:

Initialization:

$$\alpha_0(i) = \pi_i b_i(o_0)$$

where $i = 0, 1, \ldots, N-1$.

Recursion: For $t = 0, 1, \ldots, T-2$ and $j = 0, 1, \ldots, N-1$ we have

$$\alpha_{t+1}(j) = \left(\sum_{i=0}^{N-1} \alpha_t(i) a_{ij} \right) b_j(o_{t+1})$$

where $j = 0, 1, \ldots, N-1$ and $t = 0, 1, \ldots, T-2$.

Probability: We have

$$P(\mathbf{O}|\lambda) = \sum_{i=0}^{N-1} \alpha_{T-1}(i) .$$

The initialization step involves N multiplications. In the recursion step the summation involves N multiplications plus one for the out of bracket $b_j(o_{t+1})$ term. This has to be done for $j = 0$ to $N-1$ and $t = 0$ to $T-2$, making the total number of multiplications in step 2 as $(N-1)N(T+1)$. Step 3 involves no multiplications only summations. Thus the total number of multiplications is

$$N + N(N+1)(T-1)$$

i.e., of the order N^2T as compared to $2TN^T$ required for the direct method. For $N = 5$ and $T = 100$ we need about 3000 computations for the forward method as compared to 10^{72} required by the direct method - a saving of about 69 orders of magnitude. The forward algorithm is implemented in the following Java program.

```
// Forward.java

public class Forward
{
 public static void main(String[] args)
 {
 int T = 10; // number of observations
 int M = 2;  // number of observation symbols
 int N = 3;  // number of states
 int[] obser = new int[T];
 obser[0] = 0; obser[1] = 0; obser[2] = 0; obser[3] = 0;
 obser[4] = 1; obser[5] = 0; obser[6] = 1; obser[7] = 1;
 obser[8] = 1; obser[9] = 1;
 double[][] b = new double[N][M];
 b[0][0] = 0.5; b[1][0] = 0.75; b[2][0] = 0.25;
 b[0][1] = 0.5; b[1][1] = 0.25; b[2][1] = 0.75;
 double[][] A  = new double[N][N];
 A[0][0] = 1.0/3.0; A[0][1] = 1.0/3.0; A[0][2] = 1.0/3.0;
 A[1][0] = 1.0/3.0; A[1][1] = 1.0/3.0; A[1][2] = 1.0/3.0;
 A[2][0] = 1.0/3.0; A[2][1] = 1.0/3.0; A[2][2] = 1.0/3.0;
 double[] alphaold = new double[N]; double[] alphanew = new double[N];
 // initialization
 int temp = obser[0];
 for(int i=0;i<N;i++) { alphaold[i] = (1.0/3.0)*b[i][temp]; }
 // iteration
 for(int t=0;t<=T-2;t++)
 { temp = obser[t+1];
 for(int j=0;j<N;j++)
 { double sum = 0.0;
 for(int i=0;i<N;i++)
 { sum += alphaold[i]*A[i][j]; } // end for loop i
 alphanew[j] = sum*b[j][temp];
 } // end for loop j
 for(int k=0;k<N;k++) { alphaold[k] = alphanew[k]; }
 } // end for loop t
 // probability
 double P = 0.0;
 for(int i=0;i<N;i++) { P += alphanew[i]; }
```

```
 System.out.println("P = " + P);
 } // end main
}
```

In a similar manner we may define a backward variable $\beta_t(i)$ as

$$\beta_t(i) := P(o_{t+1}o_{t+2}\ldots o_{T-1}, q_t = i|\lambda)$$

where $i = 0, 1, \ldots, N-1$. Thus $\beta_i(t)$ is the probability of the observation sequence from $t+1$ to $T-1$ given the state i at time t and the model λ. It can be shown that $\beta_t(i)$ can be computed as follows

$$\beta_{T-1}(i) = 1$$

where $i = 0, 1, \ldots, N-1$. For $t = T-2, T-1, \ldots, 1, 0$ and $i = 0, 1, \ldots, N-1$ we have

$$\beta_t(i) = \sum_{j=0}^{N-1} a_{ij} b_j(o_{t+1}) \beta_{t+1}(j) .$$

Thus

$$P(O|\lambda) = \sum_{i=0}^{N-1} \pi_i b_i(o_0) \beta_0(i) .$$

The computation of $P(O|\lambda)$ using $\beta_t(i)$ also involves of the order of N^2T calculations. Hence both the forward and backward method are equally efficient for the computation of $P(O|\lambda)$.

17.5 Viterbi Algorithm

The Viterbi algorithm is an algorithm to compute the optimal (most likely) state sequence $(q_0, q_1, \ldots, q_{T-1})$ in a hidden Markov model given a sequence of observed outputs. In other words we have to find a state sequence such that the probability of occurrence of the observation sequence

$$(o_0, o_1, \ldots, o_{T-1})$$

from this state sequence is greater than that from any other state sequence. Thus the problem is to find $\mathbf{q}$ that will maximize $P(O, \mathbf{q}|\lambda)$.

In order to give an idea of the Viterbi algorithm as applied to the optimum state estimation problem a reformulation of the problem will be useful. Consider the expression for $P(O, \mathbf{q}|\lambda)$

$$P(O, \mathbf{q}|\lambda) = P(O|\mathbf{q}, \lambda) P(\mathbf{q}|\lambda) = \pi_{q_0} b_{q_0}(o_0) a_{q_0 q_1} b_{q_1}(o_1) \cdots a_{q_{T-2} q_{T-1}} b_{q_{T-1}}(o_{T-1}) .$$

We define

$$U(q_0, q_1, \ldots, q_{T-1}) := -\left(\ln(\pi_{q_0} b_{q_0}(o_0)) + \sum_{t=1}^{T-1} \ln(a_{q_{t-1} q_t} b_{q_t}(o_t)) \right) .$$

It follows that

$$P(O, \mathbf{q}|\lambda) = \exp(-U(q_0, q_1, \ldots, q_{T-1})) .$$

Consequently the problem of optimal state estimation, namely

$$\max_{q_t} P(O, q_0, q_1, \ldots, q_{T-1}|\lambda)$$

becomes equivalent to

$$\min_{q_t} U(q_0, q_1, \ldots, q_{T-1}) .$$

This reformulation now enables us to view terms like

$$-\ln(a_{q_i q_j} b_{q_j}(o_t))$$

as the cost associated in going from state q_i to state q_j at time t.

The Viterbi algorithm to find the optimum state sequence can now be described as follows: Suppose we are currently in state i and we are considering visiting state j next. We say that the weight on the path from state i to state j is

$$-\ln(a_{ij} b_j(o_t))$$

i.e., the negative of the logarithm of the probability of going from state i to state j and selecting the observation symbol o_t in state j. Here o_t is the observation symbol selected after visiting state j. This is the same symbol that appears in the observation sequence

$$O = (o_0, o_1, \ldots, o_{T-1}) .$$

When the initial state is selected as state i the corresponding weight is

$$-\ln(\pi_i b_i(o_0)) .$$

We call this the initial weight. We define the weight of a sequence of states as the sum of the weights on the adjacent states. This corresponds to multiplying the corresponding probabilities. Now finding the optimum sequence is a matter of finding the path (i.e. a sequence of states) of minimum weight through which the given observation sequence occurs.

17.6 Baum-Welch Algorithm

As described above the third problem in hidden Markov models deals with training the hidden Markov model such a way that if a observation sequence having many characteristics similar to the given one be encountered later it should be able to identify it. There are two methods that can be used:

The Segmental K-means Algorithm: In this method the parameters of the model $\lambda = (A, B, \pi)$ are adjusted to maximize $P(O, \mathbf{q}|\lambda)$, where $\mathbf{q}$ here is the optimum sequence as given by the solution to the problem 2.

Baum-Welch Algorithm: Here parameters of the model $\lambda = (A, B, \pi)$ are adjusted so as to increase $P(O|\lambda)$ until a maximum value is reached. As described before calculating $P(O|\lambda)$ involves summing up $P(O, \mathbf{q}|\lambda)$ over all possible state sequences.

17.7 Distances between Hidden Markov Models

If we want to compare two hidden Markov models then we need a measure for the distance between two hidden Markov models. Such a measure is based on the *Kullback-Leibler distance* between two probability distribution functions.

The Kullback-Leibler distance measure is given as follows: let $\rho_1(x)$ and $\rho_2(x)$ be two probability density functions (or probability mass functions) then the Kullback-Leibler distance measure can be used to find out how close the two probability distributions are.

Definition. The Kullback-Leibler distance measure $I(\rho_1, \rho_2)$ for determining how close the probability density function $\rho_2(x)$ is to $\rho_1(x)$ is

$$I(\rho_1, \rho_2) := \int_{-\infty}^{\infty} \rho_1(x) \ln\left(\frac{\rho_1(x)}{\rho_2(x)}\right) dx .$$

If $\rho_1(x)$ and $\rho_2(x)$ are probability mass functions, then

$$I(p_1, p_2) := \sum_{\text{all } x} \rho_1(x) \ln\left(\frac{\rho_1(x)}{\rho_2(x)}\right) .$$

Note that the Kullback-Leibler distance measure is not symmetric, i.e.

$$I(\rho_1, \rho_2) \neq I(\rho_2, \rho_1)$$

in general. If the objective is to simply compare ρ_1 and ρ_2 we can define a symmetric distance measure as

$$I_s(\rho_1, \rho_2) := \frac{1}{2}(I(\rho_1, \rho_2) + I(\rho_2, \rho_1)).$$

It is the use of the Kullback-Leibler distance measure which leads to the definition of the distance measure between two hidden Markov models. For hidden Markov models, the probability function is very complex, and practically it can only be computed via a recursive procedure - the forward/backward or upward/downward algorithm (Rabiner [171]). Thus there is no simple closed form expression for the Kullback-Leibler distance for these models. The Monte-Carlo method can be used to numerically approximate the integral given above.

Example. For the *Gaussian density* one can find an analytic expression for the Kullback-Leibler distance. Suppose that $\rho(\mathbf{x},\mu,\Sigma)$ is a multivariate Gaussian density defined by

$$\rho(\mathbf{x},\mu,\Sigma) := \frac{1}{\sqrt{(2\pi)^n \det(\Sigma)}} \exp\left(-\frac{1}{2}(\mathbf{x}-\mu)^T\Sigma^{-1}(\mathbf{x}-\mu)\right)$$

where $\mathbf{x},\mu \in \mathbb{R}^n$ and ^{T} means transpose. Now let

$$\rho_1(\mathbf{x}) = \rho(\mathbf{x},\mu_1,\Sigma_1), \qquad \rho_2(\mathbf{x}) = \rho(\mathbf{x},\mu_2,\Sigma_2).$$

Then

$$I(\rho_1,\rho_2) = \frac{1}{2}(\mu_1-\mu_2)^T\Sigma_2^{-1}(\mu_1-\mu_2) + \frac{1}{2}\ln\frac{\det\Sigma_2}{\det\Sigma_1} + \frac{1}{2}\mathrm{tr}(\Sigma_1\Sigma_2^{-1} - I_n)$$

and

$$I_s(\rho_1,\rho_2) = \frac{1}{2}(\mu_1-\mu_2)^T(\Sigma_1^{-1}+\Sigma_2^{-1})(\mu_1-\mu_2) + \frac{1}{2}\mathrm{tr}(\Sigma_1^{-1}\Sigma_2 + \Sigma_2^{-1}\Sigma_1 - 2I_n)\,. \quad \clubsuit$$

17.8 C++ Program

The C++ program implements the hidden Markov model, where N is the number of states ($N = 3$ in `main`), M is the total number of distinct observations (only $M = 2$ possible) and T is the number of observations.

```
// hmmt1.cpp

#include <iostream>
#include <string>
#include <cmath>       // for log
using namespace std;

class Data {
 private:
  double** transitions;
  double** emissions;
  double* pi_transitions;
  int n;
 public:
  Data(int,int); // constructor
  ~Data(); // destructor
  double get_transition(int i,int j) { return transitions[i][j]; }
  double get_emission(int i,int j) { return emissions[i][j]; }
  double get_pi_transition(int i) { return pi_transitions[i]; }
  void set_transition(int i,int j,double v)
  { transitions[i][j] = v; }
```

```
  void set_emission(int i,int j,double v) { emissions[i][j] = v; }
  void set_pi_transition(int i,double v)  { pi_transitions[i] = v; }
};

Data::Data(int n=0,int m=0)
{
 this->n = n;
 transitions = new double*[n+1];
 for(int i=0;i<n+1;i++) transitions[i] = new double[n+1];
 emissions = new double*[n+1];
 for(int i=0;i<n+1;i++) emissions[i] = new double[m+1];
 pi_transitions = new double[n+1];
}

Data::~Data()
{
 for(int i=0;i<n+1;i++) delete[] transitions[i]; delete[] transitions;
 for(int i=0;i<n+1;i++) delete[] emissions[i]; delete[] emissions;
 delete[] pi_transitions;
}

class HMM {
 private:
  int N, M, T;
  string o;
  double** alpha_table; double** beta_table;
  double* alpha_beta_table; double* xi_divisor;
  Data* current; Data* reestimated;

 public:
  HMM(int n,int m);
  ~HMM();
  void error(const string s) { cerr << "error: " << s << '\n'; }
  void init(int s1,int s2,double value)
  { current->set_transition(s1,s2,value); }
  void pi_init(int s,double value)
  { current->set_pi_transition(s,value); }
  void o_init(int s,const char c,double value)
  { current->set_emission(s,index(c),value); }
  double a(int s1,int s2)
  { return current->get_transition(s1,s2); }
  double b(int state,int pos)
  { return current->get_emission(state,index(o[pos-1])); }
  double b(int state,int pos,string o)
  { return current->get_emission(state,index(o[pos-1])); }
  double pi(int state)
  { return current->get_pi_transition(state); }
  double alpha(const string s);
```

```
  double beta(const string s);
  double gamma(int t,int i);
  int index(const char c);
  double viterbi(const string s,int *best_sequence);
  double** construct_alpha_table();
  double** construct_beta_table();
  double* construct_alpha_beta_table();
  double xi(int t,int i,int j);
  void reestimate_pi();
  void reestimate_a();
  void reestimate_b();
  double* construct_xi_divisor();
  void maximize(string training,string test);
  void forward_backward(string s);
};

HMM::HMM(int n=0,int m=0) // constructor
{
 N = n; M = m;
 current = new Data(n,m);
 reestimated = new Data(n,m);
}

HMM::~HMM() // destructor
{ delete current; delete reestimated; }

double HMM::alpha(const string s)
{
 string out;
 double P = 0.0;
 out = s;
 int T1 = out.length();
 double* previous_alpha = new double[N+1];
 double* current_alpha = new double[N+1];
 // initialization
 for(int i=1;i<=N;i++) previous_alpha[i] = pi(i)*b(i,1,out);
 // induction
 for(int t=1;t<T1;t++) {
  for(int j=1;j<=N;j++) { double sum = 0.0;
   for(int i=1;i<=N;i++) { sum += previous_alpha[i]*a(i,j); }
    current_alpha[j] = sum*b(j,t+1,out);
    }
  for(int c=1;c<=N;c++)
   previous_alpha[c] = current_alpha[c];
 }
 // termination
 for(int i=1;i<=N;i++) P += previous_alpha[i];
 delete[] current_alpha; delete[] previous_alpha;
```

```
 return P;
}

double HMM::beta(const string s)
{
 double P = 0.0;
 o = s;
 int T = o.length();
 double* next_beta = new double[N+1];
 double* current_beta = new double[N+1];
 // initialization
 for(int i=1;i<=N;i++) next_beta[i] = 1.0;
 // induction
 double sum;
 for(int t=T-1;t>=1;t--) {
  for(int i=1;i<=N;i++) { sum = 0.0;
    for(int j=1;j<=N;j++) { sum += a(i,j)*b(j,t+1)*next_beta[j]; }
    current_beta[i] = sum;
  }
  for(int c=1;c<=N;c++) next_beta[c] = current_beta[c];
 }
 // termination
 for(int i=1;i<=N;i++) P += next_beta[i]*pi(i)*b(i,1);
 delete[] next_beta; delete[] current_beta;
 return P;
}

double HMM::gamma(int t,int i)
{
 return (alpha_table[t][i]*beta_table[t][i])/(alpha_beta_table[t]);
}

int HMM::index(const char c) {
 switch(c) {
 case 'H': return 0;
 case 'T': return 1;
 default: error("no legal input symbol!");
 return 0;
 }
}

double HMM::viterbi(const string s,int best_path[])
{
 double P_star = 0.0;
 string o = s;
 int T = o.length();
 double* previous_delta = new double[N+1];
 double* current_delta = new double[N+1];
```

```
 int** psi = new int*[T+1];
 for(int i=0;i<=T;i++) psi[i] = new int[N+1];
 // initializitaion
 for(int i=1;i<=N;i++)
 { previous_delta[i] = pi(i)*b(i,1); psi[1][i] = 0; }
 double tmp, max;
 // recursion
 for(int t=2;t<=T;t++) {
  for(int j=1;j<=N;j++) { max = 0.0;
   for(int i=1;i<=N;i++) {
    tmp = previous_delta[i]*a(i,j);
     if(tmp >= max) { max = tmp; psi[t][j] = i; }
  }
  current_delta[j] = max*b(j,t);
  }
  for(int c=1;c<=N;c++)
   previous_delta[c] = current_delta[c];
 }
 // termination
 for(int i=1;i<=N;i++) {
  if(previous_delta[i] >= P_star) {
   P_star = previous_delta[i]; best_path[T] = i;
  }
 }
 // get best sequence:
 for(int t=T-1;t>=1;t--) best_path[t] = psi[t+1][best_path[t+1]];
 best_path[T+1] = -1;

 for(int i=0;i<=T;i++) delete[] psi[i];
 delete[] psi; delete[] previous_delta; delete[] current_delta;
 return P_star;
}

double** HMM::construct_alpha_table()
{
 double** alpha_table = new double*[T+1];
 for(int i=0;i<=T+1;i++) alpha_table[i] = new double[N+1];
 // initialization
 for(int i=1;i<=N;i++) alpha_table[1][i] = pi(i)*b(i,1);
 // induction
 for(int t=1;t<T;t++) {
  for(int j=1;j<=N;j++) { double sum = 0.0;
   for(int i=1;i<=N;i++) { sum += alpha_table[t][i]*a(i,j); }
    alpha_table[t+1][j] = sum*b(j,t+1);
 }
 }
 return alpha_table;
}
```

```
double** HMM::construct_beta_table()
{
 double** beta_table = new double*[T+1];
 for(int i=0;i<=T+1;i++) beta_table[i] = new double[N+1];
 // initialization
 for(int i=1;i<=N;i++) beta_table[T][i] = 1.0;
 // induction
 double sum;
  for(int t=T-1;t>=1;t--) {
   for(int i=1;i<=N;i++) { sum = 0.0;
    for(int j=1;j<=N;j++) { sum += a(i,j)*b(j,t+1)*beta_table[t+1][j]; }
    beta_table[t][i] = sum;
  }
 }
 // termination
 for(int i=1;i<=N;i++)
  beta_table[1][i] = beta_table[1][i]*pi(i)*b(i,1);
  return beta_table;
}

double* HMM::construct_alpha_beta_table()
{
 double* alpha_beta_table = new double[T+1];
 for(int t=1;t<=T;t++) {
  alpha_beta_table[t] = 0;
   for(int i=1;i<=N;i++) {
    alpha_beta_table[t] += (alpha_table[t][i]*beta_table[t][i]);
   }
 }
 return alpha_beta_table;
}

double* HMM::construct_xi_divisor()
{
 xi_divisor = new double[T+1];
 double sum_j;
 for(int t=1;t<T;t++) { xi_divisor[t] = 0.0;
  for(int i=1;i<=N;i++) { sum_j = 0.0;
   for(int j=1;j<=N;j++) {
     sum_j += (alpha_table[t][i]*a(i,j)*b(j,t+1)*beta_table[t+1][j]);
   }
   xi_divisor[t] += sum_j;
  }
 }
 return xi_divisor;
}
```

```
double HMM::xi(int t,int i,int j)
{
 return ((alpha_table[t][i]*a(i,j)*b(j,t+1)
          *beta_table[t+1][j])/(xi_divisor[t]));
}

void HMM::reestimate_pi()
{
 for(int i=1;i<=N;i++)
 { reestimated->set_pi_transition(i,gamma(1,i)); }
}

void HMM::reestimate_a()
{
 double sum_xi, sum_gamma;
 for(int i=1;i<=N;i++) {
  for(int j=1;j<=N;j++) { sum_xi = 0.0; sum_gamma = 0.0;
   for(int t=1;t<T;t++) { sum_xi += xi(t,i,j); }
    for(int t=1;t<T;t++) { sum_gamma += gamma(t,i); }
    reestimated->set_transition(i,j,(sum_xi/sum_gamma));
  }
 }
}

void HMM::reestimate_b()
{
 double sum_gamma, tmp_gamma;
 double sum_gamma_output;
 for(int j=1;j<=N;j++) {
  for(int k=0;k<M;k++) { sum_gamma = 0.0; sum_gamma_output = 0.0;
   for(int t=1;t<=T;t++) { tmp_gamma = gamma(t,j);
    if(index(o[t-1])==k) { sum_gamma_output += tmp_gamma; }
   sum_gamma += tmp_gamma;
   }
   reestimated->set_emission(j,k,(sum_gamma_output/sum_gamma));
   }
 }
}

void HMM::forward_backward(string o)
{
 T = o.length();
 alpha_table = construct_alpha_table();
 beta_table = construct_beta_table();
 alpha_beta_table = construct_alpha_beta_table();
 xi_divisor = construct_xi_divisor();
 reestimate_pi(); reestimate_a(); reestimate_b();
 for(int t=1;t<=T;t++) delete[] alpha_table[t];
```

```
 delete[] alpha_table;
 for(int t=1;t<=T;t++) delete[] beta_table[t];
 delete[] beta_table; delete[] alpha_beta_table; delete[] xi_divisor;
 Data* tmp_value = current;
 current = reestimated;
 reestimated = tmp_value;
}

void HMM::maximize(string o,string test)
{
 double diff_entropy, old_cross_entropy, new_cross_entropy;
 int c = 1;
 int t = test.length();
 old_cross_entropy = -((log10(alpha(test))/log10(2.0))/t);
 cout << "Re-estimation:\n";
 cout << " initial cross_entropy: " << old_cross_entropy << "\n";
 do { forward_backward(o);
  new_cross_entropy = -(log10(alpha(test))/log10(2.0))/t;
  diff_entropy = (old_cross_entropy - new_cross_entropy);
  old_cross_entropy = new_cross_entropy;
  c++;
 } while(diff_entropy > 0.0);
 cout << " No of iterations: " << c << "\n";
 cout << " resulting cross_entropy: " << old_cross_entropy << "\n";
}

int main(void)
{
 int t;
 HMM hmm(3,2);
 hmm.pi_init(1,0.33333333); hmm.pi_init(2,0.33333333);
 hmm.pi_init(3,0.33333333); hmm.init(1,1,0.33333333);
 hmm.init(1,2,0.33333333);  hmm.init(1,3,0.33333333);
 hmm.init(2,1,0.33333333);  hmm.init(2,2,0.33333333);
 hmm.init(2,3,0.33333333);  hmm.init(3,1,0.33333333);
 hmm.init(3,2,0.33333333);  hmm.init(3,3,0.33333333);
 hmm.o_init(1,'H',0.5);  hmm.o_init(2,'H',0.75);
 hmm.o_init(3,'H',0.25); hmm.o_init(1,'T',0.5);
 hmm.o_init(2,'T',0.25); hmm.o_init(3,'T',0.75);
 string training = "HTTHTTTHHTTHTTTHHTTHTTTHHTTHTTT"
                   "HHTTHTTTHHTTHTTTHHTTHTTTHHTTHTTTH";
 string test = "HTHTTHTHTTHTHTHTHTTHHTHTHTTHTHTTHHT";
 cout << "\nInput:  " << training << "\n\n";
 cout << "Probability (forward) : " << hmm.alpha(training) << "\n";
 cout << "Probability (backward): " << hmm.beta(training) << "\n\n";
 int* best_path = new int[256];
 cout << "Best-path-probability : "
      << hmm.viterbi(training,best_path) << "\n\n";
```

```
 cout << "Best path: ";
 for(t=1;best_path[t+1]!=-1;t++) cout << best_path[t] << ",";
 cout << best_path[t] << "\n\n";
 hmm.maximize(training,test);
 cout << " Probability (forward) : " << hmm.alpha(training) << "\n";
 cout << " Best-path-probability : "
      << hmm.viterbi(training,best_path) << "\n";
 delete best_path;
 return 0;
}
```

A C++ implementation for speech recognition has been described by Becchetti and Ricotti [17].

17.9 Application of Hidden Markov Models

The main applications of hidden Markov models are in speech recognition (Rabiner [171], Rabiner and Juang [172], Charniak [28], Jelinek [104]) and biological sequences such as DNA (Koski [120], Cristianni and Hahn [41]). We consider speech recognition. We follow in the representation mainly Rabiner [171] and Rabiner and Juang [172].

Both the phoneme and the syllable are often utilized as the primary unit of speech recognition. A syllanle consists of a nucleus (either a vowel or diphtong) plus some neighbouring consonants. A phoneme is the fundamental unit of phonology. The phonemes used for hidden Markov models. If one assigns each phoneme to a hidden Markov model one needs around 45 models for English. An additional model is also created for silence and background noise. Using this approach, a model for any word can be constructed by chaining together models for the component phonemes.

Each phoneme model will be made up of a number of states; the number of states per model is another design decision which needs to be made by the system designer. Each state in the model corresponds to some part of the input speech signal; we would like the feature vectors assigned to each state to be as uniform as possible so that the Gaussian model can be accurate. A very common approach is to use three states for each phoneme model; intuitively this corresponds to one state for the transition into the phoneme, one for the middle part and one for the transition out of the phoneme. Similarly the topology of the model must be decided. The three states might be linked in a chain where transitions are only allowed to higher numbered states or to themselves. Alternatively each state might be all linked to all others, the so called ergodic model. These two structures are common but many other combinations are possible.

When phoneme based hidden Markov models are being used, they must be concatenated to construct word or phrase hidden Markov models. For example, an hidden

Markov model for `cat` can be constructed from the phoneme hidden Markov models for `/k/` `/a/` and `/t/`. If each phoneme hidden Markov model has three states the `cat` hidden Markov model will have nine states. Some words have alternate pronunciations and so their composite models will have a more complex structure to reflect these alternatives. An example might be a model for `lives` which has two alternatives for the pronunciation of `'i'`.

While phoneme based models can be used to construct word models for any word they do not take into account any contextual variation in phoneme production. One way around this is to use units larger than phonemes or to use context dependant models. The most common solution is to use triphone models where there is one distinct phoneme model for every different left and right phoneme context. Thus there are different models for the `/ai/` in `/k-ai-t/` and in `/h-ai-t/`. A word model is made up from the appropriate context dependant triphone models: `'cat'` would be made up from the three models `[/sil-k-a/ /k-a-t/ /a-t-sil/]`.

While the use of triphones solves the problem of context sensitivity it presents another problem. With around 45 phonemes in English there are $45^3 = 91125$ possible triphone models to train (although not all of these occur in speech due to phonotactic constraints). The problem of not having enough data to effectively train these models becomes very important. One technique is state tying but another is to use only word internal triphones instead of the more general cross word triphones. Cross word triphones capture coarticulation effects accross word boundaries which can be very important for continuous speech production. A word internal triphone model uses triphones only for word internal triples and diphones for word final phonemes; `cat` would become

```
[sil /k-a/ /k-a-t/ /a-t/ sil]
```

This will be less accurate for continous speech modelling but the number of models required is smaller (none involve silence as a context) and so they can be more accurately trained on a given set of data.

There are many free parameters in an hidden Markov model, there are some techniques to reduce the number to allow better training. One is state tying and another one diagonal covariance matrices.

A single hidden Markov model contains a number of free parameters whose values must be determined from training data. For a fully connected three state model there are nine transition probabilities plus the parameters (means and covariance matrices) of three Gaussian models. If ome uses 24 input parameters (12 Mel-frequency cepstral coefficients plus 12 delta Mel-frequency cepstral coefficients) then the mean vector has 24 free parameters and the covariance matrix has $24^2 = 576$ free parameters making 609 in all. Multiply this by 45 phoneme models and there are 27,405 free parameters to estimate from training data; using context sensitive models there are many more (around 2.5 billion). With this many free parameters, a very large amount of training data will be required to get reliable statistical estimates. In addition, it is unlikely that the training data will be distributed evenly so that some models in a triphone based recogniser will recieve only a small number of training tokens while others will recieve many.

To address this problem one to shares states between triphone models. If the context sensitive triphone models consist of three states (for example) then we might assume that for all `/i/` vowel models (`/i/` in all contexts) the middle state might be very similar and so can be shared between all models. Similarly, the initial state might be shared between all `/i/` models preceeded by fricatives. Sharing states between models means that the Gaussian model associated with the state is trained on the data assigned to that state in both models: more data can be used to train each Gaussian making them more accurate. The limit of this kind of operation is to have all `/i/` models share all states, in which case we have reduced the models to context insensitive models again. In practice, states are shared between a significant number of models based on phonetic similarity of the preceding and following contexts.

The covariance matrix associated with each Gaussian inside each state measures both the amount of variance along each dimension in the parameter space and the amount of co-variance between parameters. In two dimensions this co-variance provides the orientation and 'stretch' of the ellipse shape. One simplification that can be made is to ignore the co-variance part of the matrix and only compute and use the individual dimension variances. In doing this we retain only the diagonal part of the co-variance matrix, setting all of the off-diagonal elements to zero. While this simplification does lose information it means a significant reduction in the number of parameters that need to be estimated. It is therefore used in some hidden Markov model implementations.

Choosing to build phoneme or triphone based models means that to recognise words or phrases we must make composite hidden Markov models from these subword building blocks. A model for the word `cat` can be made by joining together phoneme models for `/k/` `/a/` and `/t/`. If each phoneme model has three states then this composite model has nine states but can be treated just like any other hidden Markov model for the purposes of matching against an input observation sequence.

To be able to recognise more than one word we need to construct models for each word. Rather than have many separate models it is better to construct a network of phoneme models and have paths through the network indicate the words that are recognised. The phonemes can be arranged in a tree, each leaf of the tree corresponds to a word in the lexicon.

An example lexical tree which links together phoneme models in a network such that alternate paths through the network represent different words in the lexicon. The tree of phoneme models ensures that any path through the network corresponds to a real word. Each open circle represents a single hidden Markov model which might consist of three states. Each solid circle corresponds to a word boundary. Cases of multiple pronunciations (for `'A'`) and homonyms (`'hart'` and `'heart'`) can be seen in this network. If triphones models are being used this network would be expanded.

For connected word recognition this network can be extended to link words together into phrases such that the legal paths through the network correspond to phrases that we want to recognise. This begs the question of what phrases are to be allowed; the answer lies in what is called a language model who's job is to define possible word sequences either via definite patterns such as grammar rules or via statistical models. The language model defines the shape of the network of hidden Markov models; in anything but the simplest cases this network will be extremely complicated. While it is useful to think of a static, pre-constructed network being searched by the Viterbi algorithm, in a real recogniser the network is constructed as it is searched.

Assuming that we have designed the individual phoneme or triphone hidden Markov models then the first step is to initialise the free parameters in the models. These parameters are the transition probabilities between states and the means and covariance matrices of the Gaussian models associated with each state. The next step is to begin supervised training where we will force align the models with speech samples and update the model parameters to better fit that segmentation. If we begin with a poor set of parameters (for example, by choosing random values for each parameter) the forced aligment will be unlikely to assign appropriate phonemes to each model and so the model will be unlikely to improve itself. hidden Markov model training can be thought of as a search for the lowest point on a hilly landscape; if we begin from a point close to the lowest point we may well find it but if we begin from a point close to a higher dip, we may get stuck there and be unable to see the better solution over the horizon. The standard way to initialise the Gaussian models in each state is to use a small amount of hand segmented data and align it to each model. In a three state model, each state might be given four vectors of a twelve vector input sequence corresponding to one token. In this way the Gaussians are initialised to approximate the distributions for each phoneme.

Chapter 18

Fuzzy Sets and Fuzzy Logic

18.1 Introduction

A *classical set* X (also called crisp set) is defined as a collection of elements or objects $x \in X$ which can be finite, countable, or overcountable. Each single element can either belong to or not belong to a set A, $A \subset X$. In the former case, the statement *x belongs to A* is true, whereas in the latter case this statement is false. Such a classical set can be described in different ways: one can either enumerate (list) the elements that belong to the set; describe it analytically, for instance, by stating conditions for membership; or define the member elements by using the characteristic function, in which 1 indicates membership or 0 nonmembership.

Example. The set

$$\mathbb{N}_0 := \{ 0, 1, 2, 3, \ldots \}$$

is the set of all non-negative integers. The set is countable. The set of all integers $\mathbb{Z}$ is also countable, because it can be mapped 1-1 into $\mathbb{N}_0$ via $0 \leftrightarrow 0$, $1 \leftrightarrow 1$, $-1 \leftrightarrow 2$, $2 \leftrightarrow 3$, $-2 \leftrightarrow 4$, etc. The set $\mathbb{Z} \times \mathbb{Z}$ is also countable. ♣

Example. The set

$$A := \{ x : x \text{ is a letter in the english alphabet}, \ x \text{ is a vowel} \}$$

is given by

$$A = \{ a, e, i, o, u \} .$$

The set is finite. ♣

Example. The set of all real numbers is denoted by $\mathbb{R}$. This set is overcountable. The set of real numbers in the closed interval $[0, 1]$ is also overcountable. ♣

Both C++ using the STL with the class `set` and Java with the class `TreeSet` allow set-theoretical operations. We can find the union, intersection and difference of two finite sets. We can also get the cardinality of the finite set (i.e. the number of elements). Furthermore, we can find out whether a finite set is a subset of another finite set and whether a finite set contains a certain element.

In C++ the class `set` is a sorted associative container that stores objects of type Key. The class `set` is a simple associative container, meaning that its value type, as well as its key type, is key. It is also a unique associative container meaning that no two elements are the same. The class `set` is suited for the set algorithms

```
includes
set_union
set_intersection
set_difference,
set_symmetric_difference
```

The reason for this is twofold. First, the set algorithms require their arguments to be sorted ranges, and, since the class `set` is a sorted associative container, their elements are always sorted in ascending order. Second, the output range of these algorithms is always sorted, and inserting a sorted range into a `set` is a fast operation. The class `set` has the important property that inserting a new element into a `set` does not invalidate iterators that point to existing elements. Erasing an element from a set also does not invalidate any iterator, except of course, for iterators that actually point to the element that is being erased.

The following C++ program shows an application of this class.

```
// setsstl.cpp

#include <iostream>
#include <set>
#include <algorithm>
#include <string>
#include <iterator>
using namespace std;

int main(void)
{
 const int M = 4; const int N = 3;
 const string a[M] = { "Jones", "Miller", "Steeb", "Smith" };
 const string b[N] = { "Hardy", "Copper", "Steeb" };
 set<string> s1(a,a+M);
 set<string> s2(b,b+N);
 set<string> s3;

 cout << "union of the sets s1 and s2: " << endl;
```

```
 set_union(s1.begin(),s1.end(),s2.begin(),s2.end(),
           ostream_iterator<string>(cout," "));
 cout << endl << endl;

 cout << "intersection of the set s1 and s2: " << endl;
 set_intersection(s1.begin(),s1.end(),s2.begin(),s2.end(),
                  ostream_iterator<string>(cout," "));
 cout << endl << endl;

 cout << "difference of the sets s1 and s2: " << endl;
 set_difference(s1.begin(),s1.end(),s2.begin(),s2.end(),
                inserter(s3,s3.begin()));
 copy(s3.begin(),s3.end(),ostream_iterator<string>(cout," "));
 cout << endl << endl;

 // s2 subset of s1 ?
 bool ss = includes(s1.begin(),s1.end(),s2.begin(),s2.end());
 cout << "s2 subset of s1 ? " << ss << endl;
 // s4 subset of s2 ?
 const string c[1] = { "Hardy" };
 set<string> s4(c,c+1);
 ss = includes(s2.begin(),s2.end(),s4.begin(),s4.end());
 cout << "s4 subset of s2 ? " << ss << endl;
 // size of set (number of elements)
 cout << "s1 has " << s1.size() << " elements " << endl;
 // is set s2 empty?
 bool empty = s2.empty();
 cout << "empty = " << empty << endl;
 return 0;
}
```

In Java the interface `Set` is a `Collection` that cannot contain duplicate elements. The interface `Set` models the mathematical set abstraction. The `Set` interface extends `Collection` and contains no methods other than those inherited from `Collection`. It adds the restriction that duplicate elements are prohibited. The JDK contains two general-purpose `Set` implementations. The class `HashSet` stores its elements in a hash table. The class `TreeSet` stores its elements in a red-black tree. This guarantees the order of iteration. The union of sets is provided by the method `addAll()` and the intersection of sets is provided by the method `retainAll()`.

The following Java program shows an application of the `TreeSet` class.

```
// SetOperation.java

import java.util.*;

public class SetOperation
{
```

```
public static void main(String[] args)
{
String[] A = { "Steeb", "C++", "80.00" };
String[] B = { "Solms", "Java", "80.00" };
TreeSet S1 = new TreeSet();
for(int i=0;i<A.length;i++) S1.add(A[i]);
System.out.println("S1 = " + S1);
TreeSet S2 = new TreeSet();
for(int i=0;i<B.length;i++) S2.add(B[i]);
System.out.println("S2 = " + S2);
// union
TreeSet S3 = new TreeSet(S1);
boolean b1 = S3.addAll(S2);
System.out.println("S3 = " + S3);
System.out.println("S1 = " + S1);
// intersection
TreeSet S4 = new TreeSet(S1);
boolean b2 = S4.retainAll(S2);
System.out.println("S4 = " + S4);
System.out.println("S2 = " + S2);
// (asymmetric) set difference
TreeSet S5 = new TreeSet(S1);
boolean b3 = S5.removeAll(S2);
System.out.println("S5 = " + S5);
// test for subset
TreeSet S6 = new TreeSet(S1);
boolean b4 = S6.containsAll(S2);
System.out.println("b4 = " + b4);
// is element of set (contains)
boolean b = S1.contains("80.00");
System.out.println("b = " + b);
b = S2.contains("Steeb");
System.out.println("b = " + b);
}
}
```

The output is

```
S1 = [80.00, C++, Steeb]
S2 = [80.00, Java, Solms]
S3 = [80.00, C++, Java, Solms, Steeb]
S1 = [80.00, C++, Steeb]
S4 = [80.00]
S2 = [80.00, Java, Solms]
S5 = [C++, Steeb]
b4 = false
b = true
b = true
```

In *fuzzy logic* (Yager and Zadeh [221], Bojadziev [20], Bandemer [11], Ross [175], Zimmermann[229]) the notion of binary membership is extended to accommodate various degrees of membership on the real continuous interval $[0,1]$, where the endpoints of 0 and 1 conform to no membership and full membership, respectively, just as the indicator function does for crisp sets, but where the infinite number of values in between the endpoints can represent various degrees of membership for an element x in some set on the universe X. The sets on the universe X that can accommodate degrees of membership are termed as *fuzzy sets*. For a fuzzy set, the characteristic function allows various degrees of membership for the elements of a given set. The *universe of discourse* is the universe of all available information on a given problem. Once the universe is defined we are able to define certain events on this information space. We describe sets as mathematical abstractions of these events and of the universe itself.

Fuzzy logic and fuzzy sets are powerful mathematical tools for modelling the following: uncertain systems in industry, nature, and humanity as well as facilitators for common-sense reasoning in decision making in the absence of complete and precise information. Fuzzy logic and fuzzy sets find applications in control theory, signal processing, robotics, intelligent process control, expert systems, image processing, decision making, pattern recognition, cluster analysis, and a variety of learning methods, such as neural networks, genetic algorithms, and inductive reasoning.

Let us first give an example where fuzzy sets can be applied.

Example. If we talk about hot and cold we might classify all temperatures above 25^oC as hot and all temperatures equal to or below 25^oC as cold. This would give us a crisp binary set where any temperature would be classified as either hot or cold. Alternatively we might want to specify a degree of hotness/coldness by the following mapping

$$\begin{array}{rcccl} & & T & \leq & 0^oC & \to & \text{very cold} \\ 0^oC & < & T & \leq & 18^oC & \to & \text{moderately cold} \\ 18^oC & < & T & \leq & 25^oC & \to & \text{just right} \\ 25^oC & < & T & \leq & 30^oC & \to & \text{moderately hot} \\ & & T & > & 30^oC & \to & \text{very hot} \end{array}$$

Here we map temperatures on a crisp quintary set.

Definition. A *fuzzy set*, $\widetilde{F}$, on a collection of objects, X, is a mapping

$$\mu_{\widetilde{F}}(x) \,:\, X \to [0,\alpha]\,.$$

Here $\mu_{\widetilde{F}}(x)$ indicates the extent to which x has the attribute $\widetilde{F}$. Hence $\mu_{\widetilde{F}}(x)$ is called the *membership function*. The most common types of fuzzy sets are *normalized fuzzy sets*, $\widetilde{F}$, for which

$$\alpha \equiv \sup_{x\in X} \mu_{\widetilde{F}}(x) = 1\,.$$

A nonempty fuzzy set $\widetilde{A}$ can always be normalized by dividing $\mu_{\widetilde{A}}(x)$ by $\sup_x \mu_{\widetilde{A}}(x)$.

In this case the result of the mapping, $\mu_{\widetilde{F}}(x)$, can be interpreted as a degree of membership or as a level of confidence in the truth or compatibility of the statement.

An alternative definition of a fuzzy set can be given.

Definition. If X is a collection of objects denoted generically by x then a fuzzy set $\widetilde{A}$ in X is a set of ordered pairs

$$\widetilde{A} := \{ (x, \mu_{\widetilde{A}}(x)) \,|\, x \in X \} .$$

The function $\mu_{\widetilde{A}}$ is called the *membership function* or grade of membership (also called degree of compatibility or degree of truth) of x in $\widetilde{A}$ which maps X to the membership space M. When M contains only two points 0 and 1, $\widetilde{A}$ is nonfuzzy and $\mu_{\widetilde{A}}(x)$ is identical to the characteristic function of a nonfuzzy (crisp) set. The range of the membership function is a subset of the nonnegative real numbers whose supremum is finite. Elements with a zero degree of membership are normally not listed.

Thus we consider ordinary sets as special cases of fuzzy sets, viz. those with only 0 and 1 as membership degrees.

Definition. Two fuzzy sets $\widetilde{A}$ and $\widetilde{B}$ are equal if they have the same membership functions

$$\widetilde{A} = \widetilde{B} \qquad \Leftrightarrow \qquad \mu_{\widetilde{A}}(x) = \mu_{\widetilde{B}}(x) \quad \text{for all} \quad x \in X .$$

Example. Employees of a company might need a minimum of 16MB RAM to run the CAD package used by the company and should preferably have 64MB to run it efficiently. The degree of cost efficiency of a computer may be described by the following fuzzy set

$$\widetilde{E} = \{ (4,0), (8,0), (16,0.2), (32,0.5), (64,1), (128,0.8), (256,0.5), (512,0.2) \} .$$

Example. A vehicle should be kept close to the desired travelling speed. The degree of closeness is modeled by the continuous membership function

$$\mu_{\widetilde{F}}(x) = \exp\left(-\frac{(x-d)^2}{2d} \right)$$

where d is the desired travelling speed. A fuzzy set of speeds can be denoted by

$$\widetilde{S} = \{ (x, \mu_{(\widetilde{S})}(x)) \mid x \in [0, 200] \} .$$

Example. The fuzzy set $\widetilde{A}$ "real numbers close to 3" could be modelled by the membership function

$$\mu_A(x) = \exp(-2|x-3|) .$$

Example. The fuzzy set $\widetilde{A}$ "integers close to 15" can be expressed by the finite fuzzy set

$$\widetilde{A} = \{\ (12,0.2),\ (13,0.5),\ (14,0.8),\ (15,1.0),\ (16,0.8),\ (17,0.5),\ (18,0.2)\ \}.$$

The fuzzy set consists of 7 ordered pairs. Thus the membership function $\mu_{\widetilde{A}}$ takes the following values on $[0,1]$

$$\mu_{\widetilde{A}}(12) = 0.2, \quad \mu_{\widetilde{A}}(13) = 0.5, \quad \mu_{\widetilde{A}}(14) = 0.8, \quad \mu_{\widetilde{A}}(15) = 1.0,$$

$$\mu_{\widetilde{A}}(16) = 0.8, \quad \mu_{\widetilde{A}}(17) = 0.5, \quad \mu_{\widetilde{A}}(18) = 0.2. \qquad \clubsuit$$

In many cases the membership function is a *trapezoidal function*. Let $a < b < c < d$. Then the trapezoidal function is defined by

$$f(x) := \begin{cases} 0 & \text{if} \quad x < a \\ (x-a)/(b-a) & \text{if } x \geq a \text{ and } x < b \\ 1 & \text{if } x \geq b \text{ and } x \leq c \\ (d-x)/(d-c) & \text{if } x > c \text{ and } x < d \\ 0 & \text{if} \quad x \geq d \end{cases}$$

A C++ implementation is as follows. We assume that $a < b < c < d$.

```
// trapez.cpp

#include <iostream>
using namespace std;

double trapez(double a,double b,double c,double d,double x)
{
 if((x >= b) && (x <= c)) return 1.0;
 if((x > a) && (x < b)) return (x-a)/(b-a);
 if((x > c) && (x < d)) return (d-x)/(d-c);
 else return 0.0;
}

int main(void)
{
 double a = 1.0, b = 2.0, c = 5.3, d = 7.3;
 double x = 6.1;
 double result = trapez(a,b,c,d,x);
 cout << "x = " << x << "  " << "result = " << result << endl;
 x = 4.1;
 result = trapez(a,b,c,d,x);
 cout << "x = " << x << "  " << "result = " << result << endl;
 return 0;
}
```

Let $a, b, p, q, r \in \mathbb{R}$, $b > a$, $a < p$, $p < r$, $r < q$, $q < b$ and

$$p - a = r - p, \qquad q - r = b - q.$$

Another important membership function is given by

$$f(x; a, b, r, p, q) = \begin{cases} 0 & x \leq a \\ 2^{m-1}((x-a)/(r-a))^m & a < x \leq p \\ 1 - 2^{m-1}((r-x)/(r-a))^m & p < x \leq r \\ 1 - 2^{m-1}((x-r)/(b-r))^m & r < x \leq q \\ 2^{m-1}((b-x)/(b-r))^m & q < x < b \\ 0 & x \geq b \end{cases}$$

where m is the fuzzifier. In most cases $m = 2$. Where are the crossover points? What is the value at the centre? Is the function differentiable?

Definition. A fuzzy set $\widetilde{A}$ can be *contained* within another fuzzy set $\widetilde{B}$ and we write

$$\widetilde{A} \subseteq \widetilde{B} \text{ iff } \mu_{\widetilde{A}}(x) \leq \mu_{\widetilde{B}}(x) \quad \text{for all } x \in X .$$

$\widetilde{A}$ is strictly contained in $\widetilde{B}$, i.e. $\widetilde{A} \subset \widetilde{B}$, iff $\mu_{\widetilde{A}}(x) < \mu_{\widetilde{B}}(x)$ for all $x \in X$.

Example. The fuzzy set

$$\widetilde{A} = \{ (0.1, 0.2), (0.2, 0.4), (0.3, 0.7), (0.4, 1.0), (0.5, 0.2), (0.6, 0.1) \}$$

is contained within $\widetilde{B}$

$$\widetilde{B} = \{ (0.1, 0.3), (0.2, 0.4), (0.3, 1.0), (0.4, 1.0), (0.5, 0.6), (0.6, 0.2) \}.$$

Both, $\widetilde{A}$ and $\widetilde{B}$ are normalized fuzzy sets.

Definition. The *support for a fuzzy set*, $\widetilde{F}$, $S(\widetilde{F})$, is the crisp set of all $x \in X$ such that $\mu_{\widetilde{F}}(x) > 0$.

Definition. The *α-level set* is the crisp set of elements that belong to the fuzzy set $\widetilde{F}$ with degree of membership equal to or greater than α

$$F_\alpha := \{ x \in X \mid \mu_{\widetilde{F}}(x) \geq \alpha \} .$$

Example. In the first example the support for the set $\widetilde{E}$ is given by {16, 32, 64, 128, 256, 512 } and the 0.5-level set is given by

$$E_{0.5} = \{ 32, 64, 128, 256 \} .$$ ♣

Definition. A fuzzy set $\widetilde{F}$ is *convex* if its membership function, $\mu_{\widetilde{F}}(x)$, is convex, i.e. if

$$\mu_{\widetilde{F}}(\lambda x_1 + (1-\lambda)x_2) \geq \min\{\mu_{\widetilde{F}}(x_1), \mu_{\widetilde{F}}(x_2)\}, \qquad x_1, x_2 \in X, \quad \lambda \in [0, 1] .$$

Alternatively, a fuzzy set is convex if all α-level sets are convex.

A special property of two convex fuzzy sets, say $\widetilde{A}$ and $\widetilde{B}$, is that the intersection of these two fuzzy sets is also a convex fuzzy set.

Definition. The *crossover points* of a membership function are defined as the elements in the universe for which a particular fuzzy set $\widetilde{A}$ has the values equal to 0.5, i.e. for which $\mu_{\widetilde{A}} = 0.5$.

Definition. For a finite fuzzy set $\widetilde{A}$ the *cardinality* $|\widetilde{A}|$ is defined as

$$|\widetilde{A}| := \sum_{x \in X} \mu_{\widetilde{A}}(x) .$$

The quantity

$$\|\widetilde{A}\| := \frac{|\widetilde{A}|}{|X|}$$

is called the relative cardinality. Obviously the relative cardinality of a fuzzy set depends on the cardinality of the universe X. Thus we have to choose the same universe if we want to compare fuzzy sets by their relative cardinality. The relative cardinality can be interpreted as the fraction of elements of X being in $\widetilde{A}$, weighted by their degrees of membership in $\widetilde{A}$.

Example. Consider the discrete fuzzy set

$$\widetilde{A} = \{ (1, 0.2), (2, 0.5), (3, 0.8), (4, 1.0), (5, 0.7), (6, 0.3) \} .$$

Then

$$|\widetilde{A}| = 0.2 + 0.5 + 0.8 + 1.0 + 0.7 + 0.3 = 3.5 .$$

Its relative cardinality is

$$\|\widetilde{A}\| = \frac{3.5}{6} .$$ ♣

18.2 Operators for Fuzzy Sets

18.2.1 Logical Operators

In this section we give the basic definitions for operations on fuzzy sets. The membership function is obviously the crucial component of a fuzzy set. It is therefore not surprising that operations with fuzzy sets are defined via their membership functions. They constitute a consistent framework for the theory of fuzzy sets. They are, however, not the only possible way to extend classical set theory consistently. The main operations of fuzzy sets are:

the *intersection of fuzzy sets*

the *union of fuzzy sets*

the *complement of fuzzy sets.*

Definition. The membership function $\mu_{\widetilde{C}}(x)$ of the *intersection*

$$\widetilde{C} = \widetilde{A} \cap \widetilde{B}$$

satisfies for each $x \in X$

$$\mu_{\widetilde{C}}(x) = \min\{\mu_{\widetilde{A}}(x), \mu_{\widetilde{B}}(x)\}\,.$$

Definition. The membership function $\mu_{\widetilde{C}}(x)$ of the *union*

$$\widetilde{C} = \widetilde{A} \cup \widetilde{B}$$

satisfies for each $x \in X$

$$\mu_{\widetilde{C}}(x) = \max\{\mu_{\widetilde{A}}(x), \mu_{\widetilde{B}}(x)\}\,.$$

Definition. The membership function $\mu_{\widetilde{C}}(x)$ of the *complement* of $\widetilde{A}$

$$\widetilde{C} \not\subset \widetilde{A}$$

satisfies for each $x \in X$

$$\mu_{\widetilde{C}}(x) = \mu_{\not\subset \widetilde{A}}(x) = 1 - \mu_{\widetilde{A}}(x)\,.$$

These are extensions of the intersection, union, and complement for classical sets. In the C++ program we implement the intersection of fuzzy sets. We pass the membership function `f1()` and `f2()` to the function `min()` to find the intersection

```
// fuzzmin.cpp

#include <iostream>
#include <cmath>      // for exp(), fabs()
using namespace std;

double f1(double x) { return exp(-fabs(x-3.0)); }

double f2(double x) { return exp(-fabs(x-4.0)); }

double min(double (*f1)(double),double (*f2)(double),double x)
{ if(f1(x) >= f2(x)) return f2(x); else return f1(x); }

int main(void)
{
 double x = 3.5;
 double result = min(f1,f2,x);
```

```
  cout << "x = " << x << " " << "result = " << result << endl;
  x = 3.0;
  result = min(f1,f2,x);
  cout << "x = " << x << " " << "result = " << result << endl;
  x = 5.0;
  result = min(f1,f2,x);
  cout << "x = " << x << " " << "result = " << result << endl;
  return 0;
}
```

Example. Let $\widetilde{E}$ be the discrete fuzzy set of efficient PCs discussed above and let $\widetilde{B}$ be the fuzzy set of powerful PCs

$$\widetilde{P} = \{(4,0),(8,0),(16,0.2),(32,0.3),(64,0.4),(128,0.5),(256,0.8),(512,1.0)\}.$$

Then the fuzzy set of PCs which are efficient AND powerful is given by

$$\begin{aligned}\widetilde{A} &= \widetilde{E} \cap \widetilde{P}\\ &= \{(4,0),(8,0),(16,0.2),(32,0.3),(64,0.4),(128,0.5),(256,0.5),(512,0.2)\}\end{aligned}$$

while the union operation yields

$$\begin{aligned}\widetilde{O} &= \widetilde{E} \cup \widetilde{P}\\ &= \{(4,0),(8,0),(16,0.2),(32,0.5),(64,1),(128,0.8),(256,0.8),(512,1.0)\}.\end{aligned}$$

The fuzzy set of PCs which are NOT powerful is given by

$$\begin{aligned}\widetilde{N} &= \not\subset P\\ &= \{(4,1),(8,1),(16,0.8),(32,0.7),(64,0.6),(128,0.5),(256,0.2),(512,0)\}.\end{aligned}$$ ♣

Example. Assume that the statement: "The weather in Johannesburg is hot" is given a confidence level represented by the following membership function

$$\mu_{\widetilde{H}}(T) = \frac{1}{1+\exp(-(T-24)/2)}$$

and that the pleasant temperature range is given by

$$\mu_{\widetilde{P}}(T) = \exp\left(-\frac{(T-24)^2}{2}\right).$$

Thus for the logical AND we have "It is hot AND pleasant", and for the logical OR we have "It is hot OR pleasant". ♣

Consider the fuzzy sets $\widetilde{A}$, $\widetilde{B}$, and $\widetilde{C}$. We have the following properties

$$\widetilde{A} \cap \widetilde{A} = \widetilde{A} \qquad \text{idempotence}$$

$$\widetilde{A} \cup \widetilde{A} = \widetilde{A} \qquad \text{idempotence}$$

$$\widetilde{A} \cap \widetilde{B} = \widetilde{B} \cap \widetilde{A} \quad \text{commutativity}$$

$$\widetilde{A} \cup \widetilde{B} = \widetilde{B} \cup \widetilde{A} \quad \text{commutativity}$$

$$(\widetilde{A} \cap \widetilde{B}) \cap \widetilde{C} = \widetilde{A} \cap (\widetilde{B} \cap \widetilde{C}) \quad \text{associativity}$$

$$(\widetilde{A} \cup \widetilde{B}) \cup \widetilde{C} = \widetilde{A} \cup (\widetilde{B} \cup \widetilde{C}) \quad \text{associativity}$$

$$\widetilde{A} \cap (\widetilde{B} \cup \widetilde{C}) = (\widetilde{A} \cap \widetilde{B}) \cup (\widetilde{A} \cap \widetilde{C}) \quad \text{distributivity}$$

$$\widetilde{A} \cup (\widetilde{B} \cap \widetilde{C}) = (\widetilde{A} \cup \widetilde{B}) \cap (\widetilde{A} \cup \widetilde{C}) \quad \text{distributivity}$$

$$\overline{\overline{\widetilde{A}}} = \widetilde{A} \quad \text{double complement}$$

$$\overline{\widetilde{A} \cap \widetilde{B}} = \overline{\widetilde{A}} \cup \overline{\widetilde{B}} \quad \text{DeMorgan's law}$$

$$\overline{\widetilde{A} \cup \widetilde{B}} = \overline{\widetilde{A}} \cap \overline{\widetilde{B}} \quad \text{DeMorgan's law}$$

These properties have the same structure as those for classical sets.

Let $\emptyset$ be the empty fuzzy set. Then we have the rules

$$\widetilde{A} \cap (X \times [0,1]) = \widetilde{A} \quad \text{identity}$$
$$\widetilde{A} \cup \emptyset = \widetilde{A} \quad \text{identity}$$
$$\widetilde{A} \cap \emptyset = \emptyset \quad \text{identity}$$
$$\widetilde{A} \cup (X \times [0,1]) = X \times [0,1] \quad \text{identity}$$

18.2.2 Algebraic Operators

Besides the basic operations on fuzzy sets given above, namely intersection, union and complement we can define a large number of other algebraic operations which we now summarize.

Definition. The *algebraic sum*, $\widetilde{C} = \widetilde{A} + \widetilde{B}$, of two fuzzy sets with membership functions $\mu_{\widetilde{A}}$ and $\mu_{\widetilde{B}}$ is a fuzzy set with membership function

$$\mu_{\widetilde{A}+\widetilde{B}}(x) = \mu_{\widetilde{A}}(x) + \mu_{\widetilde{B}}(x) - \mu_{\widetilde{A}}(x)\mu_{\widetilde{B}}(x), \qquad x \in X.$$

We write

$$\widetilde{C} = \widetilde{A} + \widetilde{B} = \{ (x, \mu_{\widetilde{A}+\widetilde{B}}(x)) \mid x \in X \}.$$

Definition. The *algebraic product*, $\widetilde{C} = \widetilde{A} \cdot \widetilde{B}$, of two fuzzy sets with membership function $\mu_{\widetilde{A}}$ and $\mu_{\widetilde{B}}$ is a fuzzy set with membership function

$$\mu_{\widetilde{A} \cdot \widetilde{B}} \equiv \mu_{\widetilde{A}}(x)\mu_{\widetilde{B}}(x), \qquad x \in X.$$

We write

$$\widetilde{C} = \{ (x, \mu_{\widetilde{A} \cdot \widetilde{B}}(x)) \mid x \in X \}.$$

Definition. The *bounded sum*, $\widetilde{C} = \widetilde{A} \oplus \widetilde{B}$, of two fuzzy sets with membership function $\mu_{\widetilde{A}}$ and $\mu_{\widetilde{B}}$ is a fuzzy set with membership function

$$\mu_{\widetilde{A}\oplus\widetilde{B}} \equiv \min\{\, 1, \mu_{\widetilde{A}}(x) + \mu_{\widetilde{B}}(x)\,\}\,.$$

We write

$$\widetilde{C} = \{\, (x, \mu_{\widetilde{A}\oplus\widetilde{B}}(x)) \,|\, x \in X\,\}\,.$$

Definition. The *bounded difference*, $\widetilde{C} = \widetilde{A} \ominus \widetilde{B}$, of two fuzzy sets with membership function $\mu_{\widetilde{A}}$ and $\mu_{\widetilde{B}}$ is a fuzzy set with membership function

$$\mu_{\widetilde{A}\ominus\widetilde{B}} \equiv \max\{\, 0, \mu_{\widetilde{A}}(x) + \mu_{\widetilde{B}}(x) - 1\,\}\,.$$

We write

$$\widetilde{C} = \{\, (x, \mu_{\widetilde{A}\ominus\widetilde{B}}(x)) \,|\, x \in X\,\}\,.$$

The Cartesian product of fuzzy sets is defined as follows.

Definition. Let $\widetilde{A}_1,\ldots,\widetilde{A}_n$ be fuzzy sets in $X_1, \ldots, X_n$. The *Cartesian product* is then a fuzzy set in the product space $X_1 \times \cdots \times X_n$ with a membership function

$$\mu_{\widetilde{A}_1\times\cdots\times\widetilde{A}_n}(x) = \min_i\{\, \mu_{\widetilde{A}_i}(x_i) \,|\, x = (x_1, x_2, \ldots, x_n),\ x_i \in X_i\,\}\,.$$

Definition. The mth power of a fuzzy set $\widetilde{A}$ is a fuzzy set with the membership function

$$\mu_{\widetilde{A}^m}(x) = (\mu_{\widetilde{A}}(x))^m, \qquad x \in X\,.$$

The *Fuzzy AND* and *Fuzzy OR* operators combine the logical AND and OR operators with the arithmetic norm.

Definition. The *fuzzy AND* of two fuzzy sets $\widetilde{A}$ and $\widetilde{B}$ is defined as

$$\widetilde{C} = \widetilde{A}\widetilde{\cap}\widetilde{B}$$

with membership function

$$\mu_{\widetilde{C}}(x) = \mu_{\widetilde{A}\widetilde{\cap}\widetilde{B}} = \gamma \min\{\mu_{\widetilde{A}}(x), \mu_{\widetilde{B}}(x)\} + \frac{1}{2}(1-\gamma)(\mu_{\widetilde{A}}(x) + \mu_{\widetilde{B}}(x))$$

where γ can be varied between 0 and 1 in order to weight the logical AND against the arithmetic mean. For $\gamma = 1$ the fuzzy AND reduces to the logical AND and for $\gamma = 0$ the fuzzy AND operator reduces to the arithmetic mean.

Definition. The *fuzzy OR* of two fuzzy sets $\widetilde{A}$ and $\widetilde{B}$ is defined as

$$\widetilde{C} = \widetilde{A}\widetilde{\cup}\widetilde{B}$$

with membership function

$$\mu_{\widetilde{C}}(x) = \mu_{\widetilde{A}\widetilde{\cup}\widetilde{B}} = \gamma \max\{\mu_{\widetilde{A}}(x), \mu_{\widetilde{B}}(x)\} + \frac{1}{2}(1-\gamma)(\mu_{\widetilde{A}}(x) + \mu_{\widetilde{B}}(x)), \quad \gamma \in [0,1]\,.$$

18.2.3 Defuzzification Operators

In many practical applications we would like to obtain a crisp decision from the fuzzy analysis of the problem. For example, in a problem where a company uses fuzzy logic to decide on one of many marketing companies the result of the fuzzy analysis should be exactly one company. Similarly, in a control problem (e.g. the classic pole-balancing problem) we would like a crisp decision for the force which should be applied to the cart. The following operators are commonly used to extract a crisp decision from a fuzzy set.

Definition. The *maximum grade operator* returns that support value which has the maximum grade (degree of truth). If there is no unique support value corresponding to the maximum grade then it returns any one of these.

Definition. The *minimum grade operator* returns that support value which has the minimum grade. If there is no unique support value corresponding to the minimum grade then it returns any one of these.

In many cases a small variation in the membership function can cause a very big variation in the decision (conclusion). For example, if we have a non-convex fuzzy set, then a small variation might change the decision from the one maximum to the other. This can be problematic, especially in the case of control problems where it can be a major cause of instability. Hence, for control problems one often uses the centroid of the fuzzy set for the crisp control strategy.

Definition. The *centroid* c (or centre of mass) of a fuzzy set, $\widetilde{F}$ with membership function $\mu_{\widetilde{F}}(x)$ for $x \in X$ is defined by

$$c := \frac{\sum_{x\in X} x \cdot \mu_{\widetilde{F}}(x)}{\sum_{x\in X} \mu_{\widetilde{F}}(x)}$$

in the case of a discrete membership function and by

$$c := \frac{\int_{x\in X} x \cdot \mu_{\widetilde{F}}(x)}{\int_{x\in X} \mu_{\widetilde{F}}(x)}$$

in the case where $\mu_{\widetilde{F}}(x)$ is continuous.

For some applications it can be useful to obtain a crisp set from a fuzzy set. For example, when trading with stocks, one might want to use fuzzy logic inference to decide which stocks should be sold.

Definition. The α*-cut operator* returns a crisp set with a grade of 0 for support values with grade less than α and 1 for support values which have a grade larger than or equal to α.

Definition. The *β-cut operator* returns the crisp set with grade 1 for the β support values which have the highest grade and 0 for the remaining support values.

Example. Consider the discrete fuzzy set

$$\widetilde{F} = \{ (1, 0.85), (2, 0.7), (3, 0.1), (4, 0.44), (5, 0.87), (6, 0.2), (7, 0.19) \} .$$

The maximum grade is 5, the minimum grade is 3, the centroid is $3.34 \approx 3$, the α-cut with $\alpha = 0.7$ returns the crisp set

$$\{ (1, 1), (2, 0), (3, 0), (4, 0), (5, 1), (6, 0), (7, 0) \}$$

and the β-cut with $\beta = 0.3$ returns the crisp set

$$\{ (1, 1), (2, 1), (3, 0), (4, 0), (5, 1), (6, 0), (7, 0) \} .$$ ♣

There are a number of other defuzzification operators described in the literature, for example the weighted average method (only applies to symmetrical membership functions), the centre of sums, centre of largest area and first (or last) maxima. For details and examples we refer to the literature (Ross [175]).

18.2.4 Fuzzy Concepts as Fuzzy Sets

It is often convenient to have implemented fuzzy concepts like *large*, *small*, *near*, *greater than* and *less than* as fuzzy sets on a given universe. For example, the profit made on a product might, for arguments sake, be between 0% and 100%. We would like to implement the concept *large* as a fuzzy set. A simple way of doing this is to assume that the fuzzy set has points on a straight line with grade 1 for 100% profit and grade 0 for 0% profit. Similarly, we could define the concept, *small*, as a fuzzy set with a straight line as the membership function which gives grade 1 for 0% profit and grade 0 for 100% profit.

Definition. The concept *small* is represented by a fuzzy set, $\widetilde{S}$ with membership function

$$\mu_{\widetilde{S}}(x) = \frac{x - x_{max}}{x_{min} - x_{max}}$$

and the concept *large* is represented by a fuzzy set $\widetilde{L}$ with membership function

$$\mu_{\widetilde{L}}(x) = \frac{x - x_{min}}{x_{max} - x_{min}} .$$

Another useful concept is *near*. For example, the profit on a certain product might be near 40%. We choose to implement near, as a fuzzy set whose membership function is given by a normal distribution with width 10% of the size of the universe.

Definition. Consider a universe, X. The concept *near* p, with $p \in X$, is represented by a fuzzy set, $\widetilde{N}$, with membership function

$$\mu_{\widetilde{N}}(x) = \frac{1}{\sqrt{2\pi}\sigma} \exp\left(\frac{-(x-p)^2}{2\sigma^2} \right)$$

where the variance, σ, is chosen by default as $\sigma = 0.1(x_{max} - x_{min})$. If $\sigma \to 0$, then we find the crisp is equal to the Dirac-Delta function.

Gaussian membership functions are also used in fuzzy pattern recognition. Suppose we have a one-dimensional universe on the real line, i.e., $X = \mathbb{R}$. Consider two fuzzy sets $\widetilde{A}$ and $\widetilde{B}$ having normal Gaussian membership functions

$$\mu_{\widetilde{A}}(x) = \exp\left(\frac{-(x-a)^2}{\sigma_a^2}\right), \qquad \mu_{\widetilde{B}}(x) = \exp\left(\frac{-(x-b)^2}{\sigma_b^2}\right).$$

An inner product can be defined (Ross [175]) yielding

$$\widetilde{A} \bullet \widetilde{B} = \exp\left(\frac{-(a-b)^2}{(\sigma_a+\sigma_b)^2}\right) = \mu_{\widetilde{A}}(x_0) = \mu_{\widetilde{B}}(x_0)$$

where

$$x_0 := \frac{\sigma_a b + \sigma_b a}{\sigma_a + \sigma_b}.$$

If the two fuzzy sets are identical, then the inner product equals one.

Finally we define the fuzzy concepts *greater than* and *less than* For example, the profit on a certain product might be greater than 20%. We use a *Fermi-Dirac distribution* whose transition is at 40% to represent this fuzzy concept.

Definition. Consider a universe, X. The concept, *greater than* p, with $p \in X$ is represented by a fuzzy set, $\widetilde{G}$, with membership function

$$\mu_{\widetilde{G}}(x) = \left[1 + \exp\left(\frac{-(x-p)}{\lambda}\right)\right]^{-1}$$

where the slope λ is chosen by default as $\lambda = 0.1(x_{max} - x_{min})$. For $\lambda \to 0$ we recover the *crisp greater than* which is the step function

$$\mu_{\widetilde{G}}(x)|_{\lambda\to 0} = \begin{cases} 1 \text{ for all } x > p \\ 0 \text{ for all } x \le p \end{cases}$$

18.2.5 Hedging

A linguistic hedge or a modifier is an operation which modifies the meaning of a term. Concentration and dilation hedges are frequently used in fuzzy inference. Typical linguistic terms for concentrators are extremely and very. This shifts the emphasis towards larger support values, i.e. they increase the restrictiveness of the fuzzy set.

Definition. A *linguistic hedge* is an operation that modifies the meaning of a term or, more generally, of a fuzzy set. If $\widetilde{A}$ is a fuzzy set then the modifier m generates the (composite) term $\widetilde{B} = m(\widetilde{A})$.

Mathematical models frequently used for modifiers are

$$\text{concentration} \quad \mu_{\text{con}\widetilde{A}}(x) = (\mu_{\widetilde{A}}(x))^2$$
$$\text{dilation} \quad \mu_{\text{dil}\widetilde{A}}(x) = (\mu_{\widetilde{A}}(x))^{1/2}.$$

Example. Assume we choose the following membership function for the fuzzy set, $\widetilde{H}$, corresponding to the linguistic term *the weather is hot*

$$\mu_{\widetilde{H}}(T) = \frac{T}{40}, \quad T \in [0:40].$$

The grade of membership (degree of truth) given to various temperatures is given in the following table

T	0	5	10	15	20	25	30	35	40
$\mu_{\widetilde{H}}(T)$	0	0.125	0.25	0.375	0.5	0.625	0.75	0.825	1

This membership function will give a support of 0.5 to a temperature of 20°C, 0.75 to 30°C and 0.825 to 35°C. The set of *very hot* temperatures should give only higher temperatures significant support. The membership function for the fuzzy set *very hot*, very$\widetilde{H}$, can be obtained from the membership function of $\widetilde{H}$ by using a hedge

$$\mu_{\text{very}\widetilde{H}}(T) = (\mu_{\widetilde{H}}(T))^2$$

The grade of membership for *very hot* is given in the following table

T	0	5	10	15	20	25	30	35	40
$\mu_{\text{very}\widetilde{H}}(T)$	0	0.016	0.063	0.141	0.25	0.391	0.563	0.681	1

Only the temperatures 35°C and 40°C have a 0.6-level support for very hot while 25°C, 30°C, 35°C and 40°C all have a 0.6-level support for hot. The hedging thus restricts (or compressed) the fuzzy set around high temperatures. ♣

A commonly used mathematical expression for hedging is

$$\mu_{\text{hedge}\widetilde{H}} = (\mu_{\widetilde{H}})^{\kappa}$$

with $\kappa > 1$ for compression and $\kappa < 1$ for dilation. For example, we could use the following hedge allocation for compression and dilation:

hedge		κ
vaguely	$\longrightarrow$	0.05
slightly	$\longrightarrow$	0.25
somewhat	$\longrightarrow$	0.5
very	$\longrightarrow$	2
extremely	$\longrightarrow$	3
exactly	$\longrightarrow$	∞

Another hedge commonly used is a contrast enhancer

$$\mu_{\text{enh}\widetilde{H}} = \begin{cases} 2(\mu_{\widetilde{H}}(T))^2 & \text{for} \quad \mu_{\widetilde{H}}(T) \in [0, \frac{1}{2}] \\ 1 - 2(1 - \mu_{\widetilde{H}}(T))^2 & \text{otherwise} \end{cases}$$

18.2.6 Quantifying Fuzzyness

From information theory we know that we can define the missing information of a probability distribution by the *Shannon entropy*

$$S(x) := -\sum_{x \in X} (p(x) \log_\alpha p(x))$$

where the base of the logarithm, α, determines the units of information. If $\alpha = 2$, then the unit of information is the *bit* (binary digit). This function has a minimum of 0 if the probability distribution is a delta-function. We would expect this since in this case we know with certainty that the solution is x_0 (with probability 1) and hence there is no missing information. Similarly, if the probability distribution is flat, i.e.,

$$p(x) = \frac{1}{\text{size of } X} \quad \text{for all} \quad x$$

then the missing information is a maximum. We can define the entropy of a fuzzy set

$$\widetilde{F} = \{ (x, \mu_{\widetilde{F}}(x)) \,|\, x \in X \}$$

by

$$\mathcal{E}_\alpha(\widetilde{F}) := S_\alpha(\widetilde{F}) + S_\alpha(\not\subset \widetilde{F})$$

with

$$S_\alpha(\widetilde{F}) \equiv -\sum_{x \in X} \mu_{\widetilde{F}}(x) \log_\alpha(\mu_{\widetilde{F}}(x))$$

where α is a positive constant. Thus the entropy is a suitable measure for fuzzyness.

Example. Consider the following two discrete fuzzy sets

$$\widetilde{A} = \{(1, 0.4), (2, 0.8), (3, 1.0), (4, 0.7), (5, 0.3)\}$$

and

$$\widetilde{B} = \{(1, 0.0), (2, 0.1), (3, 1.0), (4, 0.2), (5, 0.1)\}\,.$$

Then the $\mathcal{E}_2(\widetilde{A}) = 3.46$ bits, while $\mathcal{E}_2(\widetilde{B}) = 1.66$ bits. As we would expect, $\widetilde{A}$ is much more fuzzy than $\widetilde{B}$. ♣

18.2.7 C++ Implementation of Discrete Fuzzy Sets

In the following header file `fuzzy.h` we implement the functions described above as methods within the class `Fuzzy`. The function name `near` is replaced by `closeTo` since `near` is used in some C++ compilers as keyword.

```
// fuzzy.h

#include <fstream>
#include <cmath>        // for exp, log, pow
using namespace std;
```

```
class Fuzzy
{
 public:
  Fuzzy(); // default constructor
  Fuzzy(const int size,const double& xMin=0,const double& xMax=1);
  Fuzzy(const Fuzzy& arg);
  void setDomain(const double& xMin,const double& xMax);
  void fillGrades(const double& fillValue);
  void normalize();
  void small();
  void large();
  void rectangle(const double& left,const double& right);
  void triangle(const double& left,const double& center,
                const double& right);
  void trapezoid(const double& leftBot,const double& leftTop,
                 const double& rightTop,const double& rightBot);
  // Fuzzy greater than
  void greaterThan(const double& arg,double lambda=0);
  // Fuzzy less than
  void lessThan(const double& arg,double lambda = 0);
  // Fuzzy near
  void closeTo(const double& arg,double sigma = 0);

  int supportMaxGradeIndex() const;
  int supportMinGradeIndex() const;
  double supportMaxGrade() const;
  double supportMinGrade() const;
  double minGrade() const;
  double maxGrade() const;
  double centroid() const;
  double cardinality() const;
  double relativeCardinality() const;

  Fuzzy limit(const double& ceiling);
  Fuzzy alphaCut(const double& alpha) const;
  Fuzzy betaCut(const int beta) const;

  double entropy(const double& base = 2) const;

  double xMin() const { return _xMin; };
  double xMax() const { return _xMax; };
  double domainSize () const { return _domainSize; };
  double resolution () const { return _resolution; };

  int isSubSet(const Fuzzy& arg) const;

  // grade for support i
```

```
double& operator[] (const int i);  // Grade for support i
// grade at x via linear interpolation
double operator() (const double& x) const;

Fuzzy& operator = (const Fuzzy& arg);
Fuzzy operator ! () const;                   // Logical NOT
Fuzzy operator && (const Fuzzy& arg) const; // Logical AND
Fuzzy operator || (const Fuzzy& arg) const; // Logical OR
Fuzzy operator +  (const Fuzzy& arg) const; // Bounded +
Fuzzy operator -  (const Fuzzy& arg) const; // Bounded -
Fuzzy operator %  (const Fuzzy& arg) const; // Algebraic +
Fuzzy operator *  (const Fuzzy& arg) const; // Algebraic *
Fuzzy operator &  (const Fuzzy& arg) const; // Fuzzy AND
Fuzzy operator |  (const Fuzzy& arg) const; // Fuzzy OR
Fuzzy operator <  (const Fuzzy& arg) const; // Less Than
Fuzzy operator >  (const Fuzzy& arg) const; // Greater Than
Fuzzy operator == (const Fuzzy& arg) const; // Equal To
Fuzzy operator >= (const Fuzzy& arg) const; // Greater or Equal
Fuzzy operator <= (const Fuzzy& arg) const; // Less or Equal
Fuzzy operator != (const Fuzzy& arg) const; // Not Equal To

// Fuzzy AND and Fuzzy OR
double gamma() const { return _gamma; };
void setGamma(const double& newGamma) { _gamma = newGamma; };
// weighted mean of array of fuzzy sets
// if arg weights omitted, regular mean
static Fuzzy mean(const Fuzzy* const * const sets,const int nSets,
                  const double* const weights = NULL);

Fuzzy enhanceContrast() const;
Fuzzy hedge(const double& hedgeExp) const;
// hedge constants:
static const double extremely;
static const double very;
static const double substantially;
static const double somewhat;
static const double slightly;
static const double vaguely;

friend ostream& operator << (ostream&,const Fuzzy&);
friend istream& operator >> (istream&,Fuzzy&);

public:
 double* _grades;
 int _size;
 double _gamma;   // For FUZZY AND and FUZZY OR.
 double _xMin, _xMax, _resolution, _domainSize;
 void resize(const int newSize);
```

```
 private:
  // default value for gamma parameter
  static const double _defaultGamma;
  double linearInterpolate(const double& x0,const double& y0,
                           const double& x1,const double& y1,
                           const double& x) const;
};

// static class constants
const double Fuzzy::_defaultGamma = 0.5;
// hedge exponents:
const double Fuzzy::extremely = 4.0;
const double Fuzzy::very = 2.0;
const double Fuzzy::substantially = 1.5;
const double Fuzzy::somewhat = 0.5;
const double Fuzzy::slightly = 0.25;
const double Fuzzy::vaguely = 0.03;

// Implementation
Fuzzy::Fuzzy() : _gamma(_defaultGamma),_grades(NULL),_size(0)
{ setDomain(0,1); }

Fuzzy::Fuzzy(const int size,const double& xMin,const double& xMax)
  : _gamma(_defaultGamma),_grades(NULL),_size(0)
{ resize(size); setDomain(xMin,xMax); }

Fuzzy::Fuzzy(const Fuzzy& arg)
  : _gamma(_defaultGamma),_grades(NULL),_size(0)
{ *this=arg; }

void Fuzzy::setDomain(const double& xMin,const double& xMax)
{
 _xMin = xMin; _xMax = xMax;
 _domainSize = xMax-xMin;
 if(_size != 1) _resolution = _domainSize/(_size-1.0);
 else _resolution = 1.0;
}

void Fuzzy::fillGrades(const double& fillValue)
{ for(int i=0;i<_size;i++) _grades[i] = fillValue; }

void Fuzzy::normalize()
{
 double max = maxGrade();
 for(int i=0;i<_size;i++) _grades[i] /= max;
}
```

```
void Fuzzy::large()
{
 double range = _xMax-_xMin;
 double slope = (double)1/range;
 double intercept = -slope*_xMin;
 double dx = range/(_size-1.0);
 double x = _xMin;
 for(int i=0;i<_size;i++) { _grades[i] = slope*x+intercept; x += dx; }
}

void Fuzzy::small()
{
 large();
 for(int i=0;i<_size;i++) _grades[i] = 1.0-_grades[i];
}

void Fuzzy::rectangle(const double& left,const double& right)
{
 double range = _xMax-_xMin;
 double dx = range/(_size-1.0);
 double x = _xMin;
 for(int i=0;i<_size;i++)
 {
 if((x <= left) || (x >= right)) _grades[i] = 0.0;
 else _grades[i] = 1.0;
 x += dx;
 }
}

void Fuzzy::triangle(const double& left,const double& center,
                     const double& right)
{
 double range = _xMax-_xMin;
 double dx = range/(_size-1.0);
 double x = _xMin;
 for(int i=0;i<_size;i++)
 {
 if((x <= left) || (x >= right)) _grades[i] = 0.0;
 else if(x < center)
  _grades[i] = linearInterpolate(left,0,center,1,x);
  else _grades[i] = linearInterpolate(center,1,right,0,x);
 x += dx;
 }
}

void Fuzzy::trapezoid(const double& leftBot,const double& leftTop,
                      const double& rightTop,const double& rightBot)
{
```

```
 double range = _xMax-_xMin;
 double dx = range/(_size-1.0);
 double x = _xMin;
 for(int i=0;i<_size;i++)
 {
 if((x <= leftBot) || (x >= rightBot)) _grades[i] = 0.0;
  else if(x < leftTop)
   _grades[i] = linearInterpolate(leftBot,0,leftTop,1,x);
  else if(x <= rightTop) _grades[i] = 1.0;
  else _grades[i] = linearInterpolate(rightTop,1,rightBot,0,x);
 x += dx;
 }
}

int Fuzzy::supportMinGradeIndex() const
{
 double min = 1.0;
 int minIndex = 0;
 for(int i=0;i<_size;i++)
 if(min > _grades[i]) { min = _grades[i]; minIndex = i; }
 return minIndex;
}

int Fuzzy::supportMaxGradeIndex() const
{
 double max = 0.0;
 int maxIndex = 0;
 for(int i=0;i<_size;i++)
 if(max < _grades[i]) { max = _grades[i]; maxIndex = i; }
 return maxIndex;
}

double Fuzzy::supportMinGrade() const
{ return supportMinGradeIndex()*_resolution+_xMin; }

double Fuzzy::supportMaxGrade() const
{ return supportMaxGradeIndex()*_resolution+_xMin; }

double Fuzzy::minGrade() const
{ return _grades[supportMinGradeIndex()]; }

double Fuzzy::maxGrade() const
{ return _grades[supportMaxGradeIndex()]; }

double Fuzzy::cardinality() const
{
 double cd = 0.0;
 for(int i=0;i<_size;i++) cd += _grades[i];
```

```
 return cd;
}

double Fuzzy::relativeCardinality() const
{ return cardinality()/_domainSize; }

int Fuzzy::isSubSet(const Fuzzy& arg) const
{
 int issubset = 1, i = 0;
 while((issubset) && (i<_size))
 { issubset = (arg._grades[i] >= _grades[i]); i++; }
 return issubset;
}

Fuzzy Fuzzy::limit(const double& ceiling)
{
 Fuzzy result(*this);
 for(int i=0;i<_size;i++)
  if(_grades[i] > ceiling) _grades[i] = ceiling;
 return result;
}

Fuzzy& Fuzzy::operator = (const Fuzzy& arg)
{
 if(_size != arg._size) resize(arg._size);
 for(int i=0;i<_size;i++) _grades[i] = arg._grades[i];
 setDomain(arg._xMin,arg._xMax);
 return *this;
}

double& Fuzzy::operator [] (const int i) { return _grades[i]; }

double Fuzzy::operator () (const double& x) const
{
 int iLow  = (x-_xMin)/_resolution;
 int iHigh = iLow+1;
 double xLow  = _xMin+_resolution*iLow;
 double xHigh = _xMin+_resolution*iHigh;
 return linearInterpolate(xLow,_grades[iLow],xHigh,_grades[iHigh],x);
}

Fuzzy Fuzzy::operator ! () const
{
 Fuzzy result(_size,_xMin,_xMax);
 for(int i=0;i<_size;i++) result._grades[i] = 1.0-_grades[i];
 return result;
}
```

```
Fuzzy Fuzzy::operator && (const Fuzzy& arg) const
{
 Fuzzy result(_size,_xMin,_xMax);
  for(int i=0;i<_size;i++)
   if(_grades[i] < arg._grades[i]) result._grades[i] = _grades[i];
      else result._grades[i] = arg._grades[i];
 return result;
}

Fuzzy Fuzzy::operator || (const Fuzzy& arg) const
{
 Fuzzy result(_size,_xMin,_xMax);
  for(int i=0;i<_size;i++)
     if(_grades[i] > arg._grades[i]) result._grades[i] = _grades[i];
     else result._grades[i] = arg._grades[i];
 return result;
}

Fuzzy Fuzzy::operator % (const Fuzzy& arg) const
{
 Fuzzy result(_size,_xMin,_xMax);
  for(int i=0;i<_size;i++)
   result._grades[i] = _grades[i]+arg._grades[i]
        -_grades[i]*arg._grades[i];
 return result;
}

Fuzzy Fuzzy::operator + (const Fuzzy& arg) const
{
 Fuzzy result(_size,_xMin,_xMax);
  for(int i=0;i<_size;i++)
  {
  double rslt = _grades[i]+arg._grades[i];
  if(rslt < 1.0)  result._grades[i] = rslt;
  else result._grades[i] = 1.0;
  }
 return result;
}

Fuzzy Fuzzy::operator * (const Fuzzy& arg) const
{
 Fuzzy result(_size,_xMin,_xMax);
  for(int i=0;i<_size;i++)
    result._grades[i] = _grades[i]*arg._grades[i];
 return result;
}

Fuzzy Fuzzy::operator - (const Fuzzy& arg) const
```

```
{
 Fuzzy result(_size,_xMin,_xMax);
  for(int i=0;i<_size;i++)
  {
  double rslt = _grades[i] + arg._grades[i] - 1.0;
  if(rslt > 0.0)  result._grades[i] = rslt;
  else result._grades[i] = 0.0;
  }
 return result;
}

Fuzzy Fuzzy::operator & (const Fuzzy& arg) const
{
 Fuzzy result(_size,_xMin,_xMax);
  for(int i=0;i<_size;i++)
  {
  double g1 = _grades[i], g2 = arg._grades[i];
  double rslt;
  if(g1 < g2) rslt = _gamma*g1;
  else rslt = _gamma*g2;
  rslt += 0.5*(1.0-_gamma)*(g1+g2);
  result._grades[i] = rslt;
  }
 return result;
}

Fuzzy Fuzzy::operator | (const Fuzzy& arg) const
{
 Fuzzy result(_size,_xMin,_xMax);
  for(int i=0;i<_size;i++)
  {
  double g1 = _grades[i], g2 = arg._grades[i];
  double rslt;
  if(g1 > g2) rslt = _gamma*g1;
  else rslt = _gamma*g2;
  rslt += 0.5*(1-_gamma)*(g1+g2);
  result._grades[i] = rslt;
  }
 return result;
}

Fuzzy Fuzzy::operator > (const Fuzzy& arg) const
{
 Fuzzy result(_size,_xMin,_xMax);
  for(int i=0;i<_size;i++)
  {
  if(_grades[i] > arg._grades[i]) result[i] = 1.0;
  else result[i] = 0.0;
```

```
  }
 return result;
}

Fuzzy Fuzzy::operator < (const Fuzzy& arg) const
{
 Fuzzy result(_size,_xMin,_xMax);
  for(int i=0;i<_size;i++)
  {
  if(_grades[i] < arg._grades[i]) result[i] = 1.0;
  else result[i] = 0.0;
  }
 return result;
}

Fuzzy Fuzzy::operator == (const Fuzzy& arg) const
{
 Fuzzy result(_size,_xMin,_xMax);
  for(int i=0;i<_size;i++)
  {
  if(_grades[i] == arg._grades[i]) result[i] = 1.0;
  else  result[i] = 0.0;
  }
 return result;
}

Fuzzy Fuzzy::operator >= (const Fuzzy& arg) const
{
 Fuzzy result(_size,_xMin,_xMax);
  for(int i=0;i<_size;i++)
  {
  if(_grades[i] >= arg._grades[i]) result[i] = 1.0;
  else result[i] = 0.0;
  }
 return result;
}

Fuzzy Fuzzy::operator <= (const Fuzzy& arg) const
{
 Fuzzy result(_size,_xMin,_xMax);
  for(int i=0;i<_size;i++)
  {
  if(_grades[i] <= arg._grades[i]) result[i] = 1.0;
  else  result[i] = 0.0;
  }
 return result;
}
```

```
Fuzzy Fuzzy::operator != (const Fuzzy& arg) const
{
 Fuzzy result(_size,_xMin,_xMax);
  for(int i=0;i<_size;i++)
  {
  if(_grades[i] != arg._grades[i]) result[i] = 1.0;
  else  result[i] = 0.0;
  }
 return result;
}

void Fuzzy::greaterThan(const double& arg,double lambda)
{
 double range = _xMax-_xMin;
 if(lambda <= 0.0) lambda = 0.1*range;
 double dx = range/(_size-1.0); double x = _xMin;
 for(int i=0;i<_size;i++)
 { _grades[i] = 1.0/(1.0+exp(-(x-arg)/lambda)); x += dx; }
}

void Fuzzy::lessThan(const double& arg,double lambda)
{
 double range = _xMax - _xMin;
 if(lambda <= 0.0) lambda = 0.1*range;
 double dx = range/(_size-1.0); double x = _xMin;
 for(int i=0;i<_size;i++)
 { _grades[i] = 1.0/(1.0+exp(-(arg-x)/lambda)); x += dx; }
}

void Fuzzy::closeTo(const double& arg,double sigma)
{
 double range = _xMax-_xMin;
 if(sigma <= 0.0) sigma = 0.1*range;
 double expDenom = 2.0*sigma*sigma;
 double dx = range/(_size-1.0); double x = _xMin;
 for(int i=0;i<_size;i++)
 { _grades[i] = exp(-pow(x-arg,2)/expDenom); x += dx; }
}

Fuzzy Fuzzy::mean(const Fuzzy* const * const sets,
                  const int nSets,const double* const weights)
{
 const int size = sets[0]->_size;
 Fuzzy result(sets[0]->_size,sets[0]->_xMin,sets[0]->_xMax);
 for(int ng=0;ng<size;ng++)
 {
 double rslt = 0.0;
  for(int ns=0;ns<nSets;ns++)
```

```
    if(weights) rslt += sets[ns]->_grades[ng]*weights[ns];
     else rslt += sets[ns]->_grades[ng];
   if(!weights) rslt /= nSets;
   result._grades[ng] = rslt;
  }
 return result;
}

double Fuzzy::entropy(const double& base) const
{
 double result = 0.0;
 for(int ng=0;ng<_size;ng++)
 {
 double grade = _grades[ng];
 if(grade) result -= grade*log(grade);
 grade = 1.0-grade;
 if(grade) result -= grade*log(grade);
 }
 return result/log(base);
}

double Fuzzy::centroid() const  // uses trapezium integration
{
 double numer = (_xMin*_grades[0]+_xMax*_grades[_size-1])/2.0;
 double denom = (_grades[0]+_grades[_size-1])/2.0;
 double x = _xMin;
 for(int i=1;i<=_size-2;i++)
 { x += _resolution; numer += x*_grades[i]; denom += _grades[i]; }
 return numer/denom;  // step size cancels out
}

Fuzzy Fuzzy::hedge(const double& hedgeExp) const
{
 Fuzzy result(_size,_xMin,_xMax);
 for(int i=0;i<_size;i++) result._grades[i] = pow(_grades[i],hedgeExp);
 return result;
}

Fuzzy Fuzzy::enhanceContrast() const
{
 Fuzzy result(_size,_xMin,_xMax);
 for(int i=0;i<_size;i++)
 {
 double grade = _grades[i];
  if(grade < 0.5) result._grades[i] = 2.0*pow(grade,2);
  else result._grades[i] = 1.0-2*pow(1-grade,2);
 }
 return result;
```

```
}

Fuzzy Fuzzy::alphaCut(const double& alpha) const
{
 Fuzzy result(_size,_xMin,_xMax);
 for(int i=0;i<_size;i++)
  if(_grades[i] >= alpha) result._grades[i] = 1.0;
  else result._grades[i] = 0.0;
 return result;
}

Fuzzy Fuzzy::betaCut(const int beta) const
{
 // make a copy of the set for bubble sort
 double* grades = new double[_size];
 int* indexes = new int[_size];
 for(int i=0;i<_size;i++) { grades[i] = _grades[i]; indexes[i] = i; }
 int exchanged, nFound = 0;
 // bubble sort until the beta largest elements at end of array
 do
 {
 exchanged = 0;
 for(int n=0;n<_size-nFound-1;n++)
 {
 if(grades[n] > grades[n+1])
 {
 double ddum  = grades[n+1];
 grades[n+1]  = grades[n];  grades[n] = ddum;
 int idum     = indexes[n+1];
 indexes[n+1] = indexes[n]; indexes[n] = idum;
 exchanged++;
 }
 }
 nFound++;
 } while((exchanged) && (nFound<=beta));

 Fuzzy result(_size,_xMin,_xMax);
 result.fillGrades(0);
 for(int nb=1;nb<=beta;nb++) result._grades[indexes[_size-nb]] = 1;
 delete[] indexes; delete[] grades;
 return result;
}

ostream& operator << (ostream& os,const Fuzzy& arg)
{
 os << "[";
 double x = arg._xMin;
 for(int i=0;i<arg._size;i++)
```

```
 {
 os << "(" << x << "," << arg._grades[i] << ")";
 if(i!=arg._size-1) x += arg._resolution;
 }
 os << "]";
 return os;
}

istream& operator >> (istream& is,Fuzzy& arg)
{
 int size;
 is >> size;
 arg.resize(size);
 double xMin, xMax;
 is >> xMin >> xMax;
 arg.setDomain(xMin,xMax);
 for(int i=0;i<size;i++) is >> arg._grades[i];
 return is;
}

double Fuzzy::linearInterpolate(const double& x0,
               const double& y0,const double& x1,
               const double& y1,const double& x) const
{ double m = (y1-y0)/(x1-x0); double c = y1-m*x1; return m*x+c; }

void Fuzzy::resize(const int newSize)
{
  delete[] _grades;
  _grades = new double[newSize];
  _size = newSize;
}
```

We supply 3 constructors. The default constructor `Fuzzy()` which allows the user to create an array of fuzzy sets, a constructor allowing the user to create a fuzzy set of a certain size and covering a range from `xMax` to `xMin` on the support axis. If the user does not specify the support-range then it is assumed to be $[0, 1]$ as specified by the default values of the corresponding arguments. The support range can be altered at a later stage via the method `setDomain`. The last constructor is the copy constructor.

The method `fillGrades` sets all grades equal to the supplied fill-value. The method `normalize` normalizes the fuzzy set (i.e. the largest grade will be 1). The method `rectangle` fills all the grades of the fuzzy set whose support is between `left` and `right` with 1 and all others with zero. For example, the concept *crisp faster than 100* for a universe of cruising speeds between say 100 and 200 could be defined by the rectangle

```
Fuzzy fasterThan100(201,1,200);
```

```
fasterThan100.rectangle(100,200);
```

The methods `triangle()` and `trapezoid()` can be used to define a discrete fuzzy set with triangular and trapezoidal membership functions, respectively. For example, we could define the fuzzy value for optimal cruising speed by a triangle and the range of acceptable cruising speed by a trapezoid

```
Fuzzy optimalSpeed(201,0,200), acceptableSpeed(201,0,200);
optimalSpeed.triangle(100,110,120);
acceptableSpeed.trapezoid(80,100,120,130);
```

The methods `small` and `large`, `near`, `lessThan` and `greaterThan` are a direct implementation of the corresponding fuzzy concepts.

The methods `minGrade` and `maxGrade` return the minimum and maximum grade of the fuzzy set, the methods `minGradeSupport` and `maxGradeSupport` the corresponding support value and the methods

```
minGradeSupportIndex   maxGradeSupportIndex
```

the array index for these support values. The method `centroid` is usually used for defuzzification. It returns the centroid of the fuzzy set as a crisp value. The cardinality and relative cardinality are returned by the corresponding member functions.

The `if ... then ...` rules require us to limit the degree to which a consequence is *true* to the degree that the antecedent is *true*. For this purpose we supply the member function, `limit`, which causes the set to saturate at the ceiling level. For example

```
Fuzzy optimalSpeed(201,0,200);
optimalSpeed.triangle(100,110,120);
optimalSpeed.limit(0.5);
```

The final membership function of the set `optimalSpeed` is a trapezoidal with corner points (100,0), (105,0.5), (115,0.5), (120,0). The methods `alphaCut` and `betaCut` implement the α-cut and β-cut operators. The method `entropy` estimates the entropic fuzzyness measure. By default the base of the logarithm is taken as 2.

The methods `xMin()`, `xMax()` and `domainSize()` are simple query functions, the latter returning the size of the support domain. Since this is an implementation of a discrete fuzzy set we have a finite resolution on the support domain. The resolution can be queried via the method `resolution`. The method `isSubSet` returns true, i.e. 1, if the set for which the method is called is a subset of the set supplied as argument to the method.

We supply two operators for querying the grade for a given support value. The first is the array element access operator `[]` which takes an integer as argument. It simply returns the grade for the support value at the specified array index, hence at the discretized points. This operator can also be used to change the grade at that point. The second operator is the function call operator `()`. This operator can be used to query the grade at any support value between `xMin` and `xMax`, not only at the discretized grid points. The method uses linear interpolation to obtain the grade values between grid points. The assignment operator allows us to assign one fuzzy set to another. It returns the object itself by reference to allow concatenated standard C-style assignments

```
set1 = set2 = set3 = set4;
```

Most of the remaining operators are a direct implementation of the fuzzy-set operators. The relational operators

```
< > == >= <= !=
```

each return a crisp set (the grades are all either zero or one). For example, the greater-than operator, `>`, returns a fuzzy set with grade one for all support values for which the current set, the set for which the method was called, has greater support than the set supplied as argument to the method. The grades for all other support values are zero. The remaining relational operators act in a similar fashion.

For the fuzzy-AND and fuzzy-OR operators we have to specify a value for γ. The value of γ can be queried and set via the methods `gamma` and `setGamma`.

The input/output stream operators, `>>` and `<<` are used to read a fuzzy set from an input stream (e.g. a file or the keyboard) or to an output stream (e.g. a file or the screen).

The static member function `mean()` calculates the weighted mean of an array of fuzzy sets. If the array of weights is not supplied, the standard arithmetic mean is returned. The fact that the member function is static implies that it can be called without referring to an object of the class. It is effectively a global member function with the scope of the class. We can call it as follows

```
Fuzzy::mean(setsArray,weights);
```

Note that both arguments are as constant as can be, i.e. all pointers and what they point to are declared constant.

Finally we implement linguistic hedging and contrast enhancement. We allow hedging with any continuous hedge-value. In order to introduce linguistic hedges we define static class constants for the terms

```
extremely, very, substantially, somewhat, slightly, vaguely .
```

The use could define the linguistic concept, `large` and then use hedging to introduce the concept `verylarge`:

```
Fuzzy large(50);
large.large();
Fuzzy veryLarge = large.hedge(Fuzzy::very);
```

Application 1. The following C++ program `mfuzzy.cpp` illustrates the usage of the discrete fuzzy set class `Fuzzy` in the file `fuzzy.h`. We used the following input file, `fuzzy.dat`

```
5 1 5
0.4 0.8 1.0 0.7 0.7
5 1 5
0.0 0.1 1.0 0.2 0.1
5 1 5
0.1 0.2 0.5 0.2 0.4
```

```
// mfuzzy.cpp

#include <iostream>
#include <cstdlib>
#include "fuzzy.h"
using namespace std;

int main(void)
{
 ifstream fin("fuzzy.dat");
 if(fin == NULL) { cout << "file cannot be opened"; exit(0); }
 Fuzzy set1, set2;
 fin >> set1 >> set2;
 cout << " s1 = " << set1 << endl;
 cout << "!s1 = " << !set1 << endl;
 cout << " s2 = " << set2 << endl;
 cout << "s1 && s2 = " << (set1 && set2) << endl;
 cout << "s1 || s2 = " << (set1 || set2) << endl;
 cout << "s1 +  s2 = " << (set1 +  set2) << endl;
 cout << "s1 -  s2 = " << (set1 -  set2) << endl;
 cout << "s1 &  s2 = " << (set1 &  set2) << endl;
 cout << "s1 |  s2 = " << (set1 |  set2) << endl;

 Fuzzy set3;
 fin >> set3;  set3.normalize();
 cout << " s3 = " << set3 << endl;
 cout << "s3.centroid/cardinality/relativeCardinality/entropy = "
     << set3.centroid() << " " << set3.cardinality() << " "
     << set3.relativeCardinality() << " " << set3.entropy() << endl;

 int nSets2 = 2;
 Fuzzy** sets2 = new Fuzzy*[nSets2];
 double* weights = new double[2];
 weights[0] = 0.2; weights[1] = 0.8;
 sets2[0] = &set1; sets2[1] = &set2;
 cout << "mean(s1..s2,weights[0.2,0.8]) = "
     << Fuzzy::mean(sets2,nSets2,weights) << endl;
```

```
 int nSets3 = 3;
 Fuzzy** sets3 = new Fuzzy*[nSets3];
 sets3[0] = &set1; sets3[1] = &set2; sets3[2] = &set3;
 cout << "mean(s1..s3) = " << Fuzzy::mean(sets3,nSets3) << endl;
 cout << "s3.hedge(Fuzzy::very) = "<< set3.hedge(Fuzzy::very) << endl;
 cout << "s3.enhanceContrast() = " << set3.enhanceContrast() << endl;
 cout << "s3.alphaCut(0.65) = " << set3.alphaCut(0.65) << endl;
 cout << "s3.betaCut(3) = " << set3.betaCut(3) << endl;
 cout << "s3.entropy() <=> s3.enhanceContrast().entropy() = "
      << set3.entropy() << " <=> "
      << set3.enhanceContrast().entropy() << endl;
 Fuzzy small(5,0,2);  small.small();
 Fuzzy large(5,0,2);  large.large();
 cout << "small: " << small << endl;
 cout << "large: " << large << endl;
 Fuzzy legalSpeed(7,80,140);  legalSpeed.lessThan(120);
 cout << "legalSpeed = " << legalSpeed << endl;
 Fuzzy optimumSpeed(7,80,140);  optimumSpeed.closeTo(110);
 cout << "optimumSpeed = " << optimumSpeed << endl;
 fin.close();
 return 0;
}
```

The output of the program is given below

```
 s1 = [(1,0.4)(2,0.8)(3,1)(4,0.7)(5,0.7)]
!s1 = [(1,0.6)(2,0.2)(3,0)(4,0.3)(5,0.3)]
 s2 = [(1,0)(2,0.1)(3,1)(4,0.2)(5,0.1)]
s1 && s2 = [(1,0)(2,0.1)(3,1)(4,0.2)(5,0.1)]
s1 || s2 = [(1,0.4)(2,0.8)(3,1)(4,0.7)(5,0.7)]
s1 +  s2 = [(1,0.4)(2,0.9)(3,1)(4,0.9)(5,0.8)]
s1 -  s2 = [(1,0)(2,0)(3,1)(4,0)(5,0)]
s1 &  s2 = [(1,0.1)(2,0.275)(3,1)(4,0.325)(5,0.25)]
s1 |  s2 = [(1,0.3)(2,0.625)(3,1)(4,0.575)(5,0.55)]

s3 = [(1,0.2)(2,0.4)(3,1)(4,0.4)(5,0.8)]
s3.centroid/cardinality/relativeCardinality/entropy
= 3.26087 2.8 0.7 3.38576
mean(s1..s2,weights[0.2,0.8]) = [(1,0.08)(2,0.24)(3,1)(4,0.3)(5,0.22)]
mean(s1..s3) = [(1,0.2)(2,0.433333)(3,1)(4,0.433333)(5,0.533333)]
s3.hedge(Fuzzy::very) = [(1,0.04)(2,0.16)(3,1)(4,0.16)(5,0.64)]
s3.enhanceContrast() = [(1,0.08)(2,0.32)(3,1)(4,0.32)(5,0.92)]
s3.alphaCut(0.65) = [(1,0)(2,0)(3,1)(4,0)(5,1)]
s3.betaCut(3) = [(1,0)(2,0)(3,1)(4,1)(5,1)]
s3.entropy() <=> s3.enhanceContrast().entropy() = 3.38576 <=> 2.61312

small: [(0,1)(0.5,0.75)(1,0.5)(1.5,0.25)(2,0)]
large: [(0,0)(0.5,0.25)(1,0.5)(1.5,0.75)(2,1)]
```

```
legalSpeed
= [(80,0.998729)(90,0.993307)(100,0.965555)(110,0.841131)(120,0.5)
(130,0.158869)(140,0.0344452)]
optimumSpeed = [(80,3.72665e-06)(90,0.00386592)(100,0.249352)(110,1)
(120,0.249352)(130,0.00386592)(140,3.72665e-06)]
```

Application 2. When buying a pair of shoes there are certain crisp boundary conditions determined by the anatomy of one's feet and by the balance of one's bank account. We have a fuzzy concept of what we find good-looking and what we find comfortable. The problem is thus one of finding a solution which, subject to certain crisp boundary conditions, optimizes some fuzzy objectives. Take the simple example of having to choose between N pairs of shoes, some of which are too small and some of which are too expensive. For this we use two crisp sets, `tooSmall` and `tooExpensive`. The grades of these crisp sets are either 0 or 1. We implement the fuzzy concepts of `looks` and `comfort` by the corresponding two fuzzy sets whose membership value is now a grade between zero and 1, representing our fuzzy evaluation of the looks and comfort of each pair of shoes. Consider now the code given below:

```
// shoe.cpp

#include <iostream>
#include <fstream>
#include "fuzzy.h"
using namespace std;

int main(void)
{
 ifstream infile("shoe.dat");
 Fuzzy tooSmall, tooExpensive, looks, comfort;
 infile >> tooSmall >> tooExpensive >> looks >> comfort;
 ofstream fout("shoe.out");
 fout << "Select a comfortable good-looking shoe which"
      << " is big enough and affordable:" << endl;
 fout << "==============================================="
      << "================" << endl;
 fout << "Too-small constraint: " << tooSmall
      << " (crisp set) " <<  endl;
 fout << "Too expensive constraint: " << tooExpensive
      << " (crisp set) " <<  endl << endl;
 fout << "Looks preferences: " << looks << " (fuzzy set) "
      << endl;
 fout << "Comfort preferences: " << comfort << " (fuzzy set) "
      << endl << endl;
 Fuzzy satisfyConstraints = !(tooSmall || tooExpensive);
 fout << "Constraints satisfied by : " << satisfyConstraints
      << endl;
 int nPreferences = 2;
 Fuzzy** preferences = new Fuzzy*[nPreferences];
```

```
preferences[0] = &looks;
preferences[1] = &comfort;
Fuzzy objective = Fuzzy::mean(preferences,nPreferences);
fout << "Objective = " << objective << endl << endl;
Fuzzy shoeScores = (satisfyConstraints && objective);
fout << "Shoe scores = " << shoeScores << endl;
fout << "Best shoe = " << shoeScores.supportMaxGrade()
     << " which has a grade of " << shoeScores.maxGrade()
     << "." << endl << endl;
fout << "Select a shoe which is somewhat good-looking "
     << "and very comfortable"
     << endl << "(still not too small and not too expensive):"
     << endl;
fout << "=================================================="
     << endl;
Fuzzy somewhatGoodLooking = looks.hedge(Fuzzy::somewhat);
Fuzzy veryComfortable = comfort.hedge(Fuzzy::very);
fout << "somewhatGoodLooking = " << somewhatGoodLooking << endl;
fout << "veryComfortable = " << veryComfortable << endl << endl;
preferences[0] = &somewhatGoodLooking;
preferences[1] = &veryComfortable;
objective = Fuzzy::mean(preferences,nPreferences);
fout << "Objective = " << objective << endl << endl;
shoeScores = (satisfyConstraints && objective);
fout << "Shoe scores = " << shoeScores << endl;
fout << "Best shoe = " << shoeScores.supportMaxGrade()
     << " which has a grade of " << shoeScores.maxGrade() << "."
     << endl;
infile.close();
fout.close();
return 0;
}
```

The subset of shoes which satisfies the crisp constraints is determined via

```
Fuzzy satisfyConstraints = !(tooSmall || tooExpensive);
```

We have to somehow take a weighted average of the fuzzy preferences. This is done via

```
Fuzzy objective = FuzzySet::mean(preferences,nPreferences);
```

which weigths the two preferences equally. We determine now the confidence with which we select the various shoes by taking a logical AND (in the fuzzy sense) between the constraints and the preferences, i.e. the ideal shoe must satisfy the constraints AND the preferences:

```
FuzzySet shoeScores = (satisfyConstraints && objective);
```

This is the fuzzy output of the fuzzy inference engine. At the end of the day we have to make a crisp choice between the shoes. For this purpose we have to defuzzify the output. In this simple example (which is naturally discrete) we select that particular shoe which obtains the highest confidence level, i.e. has the maximum grade. This is done by the `supportMaxGrade()` defuzzifier, which returns the support for that member of the fuzzy set which received the highest confidence level (grade):

```
int choiceOfShoe = shoeScores.supportMaxGrade();
```

An example of an input file `shoe.dat` is

```
5 0 4
0 1 0 0 0
5 0 4
0 0 0 0 1
5 0 4
0.7 1.0 0.4 0.7 0.5
5 0 4
0.4 0.0 0.8 0.6 0.9
```

In this example we have 5 pairs of shoes to choose from. Hence each of the fuzzy sets has length 5 and has a domain ranging from 0 to 4. The first set is actually a crisp set specifying that the second shoe is the only shoe which is too small. The second set specifies the crisp *too-expensive* boundary condition (only the last shoe is too expensive). The following two sets represent the fuzzy evaluation of the looks and the comfort for the various pairs of shoes.

The output `shoe.out` of the first part of the program is given below:

```
Select a comfortable, good-looking shoe which is
big enough and affordable:
==========================
Too-small constraint: [(0,0)(1,1)(2,0)(3,0)(4,0)] (crisp set)
Too expensive constraint: [(0,0)(1,0)(2,0)(3,0)(4,1)] (crisp set)

Looks preferences: [(0,0.7)(1,1)(2,0.4)(3,0.7)(4,0.5)] (fuzzy set)
Comfort preferences: [(0,0.4)(1,0)(2,0.8)(3,0.6)(4,0.9)] (fuzzy set)

Constraints satisfied by: [(0,1)(1,0)(2,1)(3,1)(4,0)]
Objective = [(0,0.55)(1,0.5)(2,0.6)(3,0.65)(4,0.7)]

Shoe scores = [(0,0.55)(1,0)(2,0.6)(3,0.65)(4,0)]
Best shoe = 3 which has a grade of 0.65.

Select a shoe which is somewhat good-looking and very comfortable
(still not too small and not too expensive):
==================================================================
somewhatGoodLooking =
[(0,0.83666)(1,1)(2,0.632456)(3,0.83666)(4,0.707107)]
```

```
veryComfortable =
[(0,0.16)(1,0)(2,0.64)(3,0.36)(4,0.81)]

Objective = [(0,0.49833)(1,0.5)(2,0.636228)(3,0.59833)(4,0.758553)]

Shoe scores = [(0,0.49833)(1,0)(2,0.636228)(3,0.59833)(4,0)]
Best shoe = 2 which has a grade of 0.636228.
```

We select a comfortable, good-looking shoe which is neither too small, nor too expensive. The "best" shoe in this case is shoe number 3 (the fourth shoe). If we use hedging to select a shoe which is somewhat comfortable, but very good looking, then the "best" shoe is shoe number 2 (the third shoe).

Next we look at a simplified version of an optimal pricing problem. We have a number of possibly overlapping or even conflicting objectives. For example, the shareholders might push for a high price for the product, while the salespersons might want a low price so that they can sell a larger quantity. Finally there might be a third objective of having a profit margin of around 30% which ensures that the company does not come into bad light with the consumer watchdogs and hence has to face bad publicity. We might also have an absolute minimum below which the producer absolutely refuses to produce and an absolute maximum above which the salespersons absolutely refuse to try and market the product. Furthermore, the objectives are to make a profit (i.e. that the retail price must be above the cost of the product) and to underbid the opposition.

Application 3. The following C++ code asks the user for the minimum and maximum price, the manufacturing cost and the competitor's price. It also asks the user to rate the various objectives according to their relative importance. It then uses fuzzy logic to determine the optimum retail price for the product.

```
// price.cpp

#include <iostream>
#include <fstream>
#include "fuzzy.h"
using namespace std;

int main(void)
{
 int minPrice, maxPrice, manufacturingCost, competitorsPrice;
 cout << "Enter minimum price (to nearest Rand): ";
 cin >> minPrice;
 cout << "Enter maximum price (to nearest Rand): ";
 cin >> maxPrice;
 cout << "Enter manufacturing cost (to nearest Rand): ";
 cin >> manufacturingCost;
 cout << "Enter competitors price (to nearest Rand):  ";
```

```
cin >> competitorsPrice;
int nPointsOnGrid = maxPrice - minPrice + 1;
const int nObjectives = 5;
Fuzzy** objectives = new Fuzzy*[nObjectives];
// Shareholders objective of a very high price:
objectives[0] = new Fuzzy(nPointsOnGrid,minPrice,maxPrice);
objectives[0]->large();
// Salesperson's objective of a low price:
objectives[1] = new Fuzzy(nPointsOnGrid,minPrice,maxPrice);
objectives[1] -> small();
// Moralists objective of a 30% profit:
objectives[2] = new Fuzzy(nPointsOnGrid,minPrice,maxPrice);
objectives[2] -> closeTo(manufacturingCost*1.3);
// Objective of making a profit:
objectives[3] = new Fuzzy(nPointsOnGrid,minPrice,maxPrice);
objectives[3] -> greaterThan(manufacturingCost);
// Objective for competitiveness:
objectives[4] = new Fuzzy(nPointsOnGrid,minPrice,maxPrice);
objectives[4] -> lessThan(competitorsPrice);
int importance[nObjectives];
cout << "Enter the importance of each of the following criteria:"
     << endl;
cout << "(0->vaguely,1->slightly,2->somewhat,3->substantially,"
     << endl
     << " 4->very,5->extremely)" << endl;
cout << "=========================================================="
     << endl;
cout << "Shareholder's desire for high price:   ";
cin >> importance[0];
cout << "Salesperson's desire for a low price: ";
cin >> importance[1];
cout << "Moralist's objective of a 30% profit:  ";
cin >> importance[2];
cout << "Retail price > manufacturing price:    ";
cin >> importance[3];
cout << "Retail price < competitors price:      ";
cin >> importance[4];

for(int nOb=0;nOb<nObjectives;nOb++)
{
switch(importance[nOb])
{
case 0: *objectives[nOb]=objectives[nOb]->hedge(Fuzzy::vaguely);
         break;
case 1: *objectives[nOb]=objectives[nOb]->hedge(Fuzzy::slightly);
        break;
case 2: *objectives[nOb]=objectives[nOb]->hedge(Fuzzy::somewhat);
        break;
```

```
 case 3: *objectives[nOb]=objectives[nOb]->
          hedge(Fuzzy::substantially); break;
 case 4: *objectives[nOb]=objectives[nOb]->hedge(Fuzzy::very);
         break;
 case 5: *objectives[nOb]=objectives[nOb]->hedge(Fuzzy::extremely);
         break;
 }
 }
 Fuzzy fuzzyPrice = (*objectives[0]) && (*objectives[1])
                  && (*objectives[2]) && (*objectives[3])
                  && (*objectives[4]);
 cout << endl << "Recommended retail price: R"
      << fuzzyPrice.supportMaxGrade() << endl;
 return 0;
}
```

The user dialogue could be as follows

```
Enter minimum and maximum price (to nearest Rand): 2000 5000
Enter manufacturing cost (to nearest Rand):        2500
Enter competitors price (to nearest Rand):         4500
Enter the importance of each of the following criteria:
(0 -> vaguely, 1 -> slightly,
 2 -> somewhat, 3 -> substantially,
 4 -> very, 5 -> extremely)
==============================================================
Shareholder's desire for high price:    4
Salesperson's desire for a low price:  3
Moralist's objective of a 30% profit:  1
Retail price > manufacturing price:     5
Retail price < competitors price:       4

Recommended retail price: R3649
```

Note that each of the fuzzy sets has 3000 elements. Assume now that the need to achieve turn-over becomes high and thus the salesperson's desire for a low price is given more importance. The shareholders still refuse to budge from rating their importance of a high price as *very important*, but the turn-over becomes now *extremely important.*

```
Enter minimum price (to nearest Rand): 2000
Enter maximum price (to nearest Rand): 5000
Enter manufacturing cost (to nearest Rand): 2500
Enter competitors price (to nearest Rand):  4500
Enter the relative importance of each of the following criteria:
(0 -> vaguely, 1 -> slightly, 2 -> somewhat, 3 -> substantially,
 4 -> very, 5 -> extremely)
==============================================================
Shareholder's desire for high price:    4
```

```
Salesperson's desire for a low price:  5
Moralist's objective of a 30% profit:  1
Retail price > manufacturing price:    5
Retail price < competitors price:      5

Recommended retail price: R3146
```

Our fuzzy inference engine suggests now a significantly lower price.

18.3 Fuzzy Numbers and Fuzzy Arithmetic

18.3.1 Introduction

We have discussed only fuzzy sets and the algebraic operations that can be performed on fuzzy sets. In real-life one usually works with numbers. Often these numbers are fuzzy, for example influenced by inaccurate measurements, noisy data, etc.. In this section we define fuzzy numbers (Dubois and Prade [50], Zimmermann [229]). and the algebraic operations between fuzzy numbers. Simple examples of fuzzy numbers are *large*, *approximately 30*, *much larger than 10*, *a few*. It would of course be convenient if we did not have to redevelop the entire algebra for fuzzy numbers. Fortunately there is the extension principle which is a general method for extending algebraic operations on standard crisp numbers to fuzzy numbers.

Definition. A *fuzzy interval*, $\widetilde{I}$, is a convex, normalized fuzzy set of the real numbers $\mathbb{R}$ whose membership function $\mu_{\widetilde{I}}(x)$ is piecewise continuous on $\mathbb{R}$.

Definition. A *fuzzy number*, $\widetilde{N}$, is a fuzzy interval with membership function, $\mu_{\widetilde{N}}(x)$, such that there exists exactly one $x_0 \in \mathbb{R}$ for which $\mu_{\widetilde{N}}(x_0) = 1$.

Definition. A positive (negative) fuzzy number, $\widetilde{N}$, has a membership function such that

$$\mu_{\widetilde{N}}(x) = 0 \quad \text{for all } x < 0 \quad (\text{for all } x > 0)\,.$$

Example. Consider

$$\widetilde{N} = \{\,(3,0.2),\,(4,0.6),\,(5,1.0),\,(6,0.6),\,(7,0.2)\,\}\,.$$

This fuzzy number could be called "approximately 5".

Example. The fuzzy set

$$\widetilde{M} = \{\,(1,0.2),\,(2,0.8),\,(3,1.0),\,(4,1.0),\,(5,0.8),\,(6,0.2)\,\}$$

is not a fuzzy number since $\mu(3) = 1.0$ and $\mu(4) = 1.0$.

Example. The fuzzy set

$$\widetilde{P} = \{\,(3,0.3),\,(4,0.7),\,(5,1.0),\,(6,0.2),\,(7,0.5),\,(8,0.1)\,\}$$

is not a fuzzy number due to the terms $(6, 0.2)$, $(7, 0.5)$, where $0.2 < 0.5$. ♣

The definition of a fuzzy number is often modified. In many cases trapezoidal membership functions are used.

18.3.2 Algebraic Operations

First we define the *extension principle* (Dubois and Prade [50], Zimmermann [229]).

Definition. Consider N fuzzy sets $\widetilde{F}_1, \ldots, \widetilde{F}_N$ in $X_1, \ldots, X_N$. Suppose g is a mapping from the product space $X \equiv X_1 \times \cdots \times X_N$ onto a universe Y,

$$y = g(x_1, x_2, \ldots, x_N).$$

Then the *extension principle* allows us to define a fuzzy set $\widetilde{G}$ in Y by

$$\widetilde{G} := \{ (y, \mu_{\widetilde{G}}(y)) \mid y = g(\mathbf{x}), \mathbf{x} = (x_1, \ldots, x_N) \in X \}$$

with

$$\mu_{\widetilde{G}}(y) = \begin{cases} \sup_{\mathbf{x} \in g^{-1}(y)} \min\{\mu_{\widetilde{F}_1}(x_1), \ldots, \mu_{\widetilde{F}_N}(x_N)\} & \text{if } g^{-1}(y) \neq \emptyset \\ 0 & \text{otherwise} \end{cases}$$

For $N = 1$, the extension principle reduces to

$$\widetilde{H} = g(\widetilde{F}) = \{ (y, \mu_{\widetilde{H}}(y)) \mid y = g(x),\ x \in X \}$$

where

$$\mu_{\widetilde{H}}(y) = \begin{cases} \sup_{x \in g^{-1}(y)} \mu_{\widetilde{F}}(x) & \text{if } g^{-1}(y) \neq \emptyset \\ 0 & \text{otherwise} \end{cases}$$

Example. Consider the fuzzy set

$$\widetilde{F} = \{ (-1, 0.5),\ (0, 0.8),\ (1, 1.0),\ (2, 0.4) \}$$

and $g(x) = x^2$. Then we find by applying the extension principle

$$\widetilde{B} = f(\widetilde{A}) = \{(0, 0.8),\ (1, 1.0),\ (4, 0.4) \}.$$

The term $(1, 1.0)$ we find from $-1 \cdot -1 = 1$ and $\sup\{ 0.5, 1.0 \} = 1.0$. ♣

Example. Consider the fuzzy sets

$$\widetilde{A} = \{ (1, 0.2), (2, 1.0), (3, 0.5), (4, 0.3) \}$$

and

$$\widetilde{B} = \{ (3, 0.9), (4, 1.0), (5, 0.4) \}.$$

Let

$$y := g(x_1, x_2) = x_1 + x_2.$$

Then the fuzzy addition of the two fuzzy sets, $\widetilde{A}$ and $\widetilde{B}$, yields a new fuzzy set, $\widetilde{C} = \widetilde{A}\widetilde{+}\widetilde{B}$ with membership function

$$\begin{aligned}
\mu_{\widetilde{C}}(y=4) &= \mu_{\widetilde{C}}(1+3) = 0.2\\
\mu_{\widetilde{C}}(y=5) &= \sup\{\mu_{\widetilde{C}}(1+4), \mu_{\widetilde{C}}(2+3)\} = \sup\{0.2, 0.9\} = 0.9\\
\mu_{\widetilde{C}}(y=6) &= \sup\{\mu_{\widetilde{C}}(1+5), \mu_{\widetilde{C}}(2+4), \mu_{\widetilde{C}}(3+3)\} = \sup\{0.2, 1.0, 0.5\} = 1.0\\
\mu_{\widetilde{C}}(y=7) &= \sup\{\mu_{\widetilde{C}}(2+5), \mu_{\widetilde{C}}(3+4), \mu_{\widetilde{C}}(4+3)\} = \sup\{0.4, 0.5, 0.3\} = 0.5\\
\mu_{\widetilde{C}}(y=8) &= \sup\{\mu_{\widetilde{C}}(3+5), \mu_{\widetilde{C}}(4+4)\} = \sup\{0.4, 0.3\} = 0.4\\
\mu_{\widetilde{C}}(y=9) &= \mu_{\widetilde{C}}(4+5) = 0.3
\end{aligned}$$

Thus we obtain the fuzzy set

$$\widetilde{C} = \widetilde{A}\widetilde{+}\widetilde{B} = \{(4,0.2), (5,0.9), (6,1.0), (7,0.5), (8,0.4), (9,0.3)\}.$$

Inspecting the membership function of $\widetilde{C}$ we note that the addition of fuzzy 2 and fuzzy 4 does indeed result in fuzzy 6. The fuzzyness of the sum is equal to the sum of the fuzzyness of the terms in the sum. Calculating the fuzzyness (entropy) S_2 of $\widetilde{A}$, $\widetilde{B}$ we obtain $S_2(\widetilde{A}) = 2.60322$ and $S_2(\widetilde{B}) = 1.43995$ bits respectively. The sum of these two is exactly equal to fuzzyness of the sum, $S_2(\widetilde{C}) = 4.04317$ bits. ♣

The extended operations on the basis of the extension principle cannot be applied directly to the fuzzy numbers with discrete support. We show this for the multiplication of fuzzy sets (Zimmermann [229]).

Example. Let

$$\widetilde{X} = \{\,(1,0.3), (2,1.0), (3,0.4)\,\}, \qquad \widetilde{Y} = \{\,(2,0.7), (3,\ 1.0), (4,0.2)\,\}.$$

Then we find

$$\begin{aligned}
\mu_{\widetilde{U}}(u=2) &= \mu_{\widetilde{U}}(1\cdot 2) = 0.3\\
\mu_{\widetilde{U}}(u=3) &= \mu_{\widetilde{C}}(1\cdot 3) = 0.3\\
\mu_{\widetilde{U}}(u=4) &= \sup\{\mu_{\widetilde{U}}(1\cdot 4), \mu_{\widetilde{C}}(2\cdot 2)\} = \sup\{0.2, 0.7\} = 0.7\\
\mu_{\widetilde{U}}(u=6) &= \sup\{\mu_{\widetilde{U}}(2\cdot 3), \mu_{\widetilde{C}}(3\cdot 2)\} = \sup\{1.0, 0.4\} = 1.0\\
\mu_{\widetilde{U}}(u=8) &= \sup\{\mu_{\widetilde{C}}(2\cdot 4)\} = \sup\{0.2\} = 0.2\\
\mu_{\widetilde{U}}(u=9) &= \mu_{\widetilde{C}}(3\cdot 3) = 0.4\\
\mu_{\widetilde{U}}(u=12) &= \sup\{\mu_{\widetilde{U}}(3\cdot 4)\} = \sup\{0.2\} = 0.2\,.
\end{aligned}$$

It follows that

$$\widetilde{U} = \widetilde{X}\,\widetilde{\cdot}\,\widetilde{Y} = \{\,(2,0.3), (3,0.3), (4,0.7), (6,1.0), (8,0.2), (9,0.4), (12,0.2)\,\}.$$

The result is obviously not a fuzzy number due to the term $(9,0.4)$.

Given a set of one-dimensional arrays of real numbers. All these one-dimensional arrays have the same length. The following C++ program `maxmin.cpp` finds first the minimum values for each of the one-dimensional arrays and then the maximum of these values. We store the set of the one-dimensional arrays as a two-dimensional array `x`, where the first index runs through the number of arrays and the second index runs through the length of each array.

```
// maxmin1.cpp

#include <iostream>
using namespace std;

int main(void)
{
 // m number of vectors, n length of each vector
 int m, n;
 m = 4; n = 3;
 int i, j;  // indices for for-loops
 double** x = NULL; x = new double*[m];
 for(i=0;i<m;i++) x[i] = new double[n];
 x[0][0] = 3.1; x[0][1] = 4.5; x[0][2] = 1.7;
 x[1][0] = 1.1; x[1][1] = 6.7; x[1][2] = 5.1;
 x[2][0] = 0.9; x[2][1] = 0.5; x[2][2] = 4.9;
 x[3][0] = 1.1; x[3][1] = 0.7; x[3][2] = 6.1;
 // store minimum value of each vector
 double* y = new double[m];
 double element;
 for(i=0;i<m;i++) { element = x[i][0];
  for(j=0;j<n;j++)
  { if(x[i][j] < element) element = x[i][j]; y[i] = element; }
 }
 double maxmin = y[0];
 for(i=1;i<m;i++) { if(y[i] > maxmin) maxmin = y[i]; }
 for(i=0;i<m;i++) { cout << "y[" << i << "] = " << y[i] << endl; }
 cout << "maxmin = " << maxmin << endl;
 for(i=0;i<m;i++) delete x[i];
 delete[] x; delete[] y;
 return 0;
}
```

The output is

```
y[0] = 1.7
y[1] = 1.1
y[2] = 0.5
y[3] = 0.7
maxmin = 1.7
```

18.3.3 LR-Representations

We can define a fuzzy set by its membership function only. However, if any fuzzy set can be represented by some arbitrary membership function, then applying algebraic operations (e.g. addition) repetitively to such fuzzy sets results in fuzzy sets with complex membership functions. For this reason the form of the membership function is often restricted to a relatively small class of membership functions. For example, fuzzy numbers are often represented by triangular fuzzy sets. Triangular fuzzy sets are a special case of the more general LR-fuzzy numbers introduced for reasons of efficiency and generality (Dubois and Prade [50], Zimmermann [229]).

Definition. A fuzzy number, $\widetilde{N}$, is of *LR-type* (L for left and R for right) if the membership function can be written in the form

$$\mu_{\widetilde{N}}(x) := \begin{cases} L\left(\frac{n-x}{\alpha}\right) & \text{for all } x < n \\ R\left(\frac{x-n}{\beta}\right) & \text{for all } x \geq n \end{cases}$$

Here L and R are the left and right spread functions, n is the mean value of $\widetilde{N}$ and α and β are the left and right spreads of the LR-membership function.

Usually the left and right spread functions, $L(z)$ and $R(z)$, are chosen to be decreasing functions which map the positive real numbers onto the interval $[0, 1]$ and for which $L(0) = R(0) = 1$.

The LR-representation of fuzzy numbers can be expanded in such a way that it can be used for fuzzy intervals.

Definition. A fuzzy interval, $\widetilde{I}$, is of *LR-type* if the membership function can be written in the form

$$\mu_{\widetilde{I}}(x) := \begin{cases} L\left(\frac{m-x}{\alpha}\right) & \text{for all } x \leq m \\ 1 & \text{for all } m \leq x \leq n \\ R\left(\frac{x-n}{\beta}\right) & \text{for all } x \geq n \end{cases}$$

As before, L and R are the left and right spread functions, and α and β are the left and right spreads. The values at which the left and right fall-off starts are m and n, respecitively.

Often fuzzy numbers are represented by triangular fuzzy sets. For a triangular fuzzy set one chooses a straight line with negative (positive) slope for the left (right) spread functions. For example

$$L(z) = R(z) = f(z) \equiv \begin{cases} 1 - z & \text{for all } z \in [0, 1] \\ 0 & \text{otherwise} \end{cases}$$

In order to represent the fuzzy number, $\widetilde{4}$, we would choose the mean to be $\mu = 4$. The left and right spreads would be determined by the application. For example, we could have $\alpha = 2$ and $\beta = 0.5$

$$\mu_{\tilde{4}}(x) = \begin{cases} 0 & \text{for all } x < 2 \\ 1 - \frac{4-x}{2} & \text{for all } x \in [2,4] \\ 1 - \frac{x-4}{0.5} & \text{for all } x \in [4, 4.5] \\ 0 & \text{for all } x > 4.5 \end{cases}$$

Naturally the left and right spread functions need not have the same functional form. For example we could have an exponential fall-off to the left and a Gaussian fall-off to the right as is illustrated in the following example.

Example. Let

$$L(z) = e^{-z}, \qquad R(z) = e^{-z^2}$$

and assume $\mu = 5$, $\alpha = 1$, $\beta = 3$.

The LR-representation for fuzzy intervals contains trapezoidal fuzzy intervals as a special case. This is illustrated in the following example.

Example. An acceptable cruising speed might be represented by a trapezoidal fuzzy set with grade 1 for speeds between 100km/h and 120km/h with a very rapid fall-off at speeds exceeding 120km/h (due to possible safety implications) and a more moderate fall-off as the cruising speed falls below 100km/h. Again we choose the same straight line for the left and right spread functions

$$L(z) = R(z) = f(z) \equiv \begin{cases} 1 - z & \text{for all } \quad z \in [0,1] \\ 0 & \text{otherwise} \end{cases}$$

Thus we choose $\mu = 100$, $\nu = 120$, $\alpha = 30$ and $\beta = 10$.

Commonly-Used Spreading Functions

When choosing suitable spreading functions, $f(z)$, for $L(z)$ and $R(z)$ one generally looks for decreasing functions which map the positive real numbers onto the interval $[0,1]$ and for which $f(0) = 1$. The following functions are the most commonly used spreading functions:

- **Straight Lines:**

$$f(z) = \begin{cases} 1 - z & \text{for all } z \in [0,1] \\ 0 & \text{otherwise} \end{cases}$$

- **Exponentials:** $f(z) = e^{-z}$
- **Gaussians:** $f(z) = e^{-z^2}$
- **Range Limited Polynomials of the form**

$$f(z) = \max\{0, (1-z)^p\}$$

- **Range Limited Polynomials of the form**

$$f(z) = \max\{0, (1 - z^p)\}$$

The following example considers once again the fuzzy set of acceptable cruising speeds, but it uses range-limited polynomials instead of straight lines for the spreading function.

Example. The fuzzy interval of acceptable cruising speed with left spreading function

$$L(z) = \max\{\, 0, (1 - z^{2.5})\,\}$$

and right spreading function

$$R(z) = \max\{\, 0, (1 - z)^{5.5}\,\}\,.$$

When using only a single spreading function within a fuzzy system, it is customary to use the notation

$$\widetilde{N} = (\mu, \alpha, \beta)_{LR}$$

for fuzzy numbers and

$$\widetilde{I} = (\mu, \nu, \alpha, \beta)$$

for fuzzy intervals.

18.3.4 Algebraic Operations on Fuzzy Numbers

The extension principle can be used to define the algebraic operations such as addition, subtraction, multiplication and division for fuzzy numbers. The definitions of the addition and subtraction operators follows exactly from the extension principle.

Definition. If $\widetilde{N}_1 = (\mu_1, \alpha_1, \beta_1)_{LR}$ and $\widetilde{N}_2 = (\mu_2, \alpha_2, \beta_2)_{LR}$ are two fuzzy numbers, then the *algebraic sum* of these two numbers is defined by

$$(\mu_1, \alpha_1, \beta_1)_{LR} \widetilde{+} (\mu_2, \alpha_2, \beta_2)_{LR} := (\mu_1 + \mu_2, \alpha_1 + \alpha_2, \beta_1 + \beta_2)_{LR}\,.$$

Definition. If $\widetilde{N} = (\mu, \alpha, \beta)_{LR}$ is a fuzzy number, then changing its sign (*unary minus*) is defined via

$$\widetilde{-}(\mu, \alpha, \beta)_{LR} := (-\mu, \beta, \alpha)_{LR}\,.$$

From the above two definitions we can define the algebraic difference between two fuzzy numbers/intervals.

Definition. If $\widetilde{N}_1 = (\mu_1, \alpha_1, \beta_1)_{LR}$ and $\widetilde{N}_2 = (\mu_2, \alpha_2, \beta_2)_{LR}$ are two fuzzy numbers, then the *algebraic difference* of these two numbers is defined by

$$(\mu_1, \alpha_1, \beta_1)_{LR} \widetilde{-} (\mu_2, \alpha_2, \beta_2)_{LR} := (\mu_1 - \mu_2, \alpha_1 + \beta_2, \beta_1 + \alpha_2)_{LR}\,.$$

Fuzzy algebra can be extended to include multiplication of fuzzy numbers. No closed form definition can be obtained from the extension principle for multiplication and division (the result cannot be represented with the same LR-functions as the operands). The expressions given below only approximate the exact result with an accuracy which improves with decreasing spreads.

Definition. If $\widetilde{N}_1 = (\mu_1, \alpha_1, \beta_1)_{LR}$ and $\widetilde{N}_2 = (\mu_2, \alpha_2, \beta_2)_{LR}$ are two fuzzy numbers, then the *algebraic product* is defined by

$$(\mu_1, \alpha_1, \beta_1)_{LR} \widetilde{\times} (\mu_2, \alpha_2, \beta_2)_{LR} := (\mu_1\mu_2, |\mu_1|\alpha_2 + |\mu_2|\alpha_1, |\mu_1|\beta_2 + |\mu_2|\beta_1)_{LR} .$$

The operation is only defined if both, $\widetilde{N}_1$ and $\widetilde{N}_2$ are either positive or negative. We can thus multiply positive numbers, negative numbers and a positive and negative fuzzy number but not a fuzzy number whose membership includes both positive and negative numbers.

It is not possible to define an inverse such that a fuzzy number times its inverse results in a crisp 1. Below we give a definition which reduces to the crisp analogue if the spreads, α and β, are zero.

Definition. If $\widetilde{N} = (\mu, \alpha, \beta)_{LR}$ is either a positive or a negative fuzzy number then we define (*multiplicative inverse*)

$$\frac{1}{\widetilde{N}} = \frac{1}{(\mu, \alpha, \beta)_{LR}} := \left(\frac{1}{\mu}, \frac{\beta}{\mu(\mu+\beta)}, \frac{\alpha}{\mu(\mu+\alpha)} \right) .$$

If we divide one fuzzy number, $\widetilde{N}_1$, by another, $\widetilde{N}_2$, we multiply $\widetilde{N}_1$ with the inverse of $\widetilde{N}_2$

$$\widetilde{N}_1 \widetilde{/} \widetilde{N}_2 = \widetilde{N}_1 \widetilde{\times} \frac{1}{\widetilde{N}_2} .$$

Example. Assume we have

$$\widetilde{N}_1 = (2, 0.5, 0.5)_{LR}, \qquad \widetilde{N}_2 = (3, 0.5, 0.5)_{LR} .$$

Then

$$\begin{aligned}
&(2, 0.5, 0.5)_{LR} \widetilde{+} (3, 0.5, 0.5)_{LR} = (5, 1, 1)_{LR} \\
&(2, 0.5, 0.5)_{LR} \widetilde{-} (3, 0.5, 0.5)_{LR} = (-1, 1, 1)_{LR} \\
&(2, 0.5, 0.5)_{LR} \widetilde{\times} (3, 0.5, 0.5)_{LR} = (6, 2.5, 2.5)_{LR} \\
&(2, 0.5, 0.5)_{LR} \widetilde{/} (3, 0.5, 0.5)_{LR} = (2/3, 0.2619, 0.2619)_{LR} .
\end{aligned}$$

To perform algebra with a mixture of crisp and fuzzy numbers is straightforward. We simply take the crisp numbers as fuzzy numbers with zero spreads and use the fuzzy algebraic operators to obtain a resultant fuzzy number.

18.3.5 C++ Implementation of Fuzzy Numbers

The C++ implementation is a straightforward implementation of an abstract data type. Below we give the implementation for triangular fuzzy numbers. To change the LR-spread functions to something other than straight lines one has to only change a single line in the private member function `LRfunc`.

The private data members are simply the mean, μ, the left spread, α and the right spread, β. The private member function `LRfunc()` defines the spread function used. In the current implementation these are simply straight lines and fuzzy numbers are thus triangular fuzzy numbers. To change the type of fuzzy number, for example to one using Gaussian or exponential spread, one only has to change the private member function `LRfunc`.

Besides the copy constructor we have a constructor taking the mean, left- and right-spread as arguments. Each of the arguments has the default value zero. Hence, if no arguments are supplied, a crisp number (zero spread) with value zero is created. If we omit the spreads but supply the mean, a crisp number with the value of the mean is created. The following six methods are query/set functions for the mean, the left-spread and the right-spread. The LR representation for fuzzy numbers is of course a continuous representation. We use the function call operator `()` to retrieve the grade of a fuzzy number for a certain support level.

Next we define the arithmetic operators, most of which are class constants (i.e. they do not modify the object for which they are called). To see this it is helpful to write the operator call as a function call.

```
x + y  ->  x.operator + (y)
```

Hence `x` is the object for which the operator method is called and since `x` remains unchanged the addition operator is a class constant. In order to allow for algebra with both, crisp and fuzzy operands we define class operators which take a crisp value as argument enabling the user to make constructs like

```
fuz1 = fuz2 + 4;
```

In order to support the reverse syntax

```
fuz1 = 4 + fuz2;
```

we have to define a global operator. The class operator is not called since the above is interpreted as

```
4.operator+(fuz2);
```

The operators take a crisp number as first argument and a fuzzy number as second argument. The global operators are global functions (not encapsulated within any class) which are friends of the class, that is, they have access to the private members of the class. Finally we supply stream access also via global friend operators `>>` and

<<.

The implementation is straightforward and requires little explanation. If we want to change the LR spread function (to something else than linear resulting in fuzzy numbers other than triangular fuzzy numbers) one only has to change a single line in the function LRfunc (the return statement):

```
double FuzzyNumber::LRfunc(const double& x) const
{
 if((x >= 0.0) && (x <= 1.0)) return 1.0-x;
 else return 0.0;
}
```

The header file FuzzyN.h is given by

```
// fuzzyN.h

#include <iostream>
using namespace std;

class FuzzyNumber
{
 public:
  FuzzyNumber(const double& mu=0.0,const double& alpha=0.0,
              const double& beta=0.0);
  FuzzyNumber(const FuzzyNumber& num);
  double mean() const;  // also conversion from fuzzy to crisp
  double leftSpread() const;
  double rightSpread() const;
  void setMean(const double& mu);
  void setLeftSpread(const double& alpha);
  void setRightSpread(const double& beta);
  double operator() (const double& x) const;
  FuzzyNumber& operator = (const FuzzyNumber& num);
  FuzzyNumber  operator + (const FuzzyNumber& num) const;
  FuzzyNumber  operator - (const FuzzyNumber& num) const;
  FuzzyNumber  operator * (const FuzzyNumber& num) const;
  FuzzyNumber  operator / (const FuzzyNumber& num) const;
  FuzzyNumber  operator - () const;   // Unary minus.
  FuzzyNumber& operator = (const double& num);
  FuzzyNumber  operator + (const double& num) const;
  FuzzyNumber  operator - (const double& num) const;
  FuzzyNumber  operator * (const double& num) const;
  FuzzyNumber  operator / (const double& num) const;
  FuzzyNumber invert() const;
  friend FuzzyNumber operator + (const double& crisp,
                                 const FuzzyNumber& fuzzy);
  friend FuzzyNumber operator - (const double& crisp,
                                 const FuzzyNumber& fuzzy);
```

```
  friend FuzzyNumber operator * (const double& crisp,
                                 const FuzzyNumber& fuzzy);
  friend FuzzyNumber operator / (const double& crisp,
                                 const FuzzyNumber& fuzzy);

  friend istream& operator >> (istream& is,FuzzyNumber& num);
  friend ostream& operator << (ostream& os,const FuzzyNumber& num);

 private:
  double _mu, _alpha, _beta;
  double LRfunc(const double& x) const;
};

// implementation
FuzzyNumber::FuzzyNumber(const double& mu,const double& alpha,
                         const double& beta)
    : _mu(mu), _alpha(alpha), _beta(beta) {};

FuzzyNumber::FuzzyNumber(const FuzzyNumber& num)
{ *this = num; }

FuzzyNumber& FuzzyNumber::operator= (const FuzzyNumber& num)
{
 _mu = num._mu; _alpha = num._alpha; _beta = num._beta;
 return *this;
}

double FuzzyNumber::mean() const { return _mu; };
double FuzzyNumber::leftSpread() const { return _alpha; };
double FuzzyNumber::rightSpread() const { return _beta; };

void FuzzyNumber::setMean(const double& mu) { _mu=mu; }
void FuzzyNumber::setLeftSpread(const double& alpha) { _alpha=alpha; }
void FuzzyNumber::setRightSpread(const double& beta) { _beta=beta; }

FuzzyNumber FuzzyNumber::operator + (const FuzzyNumber& num) const
{
 double mu = _mu+num._mu;
 double alpha = _alpha+num._alpha;
 double beta = _beta+num._beta;
 return FuzzyNumber(mu,alpha,beta);
}

FuzzyNumber FuzzyNumber::operator - (const FuzzyNumber& num) const
{
 double mu = _mu-num._mu;
 double alpha = _alpha+num._alpha;
 double beta = _beta+num._beta;
```

```
 return FuzzyNumber(mu,alpha,beta);
}

FuzzyNumber FuzzyNumber::operator * (const FuzzyNumber& num) const
{
 double mu = _mu*num._mu;
 double alpha = _mu*num._alpha+num._mu*_alpha;
 double beta = _mu*num._beta+num._mu*_beta;
 return FuzzyNumber(mu,alpha,beta);
}

FuzzyNumber FuzzyNumber::operator / (const FuzzyNumber& num) const
{ return *this*num.invert(); }

FuzzyNumber FuzzyNumber::operator- () const // unary minus.
{ return FuzzyNumber(-_mu,_alpha,_beta); }

FuzzyNumber FuzzyNumber::operator+ (const double& num) const
{ double mu = _mu+num; return FuzzyNumber(mu,_alpha,_beta); }

FuzzyNumber FuzzyNumber::operator- (const double& num) const
{ double mu = _mu-num; return FuzzyNumber(-mu,_alpha,_beta); }

FuzzyNumber FuzzyNumber::operator* (const double& num) const
{
 double mu = _mu*num;
 double alpha = num*_alpha;
 double beta = num*_beta;
 return FuzzyNumber(mu,alpha,beta);
}

FuzzyNumber FuzzyNumber::operator/ (const double& num) const
{ return *this*((double)1/num); }

double FuzzyNumber::operator() (const double& x) const
{
 if(x < _mu) return LRfunc((_mu-x)/_alpha);
 else return LRfunc((x-_mu)/_beta);
}

double FuzzyNumber::LRfunc(const double& x) const
{
 if((x >= 0.0) && (x <= 1.0)) return 1.0-x;
 else return 0.0;
}

FuzzyNumber FuzzyNumber::invert() const
{
```

```
 double mu = (double)1/_mu;
 double alpha = _beta/(_mu*(_mu+_beta));
 double beta = _alpha/(_mu*(_mu+_alpha));
 return FuzzyNumber(mu,alpha,beta);
}

// global operators
FuzzyNumber operator+ (const double& crisp,const FuzzyNumber& fuzzy)
{ return FuzzyNumber(crisp) + fuzzy; }

FuzzyNumber operator- (const double& crisp,const FuzzyNumber& fuzzy)
{ return FuzzyNumber(crisp) - fuzzy; }

FuzzyNumber operator* (const double& crisp,const FuzzyNumber& fuzzy)
{ return FuzzyNumber(crisp)*fuzzy; }

FuzzyNumber operator/ (const double& crisp,const FuzzyNumber& fuzzy)
{ return FuzzyNumber(crisp)/fuzzy; }

istream& operator >> (istream& is,FuzzyNumber& num)
{ is >> num._mu >> num._alpha >> num._beta; return is; }

ostream& operator << (ostream& os,const FuzzyNumber& num)
{
 os << "(" << num._mu << "," << num._alpha << "," << num._beta << ")";
 return os;
}
```

As an example consider the program fuzzyN.cpp

```
// fuzzyN.cpp

#include <iostream>
#include "fuzzyN.h"
using namespace std;

int main(void)
{
 FuzzyNumber num1(2.0,0.5,0.5), num2(3.0,0.5,0.5);
 cout << "num1, num2, -num2 = " << num1 << ", " << num2 << " "
      << -num2 << endl;
 const double xMin = 1.5, xMax = 3.5;
 const int nPoints = 9;
 const double dx = (xMax-xMin)/(nPoints-1.0);
 double x = xMin;
 for(int i=0;i<nPoints;i++)
 {
 cout.precision(4); cout.width(6); cout << x << " ";
 cout.precision(4); cout.width(6); cout << num1(x) << " ";
```

```
  cout.precision(4); cout.width(6); cout << num2(x) << endl;
  x += dx;
  }
  cout << "num1 + num2 = " << num1+num2 << endl;
  cout << "num1 - num2 = " << num1-num2 << endl;
  cout << "num1 * num2 = " << num1*num2 << endl;
  cout << "num1 / num2 = " << num1/num2 << endl;
  cout << "num1 + 2 = " << num1+2 << endl;
  cout << "num1 - 2 = " << num1-2 << endl;
  cout << "num1 * 2 = " << num1*2 << endl;
  cout << "num1 / 2 = " << num1/2 << endl;
  cout << "1.3 + num1 = " << 1.3+num1 << endl;
  cout << "1.3 - num1 = " << 1.3-num1 << endl;
  cout << "1.3 * num1 = " << 1.3*num1 << endl;
  cout << "1.3 / num1 = " << 1.3/num1 << endl;
  cout << "enter fuzzy number: mu, alpha, beta: ";
  FuzzyNumber num3;  cin >> num3;  cout << num3;
  return 0;
}
```

We first create two fuzzy numbers and then calculate their value on a grid using the function call operator (). Note that one can view this as a discretization process of the continuous fuzzy numbers. In the rest of the code we simply test the various algebraic operators. The output of the program is given by:

```
num1, num2, -num2 = (2,0.5,0.5), (3,0.5,0.5) (-3,0.5,0.5)
   1.5       0       0
  1.75     0.5       0
     2       1       0
  2.25     0.5       0
   2.5       0       0
  2.75       0     0.5
     3       0       1
  3.25       0     0.5
   3.5       0       0
num1 + num2 = (5,1,1)
num1 - num2 = (-1,1,1)
num1 * num2 = (6,2.5,2.5)
num1 / num2 = (0.6667,0.2619,0.2619)
num1 + 2 = (4,0.5,0.5)
num1 - 2 = (0,0.5,0.5)
num1 * 2 = (4,1,1)
num1 / 2 = (1,0.25,0.25)
1.3 + num1 = (3.3,0.5,0.5)
1.3 - num1 = (-0.7,0.5,0.5)
1.3 * num1 = (2.6,0.65,0.65)
1.3 / num1 = (0.65,0.13,0.13)
1.2 0.1 0.2
(1.2,0.1,0.2)
```

18.3.6 Applications

Fuzzy numbers have a large number of applications in engineering and science. A large number of them are described in Ross [175].

Example. Assume we want to determine the power from two fuzzy measurements, the voltage and the current. The spreads for the fuzzy measurements of the voltage and the current are determined by the estimation of the errors in the measurement. Assume, for example, that the maximum reading of the voltmeter is $V_{\max} = 200\text{V}$ and that the error at full-scale is $\epsilon = 5\%$. The spreads are then set to

$$\alpha_{\widetilde{V}} = \beta_{\widetilde{V}} = \epsilon V_{\max} = 10\text{V}\,.$$

Assuming we measure a potential difference of 120V, we have $\widetilde{V} = (120, 10, 10)_{LR}$. In the same way we obtain the fuzzy current reading $\widetilde{I}$. The power is then given by

$$\widetilde{P} = \widetilde{V} \widetilde{\times} \widetilde{I}\,.$$

Assuming

$$\widetilde{I} = (10, 0.2, 0.2)_{LR}$$

we obtain

$$\widetilde{P} = (1200, 124, 124)_{LR}\,.$$ ♣

Example. The equation of state for an ideal gas is given by $pV = NkT$. If N and T are kept constant, we have

$$pV = \text{const}\,.$$

Assume that for T and N fixed an ideal gas of fuzzy volume

$$V_1 = \{\,(0.5, 0.0),\ (0.75, 0.5)\ (1.0, 1.0),\ (1.25, 05),\ (1.5, 0.0)\,\}$$

is under a fuzzy pressure

$$p_1 = \{\,(0.5, 0.0),\ (1.75, 0.5)\ (2.0, 1.0),\ (2.25, 0.5),\ (2.5, 0.0)\,\}\,.$$

Using the extension principle one can calculate the fuzzy pressure if the volume is reduced to

$$V_2 = \{\,(0.4, 0.0),\ (0.45, 0.5)\ (0.5, 1.0),\ (0.55, 0.5),\ (0.6, 0.0)\,\}\,.$$ ♣

18.4 Fuzzy Rule-Based Systems

18.4.1 Introduction

The purpose of control is to influence the behaviour of a system by changing an input or inputs to that system according to a rule or set of rules that model how the system operates. The system being controlled may be mechanical, electrical, chemical or any combination of these. Classic control theory uses a mathematical model to define a relationship that transforms the desired state (requested) and observed state (measured) of the system into an input or inputs that will alter the future state of that system. The most common example of a control model is the PID (proportional-integral-derivative) controller. This takes the output of the system and compares it with the desired state of the system. It adjusts the input value based on the difference between the two values according to the following equation

$$\text{output} = A.e + B.\text{INT}(e)dt + C.de/dt$$

where, A, B and C are constants, e is the error term, INT(e)dt is the integral of the error over time and de/dt is the change in the error term. The major drawback of this system is that it usually assumes that the system being modelled is linear or at least behaves in some fashion that is a monotonic function. As the complexity of the system increases it becomes more difficult to formulate that mathematical model.

Fuzzy control (Babuška [9]) replaces the role of the mathematical model and replaces it with another that is build from a number of smaller rules that in general only describe a small section of the whole system. The process of inference binding them together to produce the desired outputs. That is, a fuzzy model has replaced the mathematical one. The inputs and outputs of the system have remained unchanged. The usage of fuzzy approaches for the automatic control of technical processes tries to imitate the conscious behaviour of a human operator controlling complex nonlinear processes or production plants which can hardly be modelled by mathematical-physical reflections. These circumstances disable a systematical design of a conventional model-based controller.

Examples for these kinds of processes are biotechnological production processes, chemical processes, image processing, wastewater treatment plans or autonomous vehicle steering and control.

Instead of a conventional linear or nonlinear controller a fuzzy controller is used which consists of the three elements

- fuzzyfication,
- inference and
- defuzzyfication

The first task of a fuzzy controller is the translation of the numerical input variables to linguistic variables.

Definition. Labeling the crisp value of a numerical input variable with a linguistic term and determing the corresponding grade of membership is called *fuzzification.*

Example. Assume that the output variable y can be measured in a standard control loop. In the following steps only the corresponding linguistic variable with terms like *negative big, negative* etc. will be processed. Therefore a set of membership functions for y has to be defined.

Definition. The determination of conclusions or the generation of hypotheses based on a given input state is called *inference.*

For the operation within the standard control loop this means, that the rules define the dependencies between linguistically classified input values and linguistically classified output values. The result is a variable u manipulated according to the input situation. This all occurs in an upper symbolic level first. The implementation can use a couple of operators which were partially discussed above. Mainly the inference component imitates the human operator strategies. Common inference strategies are

- the *max-prod inference*, which multiplies the whole output's membership function, and
- the *max-min inference*, which cuts the output's membership function at the top.

However the symbolic control action cannot be used for a real technical plant, since the linguistically obtained manipulated variable has to be defuzzified.

Definition. *Defuzzification* is the calculation of a crisp numerical value as the controller output based on the symbolic results.

In most cases several rules will be fired and caused by the fuzzyness different terms and therefore different control actions will be activated. But the actuator requires a crisp value which can be calculated by different approaches. The most common defuzzification methods are

- center of gravity,
- COS - center of singletons,
- maximum membership and
- center of maximum memberships.

The practical usage of the given definitions is illustrated by the following example.

Example. A fuzzy controller uses the actual system deviation $e(k)$ as an external input variable. An internal short term memory delays the deviation for one sampling period so that the previous system deviation $e(k-1)$ can also be used for the controller mapping. The membership functions are chosen equally for all linguistic variables, i.e. $e(k)$, $e(k-1)$ and the manipulated variable $u(k)$. For the sake of simplicity the inference contains only two rules:

- **IF** the system deviations $e(k)$ and $e(k-1)$ are approximately zero, **THEN** the control action $u(k)$ is approximately zero.
- **IF** the system deviations $e(k)$ and $e(k-1)$ are positive, **THEN** the control action $u(k)$ is positive.

It can easily be seen that this rulebase is incomplete because a lot of terms of the input variables and the possible combinations are not used.

A fuzzy control system usually receives a fuzzy input signal and performs some logical or arithmetic operations on that input and gives a fuzzy output. Often the original input signal is obtained from crisp measurements and has to be fuzzified. Often knowledge about the error in the measurement apparatus can be used to fuzzify the input signal. Also, for many control problems the output must again be crisp. In these cases we defuzzify the output via one of the defuzzification operators, for example by determining the centroid of the fuzzy output signal. Consider, for example, the typical fuzzy control system depicted in the figure. Usually measurements on the system result in a crisp measurement signal. This is then fuzzified, often using the error information of the measurement apparatus. The fuzzy system then uses the resultant fuzzy input signal together with approximate reasoning based on fuzzy rules to obtain a fuzzy output signal. Again, the controlled system usually requires a crisp input signal, the change in the throttle setting of the aeroplane or the amount invested in a certain stock item. To obtain this crisp signal we defuzzify the fuzzy signal with one of the defuzzification operators. For the system to work well we need a good knowledge base. The knowledge base contains the rules as well as the information regarding a suitable fuzzification and defuzzification method for the system.

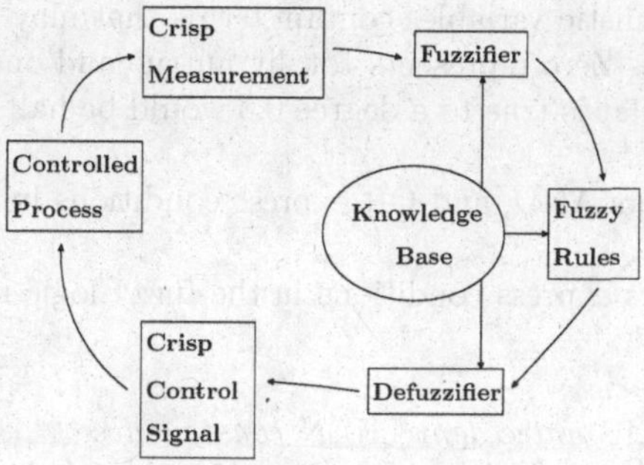

Figure 18.1: Fuzzy Control System

A simple example is temperature control. One can apply a fuzzy logic control strategy to control the temperature. A temperature control system consists of a thermostat that contains a heater resistor, a potentiometer, and a thermistor. The resistor is the temperature controlling controlling element, and the thermistor is the temperature sensing element. The resistor, activated by a pulse width modulated

(PWM) source, heats the adjacent thermistor. A potentiometer sets the desired final temperature (setpoint), and the thermistor senses the current temperature. In proportional/differential control mode one determines the value for the temperature error, T_E, and the rate of the temperature error change dT_E/dt. The fuzzy control algorithm uses these values and sends back the duty cycle for the pulse width modulation resistor. The rule base for the temperature is as follows.

Rule 1. IF the linguistic input term is COLD,
THEN the DUTY CYCLE is LARGE.

Rule 2. IF the linguistic input term is COOL,
THEN the DUTY CYCLE is MEDIUM.

Rule 3. If the linguistic input term is JUST RIGHT,
THEN the DUTY CYCLE is SMALL.

Rule 4. If the linguistic input term is HOT,
THEN the DUTY CYCLE is ZERO.

In the defuzzification process we have to associate output linguistic terms with the crisp output DUTY CYCLE values which vary from zero to 100 percent.

18.4.2 Fuzzy If-Then Rules

Fuzzy control systems use the following conditions to define and implement a rule base:

1) Input and output linguistic variables contain terms that may be true to a degree ranging from zero to one. Zero represents totally untrue and one represents totally true. A linguistic term that is true to a degree 0.5 would be half true and half false.

2) The linguistic operators AND, and OR express conditions in the rule base.

3) IF ... THEN operators express conditions in the fuzzy logic rule base. This rule is of the general form

IF *antecedent* THEN *consequence*

4) A fuzzification step links crisp input values to the fuzzy logic control.

5) A defuzzification step links output linguistic terms to crisp output values.

To implement a fuzzy control strategy to control for example the temperature we must use two way translations between the crisp and linguistic terms. To implement a fuzzy logic control process we uses the following three steps.

1) Fuzzification. One translates crisp input values into linguistic terms and assigns a degree of membership to each crisp input value. Linguistic variables are fuzzy sets which contain linguistic terms as members of the set.

2) Fuzzy Rule Inference. IF ... THEN rules that define the relationship between the linguistic variables are established. These rules determine the course of action the controller must follow. Thus most fuzzy logic application solutions use production rules to represent the relationship between linguistic variables and to derive actions from sensor inputs. Production rules consist of a precondition (IF part) and a consequence (THEN part). The IF part can consist of more than one condition linked together by linguistic conjunctions like AND and OR. The computation of fuzzy rules is called fuzzy rule inference. Aggregation uses fuzzy logic operators to calculate the result of the IF part of a production rule when the rule consists of more than one input condition. One of the linguistic conjunctions, AND or OR, links multiple input conditions. Composition uses the fuzzy logic operator, PROD, to link the input condition to the output condition. Using the conjunction AND for the minimum and OR for the maximum is often appropriate in small control applications.

3) Defuzzification. The result of the fuzzy inference is retranslated from the linguistic concept to a crisp output value.

18.4.3 Inverted Pendulum Control System

The most famous example (Ross [175], Bojadziev and Bojadziev [20]) for fuzzy control is the broom stick balancer (also called inverted pendulum control system). The inverted pendulum consists of an arm attached to a cart at a point. The arm rotates in the plane Oxy. The cart moves along the x axis forward and backward under the influence of a force F. The input parameters are the angle α and the velocity v. The output parameter is the force. The control objectives is for any pair (α, v) which specify the position and the velocity of the system to find apply a corresponding force F to the cart that the arm is balanced, i.e. does not fall over the cart.

For fuzzy control applications it is often convenient to write the fuzzy controllers output as a system of *if-then* rules. For example, we could use the following rule as part of a simple broom balancer:

IF
the angle of the broom is small positive,
THEN
set the force on its base small negative.

In Boolean logic the consequence becomes true when the antecedent is true (not necessarily vice-versa). In fuzzy logic we often work with half-truths or graded truths. Here the consequence only becomes true to the extent that the antecedent is true. We can implement this simple rule in the following way:

```
Fuzzy smallPositiveAngle(100,-45,45);
smallPositiveAngle.triangle(0,15,30);
Fuzzy smallNegativeForce(40,-3,3);
smallNegativeForce.triangle(-2,-1,0);
Fuzzy fuzzyForce
= smallNegativeForce.limit(smallPositiveAngle(angle));
```

The actual force applied to the broom must, of course, be a crisp value. For this purpose we defuzzify the fuzzy force via, for example, the centroid defuzzifier

```
double force = fuzzyForce.centroid();
```

Generally we have of course more complicated fuzzy control systems. For example, the broom balancer might be a little more successful if we use the angle and angular velocity as inputs. We might, for example, have the rule

IF
the angle of the broom is large positive,
AND
the angular velocity is large positive
THEN
set the force on its base large negative.

```
double antecedent
= min(smallPositiveAngle(angle),largePositiveVelocity(velocity));

force = largeNegativeForce.limit(antecedent).centroid();
```

In general we have many rules and many possible consequences. Consider, once more the very simple broom balancing problem. Our universe of discourse for angles could be LN, SN, Z, SP and LP for *large negative*, *small negative*, *zero*, *small positive* and *large positive* represented by fuzzy sets. We have a similar universe of discourse for the applied force and a rule base connecting the two universes. We have the primitive rule base:

IF *angle is LP* THEN *force is LN.*
IF *angle is SP* THEN *force is SN.*
IF *angle is Z* THEN *force is Z.*
IF *angle is SN* THEN *force is SP.*
IF *angle is LN* THEN *force is LP.*

Assume now the actual angle is about 25°. Then the confidence level for an LP angle is about 0.68, the confidence for an SP angle is about 0.32 and the confidence for all other fuzzy angles is zero. Hence the highest confidence level is given to the fuzzy angle LP and consequently the first rule is selected. The IF ... THEN rule limits the confidence level of the consequence (the LN force) to that of the antecedent (the LP angle). Using the centroid defuzzifier results in a crisp force of about -1.75.

18.4.4 Fuzzy Controllers with B-Spline Models

B-spline basis functions can be used for input variables and fuzzy singletons for output variables to specify linguistic terms. Product is chosen as the fuzzy conjunction, and centroid as the defuzzification method. By appropriately designing the rule base, a fuzzy controller can be interpreted as a B-spline interpolator (Zhang and Knoll [228]). Such a fuzzy controller may learn to approximate any known data sequences and to minimise a certain cost function. By choosing such a function appropriately, the learning process can be made to converge rapidly. Fuzzy controllers are universal approximators. The membership functions can be selected as triangles as described above and each pair overlaps. Triangular membership functions with a 1/2 overlap level produce a reconstruction error of zero.

To solve the problem of numerical approximation for smoothing statistical data, "basis splines" (B-Splines) can be introduced. B-splines are also used in computer aided geometric design for curve and surface representation (Hardy and Steeb [88]). Owing to their versatility based on only low-order polynomials and their straightforward computation, B-splines have become more and more popular. B-splines can also be used for neural network modelling and control.

B-spline basis functions and parametric membership functions of a linguistic variable are both convex, overlapping set functions. Splines and fuzzy controllers possess good interpolation features. The synthesis of a smooth curve with spline functions can easily be associated with the defuzzification process.

One considers the membership functions that are used in the context of specifying linguistic terms ("values" or "labels") of input variables of a fuzzy controller. In the following, basis functions of periodical Nonuniform B-Splines are summarised and compared with a fuzzy controller.

Assume x is a general input variable of a control system that is defined on the universe of discourse $[x_0, x_m]$. Given a sequence of ordered parameters (knots): $(x_0, x_1, x_2, \ldots, x_m)$, the ith normalised B-spline basis function (B-function) $N_{i,k}$ of order k is defined as

$$N_{i,k}(x) := \begin{cases} 1 \text{ for } x_i \leq x < x_{i+1} \\ 0 \text{ otherwise} \end{cases} \qquad \text{if } k = 1$$

and

$$N_{i,k}(x) := \frac{x - x_i}{x_{i+k-1} - x_i} N_{i,k-1}(x) + \frac{x_{i+k} - x}{x_{i+k} - x_{i+1}} N_{i+k,k-1}(x) \qquad \text{if } k > 1$$

with $i = 0, 1, \ldots, m - k$. The properties of B-functions are

Partition of unity:	$\Sigma_{i=0}^{m} N_{i,k}(x) = 1$.
Positivity:	$N_{i,k}(x) \geq 0$.
Local support:	$N_{i,k}(x) = 0$ for $x \notin [x_i, x_{i+k}]$.
C^{k-2} *continuity:*	If the knots $\{x_i\}$ are pairwise different from each other, then $N_{i,k}(x) \in C^{k-2}$, i.e., $N_{i,k}(x)$ is $(k-2)$ times continuously differentiable.

The B-functions are employed to specify the linguistic terms, and knots are chosen to be different from each other (periodical model). The selection of k (the order of the B-functions) determines the following factors of the fuzzy sets for modelling the linguistic terms. Thus for order $k = 1$ we have a rectangular shape, for $k = 2$ a triangular shape, for $k = 3$ a quadratic shape and for $k = 4$ a cubic shape. (Hardy and Steeb [88]).

The linguistic terms are to be defined over $[x_0, x_m]$, the universe of an input variable x of a fuzzy controller. They are referred to as *real linguistic terms.* To maintain the partition of unity for all $x \in [x_0, x_m]$, more B-functions should be added at both ends of $[x_0, x_m]$. They are called *marginal B-functions*, defining virtual linguistic terms. In the case of order 2, no marginal B-function is needed. In the case of order 3 or 4, two marginal B-functions are needed, one for the left end and another for the right end.

We define the core rules as linguistic rules that use real linguistic terms. If virtual linguistic terms appear in the premise additional rules are needed to describe the control action for these cases to maintain the output continuity at both ends of the universe of x.

Since these rules use the virtual linguistic terms that are defined by membership functions neighbouring the ends of the universe of each variable, they are called marginal rules. The output value of each marginal rule is selected just as the output value of the nearest core rule, i.e., the rule using the directly adjacent linguistic terms in its premise.

Since a multi-input-multi-output rule base is normally divided into several multi-input-single-output rule bases, we consider only the multi-input-single-output case. Under the following conditions:

- Periodical B-spline basis functions as membership functions for inputs.
- Fuzzy singletons as membership functions for outputs.
- "Product" as fuzzy conjunctions.
- "Centroid" as defuzzification method.
- Addition of virtual linguistic terms at both ends of each input variable.

- Extension of the rule base for the virtual linguistic terms by copying the output values of the nearest neighbourhood.

The computation of the output of such a fuzzy controller is equivalent to that of a general B-spline hypersurface. Consider a multi-input-single-output system with n inputs $x_1, x_2, \ldots, x_n$, rules with the n conjunctive terms in the premise are given in the form

$$\{Rule(i_1, \ldots, i_n) : \text{if}(x_1 \text{ is } N^1_{i_1,k_1}) \ \& \ \cdots \ \& \ (x_n \text{ is } N^n_{i_n,k_n}) \text{ then } y \text{ is } Y_{i_1 i_2 \ldots i_n}\}$$

where

- x_j is the jth input $(j = 1, \ldots, n)$.
- k_j is the order of the B-spline basis functions used for z_j.
- $N^j_{i_j,k_j}$ is the ith linguistic term of x_j defined by B-spline basis functions.
- $i_j = 0, \ldots, m_j$ represents how fine the jth input is fuzzy partitioned.
- $Y_{i_1,i_2 \ldots i_n}$ is the control vertex (de Boor points) of $Rule(i_1, i_2, \ldots, i_n)$.

Then the output y of a multi-input-single-output fuzzy controller is

$$y = \frac{\Sigma^{m_1}_{i_1=0} \cdots \Sigma^{m_n}_{i_n=0} (Y_{i_1,\ldots,i_n} \Pi^n_{j=1} N^j_{i_j,k_j}(x_j))}{\Sigma^{m_1}_{i_1=0} \cdots \Sigma^{m_n}_{i_n=0} \Pi^n_{j=1} N^j_{i_j,k_j}(x_j)} = \sum_{i_1=0}^{m_1} \cdots \sum_{i_n=0}^{m_n} (Y_{i_1,\ldots,i_n} \prod_{j=1}^{n} N^j_{i_j,k_j}(x_j)).$$

This is called a general nonuniform B-splines hypersurface. It has the properties

- If the B-functions of order $k_1, k_2, \ldots, k_n$ are employed to specify the linguistic terms of the input variables $x_1, x_2, \ldots, x_n$ it can be guaranteed that the output variable y is $(k_j - 2)$ times continuously differentiable with respect to the input variables $x_j, j = 1, \ldots, n$.
- If the input space is partitioned fine enough and at the correct positions, the interpolation with the B-spline hypersurface can reach a given precision.

18.4.5 Application

Let us have a look at an approximate reasoning example. Consider the following rules for analyzing a computer code for the computer language used.

IF
vaguely many extends,
THEN
Java.
IF
more than n semicolons,

AND
somewhat many curly brackets
AND NOT
many begin
AND NOT
Java
THEN
C++

Again, we can use the discrete fuzzy set class to implement this fuzzy inference problem.

```
// language.cpp

#include <iostream>
#include "Fuzzy.h"
using namespace std;

template <class T>
T min4(const T& x1,const T& x2,const T& x3,const T& x4)
{
 T min = x1;
 if(min > x2) min = x2;
 if(min > x3) min = x3;
 if(min > x4) min = x4;
 return min;
}

int main(void)
{
 const int pMin = 0; const int pMax = 10;
 const int nDiscr = 200;
 Fuzzy enoughSemicolons(nDiscr,pMin,pMax);
 enoughSemicolons.greaterThan(5);
 Fuzzy somewhatManyCurlyBrackets(nDiscr,pMin,pMax);
 somewhatManyCurlyBrackets.large();
 somewhatManyCurlyBrackets
  = somewhatManyCurlyBrackets.hedge(Fuzzy::somewhat);
 Fuzzy manyBegin(nDiscr,pMin,pMax);
 manyBegin.large();
 Fuzzy slightlyManyExtends(nDiscr,pMin,pMax);
 slightlyManyExtends.large();
 slightlyManyExtends = slightlyManyExtends.hedge(Fuzzy::slightly);
 double nSemicolon, nCurlyBracket, nBegin, nExtend;
 cout << "Enter no of ; "; cin >> nSemicolon;
 cout << "Enter no of { "; cin >> nCurlyBracket;
 cout << "Enter no of begin "; cin >> nBegin;
 cout << "Enter no of extend "; cin >> nExtend;
```

```
 double java = (slightlyManyExtends)(nExtend);
 double notJava = (!slightlyManyExtends)(nExtend);
 double cpp = min4(enoughSemicolons(nSemicolon),
                   somewhatManyCurlyBrackets(nCurlyBracket),
                   (!manyBegin)(nBegin),notJava);
 cout << "Confidence that program is Java: " << java << endl;
 cout << "Confidence that program is C++ : " << cpp  << endl;
 return 0;
}
```

We can run the program now for various inputs for the frequency of occurance of the program elements, *semicolon*, *curly bracket*, *begin* and *extend* (on some arbitrary scale): Below is the output for 2 different trial runs:

```
Enter no of   ;   10
Enter no of   {   10
Enter no of   begin  0
Enter no of   extend 0
Confidence that program is Java: 0
Confidence that program is C++ : 0.993307

Enter no of   ;  20
Enter no of   {  15
Enter no of   begin  5
Enter no of   extend 10
Confidence that program is Java: 1
Confidence that program is C++ : 0
```

18.5 Fuzzy C-Means Clustering

The aim of this clustering technique (Lin et al [129]) is to group a given set of data consisting of n vectors in the vector space $\mathbb{R}^k$

$$X = \{ \mathbf{x}_0, \mathbf{x}_1, \ldots \mathbf{x}_{n-1} \}$$

into a number of clusters c, so that the data in the same group are as similar as possible and the data in different groups are as dissimilar as possible. Fuzzy C-Means achieves this classification by minimising the objective function

$$J(U,V) = \sum_{i=0}^{n-1} \sum_{j=0}^{c-1} (\mu_{ij})^m \|\mathbf{x}_i - \mathbf{v}_j\|^2$$

where μ_{ij} represents the membership degree of data $\mathbf{x}_i$ to the cluster centre $\mathbf{v}_j$. The quantity μ_{ij} satisfies the following conditions

$$\mu_{ij} \in [0,1] \quad \text{for } i = 0,1,\ldots,n-1, \; j = 0,1,\ldots,c-1$$

and

$$\sum_{j=0}^{c-1} \mu_{ij} = 1, \quad \text{for } i = 0, 1, \ldots, n-1 .$$

Furthermore $\|\mathbf{x}_i - \mathbf{v}_j\|$ is the Euclidean distance between $\mathbf{x}_i$ and $\mathbf{v}_j$. The parameter m is used to control the fuzziness of membership of each datum, $m > 1$. There is no theoretical basis for the optimal selection of the exponent m, but a value of $m = 2.0$ is usually chosen. The $n \times c$ matrix $U = (\mu_{ij})$ is called a *fuzzy partition matrix* and $V = \{\mathbf{v}_0, \mathbf{v}_1, \ldots, \mathbf{v}_{c-1}\}$ is a set of cluster centres (of course also vectors in $\mathbb{R}^k$). The fuzzy C-mean clustering algorithm attempts to partition a finite collecetion of elements of the set X into a collection of c fuzzy clusters with respect to some given criterion. Thus the fuzzy C-means clustering needs an a priori assumption of the number of clusters. The fuzzy C-means clustering algorithm is as follows:

1) Initialize the membership matrix $U = (\mu_{ij})$ with random values so that

$$\mu_{ij} \in [0,1] \quad \text{for } i = 0, 1, \ldots, n-1, \quad j = 0, 1, \ldots, c-1$$

and

$$\sum_{j=0}^{c-1} \mu_{ij} = 1, \quad \text{for } i = 0, 1, \ldots, n-1 .$$

2) Compute the fuzzy centres $\mathbf{v}_j$ using

$$\mathbf{v}_j = \frac{\sum_{i=0}^{n-1} (\mu_{ij})^m \mathbf{x}_i}{\sum_{i=0}^{n-1} (\mu_{ij})^m}$$

for $j = 0, 1, \ldots, c-1$.

3) Calculate the new distances

$$d_{ij} = \|\mathbf{x}_i - \mathbf{v}_j\|$$

for $i = 0, 1, \ldots, n-1$ and $j = 0, 1, \ldots, c-1$.

4) Update the fuzzy membership matrix $U = (\mu_{ij})$ according to: if $d_{ij} \neq 0$, then

$$\mu_{ij} = \frac{1}{\sum_{\ell=0}^{c-1} (d_{ij}/d_{i\ell})^{2/(m-1)}}$$

else

$$\mu_{ij} = 1 .$$

5) Repeat step 2) to 4) until the minimum J value is achieved.

In the C++ program we consider four vectors in $\mathbb{R}^3$, $m = 2$ and two clusters, i.e. $c = 2$. We do 20 iterations. It is left as an exercise to calculate $J(U, V)$ at each time step and consequently obtain a criteria for convergence.

```
// clustering.cpp

#include <iostream>
#include <cmath>     // for sqrt
using namespace std;

double distance(double* w1,double* w2,int k)
{
 double r = 0.0;
 for(int i=0;i<k;i++) { r += (w1[i]-w2[i])*(w1[i]-w2[i]); }
 return sqrt(r);
}

int main(void)
{
 int n = 4; // number of vectors
 int k = 3; // length of vectors
 double** X = new double*[n];
 for(int i=0;i<n;i++) { X[i] = new double[k]; }
 X[0][0] = 1.0; X[0][1] = 1.0; X[0][2] = 0.0;
 X[1][0] = 0.0; X[1][1] = 1.0; X[1][2] = 1.0;
 X[2][0] = 1.2; X[2][1] = 0.9; X[2][2] = 0.1;
 X[3][0] = 0.1; X[3][1] = 1.1; X[3][2] = 0.95;

 int c = 2; // number of clusters
 double** V = new double*[c];
 for(int i=0;i<c;i++) { V[i] = new double[k]; }

 double** U = new double*[n];
 for(int i=0;i<n;i++) { U[i] = new double[c]; }
 U[0][0] = 0.1; U[0][1] = 0.9; U[1][0] = 0.4; U[1][1] = 0.6;
 U[2][0] = 0.5; U[2][1] = 0.5; U[3][0] = 0.7; U[3][1] = 0.3;

 double** D = new double*[n];  // distances
 for(int i=0;i<n;i++) { D[i] = new double[c]; }

 // compute the fuzzy centres
 int count = 0;
 while(count < 50)
 {
 for(int j=0;j<c;j++)
 {
 for(int l=0;l<k;l++)
 { double t1 = 0.0; double t2 = 0.0;
 for(int i=0;i<n;i++)
 { t1 += U[i][j]*U[i][j]*X[i][l]; t2 += U[i][j]*U[i][j]; }
 V[j][l] = t1/t2;
 } // end for loop l
```

```
} // end for loop j

for(i=0;i<n;i++)
{ for(j=0;j<c;j++) { D[i][j] = distance(X[i],V[j],k); } }
for(int i=0;i<n;i++)
{
for(int j=0;j<c;j++)
{ double t3 = 0.0;
if(D[i][j]==0.0) U[i][j] = 1.0;
else
for(int p=0;p<c;p++) { t3 += D[i][j]*D[i][j]/(D[i][p]*D[i][p]); }
U[i][j] = 1.0/t3;
}
}
count++;
} // end while

for(int i=0;i<n;i++)
 for(int j=0;j<c;j++)
  cout << "U[" << i << "][" << j << "] = " << U[i][j] << endl;
for(int j=0;j<c;j++)
 for(int i=0;i<k;i++)
  cout << "V[" << j << "][" << i << "] = " << V[j][i] << endl;
// free memory
for(int i=0;i<n;i++) { delete[] X[i]; } delete[] X;
for(int i=0;i<c;i++) { delete[] V[i]; } delete[] V;
for(int i=0;i<n;i++) { delete[] U[i]; } delete[] U;
for(int i=0;i<n;i++) { delete[] D[i]; } delete[] D;
return 0;
}
```

Another strategy of the fuzzy clustering method is called the penalized fuzzy C-means ($PFCM$) algorithm owing to the addition of a penalty term. It is a fuzzy C-means algorithm of generalized type depending upon the penalized term in accordance with the value of w. The penalized fuzzy C-means algorithm is more meaningful and effective than the fuzzy C-means method. The objective function is given by

$$\begin{aligned} J_{PFCM} &= \frac{1}{2}\sum_{j=1}^{c}\sum_{i=1}^{n}\mu_{i,j}^{m}\|x_i - w_j\|^2 - \frac{1}{2}v\sum_{j=1}^{c}\sum_{i=1}^{n}\mu_{i,j}^{m}\ln\alpha_j \\ &= J_{FCM} - \frac{1}{2}v\sum_{j=1}^{c}\sum_{i=1}^{n}\mu_{i,j}^{m}\ln\alpha_j \end{aligned}$$

where α_j is a proportional constant of class j and v (≥ 0) is a constant. When $v = 0$, J_{PFCM} equals J_{FCM}. The penalty term

$$-\frac{1}{2}v\sum_{j=1}^{c}\sum_{i=1}^{n}\mu_{i,j}^{m}\ln\alpha_j$$

is added to the objective function and α_j, w_j, and μ_{ij} are defined as

$$\alpha_j := \frac{\sum_{i=1}^{n} \mu_{i,j}^m}{\sum_{j=1}^{c}\sum_{i=1}^{n} \mu_{i,j}^m}, \quad w_j := \frac{\sum_{i=1}^{n} \mu_{i,j}^m x_i}{\sum_{i=1}^{n} \mu_{i,j}^m}, \quad j = 1, 2, \ldots, c$$

and

$$\mu_{i,j} := \left(\sum_{l=1}^{c} \frac{(\|x_i - w_j\|^2 - v \ln \alpha_j)^{1/(m-1)}}{(\|x_i - w_l\|^2 - v \ln \alpha_l)^{1/(m-1)}} \right)^{-1}$$

where $i = 1, 2, \ldots, n$, $j = 1, 2, \ldots, c$. Then the steps in the penalized fuzzy C-means algorithm are

1) Randomly set cluster centroids w_j ($2 \leq j \leq c$), fuzzification parameter m ($1 \leq m < \infty$), and the value $\epsilon > 0$. Give a fuzzy C-partition $U(0)$.

2) Compute the $\alpha_j(t)$, $w_j(t)$ with $U(t-1)$ using the second and third equations. Calculate the membership matrix $U = (\mu_{i,j})$ with $\alpha_j(t)$, $w_j(t)$ using the fourth equation.

3) Compute $\Delta := \max(|U(t+1) - U(t)|)$. If $\Delta > \epsilon$, then go to step 2; otherwise go to step 4.

4) Find the results for the final class centroids.

18.6 T-Norms and T-Conorms

A *t-norm* (triangular norm) is a function $T : [0,1] \times [0,1] \to [0,1]$ which satisfies the following properties (Klement et al [114], Hájek [80])

$$\begin{aligned} T(x,y) &= T(y,x) \quad \text{commutativity} \\ T(x,y) &\leq T(u,v) \quad \text{if } x \leq u \text{ and } y \leq v \\ T(x,T(y,v)) &= T(T(x,y),v) \quad \text{associativity} \\ T(x,1) &= x \quad \text{identity element} \end{aligned}$$

From the last condition and commutativity it follows that $T(1,y) = 1$. Monotony expresses the assumption that increasing the truth degree of a conjunct should not decrease the truth degree of the conjunction. This also implies that $T(x,0) = T(0,y) = 0$. Note that t-norms are not necessarily continuous.

Example. The t-norm

$$T(x,y) := \max\{\, x, y \,\}$$

is called the minimum t-norm. ♣

Example. The t-norm

$$T(x,y) := xy$$

is called the product t-norm. ♣

Example. The t-norm

$$T(x,y) := \max\{\,0, x+y-1\,\}$$

is called the Łukasiewicz norm. ♣

Example. The t-norm

$$T(x,y) := \begin{cases} 0 & \text{if} \quad x=y=0 \\ \frac{xy}{x+y-xy} & \text{otherwise} \end{cases}$$

is called the Hamacher product. ♣

T-conorms (also called S-norms) are dual under the order-reversing operation which assigns $1-x$ to x on $[0,1]$. Given a t-norm the complementary conorm is defined by

$$\perp(x,y) = 1 - T(1-x, 1-y)\,.$$

This is generalization of De Morgn's laws. From the properties of the T-norm it follows that

$$\begin{aligned} \perp(x,y) &= \perp(y,x) \quad \text{commutativity} \\ \perp(x,y) &\leq \perp(u,v) \quad \text{if } x \leq u \text{ and } y \leq v \\ \perp(x, \perp(y,u)) &= \perp(\perp(x,y),u) \quad \text{associativity} \\ \perp(x,0) &= x\,. \end{aligned}$$

It follows that $\perp(x,1) = 1$ for all $x \in [0,1]$.

Example. The t-conorm

$$\perp(x,y) := \max\{\,x,y\,\}$$

is called the maximum t-conorm. ♣

Example.

$$\perp(x,y) := x + y - xy$$

is called the probabilistic sum. ♣

Example. The t-conorm

$$\perp(x,y) := \min\{\,x+y, 1\,\}$$

is called the the bounded sum. ♣

Example. The t-conorm

$$\perp(x,y) := \frac{x+y}{1+xy}$$

is called is called the Einstein sum in analogy to the addition theorem of velocities in special relativity. ♣

18.7 Fuzzy Logic Networks

The algebraic *Reed-Muller expansion* of boolean functions (Sasao [180]) is one of the fundamental approaches in the design of digital systems, especially when dealing with the VLSI technology. This algebraic representation is also useful in error detection and error correction models. Let $f(x_1, x_2, \ldots, x_n)$ be a boolean function. We define with respect to x_j $(j = 1, 2, \ldots, n)$

$$f_{x_j}(x) := f(x_1, \ldots, x_{j-1}, 1, x_{j+1}, \ldots, x_n)$$
$$f_{\bar{x}_j}(x) := f(x_1, \ldots, x_{j-1}, 0, x_{j+1}, \ldots, x_n)$$
$$\frac{\partial f}{\partial x_j} := f_{x_j}(x) \oplus f_{\bar{x}_j}(x)$$

as the positive cofactors of f, negative cofactor of f, and the boolean derivative of f. Then the Reed-Muller expansion (also called Davio expansion) is given by

$$f = f_{\bar{x}_j} \oplus \left(x_j \cdot \frac{\partial f}{\partial x_j} \right)$$

where $\oplus$ is the XOR operation and $\cdot$ is the AND operation. The complement of the variable is denoted by an overbar, that is $\bar{x} = 1 - x$. Note that $(a \oplus b) \cdot c \neq a \oplus (b \cdot c)$ in general, for example for $a = 1$, $b = 0$, $c = 0$.

Example. Consider the sum-of-product form of the boolean function

$$f(x_1, x_2, x_3, x_4) = \bar{x}_1 \cdot x_3 + \bar{x}_1 \cdot x_2 \cdot x_4 + x_1 \cdot \bar{x}_2 \cdot \bar{x}_3 + x_1 \cdot \bar{x}_3 \cdot \bar{x}_4 \,.$$

This can be written in the (obviously simpler) form

$$f(x_1, x_2, x_3, x_4) = x_1 \oplus x_3 \oplus (x_2 \cdot \bar{x}_3 \cdot x_4) \,.$$ ♣

A generalization of the Reed-Muller algebraic representation applied to multivalued (fuzzy) functions is as follows (Pedrycz and Succi [161]). One has to define fuzzy exclusive-OR functions that are a cornerstone of such a representation. One can develop a logic-based architecture of fuzzy neural networks, called fXOR networks here, that are capable of realizing this type of mapping. In contrast to the standard way of approximation of fuzzy functions that is realized via a generalized sum of minterms and becomes a generalized Shannon representation model, the approach leads to a compact representation and features several useful learning properties.

One uses t- and s-norms described above. The $t-$ and $s-$norms are treated as two general classes of logic connective (logic operators). All variables assume values in the unit interval.

The fuzzy XOR neuron (fXOR) is a generalization of the standard XOR operation (gate) used in digital (boolean) systems. An n-input single output fXOR is governed by the expression

$$y = \text{fXOR}(\mathbf{x}, \mathbf{w})$$

where $\mathbf{x}$ and $\mathbf{w}$ are elements in the n-dimensional unit hypercube, i.e. $\mathbf{x}, \mathbf{w} \in [0,1]^n$. The underlying logic static transformation is realized through the use of s- and t-norms. We have

$$y = \text{fXOR}(\mathbf{x}, \mathbf{w}) = \bigoplus_{i=1}^{n} (x_i \, s \, w_i) \equiv (x_1 \, s \, w_1) \oplus (x_2 \, s \, w_2) \oplus \cdots \oplus (x_n \, s \, w_n)$$

where the generalized exclusive-OR operation given above ($\oplus$) is defined by the $s-t$ composition

$$a \oplus b := (\bar{a} \, t \, b) \, s \, (a \, t \, \bar{b}), \quad a, b \in [0,1] \, .$$

The expression of the fXOR neuron is given with x_j and w_j being the coordinates of the fuzzy sets, $j = 1, 2, \ldots, n$ and t- and s-describing triangular norms and conorms. The above convolution of $\mathbf{x}$ and the weight vector (vector of the connections) $\mathbf{w}$ is just a $t-s$ composition of two fuzzy sets. Thus it follows the fundamentals of the calculus of fuzzy relational equations.

By studying the characteristics of the neuron (for $n = 1$), one finds that for different inputs we achieve higher values of the output. The more similar the inputs, the lower the output of the fXOR neuron. In the binary (two-valued) case we end up having the standard characteristics of the XOR function.

The meaning of the connections becomes obvious by studying the properties of the $t-s$ composition. We find that the connections help quantify the relationships between the input variables and the corresponding output. The higher the value of the connection, the less intensive (visible) the impact of the corresponding input variable on the output of the neuron. For $w_j = 1$ this impact is totally eliminated. On the other hand, for $w_j = 0$, the impact is the most evident.

The fuzzy neural network generalizes the algebraic representation of fuzzy functions and is in analogy to what occurs in the Reed-Muller expansion of Boolean functions. Two types of fuzzy neurons are: the fXOR neurons and the AND neurons. The network exhibits a single hidden layer consisting of AND neurons that is followed by the output layer of the fXOR neurons. The role of the AND neurons is to build a logical AND aggregation of the input variables (appearing here in a direct as well as complemented format), that is $x_1, x_2, \ldots, x_n$. The expression governing this is given by

$$z = \text{AND}(\mathbf{x}; \mathbf{v})$$

where $\mathbf{v}$ is a vector of the connections (weights) of this neuron. Considering individual variables, we can write

$$z = \mathop{\mathbf{T}}_{j=1}^{n} (x_j \; s \; v_j) \; t \; \mathop{\mathbf{T}}_{j=1}^{n} (\bar{x}_j \; s \; v_{n+j}) \, .$$

In view of the $t-s$ composition we find that the higher the value of the connection, the less impact is reported for the corresponding variable. If $v_j = 1$ then the associated variable (x_j) does not impact the output (z) as it has been totally eliminated.

When we are confined to boolean gates, the AND neuron generalizes a well-known AND-gate.

18.8 Fuzzy Hamming Distance

The Hamming distance quantifies the extent to which two bitstrings of the same length differ. An application is in the theory of error-correcting codes, where the Hamming distance measured the error introduced by the noise over a channel when a message is sent between its source and destination. However the distance does not distinguish whether a discrepancy of 1 bit between a target and source is separated by one or many positions. Consider the following three bitstrings of length 10

$$b_0 = 1100100000, \quad b_1 = 1100010000, \quad b_2 = 1100000001$$

The Hamming distance bewteen b_0 and b_1 is 2 and the Hamming distance between b_0 and b_2 is also 2. However one would consider b_1 a better match to b_0 than b_2 to b_0. b_1 could be mapped into b_0 by shifting the bit 1 at position 4 one position to the left. To map b_2 into b_0 we have to shift the 1 bit at position 0 five places to the left. This should be taken into account for a definition of a *fuzzy Hamming distance.*

One can define a fuzzy Hamming distance as follows (Bookstein et al [21]). One defines three operations in the bistrings

1) Insertion: flip a bit from 0 to 1
2) Deletion: flip a bit from 1 to 0
3) Shift: Move a 1 bit from position i to position j

Both insertion and deletion are edit operations. Each operation has constant cost. One could set the cost for insertion and deletion, c_i and c_d respectively, to 1 for both cases. The shift operation allows us to transfer a 1-bit in bitstring B_S to a nearby 1-bit in bitstring B_T at less cost than deleting the 1-bit in B_S and inserting it in B_T. The shift operation is an abstraction of the concrete task of attempting to match a 1-bit in a target and missing, but getting closer – it thus captures, for the measure, the notion of neighbouring bit-sites. The cost for shift, c_s, should be proportional to the difference in bit positions, i and j. Thus $c_i = c_d = 1$ and $c_s = \text{abs}(i - j) * k$, where k is arbitrarily chosen so that at some threshold point it becomes cheaper to insert and delete. A choice could be $k = 0.15$. The fuzzy Hamming distance is then defined as the optimal cost to change bitstring b_1 to bitstring b_2 using only these operations. Since the optimal cost can be written as a minimum of the cost of a current operation plus the cost of obtaining the given bitstring, dynamic programming can be applied. The fuzzy Hamming distance is a metric.

Closely related is the *Levenshtein distance* (also called the *edit distance*). The Levenshtein distance (LD) (Gusfield [78]) is a measure of the similarity between two strings not necessarily of the same length. One refers to the two strings as source string `s` and target string `t`. The distance is the number of deletions, insertions, or substitutions required to transform the source string `s` into the target string `t`. Obviously if the strings `s` and `t` are the same then $LD(s,t) = 0$.

Example. If we have the source string `test` and the target string `tent`, then $LD(s,t) = 1$. ♣

Example. Consider the strings `abc` and `bca`. Then we have

$$LD(\text{“}abc\text{”}, \text{“}bca\text{”}) = 2\,.$$

On the other hand the Hamming distance would be 3. ♣

The Levenshtein distance is a *metric* and therefore has the properties

$$LD(s,t) \geq 0$$

and (triangle inequality)

$$LD(s,t) \leq LD(s,r) + LD(r,t)\,.$$

The algorithm to find the Levenshtein distance is

1) Set n to be the length of the source string `s`. Set m to be the length of the target string `t`. If $n = 0$, return m and exit. If $m = 0$, return n and exit.

2) Construct a $(m+1) \times (n+1)$ matrix with rows $0 \ldots m$ and columns $0 \ldots n$. Initialize the first row to $0 \ldots n$. Initialize the first column to $0 \ldots m$.

3) Examine each character in the source string `s` $(i = 1, \ldots, n)$. Examine each character of the target string `t` $(j = 1, \ldots, m)$.

3a) If `s[i]==t[j]`, the cost is 0. If `s[i]!=t[j]`, the cost is 1.

3b) Set cell `d[i][j]` of the matrix equal to the minimum of:
i. The cell immediately above plus 1, i.e. `d[i-1][j]+1`
ii. The cell immediately to the left plus 1, i.e. `d[i][j-1]+1`
iii. The cell diagonally above and and to the left plus the cost, i.e. `d[i-1][j-1]+cost`.

4) After the step 3) is complete, the Levenshtein distance between `s` and `t` is in the matrix element $d[n][m]$.

A C++ implementation is

```
// levenshtein.cpp

#include <iostream>
```

```
#include <string>
using namespace std;

int min(int a,int b,int c)
{
 int m = a;
 if(b < m) { m = b; }
 if(c < m) { m = c; }
 return m;
}

int LD(string s,string t,int n,int m,int** D)
{
 int cost; // cost
 // step 1
 if(n==0) return m;
 if(m==0) return n;
 // step 2
 for(int p=0;p<=n;D[p][0]=p++);
 for(int q=0;q<=m;D[0][q]=q++);
 // step 3
 for(int i=1;i<=n;i++) {
  for(int j=1;j<=m;j++) {
   cost = (t.substr(j-1,1)==s.substr(i-1,1) ? 0 : 1);
    D[i][j] = min(D[i-1][j]+1,D[i][j-1]+1,D[i-1][j-1]+cost);
 }
 }
 return D[n][m];
}

int main(void)
{
 string s = "010101010101"; string t = "101010101010";
 int n = s.length(); int m = t.length();
 int** D = NULL; D = new int*[n+1];  // memory allocation
 for(int k=0;k<=n;k++) D[k] = new int[m+1];
 int distance = LD(s,t,n,m,D);
 cout << "distance = " << distance << endl;
 for(int l=0;l<=n;l++) delete D[l]; delete[] D;
 return 0;
}
```

Exercise. Find the Levenshtein distance between the bitstrings

0000000011111111
1111111100000000

and between the bitstrings

```
1010101010101010
0101010101010101
```

Compare to the Hamming distance.

18.9 Fuzzy Truth Values and Probabilities

What is the relationship between fuzzy truth values and probabilities? This question has to be answered in two ways: how does fuzzy theory differ from probability theory mathematically, and second, how does it differ in interpretation and application.

At the mathematical level, fuzzy values are commonly misunderstood to be probabilities, or fuzzy logic is interpreted as some new way of handling probabilities. However this is not the case. A minimum requirement of probabilities is "additivity", loosely speaking that is that they must add up to one, or the integral of the density curve must be one.

However, this does not hold in general with membership grades. While membership grades can be determined with probability densities in mind, there are other methods as well, which have nothing to do with frequencies or probabilities.

Owing to this, fuzzy researchers have gone to great pains to distance themselves from probabilities. All probability distributions are fuzzy sets. As fuzzy sets and logic generalize Boolean sets and logic, they also generalize probability.

From a mathematical perspective, fuzzy sets and probabilities exist as part of a greater Generalized Information Theory which includes many formalisms for representing uncertainty (including random sets, Demster-Shafer evidence theory, probability intervals, possibility theory, general fuzzy measures, interval analysis, etc.). Furthermore, one can also speak about random fuzzy events and fuzzy random events.

Semantically, the distinction between fuzzy logic and probability theory has to do with the difference between the notions of probability and a degree of membership. Probability statements are about the likelihoods of outcomes: an event either occurs or does not. However, with fuzziness, one cannot say unequivocally whether an event occured or not, and instead we are trying to model the "extent" to which an event ocurred.

Bibliography

[1] Abraham R. and Marsden J. E., *Foundations of Mechanics*, second edition, Benjamin/Cummings, 1982

[2] Ahuja R. K., Magnanti T. L. and Orlin J. B., *Network Flows*, Prentice Hall, 1993

[3] Antoine J.-P., Murenzi R., Vandergheynest P. and Ali S. T., *Two-dimensional wavelets and their relatives*, Cambridge University Press, 2004

[4] Arbib M. A. (Editor), *The Handbook of Brain Theory and Neural Networks*, second edition, MIT Press, 2003

[5] Arnold V. I., *Mathematical Methods of Classical Mechanics*, Springer, 1979

[6] Arnold V. I. and Avez A., *Ergodic Problems of Classical Mechanics*, Addison-Wesley, 1989

[7] Arrowsmith D. K. and Place C. M., *An Introduction to Dynamical Systems*, Cambridge University Press, 1990

[8] Bac Fam Quang and Perov V. L., New evolutionary genetic algorithms for NP-complete combinatorial problems, *Biological Cybernetics* **69**, 229-234, 1993

[9] Babuška R., *Fuzzy Modeling for Control*, Kluwer, 1998

[10] Baker G. L. and Gollub J., *Chaotic Dynamics : An Introduction*, Cambridge University Press, 1990

[11] Bandemer H. and Gottwald S., *Fuzzy Sets, Fuzzy Logic, Fuzzy Methods with Applications*, John Wiley, 1995

[12] Barnsley M., *Fractals Everywhere*, Academic Press, 1988

[13] Barnsley M., *Fractals Everywhere*, Morgan Kaufmann, 2000

[14] Bauer M. and Martienssen W., Lyapunov exponents and dimensions of chaotic neural networks, *J. Phys. A: Math. Gen.* **24**, 4557-4566, 1991

[15] Bazaraa M. S. and Shetty C. M., *Nonlinear Programming*, Wiley, 1979

[16] Beardon A. F., *Iteration of Rational Functions*, Springer, 1991

[17] Becchetti C. and Ricotti L. P., *Speech Recognition: Theory and C++ Implementation*, Wiley, 2002

[18] Bellman R., The stability of solutions of linear differential equations, *Duke Math. J.* **10**, 643-647, 1943

[19] *Wavelets: Mathematics and Applications*, Edited by John J. Benedetto and Michael W. Frazier, CRC Press, 1994

[20] Bojadziev G. and Bojadziev M., *Fuzzy Sets, Fuzzy Logic, Applications*, World Scientific, Singapore, 1995

[21] Bookstein A., Klein S. T. and Raita T., Fuzzy Hamming distance in *Combinatorial pattern machting*, Springer, 2006

[22] Bowen R., *Methods of Symbolic Dynamics*, Springer, 1975

[23] Broomhead D. S., Huke J. P. and Muldoon M. R., Linear filters and nonlinear systems, *J. Roy. Stat. Soc.* **B 54**, 373-382, 1992

[24] Brown R. and Kocarev L., A unifying definition of synchronization for dynamical systems, *Chaos* **10**, 344-349, 2000

[25] Buhmann M. D., *Radial Basis Functions: Theory and Implementations*, Cambridge University Press, 2003

[26] Cartan H., *Differential Forms*, Dover Books on Mathematics, 1950

[27] Chaitin G. J., *Information, Randomness and Incompleteness*, World Scientific, Singapore, 1987

[28] Charniak E., *Statistical Language Learning*, MIT Press, 1993

[29] Chechkin A., Gonchar V., Klafter J. and Metzler, Fundamentals of Lévy Flight Processes, *Adv. Chem. Phys.* **133B**, 439-451, 2006

[30] Chen G. and Dong X., On Feedback Control of Chaotic Continuous-Time Systems, *IEEE Trans. Circuits Syst.* **40**, 591-601, 1993

[31] Chen L. and Aihara K., Chaotic simulated annealing by a neural network model with transient chaos, *Neural Networks* **8**, 915-930, 1995

[32] Cherbit G. (editor), *Fractals : Non-integral dimensions and applications*, Wiley, 1991

[33] Chow S. N. and Hale J. K., *Methods of Bifurcation Theory*, Springer, 1982

[34] Chui C. K., *An Introduction to Wavelets*, Academic Press, 1992

[35] Cichocki A. and Unbehauen R., *Neural Networks for Optimization and Signal Processing*, Wiley, 1993

[36] Cohen A., Daubechies I. and Feauveau J.-C., Biorthogonal bases of compactly supported wavelets, *Comm. Pure and Appl. Math.* **45**, 485-560, 1992

[37] Collet P. and Eckmann J. P., *Iterated Maps on the interval as Dynamical Systems*, Birkhäuser, Boston, 1980

[38] Collins J. J. and Stewart I., Hexapodal gaits and coupled nonlinear oscillators models, *Biol. Cybern.* **68**, 287-298, 1993

[39] Collins J. J. and Richmond S. A., Hard-wired central pattern generators for quadrupedal locomotion, *Biol. Cybern.* **71**, 375-385, 1994

[40] Contopolous G., Galgani L. and Giorgilli A., On the Number of Isolating Integrals in Hamiltonian Systems, *Phys. Rev. A*, **18**, 1183-1190, 1978

[41] Cristianni N. and Hahn M. W., *Introduction to Computational Genomics*, Cambridge University Press, 2007

[42] Cronje G. A. and Steeb W.-H., Genetic Algorithms in a Distributed Computing Enviroment Using PVM, *International Journal of Modern Physics C* **8**, 327-344, 1997

[43] Cybenko G., Approximation by superpositions of a sigmoidal function, *Mathematics of Control, Signals and Systems* **2**, 303-314, 1989

[44] Daubechies I., *Ten Lectures on Wavelets*, SIAM, 1992

[45] Davis H. T., *Introduction to Nonlinear Differential and Integral Equations*, Dover Publications, New York, 1962

[46] Deslauriers G., Dubuc S. and Lemire D., Une famille d'ondelettes biorthogonales sur l'intervalle obtenue par un schéma d' interpolation itérative, *Ann. Sci. Math. Québec* **23**, 37-48, 1999

[47] Devaney R. L., *An Introduction to Chaotic Dynamical Systems*, Second Edition, Addison-Wesley, 1989

[48] Doob J. L., *Stochastic Processes*, Wiley, 1953

[49] Dorbec P. and Gajardo A., Lanton's fly, *J. Phys. A: Math. Theor.* **41**, 405101, 2008

[50] Dubois D. and Prade H., *Fuzzy Sets and Systems: Theory and Applications*, Academic Press, 1980

[51] Eckmann J.-P. and Ruelle D., Ergodic Theory of Chaos and Strange Attractors, *Reviews of Modern Physics* **57**, 617-655, 1985

[52] Eckmann J.-P., Kamphorst S. O., Ruelle D. and Ciliberto S., Liapunov Exponents from Time Series, *Physical Review A* **34**, 4971-4979, 1986

[53] Edgar G. A., *Measure, Topology, and Fractal Geometry*, Springer, 1990

[54] Elizondo D. A., Birkenhead R., Góngora M., Taillard E. and Luyima P., Analysis and test of efficient methods for building recursive deterministic perceptron neural networks, *Neural Networks* **20**, 1095-1108, 2007

[55] *Wavelets: Theory and Applications*, Edited by Gordon Erlebacher, M. Pousuff Hussaini, Leland M. Jameson, Oxford University Press, 1996

[56] Falconer K. J., *The Geometry of Fractal Sets*, Cambridge University Press, 1985

[57] Fausett L., *Fundamentals of Neural Networks : Architecture, Algorithms and Applications*, Prentice Hall, 1994

[58] Ferreira C., *Gene Expression Programming: Mathematical Modeling by an Artificial Intelligence*, 2nd edition, Springer, 2006

[59] Ferreira C., Gene Expression Programming: A New Adaptive Algorithm for Solving Problems, http://xxx.lanl.gov, cs.AI/0102027

[60] Fisher R. A., The use of multiple measurements in taxonomic problems, *Annals of Eugenics* **7**, 179-188, 1936

[61] Flanders H., *Differential Forms with Applications to the Physical Sciences*, Academic Press, 1963

[62] Frisch U., Hasslacher B. and Pommeau Y., Lattice gas methods for partial differential equations, *Phys. Rev. Lett.* **56**, 1505-1508, 1986

[63] Frieß T.-T., Christianni N. and Campbell C., The Kernel-Adatron Algorithm: A fast and simple learning procedure for support vector machines, Proceedings of the 15th International Conference on Machine Learning, pp. 188-196, Morgan Kaufmann, 1998

[64] Fröberg C. E., *Numerical Mathematics: Theory and Computer Applications*, Benjamin-Cummings, 1985

[65] Funahashi K.-I., On the approximate realization of continuous mappings by neural networks, *Neural Networks* **2**, 183-192, 1989

[66] Gajardo A., Moreira A. and Goles E., Complexity of Langton's ant, *Discrete Appl. Maths.* **117**, 41-50, 2002

[67] Gao T., Chen Z., Yuan Z. and Chen G., A hyperchaos generated from Chen's system, *Int. J. Mod. Phys. C* **17**, 471-478, 2006

[68] Gardner M., Mathematical Games: The fanatastic combinations of John Conway's new solitaire game 'Life', *Scientific American* **223**, 120-123, 1970

[69] Goldberg D. E., *Genetic Algorithms in Search, Optimization and Machine Learning*, Addison-Wesley, 1989

[70] Goldberg D. E. and Lingle R., Alleles, Loci, and the TSP, in Greffenstette, J. J. (Editor), *Proceedings of the First International Conference on Genetic Algorithms*, Lawrence Erlbaum Associates, Hillsdale, NJ, 1985

[71] González-Miranda J.M., *Synchronization and Control of Chaos*, Imperial College Press, 2004

[72] Grassberger P. and Procaccia I., Characterization of strange attractors, *Physical Review Letters* **50**, 346-349, 1983

[73] Gröbner W. (editor), *Contributions to the method of Lie series*, Bibliographisches Institut, Mannheim, 1967

[74] Gronwall T. H., Note on the derivative with respect to a parameter of the solutions of a system of differential equations, Ann. of Math. **20**, 292-296, 1919

[75] Guckenheimer J. and Holmes P., *Nonlinear Oscillations, Dynamical Systems and Bifurcations of Vector Fields*, Springer, 1983

[76] Guillemin V. and Sternberg S., *Symplectic Techniques in Physics*, Cambridge University Press, 1984

[77] Gumowski I. and Mira C., *Recurrences and Discrete Dynamical Systems*, Springer, 1980

[78] Gusfield D., *Algorithms on strings, trees and sequences: computer science and computational biology*, Cambridge University Press, 1997

[79] Hairer E., Norsett S. P. and Wanner G., *Solving Ordinary Differential Equations I*, 2nd revised edition, Springer, 1993

[80] Hájek P., *Metamathematics of Fuzzy Logic*, Kluwer, 1998

[81] Hao Bai-Lin, *Elementary Symbolic Dynamics and Chaos in Dissipative Systems*, World Scientific, Singapore, 1989

[82] Hao Bai-Lin, Symbolic Dynamics and Characterization of Complexity, *Physica D* **51**, 161-176, 1991

[83] Hardy Y., Steeb W.-H. and Villet C. M., Stabilization of Chaotic Systems with Phase Coupling, *Z. Naturf.* **55a**, 847-850, 2000

[84] Hardy Y. and Steeb W.-H., *Classical and Quantum Computing with C++ and Java Simulations*, Birkhauser Verlag, Basel, 2001

[85] Hardy Y., Steeb W.-H. and Stoop R., Genetic algorithms, floating point numbers and applications, *Int. J. of Modern Physics C* **16**, 1801-1804, 2005

[86] Hardy Y. and Sabatta D., Encoding, symbolic dynamics, cryptography and C++ implementations, *Phys. Lett. A* **366**, 575-584, 2007

[87] Hardy Y., Kiat Shi Tan and Steeb W.-H., *Computer Algebra with SymbolicC++*, World Scientific, Singapore, 2008

[88] Hardy A. and Steeb W.-H., *Mathematical Tools in Computer Graphics with C# Implementations*, World Scientific, Singapore, 2007

[89] Hassoun M. H., *Fundamentals of Artificial Neural Networks*, The MIT Press, 1995

[90] Haykin S., *Neural Networks*, Macmillan College Publishing Company, 1994

[91] Hebb D., *The organization of behaviour*, Wiley, 1949

[92] Hedlund G. A., Endomorphisms and automorphisms of the shift dynamical system, *Math. Systems Theory* **3**, 320-371, 1969

[93] Hernández E. and Weiss G., *A First Course on Wavelets*, CRC Press, 1996

[94] Herz A., Sulzer B., Kühn R. and van Hemmen J., The Hebb rule: storing static and dynamic objects in an associative neural network, *Europhysics Letters* **7**, 663-669, 1988

[95] Higuchi T., Approach to an irregular time series on the basis of fractal theory, *Physica D* **31**, 277-283, 1988

[96] Hirsch M. W. and Smale S., *Differential Equations, Dynamical Systems, and Linear Algebra*, Academic Press, 1974

[97] Holland J. H., *Adaptation in Natural and Artificial Systems*, University of Michigan Press, Ann Arbor, 1975

[98] Holmgren R. A., *A First Course in Discrete Dynamical Systems*, Springer, 1994

[99] Hopfield J. J., Neural networks and physical systems with emergent collective computational abilities, *Proceedings of the National Academy of Sciences of the USA* **79**, 2554-2558, 1982

[100] Hornik K., Stinchcombe M. and White H., Multilayer feedforward networks are universal approximators, *Neural Networks* **2**, 359-366, 1989

[101] Hübler A. W., Adaptive Control of Chaotic Systems, *Helvetica Physica Acta* **62**, 343-346, 1989

[102] Hurst H. E., Long Term Storage Capacity of Reservoirs, *Trans. Am. Soc. Civil Eng.* **116**, 770-776, 1951

[103] Iooss G. and Joseph D. D., *Elementary Stability and Bifurcation Theory*, second edition, Springer, 1990

[104] Jelinek F., *Statistical Methods for Speech Recognition*, The MIT Press, 1997

[105] Jensen A. and La Cour-Harbo A., *Ripples in mathematics: The discrete wavelet transform*, Springer, 2001

[106] Jordan D. W. and Smith P., *Nonlinear Ordinary Differential Equations*, Clarendon Press, Oxford, 1985

[107] Julia G. M., Mémoire sur literation des fonctions rationnelles, *Journal de Mathématiques Pures et Appliquées* **8**, 47-245, 1918

[108] Kahan W., IEEE Standard 754 for binary floating-point arithmetic, Lectures Notes on the Status of IEEE 754, Work in Progress, 1996

[109] Kamp Y. and Hasler M., *Recursive Neural Networks for Associative Memory*, Wiley, 1990

[110] Kaneko K. and Konishi T., Diffusion in Hamiltonian dynamical systems with many degrees of freedom, *Phys. Rev. A* **40**, 6130-6133, 1989

[111] Kantz H., A robust method to estimate the maximal Lyapunov exponent of a time series, *Physics Letters A* **185**, 77-87, 1994

[112] Kapitaniak T., *Controlling Chaos*, Harcourt Brace, 1996

[113] Kawakami H., Bifurcation of Periodic Responses in Forced Dynamics, *IEEE Trans. Circuits Syst.* CAS-31, 248-256, 1984

[114] Klement E. P., Mesiar R. and Pap E., *Triangular norms*, Kluwer, 2000

[115] Kluiving R., Capel H. W. and Pasmanter R. A., Phase-Transition Like Phenomenon in a Piecewise Linear Map, *Physica A* **164**, 593-624, 1990

[116] Knapp H. and Wanner G., On the numerical treatment of ordinary differential equations, Contributions to the Method of Lie series, Chapter II, pp. 43-97, Bibliographisches Institut, Mannheim, 1967

[117] Koch H. von, Sur une courbe continue sans tangente, obtenue par une construction géométrique élémentair, *Archiv för Matemat. Astron. och Fys.* **1**, 681-702, 1904

[118] Kohonen T., *Self-Organization and Associative Memory*, third edition, Springer, 1989

[119] Kohonen T., *Self-Organizing Maps*, third edition, Springer, 2001

[120] Koski T., *Hidden Markov Models for Bioinformatics*, Springer, 2001

[121] Kowalski K. and Steeb W.-H., *Nonlinear Dynamical Systems and Carleman Linearization*, World Scientific, Singapore, 1991

[122] Kuhn H. W. and Tucker A. W., Nonlinear Programming, Proceedings of the 2nd Berkeley Symposium, Berkeley: University of California Press, 481-492, 1951

[123] Kuznetsov Y. A., *Elements of Applied Bifurcation Theory*, second edition, Springer, 1998

[124] Lai Y.-C., Encoding Digital Information using Transient Chaos, *Int. J. Bifur. Chaos* **10**, 787-795, 2000

[125] Lai Y.-C. and Grebogi C., Synchronization of chaotic trajectories using control, *Phys. Rev.* **47**, 2357-2362, 1993

[126] Langton C. G., Studying artificial life with cellular automata, *Physica D* **22**, 120-149, 1986

[127] Lempel A. and Ziv J., On the Complexity of Finite Sequences, *IEEE Transactions on Information Theory* **IT-22**, 75-81, 1976

[128] Lichtenberg A. J. and Lieberman M. A., *Regular and Stochastic Motion*, Springer, 1983

[129] Lin J.-S., Cheng K.-S. and Mao C.-W., Segmentation of Multispectral Magnetic Resonance Image Using Penalized Fuzzy Competitive Learning Network, *Computers and Biomedical Research* **29**, 314-326, 1996

[130] Lindenmayer A., Mathematical models for cellular interaction in development, *J. Theoret. Biology* **18**, 280-315, 1968

[131] Liu Y., You Z. Cao L., A functional neural network computing some eigenvalues and eigenvectors of a special real matrix, *Neural Networks* **18**, 1293-1300, 2005

[132] Lorenz E. N., Deterministic Non-Periodic Flow, *J. Atmos. Sci.* **20**, 130-141, 1963

[133] Lopez-Ruiz R. and Perez-Garcia C., Dynamics of maps with a global multiplicate coupling, *Chaos, Solitons and Fractals* **1**, 511-528, 1991

[134] Lopez-Ruiz R. and Perez-Garcia C., Dynamics of two logistic maps with a multiplicate coupling, *Int. J. of Bifurcation and Chaos* **2**, 421-425, 1992

[135] Luo Fa-Long and Unbehauen R., *Applied Neural Networks for Signal Processing*, Cambridge University Press, 1997

[136] MacKay R. S. and Meiss J. D. (Editors), *Hamiltonian Dynamical Systems*, Adam Hilger, Bristol, 1987

[137] Mandelbrot B. B. and Wallis J. R., Noah, Joseph and operational hydrology, *Water Resources Research* **4**, 909-918, 1968

[138] Mandelbrot B. B. and Van Ness J. W., Fractional Brownian motions, fractional noises and applications, *SIAM Review* **10**, 422-437, 1968

[139] Mandelbrot B. B., *The Fractal Geometry of Nature*, Freeman, New York, 1982

[140] Maniezzo V., Genetic Evolution of the Topology and Weight Distribution of Neural Networks, *IEEE Transactions of Neural Networks* **5**, 39-53, 1994

[141] Marek M. and Schreiber I., *Chaotic behaviour of deterministic dissipative systems*, Cambridge University Press, 1991

[142] Marsden J. E. and McCracken M., *The Hopf Bifurcation and its Applications*, Springer, 1976

[143] Matsumoto T., Chua L. O. and Kobayashi K., Hyperchaos: Laboratory Experiment and Numerical Confirmation, *IEEE Trans. Circuits Syst.* vol. CAS-33, 1143-1147, 1986

[144] McLaren D. I. and Quispel G. R. W., Integral-preserving integrators, *J. Phys. A: Math. Gen.* **37**, L489-L495, 2004

[145] Michalewicz Z., *Genetic Algorithms + Data Structure = Evolution Programs*, Third Edition, Springer, 1996

[146] Mika S., Rätsch G., Weston J., Schölkopf B. and Müller K.-R., Fisher discriminant analysis with kernels. In *Neural Networks for Signal Processing:* Vol. IX, 41-48, New York: IEEE Press, 1999

[147] Mika S., Rätsch G., Weston J., Schölkopf B., Smola A. J. and Müller K.-R., Invariance features extraction and classification in feature space. In S. A. Solla, T. K. Keen and K.-R. Müller (editors) Advances in neural information processing systems: Vol. 12, pp. 526-532, MIT Press, 2000

[148] Minskey M. L. and Papert S. A., *Perceptrons*, MIT Press, 1969

[149] Mira C., *Chaotic Dynamics: From the One-dimensional Endomorphism to the Two-dimensional Diffeomorphism*, World Scientific, Singapore, 1987

[150] Mitchell M., *An Introduction to Genetic Algorithms*, MIT Press, 1966

[151] Mosekilde E., Maistrenko Y. and Postnov D., *Chaotic Synchronization*, World Scientific, Singapore 2002

[152] Munkres J., *Topology*, 2nd edition, Prentice Hall, 1999

[153] Newhouse S. E., Ruelle D. and Takens F., Occurrence of strange axiom A attractors near quasi-periodic flow on T^m, $m > 3$, *Communications in Mathematical Physics* **64**, 35-40, 1979

[154] Oltean M., "Multi Expression Programming", Technical Report, Babes-Bolyai University, Romania

[155] Oltean M., Evolving Evolutionary Algorithms with Patterns, *Soft Computing* **11**, 503-518, 2007

[156] Ott E., Grebogi C. and Yorke J. A., *Controlling Chaos*, University of Maryland, Laboratory for Plasma Research, 1989

[157] Ott E., Grebogi C. and Yorke J. A., Controlling Chaos, *Phys. Rev. Lett.* **64**, 1196-1199, 1990

[158] Parlitz U. and Lauterborn W., Resonances and Torsion Numbers of Driven Dissipative Nonlinear Oscillators, *Z. Naturforsch.* **41a**, 606-614, 1986

[159] Parlitz U. and Lauterborn W., Period-doubling cascades and devil's staircases of a driven van der Pol oscillators, *Phys. Rev. A* **36**, 1428-1436, 1987

[160] Pecora L. M. and Carroll T. L., Synchronization in Chaotic Systems, *Phys. Rev. Lett.* **64**, 821-824, 1990

[161] Pedrycz W. and Succi G., fXOR fuzzy logic networks, *Soft Computing* **7**, 115-120, 2002

[162] Peitgen H.-O. and Richter P. H., *The Beauty of Fractals*, Springer, 1986

[163] Peitgen H.-O., Jürgens H. and Saupe D., *Chaos and Fractals: new frontiers in science*, 2nd edition, Springer, 2004

[164] Penna T. J., A bit-string model for biological aging, *J. Stat. Phys.* **78**, 1629, 1995

[165] Peters E., *Fractal Market Analysis: Applying Chaos Theory to Investment and Economics*, Wiley, 1994

[166] Pikovsky A., Rosenblum M. and Kurths J., *Synchronization: A Universal Concept in Nonlinear Sciences*, Cambridge University Press, 2001

[167] Polymilis C., Servizi G., Skokos Ch., Turchetti and Vrahatis M. N., Locating periodic orbits by topological degree theory, arXiv:nlln 0211044v1

[168] Protter M. H., *Basic elements of real analysis*, Springer, 1998

[169] Pyragas K., Continuous Control by Self-Controlling Feedback, *Phys. Lett. A* **170**, 421-428, 1992

[170] Pyragas K., Control of Chaos via Unstable Delayed Feedback-Controller, *Phys. Rev. Lett.* **86**, 2265-2268, 2001

[171] Rabiner L., A tutorial on Hidden Markov Models and selected applications in speech recognition, *Proceedings of the IEEE* **77**, 257-286, 1989

[172] Rabiner L. and Juang B.-H., *Fundamentals of Speech Recognition*, Prentice Hall, 1993

[173] Rojas R., *Neural Networks*, Springer, 1996

[174] Rosenstein M. T., Collins J. J. and De Luca C. J., A practial method for calculating largest Lyapunov exponents from small data sets, *Physica D* **65**, 117-134, 1993

[175] Ross T. J., *Fuzzy Logic with Engineering Applications*, McGraw-Hill, New York, 1995

[176] Rössler O. E., An equation for hyperchaos, *Phys. Letters A* **71**, 155-158, 1979

[177] Ruelle D., *Chaotic Evolution and Strange Attractors*, Cambridge University Press, 1989

[178] Saadi K., Talbot N. L. C. and Cawley G. C., Optimally regularised kernel Fisher discriminant classification, *Neural Networks* **20**, 832-841, 2007

[179] Sano M. and Sawada Y., Measurement of the Lyapunov Spectrum from Chaotic Time Series, *Phys. Rev. Lett.* **55**, 1082-1085, 1985

[180] Sasao T., *Logic Synthesis and Optimization*, Kluwer, 1993

[181] Sato S., Sano M. and Sawada Y., Practical methods of measuring the generalized dimension and the largest Lyapunov exponent in high dimensional chaotic systems, *Progess in Theoretical Physics* **77**, 1-5, 1987

[182] Scheid F,. *Numerical Analysis*, Schaum's Outline Series, McGraw Hill, 1968

[183] Schöll E. and Schuster H. G. (Editors), *Handbook of Chaos Control*, Wiley-VCH, second edition, 2008

[184] Schuster H. G., Martin S. and Martienssen W., New Method for Determing the Largest Liapunov Exponent of Simple Nonlinear Systems, *Phys. Rev. A* **33**, 3547-3549, 1986

[185] Seydel R., *From Equilibrium to Chaos*, Elsevier, New York, 1988

[186] Shlesinger M. G., Zaslavsky G. M. and Klafter J., Strange kinetics, *Nature* **363**, 31-37, 1993

[187] Sparrow C., *The Lorenz Equations: Bifurcations, Chaos and Strange Attractors*, Springer, 1982

[188] Smale S., On the mathematical foundations of electrical circuits theory, *J. Differential Geometry* **7**, 193-210, 1972

[189] Steeb W.-H., *A Handbook of Terms Used in Chaos and Quantum Chaos*, Bibliographisches Institut, Mannheim, 1991

[190] Steeb W.-H., *Chaos und Quantenchaos in Dynamischen Systemen*, Bibliographisches Institut, Mannheim, 1994

[191] Steeb W.-H. and Louw J., *Chaos and Quantum Chaos*, World Scientific, Singapore, 1986

[192] Steeb W.-H., Louw J. A. and Villet C. M., Chaos in Yang-Mills equations, *Phys. Rev. D* **33**, 1174-1176, 1986

[193] Steeb W.-H., Solms F., Tan Kiat Shi, and Stoop R., Cubic Map, Complexity and Ljapunov Exponent, *Physica Scripta* **55**, 520-522, 1997

[194] Steeb W.-H., *Matrix Calculus and Kronecker Product with Applications and C++ Programs*, World Scientfic, Singapore, 1997

[195] Steeb W.-H., Hardy Y., Hardy A., and Stoop R., *Problems and Solutions in Scientific Computing with C++ and Java Simulations*, World Scientific, Singapore 2004

[196] Steeb W.-H. and Andrieu E. C., Ljapunov Exponents, Hyperchaos and Hurst Exponent, *Z. Naturf.* **60a**, 252-254, 2005

[197] Steeb W.-H., *Problems and Solutions in Introductory and Adanced Matrix Calculus*, World Scientific, Singapore 2006

[198] Steeb W.-H., *Continuous Symmetries, Lie Algebras, Differential Equations and Computer Algebra*, second edition, World Scientific, Singapore 2007

[199] Steeb W.-H. and Hardy Y., A Chaotic Map and Data Transmission, *Z. Naturf.* **65a**, 613-614, 2010

[200] Stoop R. and Meier P.F., Evalution of Lyapunov exponents and scaling functions from time series, *J. Opt. Soc. Am. B* **5**, 1037-1045, 1988

[201] Suhchik M., Tsimring L. S. and Volkovskii, Performance Analysis of Correlation-based Communication Schemes utilizing Chaos, *IEEE Trans. Circuits I* **47**, 1684-1691, 2000

[202] Sznajd-Weron K. and Sznajd J., Opinion Evolution in Closed Comminity, *Int. J. Mod. C* **11**, 1157-1165, 2000

[203] Tan Z., Hepburn B. S., Tucker C. and Ali M. K., Pattern Recognition Using Chaotic Neural Networks, *Discrete Dynamics in Nature and Society* **2**, 243-247, 1998

[204] Toda M., *Theory of Nonlinear Lattices*, second edition, Springer, 1988

[205] Trotter H. F., On the product of semi-groups of operators, *Proceedings of the American Mathematical Society* **10**, 545-551, 1959

[206] Ueda Y., Akamatsu N. and Hayashi C., Computer Simulations and Non-Periodic Oscillations, *Trans. IEICE Japan* **56A**, 218-225, 1973

[207] Ueda Y. and Akamatsu N., Chaotically Transitional Phenomena in the Forced Negative-Restistance Oscillator, *IEEE Trans. Circuits Syst.* CAS-28, 217-224, 1981

[208] Uezu T., Topology in Dynamical Systems, *Phys. Lett.* **93A**, 161-166, 1983

[209] Van Wyk A. and Steeb W.-H., *Chaos in Electronics*, Kluwer, 1997

[210] Vapnik V., *Statistical Learning Theory*, Wiley, 1998

[211] Vapnik V., *The nature of statistical learning theory*, second edition, Springer, 2000

[212] Verlet L., Computer Experiments on Classical Fluids I. Thermodynamical Properties of Lennard-Jones Molecules, *Phys. Rev.* **159**, 98-104, 1967

[213] von Westenholz C., *Differential Forms in Mathematical Physics*, North-Holland, 1978

[214] Wang C. and Ge S. S., Adaptive synchronization of uncertain chaotic systems via backstepping design, *Chaos, Solitons and Fractals* **12**, 1199-1206, 2001

[215] Wang L. (editor), *Support vector machine: theory and applications*, Springer, 2005

[216] Walters P., *An Introduction to Ergodic Theory*, Springer, 1982

[217] Wolf A., Swift J. B., Swinney H.L., and Vastano J. A., Determining Lyapunov Exponents From Time Series, *Physica D* **16**, 258-317, 1985

[218] Wolfram S., *Theory and Applications of Cellular Automata*, World Scientific, Singapore, 1986

[219] Wolfram S., *A New Kind of Science*, Champaign, IL., Wolfram Media, 2002

[220] Xiang-Jun W., Jing-Sen L. and Guan-Rong C., Chaos synchronization of Rikitake chaotic attractor using the passive control technique, *Nonlinear Dynamics* **53**, 45-53, 2007

[221] Yager R. R. and Zadeh L. A. (Editors), *Fuzzy Sets, Neural Networks, and Soft Computing*, Van Nostrand Reinhold, New York, 1994

[222] Yamaguchi Y., Tanikawa K. and Mishima N., Fractal basin boundary in dynamical systems and the Weierstrass-Takagi functions, *Phys. Lett. A* **128**, 470-478, 1988

[223] Yamamoto T. and Kaneko K., Helium atom as a Classical Three-Body Problem, *Phys. Rev. Lett.* **70**, 1928-1930, 1993

[224] Yoshida H., Construction of higher order symplectic integrators, *Physics Letters A* **150**, 262-268, 1990

[225] Young L.-S., Dimension, entropy and Lyapunov exponents, *Ergodic Theory and Dynamical Systems* **2**, 109-124, 1982

[226] Zaslavskii G. M., Sagdeev R.Z., Chaikovskiii D. K. and Chernikov A. A., Chaos and two-dimensional random walk in periodic and quasiperiodic fields, *Zh. Eksp. Teor. Fiz.* **97**, 1723-1733, 1989

[227] Zhang H., Liu D. and Wang Z., *Controlling Chaos: Suppression, Synchronization and Chaotification*, Springer, 2009

[228] Zhang J. and Knoll A., Constructing Fuzzy Controlles with B-Spline Methods - Principles and Applications, *International Journal of Intelligent Systems* **13**, 257-286, 1998

[229] Zimmermann H.-J., *Fuzzy Set Theory and its Applications*, second edition, Kluwer, 1994

[230] Zizza F., Differential Forms for Constrained Max-Min Problems: Eliminating Lagrange Multipliers, *The College Mathematical Journal* **29**, 387-396, 1998

Index